Full-Field Measurements and Identification in Solid Mechanics

Full-Field Measurements
and
Identification in Solid Mechanics

Edited by
Michel Grédiac
François Hild

Series Editor
André Pineau

iSTE

ⓦWILEY

First published 2013 in Great Britain and the United States by ISTE Ltd and John Wiley & Sons, Inc.

ISTE Ltd
27-37 St George's Road
London SW19 4EU
UK

www.iste.co.uk

John Wiley & Sons, Inc.
111 River Street
Hoboken, NJ 07030
USA

www.wiley.com

Library of Congress Control Number: 2012946445

British Library Cataloguing-in-Publication Data
A CIP record for this book is available from the British Library
ISBN: 978-1-84821-294-7

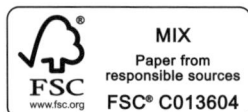

MIX
Paper from
responsible sources
FSC
www.fsc.org   FSC® C013604

Printed and bound in Great Britain by CPI Group (UK) Ltd., Croydon, Surrey CR0 4YY

Table of Contents

Foreword

From May through to early July 2011, I had the good fortune to visit colleagues at LMT-Cachan in Paris, France and see first-hand developments that they have recently been pioneering in the integration of full-field experimental measurement techniques with solid models. After attending the CNRS summer school entitled "Identification Procedures Using Full-Field Measurements: Applications in Mechanics of Materials and Structures", it was clear that faculty members are not only carrying out cutting-edge research, but also that they are translating their novel developments into academic course material at a remarkable rate. Over an intense one-week period, topics such as the reciprocity gap method, equilibrium gap method, constitutive equation gap method, virtual fields method, thermal source identification and many other similar topics were taught and applications completed for each topic.

This book on full-field measurements and identification in solid mechanics, edited by two world-class scientific investigators, Prof. Michel Grédiac and Prof. François Hild, is a natural and timely presentation of cutting-edge developments by experts in the field. Initially presenting various full-field measurement techniques used in solid mechanics, including photoelasticity, grid methods, deflectometry, holography, speckle interferometry and digital image correlation, the authors move smoothly into the evaluation of strains and the use of the measurements in subsequent parameter identification techniques to determine material properties. Since parametric identification techniques require a close coupling of theoretical models and experimental measurements, the authors focus on specific modeling approaches that include (a) finite element model updating, (b) the equilibrium gap method, (c) the constitutive equation gap method, (d) the virtual field method, and (e) the reciprocity gap method. In the latter part of the treatise, the authors discuss two particular applications of selected methods that are of special interest to many investigators: the analysis of localized phenomenon and connections between microstructure and constitutive laws. The final chapter highlights infrared measurements and their use in the mechanics of materials.

Taken as a whole, this book is a rich source of information on a rapidly developing and scientifically challenging area, the integration of experimental measurements with physically relevant models for parameter identification in materials and structures. Written and edited by knowledgeable scientists proficient in their areas of expertise, the monograph will be a valuable resource for all students, faculty and scientists seeking to expand their understanding of an important, growing research area.

Prof. Michael A. Sutton
University of South Carolina
USA
October 2012

Introduction

Non-contact full-field measurement techniques are increasingly being used in the experimental mechanics community. Such systems involve cameras, dedicated image-processing software and, in some cases, various types of more or less sophisticated optical setups. In all these cases, the goal is to measure the spatial distributions of various types of physical quantity, such as displacements, strains and temperatures, on the surface of specimens subjected to a given load, and even in their bulk in some cases. These fields can subsequently be postprocessed to identify parameters for material models.

In this context, the aim of this book is twofold. First, it proposes to describe the main features of the most popular types of full-field displacement and strain measurement techniques, which often remain poorly understood by engineers and scientists in the experimental mechanics field. It also seems relevant to closely associate the use of such types of data for material characterization purposes with the presentation of the measurement techniques themselves. Second, it analyzes numerical procedures that enable researchers and engineers to identify parameters governing constitutive equations. Any new user of full-field measurement techniques is often surprised by the wealth of data provided by such systems compared to classic measurement means, such as displacement transducers or strain gauges, which only provide a limited amount of data for comparison. This raises the question of the use of these data in a wise and rational manner. In particular, the fact that quasi-continuous information, rather than isolated measurements, must be processed requires a sound theoretical framework as well as robust numerical tools. Hence, controlling any identification procedure based on full-field measurements and assessing its global performance requires a clear overview simultaneously of purely experimental and theoretical aspects.

It is worth mentioning that some measurement techniques are not really recent because their fundamentals were described several decades ago. Their diffusion was,

Chapter written by M. GRÉDIAC and F. HILD.

however, strongly hindered for a long time because of the tedious procedures used at that time for storing, handling and processing the images they provide. In addition, the emergence and outstanding success of the finite element method attracted the majority of the community toward purely numerical problems, thus leaving experimental issues as secondary.

Two combined events have progressively contributed toward changing this situation. First, camera technology dramatically evolved in the early 1980s, especially with the advent of the charge-coupled device (CCD) and complementary metal-oxide-semiconductor (CMOS) sensors. Second, such cameras can be connected to personal computers, whose capabilities also began to increase at more or less the same time. Combined, these two events caused the above-mentioned drawbacks concerning image handling and processing to disappear gradually, and gave rise to the revival of "old" experimental techniques as well as the emergence of new techniques such as digital image correlation, thus paving the way for a new research field called photomechanics.

The first contributions naturally dealt with issues related to the actual performances of such techniques and their successive and numerous improvements, which were partly the logical consequence of advances in camera or computer technology. Many studies were also devoted to the use of full-field measurement techniques as even more powerful tools for studying particular problems in the mechanics of materials and structures. These studies generally share a common feature, namely the fact that local events are detected and studied. This was quite new for an experimental mechanics community accustomed to classic measurement instruments such as strain gauges or displacement transducers. Except in some particular cases, such devices are generally unable to give a clear understanding of the heterogeneous strain fields that occur in many situations. However, this type of information is very useful to obtain more insights into the global response of structures or tested samples.

A noteworthy aspect of heterogeneous strain fields is the fact that the number of constitutive parameters governing them is generally greater than those driving homogenous strain fields. Hence, pushing forward this idea of analyzing full-field measurements to detect specific phenomena, having displacement or strain fields available, measuring the applied load and knowing *a priori* the geometry of the specimen opens the way for the identification of the parameters of constitutive equations using this type of information. Because heterogeneous strain fields are being processed, the problem here is that no direct link generally exists between displacement or strain components measured at a given point, applied load and the constitutive parameters sought. Therefore, it is necessary to develop or use specific numerical tools that will allow us to tackle this issue, thus leading to the establishment of new links between experimental and computational mechanics. This also fully justifies the fact that this book closely associates the presentation of full-field measurement techniques with numerical strategies used to identify constitutive parameters by processing the measured fields they provide.

In this context, two main parts can be distinguished in this book.

The first part mainly deals with the description of full-field measurement techniques used in experimental solid mechanics. Metrological issues related to such techniques are addressed in Chapter 1. It is followed by Chapter 2 devoted to one of the oldest techniques: photoelasticity. Four techniques suitable for full-field displacement measurements are then described in the following chapters, namely the grid method (Chapter 3), holography (Chapter 4), speckle interferometry (Chapter 5) and digital image correlation (Chapter 6). Because the raw quantities provided by these techniques are generally displacement fields, Chapter 7 specifically deals with strain evaluation, therein closing the first part.

Identification techniques suitable for full-field measurements are introduced and discussed in the second part of the book. These techniques are the finite element model updating method (Chapter 9), the constitutive equation gap method (Chapter 10), the virtual fields method (Chapter 11), the equilibrium gap and the reciprocity gap methods (Chapters 12 and 13). First, after Chapter 8 in which these different techniques are introduced and compared, the next chapters address them in turn. Two chapters then deal with some particular applications, namely the analysis of localized phenomena (Chapter 14) and the link between microstructures and constitutive equations (Chapter 15). The final chapter (Chapter 16) deals with infrared measurements for which both experimental and identification issues are addressed at the same time to take into account the specificities of this technique.

It must be emphasized that it would have been unrealistic, even dangerous, to rank the different techniques presented in this book in terms of performance. In general, each of these techniques has its own advantages and limitations. Hence, any potential user should rather consider that they form a panel of complementary measurement and identification tools rather than competing techniques. In this context, having access to all the information on these topics should help any user to make his/her own choices in a given context. We hope that this book will be useful for this purpose.

To conclude the introduction, let us note that this book was initially written in the language of Molière [GRÉ 11], and was the result of many discussions within the "full-field measurements and identification in solid mechanics" research network (GDR 2519), created under the auspices of the French Research Council (CNRS) in 2003. Many of the contributing authors of the book are still affiliated with this network.

Bibliography

[GRÉ 11] GRÉDIAC M., HILD F. (eds), *Mesures de champs et identification en mécanique des solides*, Traité MIM, Hermes-Lavoisier, Paris, France, 2011.

Chapter 1

Basics of Metrology
and Introduction to Techniques

1.1. Introduction

Full-field optical methods for kinematic field measurement have developed tremendously in the last two decades due to the evolution of image acquisition and processing. Infrared (IR) thermography has also dramatically improved due to the extraordinary development of IR cameras. Because of their contactless nature, the amount of information they provide, their speed and resolution, these methods have enormous potential both for the research lab in the mechanics of materials and structures and for real applications in industry.

As for any measurement, it is essential to *assess* the obtained result. This is the area of metrology. The ultimate goal is to provide the user with as much information as possible about the measurement *quality*. We deal with a specific difficulty for the quality assessment of the optical methods that arises precisely from their full-field nature. The metrology community is far more familiar with point-wise or average scalar measurements (length, temperature, voltage, etc.). Currently, the metrology of full-field optical methods is not yet fully settled. However, the wide dissemination of these techniques will only efficiently occur when users have a clear understanding of how they can characterize the measurement performances of the equipment that vendors put on the market.

First, the goal of this chapter is to present some basic elements and concepts of metrology. It is by no means exhaustive, and only aims at presenting the basics of the

Chapter written by André Chrysochoos and Yves Surrel.

domain in a simple way, so that the researchers, users, developers and vendors can exchange information based on well-established concepts. Second, we will rapidly present the different optical techniques based on their main characteristics (how information is encoded, interferential or not, etc.).

It should be noted that optical measurement techniques exhibit a non-negligible amount of complexity. Figure 1.1 outlines the typical structure of a measurement chain that leads from a physical field to a numerical measurement field using a camera. It can be seen that there are many steps required to obtain the final result that the user is interested in, and numerous parameters that may impair the result and effects are involved at each step. Most importantly, there are usually numerical postprocessing stages that are often "black boxes" whose metrological characteristics or impact may be difficult or even impossible to obtain from the supplier of the equipment in the case of commercial systems.

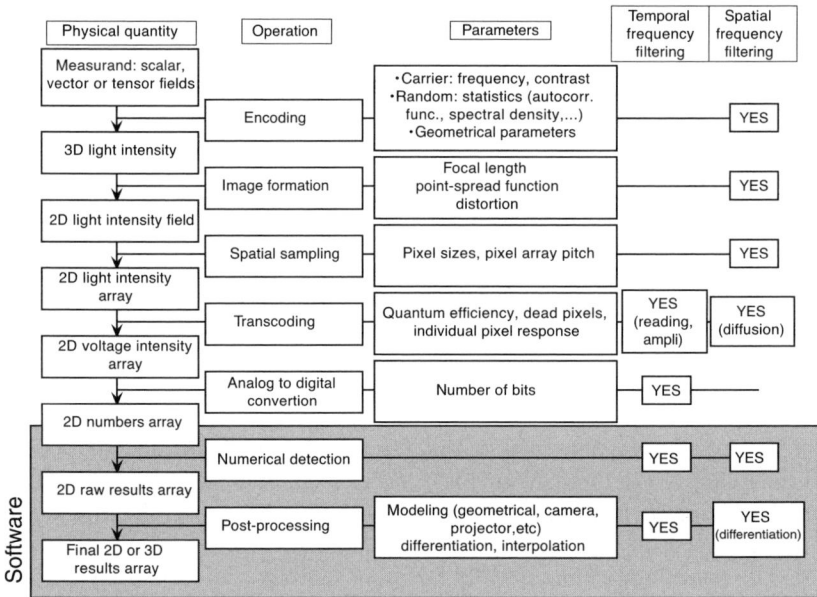

Physical quantity	Operation	Parameters	Temporal frequency filtering	Spatial frequency filtering
Measurand: scalar, vector or tensor fields				
	Encoding	·Carrier: frequency, contrast ·Random: statistics (autocorr. func., spectral density,...) ·Geometrical parameters		YES
3D light intensity				
	Image formation	Focal length point-spread function distortion		YES
2D light intensity field				
	Spatial sampling	Pixel sizes, pixel array pitch		YES
2D light intensity array				
	Transcoding	Quantum efficiency, dead pixels, individual pixel response	YES (reading, ampli)	YES (diffusion)
2D voltage intensity array				
	Analog to digital convertion	Number of bits	YES	
2D numbers array				
	Numerical detection		YES	YES
2D raw results array				
	Post-processing	Modeling (geometrical, camera, projector,etc) differentiation, interpolation	YES	YES (differentiation)
Final 2D or 3D results array				

(Software spans from "2D numbers array" through "Final 2D or 3D results array")

Figure 1.1. *Outline of the many steps involved in a measurement chain using a camera*

1.2. Terminology: international vocabulary of metrology

1.2.1. *Absolute or differential measurement*

In any scientific domain, terminology is essential. Rather than enumerating the main terms to use (precision, sensitivity, resolution, etc.), let us try to adopt the final user point of view. What are the questions he generally asks, and in which context? There are, in fact, not so many questions, and each of them leads naturally to the relevant metrological term(s):

1) Is the obtained result "true", "exact" and "close to reality"? How to be confident in the result?

2) Is the equipment "sensitive"? Does it see small things?

These two questions lie behind the separation of metrology into two distinct domains, within which the metrological approach will be different: *absolute* measurement and *differential* measurement.

1.2.1.1. *Absolute measurement*

Here, we seek the "true" value of the measurand (the physical quantity to measure), for example to assess that the functional specifications of a product or system are met. Dimensional metrology is an obvious example. The functional quality of a mechanical part will most often depend on the strict respect of dimensional specifications (e.g. diameters in a cylinder/piston system). The user is interested in the deviation between the obtained measurement result and the true value (the first of the two questions above). This deviation is called the measurement *error*. This error is impossible to know, and here is where metrology comes in. The approach used by metrologists is a *statistical* one. It will consist of evaluating the statistical distribution of the possible errors, and characterizing this distribution by its width, which will represent the average (typical) deviation between the measurement and the true value. We generally arrive here at the concept of measurement *uncertainty*, which gives the user information about the amount of error that is likely to occur.

It is worth mentioning that some decades ago, the approach was to evaluate a *maximum* possible deviation between the measurement and the real value. The obtained uncertainty values were irrelevantly overestimated because the uncertainty was obtained by considering that all possible errors were synergetically additive in an unfavorable way. Today, we take into account that statistically independent errors tend to average each other out to a certain extent. In other words, it is unlikely (and this is numerically evaluated) that they act coherently together to impair the result in the same direction. This explains the discrepancy between the uncertainty evaluation formulas that can be found in older treatises and those used today. As an example, if $c = a + b$, where a and b are independent measurements having uncertainties Δa and Δb, the older approach (maximum upper bound) would yield $\Delta c = \Delta a + \Delta b$, but the more recent approach gives $\Delta c = \sqrt{\Delta a^2 + \Delta b^2}$, these two values being in a non-negligible ratio of $\sqrt{2} = 1.414$ in the case $\Delta a = \Delta b$.

1.2.1.2. *Differential measurement*

Now, let us consider that the user is more interested in the deviation from a "reference" measurement. In the mechanics of materials field, the reference measurement will most likely be the measurement in some initial state, typically before loading. It is basically interesting to perform measurements in a differential way, and this is done for two reasons:

– As the measurement conditions are often very similar, most *systematic errors* will compensate each other when subtracting measurement results to obtain the deviation; typically, optical distortion (image deformation) caused by geometrical aberrations of the camera lens will be eliminated in a differential measurement.

– Measurement uncertainty is only about the *difference*, which implies that interesting results can be obtained even with a system of poor quality; let us take an example of a displacement measurement system exhibiting a systematic 10% error. Absolute measurements obtained with this system would probably be considered unacceptable. However, when performing differential measurements, this 10% error only affects the difference between the reference and actual measurements. Consider the following numerical illustration: with a first displacement of 100 μm and a second displacement of 110 μm; this hypothetical system would provide measured values of 90 and 99 μm, roughly a 10 μm error each time, probably not acceptable in an absolute positioning application, for example. But the measurement differential displacement is 9 μm, which is only a 1 μm error.

Regarding this second example, we should emphasize the fact that it is *not possible to use relative uncertainty* (percentage of error) with differential measurements because this relative uncertainty, ratio between the uncertainty and the measurand, is related to a very small value, the difference of measurements, that is nominally zero. Hence, "percentage of error" is strictly meaningless in the case of differential measurements. Only absolute uncertainty is meaningful in this case.

1.2.2. *Main concepts*

We will consider in this section the main concepts of metrology. The corresponding terms are *standardized* in a document (*International Vocabulary of Metrology*, abbreviated as "VIM", [VIM 12]) which can be downloaded freely from the BIPM website.

1.2.2.1. *Measurement uncertainty*

The VIM definition is as follows:

> non-negative parameter characterizing the dispersion of the quantity values being attributed to a measurand, based on the information used.

The underlying idea is that the measurement result is a random variable, which is characterized by a probability density function centered at a certain statistical average value. This probability density function is typically bell-shaped. Note # 2 in the VIM states:

> the parameter may be, for example, a standard deviation called standard measurement uncertainty (or a specified multiple of it), or the half-width of an interval, having a stated coverage probability.

So the idea is really to *characterize the width* of a statistical distribution. There is obviously a conventional feature to perform this characterization. We may choose the width at half the maximum, the width at $1/e$, the usual standard deviation, etc.

In industry, the need to assess the quality of a measurement instrument can be slightly different: it is most often necessary to be "sure" that the measurement deviation will not exceed some prescribed value. So, the chosen width will include as much as possible the total spreading width of the function. Of course, no real 100% certainty can be obtained, but we can obtain *confidence intervals* at $x\%$, which are intervals having a probability of $x\%$ to include the real measurand value. For example, for a Gaussian distribution of the measurement result probability, an interval of $\pm 3\sigma$ has a probability of 99.5% to englobe the true value. Hence, the concept of *expanded measurement uncertainty*, whose VIM definition is:

> product of a combined standard measurement uncertainty and a factor larger than the number one.

The "combined standard measurement uncertainty" is the final uncertainty obtained when all sources of uncertainty have been taken into account and merged. In general, this is what is referred to when we speak of "uncertainty" in short.

Measurement uncertainty is the key notion to use when dealing with absolute measurement.

It should be noted that the two major uncertainty sources are those related to noise, introducing random fluctuations of null average value, and those related to *systematic errors* (calibration errors, poor or insufficient modeling, external permanent biasing influences, etc.). With full-field optical methods, noise can easily be characterized (see section 1.2.2.2).

1.2.2.2. *Resolution*

The VIM definition is as follows:

> smallest change in a quantity being measured that causes a perceptible change in the corresponding indication.

This is the key notion to use when performing a differential measurement. In practice, with full-field optical methods, noise (optical, electronic, etc.) will be the factor limiting the resolution. A practical definition of resolution that we can propose is (non-VIM):

> change in the quantity being measured that causes a change in the corresponding indication greater than one standard deviation of the measurement noise.

As it happens, with techniques that use cameras, it is easy to *measure* the measurement noise because a large number of pixels are available to obtain significant statistics. To evaluate the measurement noise, it suffices to make two consecutive measurements of the same field, provided, of course, that it does not vary significantly in the meantime. Let us denote by $m_1(x, y)$ and $m_2(x, y)$ two successive measurements at point (x, y). These values are the sum of the "signal" s (the measurand value) and the noise b, as:

$$\begin{cases} m_1(x, y) = s(x, y) + b_1(x, y) \\ m_2(x, y) = s(x, y) + b_2(x, y) \end{cases} \qquad [1.1]$$

where the signal s is unchanged, but the noise has two different values. The subtraction of the two results yields:

$$\Delta m(x, y) = m_2(x, y) - m_1(x, y) = b_2(x, y) - b_1(x, y) \qquad [1.2]$$

Assuming *statistical independence* of the noise between the two measurements, which is often the case[1], we can evaluate the standard deviation of the noise b (we *always* assume that the noise is stationary, i.e. its statistical properties do not change in time) by using the theorem that states that the variance (square of the standard deviation) of the sum of two statistically independent random variables is equal to the sum of their variances, which gives:

$$\sigma^2(\Delta m) = \sigma^2(b_2 - b_1) = \sigma^2(b_2) + \sigma^2(-b_1) = 2\sigma^2(b) \qquad [1.3]$$

An estimator of the statistical variance of the noise difference $b_2 - b_1$ is obtained by calculating, for all the pixels of the field, the arithmetic variance:

$$\tilde{\sigma}^2(\Delta m) = \frac{1}{N_{\text{pixels}}} \sum_{\text{pixels}} [\Delta m(x, y) - \overline{\Delta m}]^2 \qquad [1.4]$$

where $\overline{\Delta m}$ is the arithmetic average of Δm within the field:

$$\overline{\Delta m} = \frac{1}{N_{\text{pixels}}} \sum_{\text{pixels}} \Delta m(x, y) \qquad [1.5]$$

---

1 Counter example: atmospheric turbulence has typical correlation times of the order of one or many seconds, possibly larger than the interval between successive measurements if we do not take great care.

We thus obtain for the resolution r:

$$r = \sigma_b \approx \frac{1}{\sqrt{2}} \tilde{\sigma}(\Delta m) \qquad\qquad [1.6]$$

For the sake of mathematical rigor, we have chosen to denote by an ~ sign the real statistical values, and not their *estimators* obtained from arithmetic averages over a large number of pixels. This number being typically much larger than 10,000 (which corresponds to a sensor definition of 100×100 pixels, a very low definition), we can ignore the difference and consider the estimators to be faithful.

Let us insist on the fact that this estimate of the noise level is very easy to perform, and should be systematically performed during full-field measurements.

Characterizing resolution using the noise level is common in IR thermography. An IR camera is characterized with *noise equivalent temperature difference* (NETD), which is nothing other than the output noise level converted into the input, that is converted into the temperature difference that would cause a reading equal to one standard deviation of the noise. Temperature variations will begin to "come out of the noise" when they reach the order of magnitude of the NETD. To change an output noise level into an input level, we use *sensitivity*.

1.2.2.3. *Sensitivity*

The VIM definition is as follows:

> quotient of the change in an indication of a measuring system and the corresponding change in a value of a quantity being measured.

This corresponds to the coefficient s that maps a small variation of the measurand e onto the corresponding variation of the measurement reading (the output m)

$$s = \frac{\partial m}{\partial e} \qquad\qquad [1.7]$$

Sensitivity is not always relevant for optical field measurements. As an example, correlation algorithms directly provide their results as displacement values; their sensitivity is equal to one. On the contrary, the grid method presented later in this book provides a phase after the detection step. The relationship between the phase Φ and the displacement u is:

$$\Phi = 2\pi \frac{u}{p} \qquad\qquad [1.8]$$

where p is the grid pitch because the phase varies by 2π when the displacement is equal to one grid period. Thus, the sensitivity according to the VIM is:

$$s = \frac{2\pi}{p} \qquad\qquad [1.9]$$

This notion is rarely used in optical field methods. However, in IR thermography, it remains fully meaningful and represents the voltage or current variation at the sensor level relative to the received irradiation.

1.2.2.4. *Repeatability*

The VIM definition is as follows:

> measurement precision under a set of repeatability conditions of measurement

to which we add the definition of *measurement precision*

> closeness of agreement between indications or measured quantity values obtained by replicate measurements on the same or similar objects under specified conditions

and the definition of *repeatability conditions*:

> condition of measurement, out of a set of conditions that includes the same measurement procedure, same operators, same measuring system, same operating conditions and same location, and replicate measurements on the same or similar objects over a short period of time.

Actually, the underlying idea is fairly simple: repeatability is supposed to characterize the *measurement instrument alone*, having excluded all external influences: same operator, same object under measurement, short delay between successive measurements, etc. In short, repeatability tests will characterize the noise level of the measurement instrument.

When using IR thermography, it is essential to evaluate repeatability only after the radiometric system has reached its thermal equilibrium. The non-stationary feature of the parasitic radiations coming from the optical elements and from the rest of the camera may significantly bias the results. It is noteworthy that the time to thermal equilibrium can be a few hours, even in a stabilized environment. This is a noticeable constraint [BIS 03].

The "dual" notion, which supposes, on the contrary, different operators, times, locations and even measurement systems, is *reproducibility*.

1.2.2.5. *Reproductibility*

The VIM definition is as follows:

> measurement precision under reproducibility conditions of measurement

to which we add the definition of the *reproducibility conditions*:

> condition of measurement, out of a set of conditions that includes different locations, operators, measuring systems and replicate measurements on the same or similar objects.

To emphasize the fact that a lot can be changed in this case, Note 1 in the VIM states:

> the different measuring systems may use different measurement procedures.

As can be seen, as little as possible should change when performing repeatability tests (in short, noise evaluation), and as much as possible may change when doing reproductibility tests (in short, all possible error sources can be involved).

1.2.2.6. *Calibration*

The VIM definition is as follows:

> operation that, under specified conditions, in a first step, establishes a relation between the quantity values with measurement uncertainties provided by measurement standards and corresponding indications with associated measurement uncertainties and, in a second step, uses this information to establish a relation for obtaining a measurement result from an indication.

Thus, calibration enables, from artifacts producing known values of the measurand, the identification of the transfer function of the measurement system. As an example, for an instrument having a linear response, measuring a single standard allows us to identify the slope of the straight line that maps the measurand to the output indication. Note in passing that this slope is nothing other than the sensitivity of the instrument according to the VIM definition presented in section 1.2.2.3.

A measurement standard is a

> realization of the definition of a given quantity, with stated quantity value and associated measurement uncertainty, used as a reference.

In IR thermography, the ideal measurement standard is the *black body*, a radiating object with unity emissivity. For cameras, it is practical to use plane black bodies that present isothermal surfaces (e.g. within ± 0.02 K). In this case, temperatures and radiated energies unequivocally correspond. Classic calibration procedures use two black body temperatures that include the temperature range of the scene that is to be measured. The procedure involves a non-uniformity correction (NUC), which sets the offsets of all matrix elements such that the represented thermal scene appears uniform and detects and replaces the bad pixels (bad pixel replacement (BPR)); bad pixels are those which, according to a user-prescribed criterion, provide an output signal that deviates too far from the average. Recently, this calibration step, which is convenient and fast, has tended to be replaced by a pixel-wise calibration, which is longer and more costly but offers improved precision over a wider measurement range [HON 05]. The interested reader will find in Chapter 16 more information on this subject.

1.2.2.7. *Illustration*

Let us recall here an elementary result. If we average the results of N nominally identical measurements (in other words, measurement performed in *repeatability conditions*, section 1.2.2.4), we obtain something that can also be considered a random variable, because another set of N measurements would provide a different value for the average. The classic result is that the standard deviation of the fluctuations of the average of N measurements is equal to the standard deviation of the fluctuations of a single measurement divided by $\sqrt{N}$.

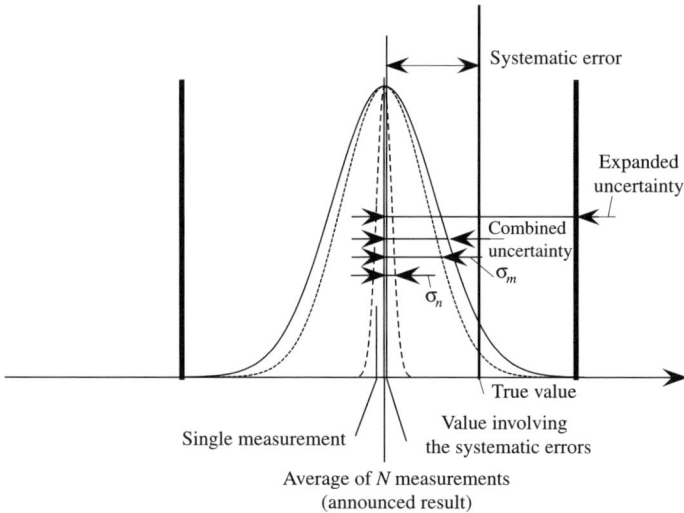

Figure 1.2. *Measurement, error and uncertainty*

Figure 1.2 shows some of the notions presented previously. We have represented on the horizontal axis different values related to one or more measurements of the same quantity, with their uncertainties, namely:

– The true value of the measurand (it is permissible to speak of the "true value", explicitly referred to in the VIM in section 2.11).

– The true value biased by the systematic errors; measurement values will fluctuate around this value.

– The result of a single measurement; many measurements will provide values that will fluctuate according to some probability density function of standard deviation σ_n, where n stands for noise.

– A value resulting from the averaging of N measurement results; many such averages will fluctuate with a standard deviation that is $\sqrt{N}$ smaller than for a single measurement.

– The standard deviation σ_m, where m stands for "model", which corresponds to the probability density function of systematic errors that could be determined by a proper modeling and investigation of the physics of the instrument.

– The combined uncertainty $\sqrt{\sigma_m^2 + \sigma_b^2}$, which takes into account the uncertainty related to systematic errors, as well as the noise.

– The expanded uncertainty that can be used as a confidence interval.

In practice, it is often the uncertainty component not related to noise (the systematic errors) that is the most difficult to evaluate.

1.3. Spatial aspect

Full-field optical measurement methods have a special aspect for metrology: the fact that they provide a spatial field instead of a single scalar. Concepts adapted to this spatial nature are lacking in the VIM. The most important is *spatial resolution*, which depends on numerous factors, especially all the digital postprocessing that can take place after the raw measurements are obtained, for example, to decrease the amount of noise. Before presenting the possible definitions, let us recall some basic results concerning spatial frequency analysis.

1.3.1. *Spatial frequency*

The relevant concepts here are commonly used by optical engineers when dealing with *image forming*. Indeed, the process of forming an image from an object using an imaging system can be seen as the transfer of a signal from an input (object irradiance) to an output (image intensity). This transfer will change some features of the signal, according to some *transfer function*. In a way similar to an electronic filtering circuit, which is characterized by a gain curve as a function of (temporal) frequency, an image-forming system can be characterized by the way it will decrease (there is never any amplification) the *spatial frequencies*. In lens specifications found in catalogs, we can sometimes find "gain curves" (called optical transfer function) that correspond exactly to gain curves characterizing electronic amplifiers or filters.

The difficulty of spatial frequency is related to its vectorial aspect because we have to deal with functions defined on a two-dimensional (2D) domain, for example the irradiance $L(x, y)$ of the object to be imaged. When dealing with a time function $f(t)$, the variable t is defined over a unidimensional space, and the Fourier conjugated variable, the frequency, is a scalar variable. With a function defined on a plane, $f(x, y)$ or $f(\vec{r})$ with vector notation, the Fourier variable is of dimension 2: (f_x, f_y) or $\vec{f}$.

To understand the notion of spatial frequency, the simplest way is to consider an elementary spatial periodic function, for example $z(x, y) = \cos[2\pi(f_x x + f_y y)]$, where f_x and f_y are constants. This "corrugated iron sheet" function is shown in Figure 1.3.

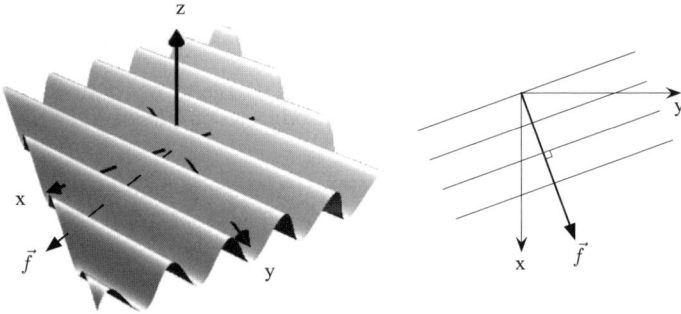

Figure 1.3. *Function* $\cos[2\pi(f_x x + f_y y)]$

We can rewrite its definitive equation as $z(x, y) = \cos(2\pi \vec{f}.\vec{r})$, where $\vec{r}$ is the vector of components (x, y) and $\vec{f}$ is the *spatial frequency vector* of components (f_x, f_y). It is easy to convince ourselves that:

– The direction of vector $\vec{f}$ is perpendicular to the function $z(x, y)$ isovalue lines.

– The norm (length) of $\vec{f}$ is equal to the reciprocal of the function period p: $||\vec{f}|| = 1/p$.

The history of instrumental optics has been a race to obtain instruments that transmit as much as possible all the spatial frequencies present in an object, in the same way as audio electronics developed to transmit with high fidelity all the acoustic frequencies present in voice or music signals. In the optical domain, test objects are often targets made up of equidistant straight lines that implement "single frequency" objects[2] (Figure 1.4). The investigation of these target images, describing whether the image-forming system has correctly transmitted ("resolved") the lines, allows us to determine the instrument *cutoff frequency* beyond which the lines will

---

2 We ignore here the fact that sharp-edged lines have in their frequency spectrum many harmonics with frequencies that are integer multiples of the basic frequency.

not be distinguished one from the other. We can also see in Figure 1.4 trumpet-shaped patterns that correspond to increasing spatial frequencies, which is the spatial implementation of what is known as "chirps" in signal processing.

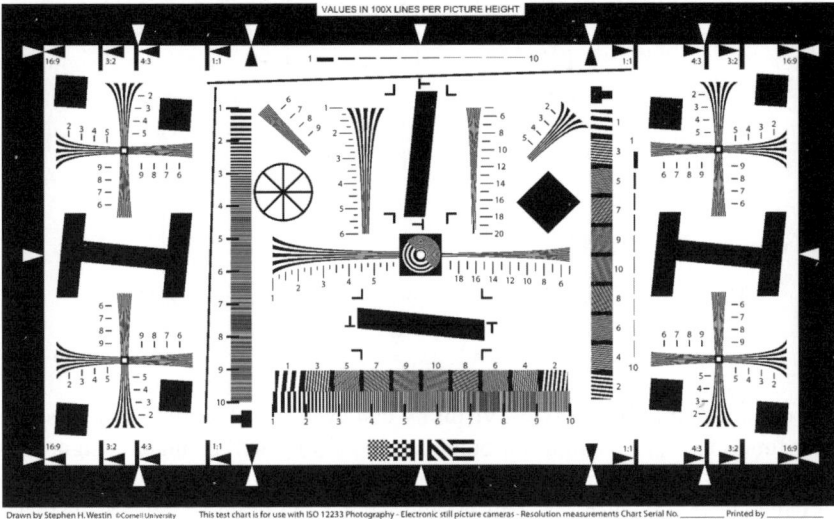

Figure 1.4. *Mire ISO 12233*

We can suppose that Fourier analysis is perfectly well adapted to deal with all these notions, and that the optical transfer function is nothing more than the filtering function in the Fourier plane (the spatial frequency plane) where the spatial frequencies present in the object are represented.

Figure 1.5 shows the process of image forming, starting from a sinusoidal grid object. The instrument gain (or *optical transfer function*) is represented in the central box, at the top.

Light intensity is always non-negative, so a constant average value has to be superimposed onto the sinusoidal variation, yielding the irradiance:

$$I(x) = A \left[1 + \gamma \cos \left(\frac{2\pi x}{p} \right) \right]$$
[1.10]

Visibility (or contrast) is the dimensionless ratio of the modulation and the average. Visibility is unity when "blacks are perfectly black", that is when the irradiance minimum reaches zero. The change in the irradiance signal by a system having the transfer function $G(f)$ that attenuates non-zero frequencies causes a loss of visibility. For a spatial frequency f_0, as represented in Figure 1.5, visibility loss is

$$\frac{\gamma'}{\gamma} = \frac{G(f_0)}{G(0)}$$
[1.11]

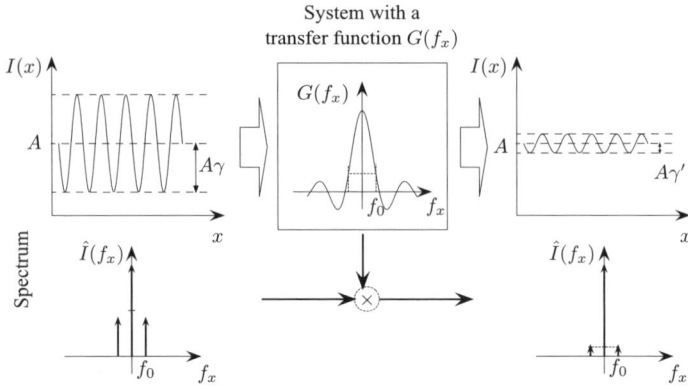

Figure 1.5. *Transmission of sinusoidal grid irradiance through an image-forming system*

For certain spatial frequencies, visibility can even vanish completely. Also, if the transfer function $G(f)$ oscillates around zero, visibility will change its sign, causing a phase lag of π of the output signal: minima and maxima will exchange their places; this is called *visibility inversion*. This effect is shown in Figure 1.6, where the visibility changes its sign when crossing the zones where it vanishes.

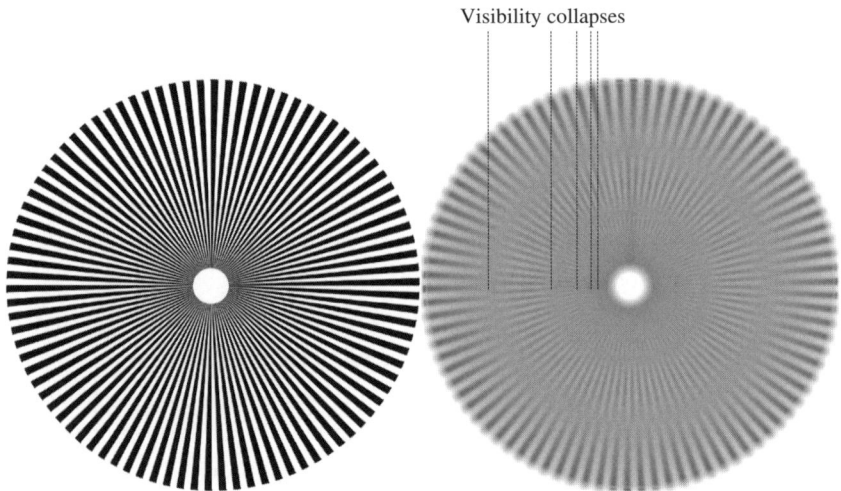

Figure 1.6. *Contrast inversions*

Historically, of course, this mathematical background was not available. The spatial resolution of an image-forming system was defined from the minimum distance between the images of two neighboring points that allowed the points to be distinguished ("resolved"). We recognize the spatial equivalent of the resolution of a measuring instrument as the "smallest distance between two points whose images can be separated". We can state this, in other words: "the smallest change in position that causes a perceptible change in the corresponding image", which is the copy *mutatis mutandis* of the VIM definition. This concept relates to what is known in optical engineering as the Rayleigh criterion.

Considering the images of grids having various frequencies instead of considering neighboring points is the dual approach (in the Fourier sense). In this case, we seek the highest frequency that is transmitted by the instrument without perceptible degradation (e.g. without loss greater than 50%). The reciprocal of this spatial frequency is a geometrical wavelength that can be considered as the spatial resolution. This explains why the following definition can be found in [AST 08]:

– "Optical data bandwidth: Spatial frequency range of the optical pattern (e.g. fringes, speckle pattern, etc.) that can be recorded in the images without aliasing or loss of information". We can recognize the notion of performance of an image-forming system, characterized by a cutoff frequency f_c. The bandwidth is the interval $[0, f_c]$.

– "Spatial resolution for optical data: One-half of the period of the highest frequency component contained in the frequency band of the encoded data". This is the quantity $1/2f_c$. The factor 2 is something of a convention and has no precise justification. However, there is always a conventional aspect in these definitions as the considered quantities vary in a continuous way. The above-mentioned Rayleigh criterion also has a conventional aspect. A factor of 2.5 or 1.5, for example, might have been used without the possibility of justifying that choice any more clearly.

For image correlation techniques, the minimum spatial resolution is the size of the subimages used to calculate the correlation, because we obtain only one measurement per subimage position, for the whole subimage area. Of course, the subimage can be (and is, in general) moved pixel-by-pixel to obtain a displacement field populated with as many pixels as the original image, but this does not change the spatial resolution, as all measurements corresponding to a distance smaller than the subimage size will be correlated and not independent.

To end this section, let us emphasize that we should not use the term "resolution" to designate the number of pixels of the camera sensor, but *definition*. Remember that we speak of high-definition TV (HDTV) to evoke a very large number of pixels in the image. High resolution is obtained with a high magnification of the imaging system. Thus, we can have a very good resolution with a low definition, with a sensor that does not have a very high number of pixels but covers a very small field of view[3].

---

3 This is a little theoretical; in practice, of course, a high definition is usually chosen when a high resolution is available.

1.3.2. *Spatial filtering*

A very common operation for reducing the spatial noise that is present in the image is spatial filtering, which consists of replacing every pixel in the image by an average, weighted or not, performed with respect to its neighbors. This operation is very well described using the Fourier transform, and it is easy to calculate the consequences on spatial resolution.

This kind of filtering is basically a convolution operation. For the sake of simplicity, let us use a one-dimensional description with continuous variables, as sampling introduces some complexity without changing the basic concepts. The simplest filtering is the moving average, where each point of a signal $g(x)$ is replaced by the signal average over a distance L around this point:

$$\overline{g}(x) = \int_{x-\frac{L}{2}}^{x+\frac{L}{2}} g(u)du \qquad\qquad [1.12]$$

This expression can be described as a convolution product with a rectangle function $\Pi(x)$ defined as:

$$\Pi(x) = \begin{cases} 0 & \text{if} \quad x < -1/2 \quad \text{or} \quad x > 1/2 \\ 1 & \text{else} \end{cases} \qquad\qquad [1.13]$$

Equation [1.12] can, indeed, be rewritten as:

$$\overline{g}(x) = \int_{-\infty}^{+\infty} g(u)\Pi\left(\frac{x-u}{L}\right) du = (g * \Pi_L)(x) \qquad\qquad [1.14]$$

where $*$ denotes a convolution product and $\Pi_L(x) = \Pi(x/L)$. The effect of such a filtering can be seen in the Fourier space (or frequency space), by taking the Fourier transform of the previous equation:

$$\hat{\overline{g}}(f_x) = \hat{g}(f_x) \times \hat{\Pi}_L(f_x) \qquad\qquad [1.15]$$

which gives:

$$\hat{\overline{g}}(f_x) = \hat{g}(f_x) \times L\,\text{sinc}(Lf_x) \qquad\qquad [1.16]$$

where sinc is the "cardinal sine" function: $\text{sinc}(u) = \sin(\pi u)/\pi u$. The initial frequency spectrum (the Fourier transform of signal g) is attenuated by the transfer function, in this case $\text{sinc}(Lf_x)$ that reaches its first zero for $Lf_x = 1$, or $f_x = 1/L$ (Figure 1.7).

Gaussian filtering is also used very frequently:

$$\overline{g}(x) = \int_{-\infty}^{+\infty} g(u)G_\sigma(x-u)du \qquad\qquad [1.17]$$

where

$$G_\sigma(x) = \frac{1}{\sigma\sqrt{2\pi}} \exp\left(-\frac{x^2}{2\sigma^2}\right) \qquad\qquad [1.18]$$

is a normalized Gaussian function having a standard deviation σ. Its Fourier transform is:

$$\hat{G}_\sigma(f_x) = \exp(-2\pi^2\sigma^2 f_x^2) = \sigma'\sqrt{2\pi}G_{\sigma'}(f_x) \qquad\qquad [1.19]$$

where $\sigma' = 1/2\pi\sigma$. It is also a Gaussian function, which has the property of decreasing very rapidly when its variable increases. This provides interesting properties for noise filtering because spatial noise is mostly present at high frequencies (typically, noise changes from one pixel to another, which corresponds to the highest frequency in a sampled image).

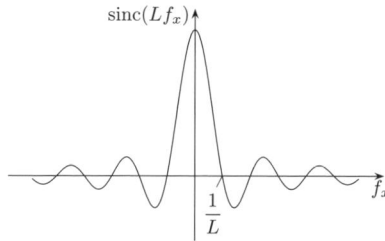

Figure 1.7. *Cardinal sine function* $\text{sinc}(Lf_x)$, *which is the transfer function of a moving average over a length* L

More generally, any (linear) filtering can be described as a convolution with a certain function that is usually called the filtering *kernel*.

Numerically, linear filtering is efficiently implemented in the Fourier plane because efficient discrete Fourier transform algorithms (the so-called Fast Fourier Transform – FFT) are available. In the Fourier plane, the convolution is replaced by a simple multiplication with the transfer function, the Fourier transform of the filtering kernel. However, these FFT algorithms can only be used if the image topology is rectangular, in other words, when no invalid pixels or holes exist in the image. Holes are zones with no measurements and cannot be simply replaced by zero. If holes are present, some kind of interpolation has to be used to fill them so that no discontinuity arises along their edges, because these possible discontinuities will introduce adverse effects in the signal spectrum. This is not a trivial operation, and it may finally be simpler to implement the convolution.

Successive filterings correspond to successive convolution products. As the convolution product is associative, this can be considered as a single filtering using a kernel that is the convolution product of all the different kernels used. As an example, if three filterings having kernels $K_1(x)$, $K_2(x)$ and $K_3(x)$ are used:

$$\overline{g}(x) = \{[(g * K_1) * K_2] * K_3\}(x) = [g * (K_1 * K_2 * K_3)](x) \qquad [1.20]$$

we obtain something equivalent to a single filtering with kernel $(K_1 * K_2 * K_3)(x)$. It can be shown [ROD 78] that with a high number of successive filterings with the *same* filter, we tend toward Gaussian filtering; this is a consequence of what is known as the central-limit theorem.

As for spatial resolution, each filtering widens an object point, since the result of the filtering of a point located at x_0 and described by a Dirac function $\delta(x - x_0)$ is simply the filtering kernel translated at the same point, because:

$$\delta(x - x_0) * K(x) = K(x - x_0) \qquad [1.21]$$

It can also be shown [ROD 78] that the "width" Δ_{f*g} of the convolution product of two functions f and g is the quadratic sum of the widths of each function:

$$\Delta_{f*g} = \sqrt{\Delta_f^2 + \Delta_g^2} \qquad [1.22]$$

This equation allows us to calculate how the spatial resolution evolves as a function of successive spatial filterings.

To end this section, let us mention that this approach for the study of spatial resolution by the frequency transfer function was the basis of the work performed by the GDR 2519 research group, reported in [BOR 09], with respect to image correlation.

1.4. Classification of optical measurement techniques

In this section, we propose a systematic classification of different optical techniques for the measurement of kinematic fields. This classification is based on a limited number of basic key concepts.

In this context, infrared thermography has a slightly special position. It enables the measurement of a surface temperature, and not a kinematic value such as displacement, strain, slope and curvature.

In fact, the signal output by infrared sensors depends on the irradiance received from emitting bodies, and its transformation into a temperature is not a trivial task. The signal-to-energy relationship first involves Stefan–Boltzmann's and Planck's laws. Then, ignoring any parasitic radiation, translating received energies into the temperature of the target is only possible if the target has a known emissivity, which is a necessary condition. Other aspects of course need to be taken into account to ensure the accuracy of the temperature measurement. They are presented in Chapter 16 of this book.

1.4.1. *White light measurement methods*

In this family, the measurand is encoded in the spatial variation of light intensity, this variation being obtained in a non-interferometric way; in other words, we are excluding everything related to interferometric fringes. Purely optical aspects may be restricted to the image-forming process: measuring displacements by using image correlation has nothing really "optical" in nature, but is only based on a geometrical phenomenon that is present in images.

1.4.1.1. *Encoding techniques*

The measurand encoding, shown in Figure 1.1, is mainly of two kinds:

Encoding through a random signal: the signal is characterized by its local random variation, acting as a signature; the receiving system will have to identify this signature to complete the measurement; in this category lie all the methods using image correlation, including the so-called (laser) speckle correlation; these methods are presented in Chapter 6.

Encoding through a periodic signal, more precisely through the modulation of the phase (or, equivalently, of the frequency) of a spatial sinusoidal signal, called a "carrier"; as an example, we can cite the displacements of the lines of a grid deposited on a substrate, or the displacements of the lines of a grid reflected by a mirror-like surface where the local slopes are not uniform; these methods are presented in Chapter 3.

It should be noted that *each technique exists under both forms* (random encoding or phase encoding), even if the terminology does not always help to recognize this fact. There are of course major differences in performance resulting from these different encoding approaches:

– The practical realization of random encoding, which is essentially noise, is *difficult to fully characterize*, using complex notions such as statistical moments (there are plenty of such "moments") and average power spectral density (PSD); the correlation method for in-plane displacement measurements often requires paint to be sprayed on to the body under examination to produce a "speckle", a random contrasted pattern; it is obvious that this manual process cannot easily be made repeatable, and the "quality" of the obtained pattern is difficult to characterize.

– On the other hand, encoding through the modulation of a spatial periodic carrier can be fully characterized by its frequency, its local phase, its local average level, its local contrast and its harmonic content.

– Random encoding *is not quantitative by itself*; the information lies within a local contrast morphology or signature that the detection system has to identify.

– On the contrary, encoding by the phase modulation of a carrier is *quantitative* because the signal (e.g. the displacement of a grid line) corresponds to a number that is the amount of phase modulation. Phase detection techniques are very well established and efficient (they mostly rely on Fourier analysis) and allow easy characterization.

– A drawback of phase encoding is its *periodicity*: the same code value periodically repeats as the signal increases, which leads to *ambiguities*. Removing these ambiguities corresponds to what is called *phase unwrapping* that suppresses the 2π jumps appearing in the detection process that only outputs the phase modulo 2π. Phase unwrapping is, in fact, simply *numbering the fringes* (or lines, depending on the method).

– In general, correlation methods are much easier and cheaper to implement; either the part under test is sufficiently textured to allow the use of digital image correlation, or it suffices to spray paint on it. This is much simpler than gluing or engraving or in some way depositing a grid onto the surface, not to mention the fact that grids can only be placed on flat or cylindrical surfaces.

1.4.1.2. *Examples*

Almost all white light techniques consist of measuring a position or a displacement in an image. A simple geometrical analysis of the measurement system allows us to understand the measurement principle.

1) Measurement of in-plane displacements: here, the object undergoing displacements is simply observed using a camera. The displacements of marked points of the object correspond to displacements of image points. The primary output of the technique is the displacement field of the image points. The available techniques are the following.

i) Grid technique: a grid has to be deposited on the surface; the local displacements will modulate the phase of the grid acting as a periodic carrier. Its main drawback is the required surface preparation.

ii) *Moiré*[4] is only an additional layer on the grid technique, since the nature of the encoding is the same. It consists of decreasing the spatial frequency of the carrier using a beat phenomenon, exactly as stroboscopy decreases the temporal frequency of a vibration.

iii) Deflectometry: in this technique, derived from the previous one, the object reflects a fixed regular grid pattern. Distortion of the grid image is caused by non-uniformity of the surface slopes. Thus, this technique enables slope fields to be measured. It is necessary that the surface be at least partially reflective, in which case

---

4 The first letter should not be capitalized within a sentence, as it is a common noun and not the name of an inventor.

the technique is quite easy to implement since no surface preparation is required. If not, a revealing fluid can be wiped onto the surface to make it reflective.

iv) Digital image correlation: this is the speckle implementation of the same concept, and the surface points are marked using the surrounding local random contrast pattern.

2) Shape measurement: Shape measurement techniques derive from displacement measurement techniques. The fundamental geometrical arrangement (Figure 1.8) is the same for stereocorrelation, where two cameras observe the object from two different points of view, and for fringe projection (or structured light projection). If the position in space of the optical centers C_1 and C_2 is known, knowing the position of points M_1' and M_2' allows the position of point M to be determined. The difference between the two techniques lies in the trajectory and nature of the information:

i) With stereocorrelation, the information starts from point M and propagates toward points M_1' and M_2', located on two camera sensors; the analysis software has to recognize the local signature around M in the images to pair M_1' and M_2'.

ii) With fringe projection, the information (which is quantitative as it is a fringe phase, proportional to the abscissa or ordinate of M_1') starts from M_1', which is most often a video projector pixel, and propagates toward M and afterward M_2'. The analysis software has only to detect the phase and unwrap it[5] to know exactly the emitting point because its coordinates are known from the phase.

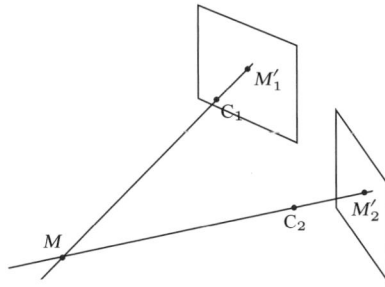

Figure 1.8. *Shape measurement using stereocorrelation or fringe projection*

1.4.2. Interference methods

The temporal phase $\psi(t) = \omega t + \phi$ of an electromagnetic wave in the visible domain, such as the one emitted by an helium–neon ($\lambda = 633$ nm) or YAG ($\lambda = 532$ nm) laser, vibrates at a frequency around one hundred terahertz (10^{14} Hz).

---

5 In this technique, unwrapping should also preserve an absolute phase origin.

There does not exist any sensor capable of recording temporal variations at such high frequencies, and they are only sensitive to the average light power value. This is why the interference phenomenon is useful to obtain phase information. The nature of interference is a beat phenomenon between two periodic signals of same frequency, as in stroboscopy or moiré, that will remove the term ωt present in ψ. The formula that gives the light intensity in two-wave interferences is:

$$I = I_0(1 + \gamma \cos \Phi) \tag{1.23}$$

where γ is the visibility (or contrast) and the phase Φ of the interference fringes is:

$$\Phi = \psi_2 - \psi_1 = \phi_2 - \phi_1 = 2\pi \frac{\delta}{\lambda} \tag{1.24}$$

where δ is the optical path difference between the two interfering beams and λ is the wavelength in the propagation medium, most often air. In fact, an interferometric measurement is always performed between an initial state that acts as the reference (due to imperfections, we never have an initial state without any interference fringes) and a final state. It is therefore always an *interference fringe phase variation*:

$$\Delta\Phi = (\phi_{2,\text{final}} - \phi_{1,\text{final}}) - (\phi_{2,\text{initial}} - \phi_{1,\text{initial}}) \tag{1.25}$$

which is used (double difference).

1.4.2.1. *Light-surface interactions*

Depending on the way the incident light, having smooth (plane or spherical) or rough (random) wave fronts, interacts with the surface under examination, different interferometry techniques can be used. The different possible interactions are outlined in Figure 1.9.

– In the first case, the surface is a mirror. All the energy is reflected in a single direction, following Snell's law, and the wave front remains smooth.

– In the second case, the surface is rough (diffusive) and the energy spreads in all directions. The wave front shape becomes very complicated, which is the reason why they are referred to as "random"; this will create spatial noise in an image obtained by an imaging system placed downstream. This spatial noise is called speckle.

– In the last case, a diffraction grating[6] is deposited (in general, molded) onto the surface so that the first diffracted order is normal to the surface. In this case, a noticeable amount of energy is re-emitted along the normal direction, and the wave front remains smooth.

---

6 A grating is a very dense pattern of straight parallel lines. A grating has the property of splitting an incident ray into many different rays (called diffraction orders) having different directions.

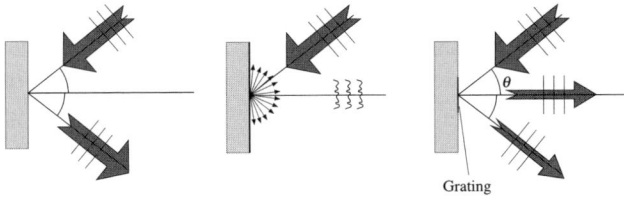

Figure 1.9. *Different interaction modes between incident light and a surface:*
specular (mirror-like) reflection, diffusion, diffraction

1.4.3. *Sensitivity vector*

When a surface point that is illuminated by a monochromatic wave undergoes a small displacement characterized by the vector $\vec{u}$, the phase variation of the beam coming from the light source and re-emitted toward the observer T is calculated from the variation in the optical path. It is easily shown that this phase variation $\Delta\phi$ is

$$\Delta\phi = \vec{g}.\vec{u} \qquad\qquad [1.26]$$

where the difference between the illumination and observation wave vectors $\vec{g} = \vec{k}_i - \vec{k}_o$ is called the *sensitivity vector*. Its direction indicates *the displacement component the measurement will be sensitive to*. In the case of a mirror-like reflection, the sensitivity vector is normal to the surface; thus, only out-of-plane displacements can be measured (Michelson interferometry).

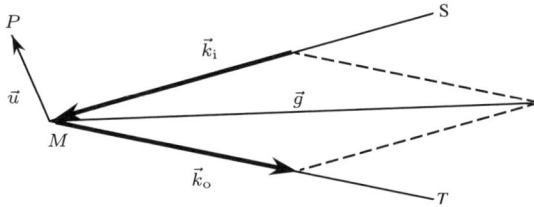

Figure 1.10. *Sensitivity vector*

1.4.4. *Synthetic sensitivity vectors*

After a measurement, an interference fringe phase variation field is obtained, which is locally proportional to the sensitivity vector $\vec{g}$. If different mesurements of the same kinematic field can be made with different illumination conditions, different phase variations for the same displacement are obtained:

$$\begin{cases} \Delta\phi_0 &= \vec{g}_0.\vec{u} \\ \Delta\phi_1 &= \vec{g}_1.\vec{u} \\ &\cdots \\ \Delta\phi_{N-1} &= \vec{g}_{N-1}.\vec{u} \end{cases} \qquad\qquad [1.27]$$

With a simple linear combination of these equations, that is of the obtained phase images, the displacement vector appears factorized with the corresponding linear combination of sensitivity vectors, which is called a *synthetic sensitivity vector*:

$$\sum_{i=0}^{N-1} \alpha_i \Delta \phi_i = \left(\sum_{i=0}^{N-1} \alpha_i \vec{g}_i \right) . \vec{u} = \vec{G} . \vec{u} \qquad [1.28]$$

Such a linear combination may also appear in an analog way with a setup that uses *simultaneously* different illumination directions, as in interferometric moiré (see section 1.4.5.2).

1.4.5. *The different types of interferometric measurements*

These simply correspond to the various possible interactions of the incident beam with the surface presented in Figure 1.9 and to the different ways of choosing the interfering beams. In the following, we will use the term "object beam" to designate the incident beam; it is the beam that illuminates the object.

1.4.5.1. *Interference between the object beam and a fixed reference beam*

In this case, we have for the double difference formula [1.25]:

$$\Delta \Phi = \Delta \phi_{\text{obj}} - \Delta \phi_{\text{ref}} = \Delta \phi_{\text{obj}} = \vec{g} . \vec{u} \qquad [1.29]$$

so it can be seen that the measured quantity is the *displacement component along the sensitivity vector* $\vec{g}$. For example, the measurement of the approximate out-of-plane displacement using interferometry in diffuse light (speckle interferometry) is shown in Figure 1.11, where the measured displacement component u_g is aligned along the bisector of the illumination and observation directions. It can be seen that there is some degree of approximation, as this bissector is not exactly normal to the surface. Existing commercial systems may or may not take this effect into account.

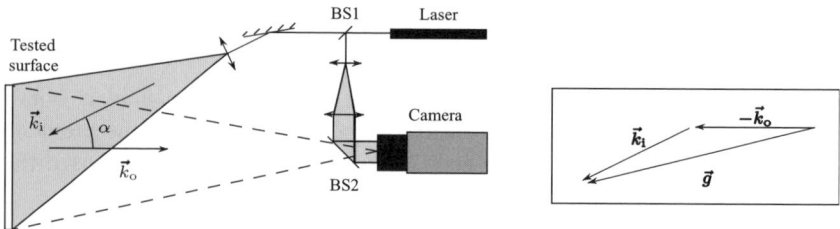

Figure 1.11. *Interferometry in diffuse light for the measurement of out-of-plane displacements ("out-of-plane speckle" setup)*

1.4.5.2. *Interference between two object beams coming from the same point*

The two beams come from the same object point, but with different incident illumination directions, so the sensitivity vectors are different. Then, the double difference formula [1.25] is written:

$$\Delta\Phi = \Delta\phi_2 - \Delta\phi_1 = (\vec{g_2} - \vec{g_1}).\vec{u} = \vec{G}.\vec{u} \qquad [1.30]$$

Thus, we obtain, in an "analog" way, a synthetic sensitivity vector. As a typical example, we have what is often referred to as interferometric moiré, but is more relevantly called *grating interferometry*, outlined in Figure 1.12.

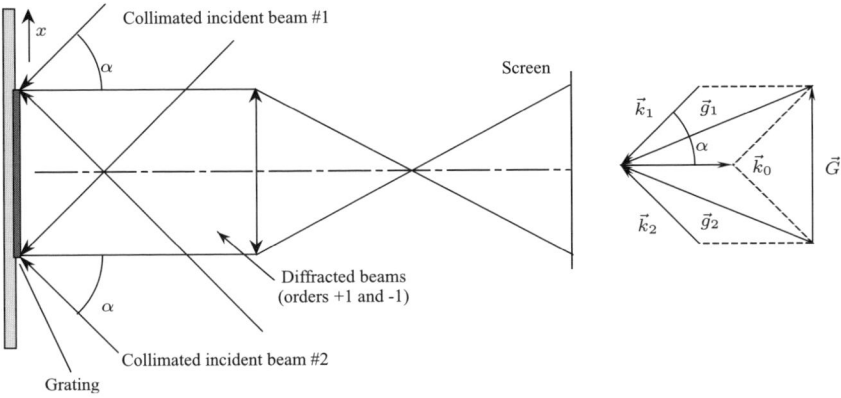

Figure 1.12. *Measurement of in-plane displacements: grating interferometry (also called interferometric moiré); corresponding sensitivity vector*

Regarding grating interferometry, the main drawback is the difficulty in depositing the grating on the surface. A typical maximum value of the grating size that can be placed is 40–50 mm. The main advantage is the absence of speckle noise in the image, leading to excellent spatial resolution for the measurements.

The same approach can be used with diffuse light, as indicated in Figure 1.13. The major difference (not speaking of the absence of the complicated process of depositing a grating onto the surface) is that the incident beams are not collimated in general, in order to investigate large surfaces.

1.4.5.3. *Interference between object beams coming from two contiguous points (differential interferometry, shearography)*

A splitting system placed in the imaging arm of the setup enables us to obtain on the sensor two images of the object, laterally shifted by a small quantity $\delta\vec{l}$ (Figure 1.14). The double difference formula [1.25] becomes in this case:

$$\Delta\Phi = \Delta\phi_2 - \Delta\phi_1 = \vec{g}.(\vec{u} + \delta\vec{u}) - \vec{g}.\vec{u} = \vec{g}.\delta\vec{u} \qquad [1.31]$$

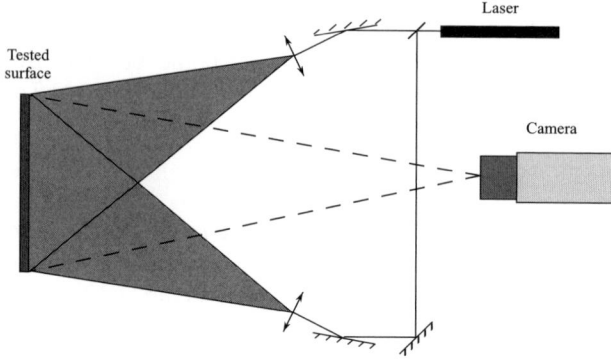

Figure 1.13. *Measurement of in-plane displacements*
("in-plane speckle setup")

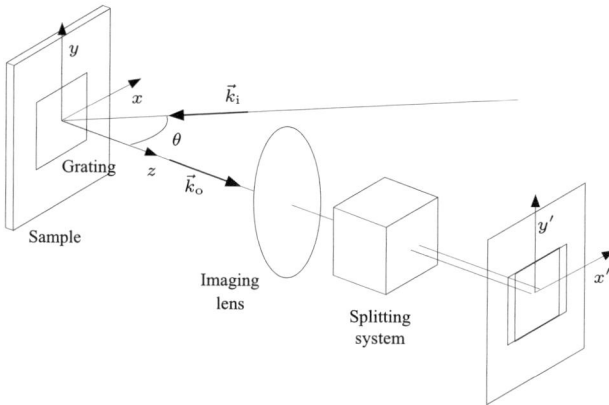

Figure 1.14. *Differential grating interferometry, or*
"shearography" if no grating is present

What is measured is then a *differential displacement*, which is very interesting for mechanical engineering, since the differential displacement basically relates to strain. With different illumination directions, we can acquire different phase fields that can be combined in order to obtain synthetic sensitivity vectors enabling the measurement of in-plane and out-of-plane displacements. Once again, grating interferometry or interferometry in diffuse light can be used (in other words, differential interferometry can be used with or without a grating), with the already mentioned drawbacks and advantages. The term "shearography" [HUN 07] is in general reserved for realizations in diffuse light.

Regarding terminology, it should be noted that the term "shear" which appears in many papers dealing with differential interferometry (and from which "shearography" is derived) designates the lateral displacement introduced by the splitting component in the imaging system. It has nothing to do with "shear" as the non-diagonal strain component as understood by mechanical engineers. When aligning the lateral shift direction with the x and y axes, and using different illumination directions, all the derivatives of all component displacements can be obtained. Thus, we can obtain all the in-plane strain components and all the slope[7] (out-of-plane displacement gradient) components.

To conclude, let us emphasize a very interesting feature of differential interferometry: it is basically insensitive to vibrations, as vibrations introduce rigid body displacements but not displacement gradients, making the technique very usable in industrial environments. This is why shearography is now becoming accepted as a non-destructive technique in the aeronautical industry.

1.4.6. *Holography, digital holography*

Holography is not, strictly speaking, a measurement technique; it is a *recording* technique for an electromagnetic field, both in amplitude and in phase, so that it can be reconstructed afterward. The principle is to make the field interfere with a so-called "reference" beam, and to record the resulting interference fringes. The recording used to be photographic, with special high-resolution photographic plates, but is now mostly digital.

Holography offers the possibility to create interferences between waves that do not exist simultaneously, for example waves diffused by an object before and after deformation. Thus, holography can be seen as a complement to all the interferometric techniques presented before.

It should be noted that there is no real difference between digital holography and the interferometric setup using a smooth reference beam like the one presented in Figure 1.11, as it is in both cases the recording of interferences between an object and a reference beam.

The only difference is that in cases called "speckle interferometry", the recording takes place in a location where the object has already been imaged, so the object reconstruction required in digital holography is not necessary. When we remove the camera lens shown in Figure 1.11 and reconstruct the object from the recorded interference fringes, we perform digital holography.

---

7 We are speaking here of slopes introduced by mechanical loading.

1.4.7. *Conclusion*

We have presented in this section an approach to classify a large number of optical techniques in terms of their fundamental features (white light or interferometric techniques, random encoding or phase modulation, nature of the measurand). Table 1.1 proposes a very short synthesis, briefly indicating their performances. Of course, such a table is only a very coarse presentation of reality. However, it is hoped that it may be useful for a first approach to the choice of a full-field optical measurement technique.

White light			
Measurand	Random encoding	Phase modulation	Remarks
In-plane disp.	Image correlation	Grid (with or without moiré)	Coupling with out-of-plane disp.
Shape	Stereocorrelation	Fringe projection (structured light)	Transverse calibration of camera required
3D disp.	Stereocorrelation with deposited speckle	Impossible	*id*.
Slopes	Not used	Deflectometry	Coupling with shape
Interferometry			
Measurand	Reflected light	Diffuse light	Diffracted light
In-plane disp.	Impossible	In-plane speckle	Interferometric moiré
Out-of-plane disp.	Michelson interferometry	Out-of-plane speckle	Not used, but a setup is conceivable
Differential setup (slopes, strain)	For example Nomarski microscopy	Shearography	Grating differential interferometry
Performances			
	White light, random	White light, phase mod.	Interferometry
Simplicity	++	+	−
Cost	−	−	+
Performances	−	−+	++

Table 1.1. *Synthesis*

1.5. Bibliography

[AST 08] Standard Guide for Evaluating Non-Contacting Optical Strain Measurement Systems, ASTM International, 2008.

[BIS 03] BISSIEUX C., PRON H., HENRY J.-F., "Pour de véritables caméras matricielles de recherche", *Contrôles Essais Mesures*, vol. 2, pp. 39–41, January 2003.

[BOR 09] BORNERT M., BRÉMAND F., DOUMALIN P., DUPRÉ J.-C., FAZZINI M., GRÉDIAC M., HILD F., MISTOU S., MOLIMARD J., ORTEU J.-J., ROBERT L., SURREL Y., VACHER P., WATTRISSE B., "Assessment of digital image correlation measurement errors: methodology and results", *Experimental Mechanics*, vol. 49, no. 3, pp. 353–370, June 2009.

[HON 05] HONORAT V., MOREAU S., MURACCIOLE J.-M., WATTRISSE B., CHRYSOCHOOS A., "Calorimetric analysis of polymer behaviour using an IRFPA camera", *International Journal on Quantitative Infrared Thermography*, vol. 2, no. 2, pp. 153–172, 2005.

[HUN 07] HUNG M.Y.Y., CHEN Y.S., NG S.P., SHEPARD S.M., HOU Y., LHOTA J.R., "Review and comparison of shearography and pulsed thermography for adhesive bond evaluation", *Optical Engineering*, vol. 46, 2007.

[ROD 78] RODDIER F., *Distributions et transformation de Fourier*, McGraw-Hill, Paris, 1978.

[VIM 12] International Vocabulary of Metrology–Basic and General Concepts and Associated Terms, JCGM, 2012, available at http://www.bipm.org/en/publications/guides/vim.html.

Chapter 2

Photoelasticity

2.1. Introduction

Photoelasticity is one of the oldest techniques in photomechanics. It is essentially based on the scientific discoveries of the 19th Century, especially studies on light propagation in transparent media. During the 20th Century, it was mainly used to visualize the distribution of stress on experimental models, in order to design and optimize the shape of complex mechanical structures (in aeronautics, automotive and civil engineering) [KOB 87]. Many experimental analysis techniques were then developed for both research and industrial purposes. Since the advent of computer software, the use of photoelasticity in industrial design has become less prominent. Its main use is now the validation of computer codes, since there is always a need to test the difference between experimentation and modeling. This is facilitated by the development of three-dimensional tools and automatic procedures for fringe pattern analysis developed using image processing.

This chapter attempts to give an indepth overview of photoelasticity: its theoretical foundations in the first three sections and its use in plane stress (sections 2.5 and 2.6); but also recent advances, including the latest techniques for automatic analysis (sections 2.7 and 2.8), and three-dimensional photoelasticity (section 2.9).

Chapter written by Fabrice BRÉMAND and Jean-Christophe DUPRÉ.

2.2. Concept of light polarization

Let us recall that, according to electromagnetic theory, light is a wave composed of an electric field E and a magnetic field H. These two fields are perpendicular to the direction of light propagation.

The polarization state of light is given by the direction of the vector representing the electric or magnetic field. For example, if the electric field E has a constant direction during propagation, then the light wave is linearly polarized. If the orientation of the field changes, the wave has elliptical or circular polarization (Figure 2.1).

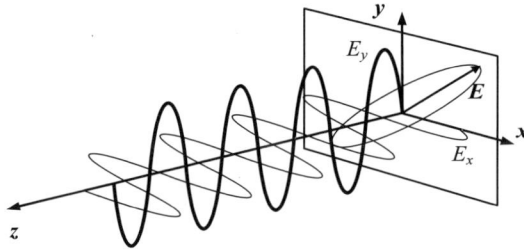

Figure 2.1. *Representation of the electric field components of a wave with elliptical polarization*

A monochromatic wave propagating along the direction z, defined by its wave vector k, its angular frequency ω and its amplitude A, can be written as:

$$E = A\cos(\omega t - k.z)$$
[2.1]

In the plane *(x, y)* this expression becomes:

$$\begin{cases} E_x = A_x \cos(\omega t - k.z + \phi_x) \\ E_y = A_y \cos(\omega t - k.z + \phi_y) \end{cases}$$
[2.2]

The phase shift $\phi = \phi_y - \phi_x$ can describe two particular cases:

$$\begin{cases} if \ \phi = n\,\pi \ \left(n \ integer\right) \ the \ polarization \ is \ linear \\ if \ \phi = \pm\dfrac{\pi}{2} + n\,\pi \ and \ A_x = A_y \ the \ polarization \ is \ circular \end{cases}$$
[2.3]

Polaroid filters are currently the most widely-used as polarizers. They are made of a stretched polyvinyl alcohol film. After treatment with a liquid rich in iodine, these filters contain some elongated and parallel molecular chains. These chains become conductive and they absorb the electric field when it is parallel to the direction of the chains, allowing the transmission of the perpendicular component. Polarization can also be performed by reflection and scattering. We will see the use of diffusion phenomenon as a polarizer later.

2.3. Birefringence phenomenon

The birefringence phenomenon is a physical property of some transparent materials that are not optically isotropic, such as quartz crystals. In fact, this results in the refractive index varying with the direction of light propagation. A polarized light vector passing through a birefringent plate is split into two perpendicular components propagating at speeds v_1 and v_2. If $v_1 > v_2$, the 1-axis is called the fast axis and the 2-axis is called the slow axis (see Figure 2.2).

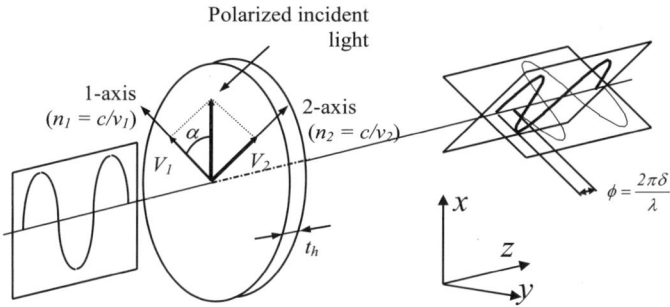

Figure 2.2. *Effect of a thin birefringent plate on polarized light (with c being the speed of light in a vacuum)*

The optical path difference δ between the two rays (also called birefringence) is given by relationship [2.4], where t_h represents the thickness of the plate:

$$\delta = (n_2 - n_1)\, t_h \tag{2.4}$$

The phase shift between the two components (defined in the previous section, equation [2.3]), called the angular birefringence, is defined as follows if λ is the wavelength of the light:

$$\phi = 2\pi\, \delta / \lambda \tag{2.5}$$

For any value of ϕ, the output light is elliptical. Following relationship [2.3], light can be also linear or circular.

Two important and special cases can be observed:

– Quarter-wave plate, when the optical path difference is $\lambda/4$. According to relationship [2.5], the phase shift introduced by this plate is $\pi/2$. If polarized light arrives with a polarization angle of $\pi/4$ with respect to one of the axes of the plate $(Ax = Ay)$, then the output light is circular.

– Half-wave plate. When the optical path difference is $\lambda/2$, a similar analysis gives a phase shift of π. If polarized light arrives with a polarization angle of θ, the half-wave plate creates an orientation shift of 2θ. This plate makes it possible to rotate the direction of polarization.

2.4. The law of optico-mechanics

The phenomenon of birefringence may appear on optically isotropic materials such as glass or certain polymers (epoxy, polycarbonate, etc.) that have the property of becoming birefringent when subjected to stress. Initially, the material that is considered optically isotropic is characterized by a refractive index n_0 for all directions. Under stress this medium is characterized by an ellipsoid of refractive indices. This phenomenon was discovered by Brewster in 1816. The ellipsoid of refractive indices is such that its principal directions coincide with the principal directions of stresses. In the 19th Century, Maxwell determined the laws linking the principal optical indices to the principal stresses:

$$n_1 = n_0 + C_1\,\sigma_1 + C_2\,(\sigma_2 + \sigma_3)$$
$$n_2 = n_0 + C_1\,\sigma_2 + C_2\,(\sigma_3 + \sigma_1)$$
$$n_3 = n_0 + C_1\,\sigma_3 + C_2\,(\sigma_1 + \sigma_2)$$

[2.6]

where C_1 and C_2 are called the absolute photoelastic constants of the material.

Independently of Maxwell, Neumann determined the laws linking the principal refractive indices to the principal strains.

In particular, for plane stress problems, we have:

$$n_1 = n_0 + C_1\,\sigma_1 + C_2\sigma_2$$
$$n_2 = n_0 + C_1\,\sigma_2 + C_2\sigma_1$$

[2.7]

We can note that for photoelastic materials, variations in the refractive indices $(n_1 - n_0)/n_0$ and $(n_2 - n_0)/n_0$ are in the range of 10^{-3}.

By eliminating n_0 from relations [2.7], we obtain:

$$n_2 - n_1 = (C_1 - C_2)(\sigma_2 - \sigma_1) = C(\sigma_2 - \sigma_1)$$

[2.8]

where C is called the stress-optics or photoelastic constant of the material. C has the dimension of the inverse of a stress and is expressed in Brewsters (Bw) as:

$$1\ \text{Bw} = 10^{-12}\ \text{m}^2/\text{N} = 10^{-12}\ \text{Pa}^{-1}$$

[2.9]

The following table gives the magnitudes of photoelastic constants for some common materials:

Material	C in Bw
Photoelastic glass	2
Plexiglass	4
Epoxy	55
CR39	35
Polyester	82
Polyurethane	3,200

Table 2.1. *Photoelastic constants for common materials*

2.5. Several types of polariscopes

2.5.1. *Plane polariscope*

This is an experimental set-up formed by a light source giving a plane wave, two Polaroids (the one closest to the light source being the polarizer and the one nearest the observer being the analyzer), a measuring device (photomultiplier, photographic film, eye, digital camera) and the model to be analyzed.

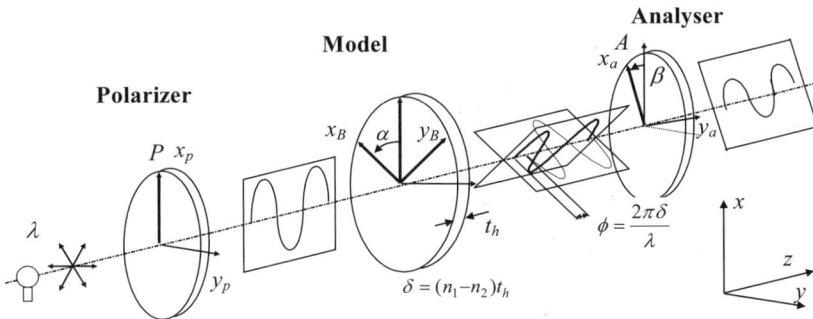

Figure 2.3. *Principle of a plane polariscope*

Ox_B and Oy_B are the principal directions of stress in the loaded model. At the entry of the model, considered the origin of the z-axis (Figure 2.3), there is a plane-polarized wave along direction Ox_P.

$$E = A \begin{bmatrix} \cos(\omega t) \\ 0 \end{bmatrix}_{x_p y_p}$$

[2.10]

Relationship [2.10] can be represented as follows by projection in the directions of principal stress:

$$E = A \begin{bmatrix} cos(\alpha)cos(\omega t) \\ -sin\,\alpha\,cos(\omega t) \end{bmatrix}_{x_B y_B}$$
[2.11]

At the output of the model with a thickness of t_h, one of the components emerges with an angular retardation ϕ:

$$E = A \begin{bmatrix} cos(\alpha)cos(\omega t) \\ -sin(\alpha)cos(\omega t - \phi) \end{bmatrix}_{x_B y_B}$$
[2.12]

To simplify calculation, from now on we will use a complex notation:

$$E = A \begin{bmatrix} cos(\alpha)e^{j\omega t} \\ -sin(\alpha)e^{j(\omega t - \phi)} \end{bmatrix}_{x_B y_B}$$
[2.13]

Between the model and the analyzer, an equal phase shift occurs on the two waves, which can be eliminated by a new change of origin. By projecting along the axes of the analyzer, we obtain:

$$E = Ae^{j\omega t} \begin{bmatrix} cos(\beta - \alpha)cos(\alpha) - sin(\beta - \alpha)sin(\alpha)e^{-j\phi} \\ -sin(\beta - \alpha)cos(\alpha) - cos(\beta - \alpha)sin(\alpha)e^{-j\phi} \end{bmatrix}_{x_A y_A}$$
[2.14]

Since the analyzer only allows the component along Ox_A of the polarizer, it only retains:

$$E = Ae^{j\omega t} \begin{bmatrix} cos(\beta - \alpha)cos(\alpha) - sin(\beta - \alpha)sin(\alpha)e^{-j\phi} \\ 0 \end{bmatrix}_{x_A y_A}$$
[2.15]

The light intensity recorded by a measurement device is then given by:

$$I = EE^* = I_0 \left[cos^2\beta - sin\,2\alpha\,sin\,2(\alpha - \beta)sin^2\frac{\phi}{2} \right]$$
[2.16]

with $I_0 = A^2$ as the intensity of the background.

By varying angle β, we find the expressions of light obtained by a parallel arrangement plane polariscope ($\beta = 0$) [2.17] or by a crossed arrangement plane polariscope ($\beta = \pi / 2$) [2.18]:

$$\text{Parallel arrangement } I = I_0 \left[1 - sin^2 \, 2\alpha \, sin^2 \, \frac{\phi}{2} \right] \qquad [2.17]$$

$$\text{Crossed arrangement } I = I_0 \left[sin^2 \, 2\alpha \, sin^2 \, \frac{\phi}{2} \right] \qquad [2.18]$$

The following example (see Figure 2.4) shows the image obtained through a plane polariscope with a crossed arrangement for an epoxy disk subjected to diametral compression. The analysis of light zones (called white fringes) and dark zones (called black fringes) provides knowledge of the principal differences in stress.

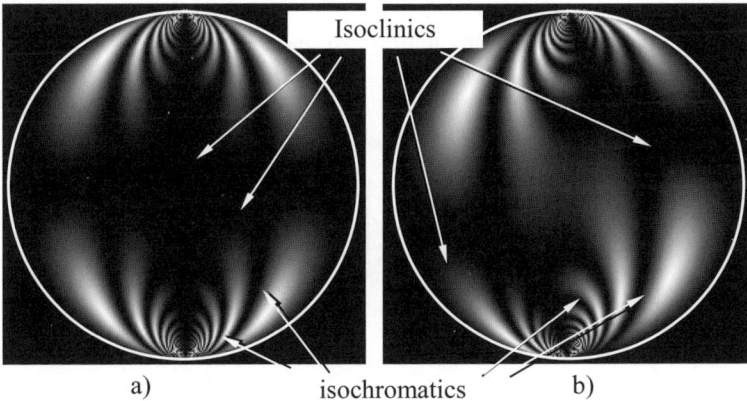

Figure 2.4. *Dark field plane polariscope for a disk under diametral compression for two orientations: a) $\alpha=0°$ and b) $\alpha=15°$*

Outside the model or in unstressed conditions, we obtain $I = 0$, so this crossed plane polariscope is also called a dark field.

Thus, in such a polariscope described in relation [2.18], the light intensity I is zero (dark fringes) when:

$$- sin \, 2\alpha = 0 \Rightarrow \alpha = k \, \pi/2$$

These lines are called *isoclinics*; they are independent of the loading intensity and the wavelength of the light. They can change either by rotating the model along the z-axis inside the polariscope or, more easily, by rotating both the polarizer and analyzer. They indicate the location of points where one of the principal orientations of stress is parallel to the direction of polarization.

$$- sin \ \phi/2 = 0 \Rightarrow \phi = 2\pi k \Rightarrow k = t \ C \ (\sigma_2 - \sigma_1)/\lambda.$$

These lines are called *isochromatics*; they represent equal maximum shear stress $(\sigma_2 - \sigma_1 = cte)$ and they are dependent on the intensity of the load applied and the wavelength used. Along an isochromatic, the optical path difference is equal to an integer number of wavelengths. These lines are called isochromatics because when observed with white light (large spectrum) they are the same color.

2.5.2. *Circular polariscope*

A circular polariscope comprises a light source giving a plane wave, a circular polarizer (see section 2.3), a quarter-wave plate, the model studied, a second quarter-wave plate, the analyzer and a recording device. We can obtain different arrangements depending on the orientation γ of the second quarter-wave plate.

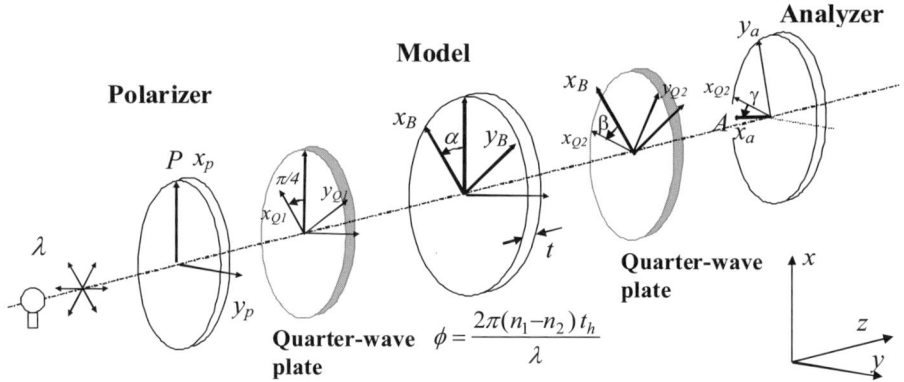

Figure 2.5. *Scheme of a circular polariscope*

A similar approach to that of the plane polariscope allows us to calculate the following expression for the output light intensity:

$$I = \frac{I_0}{2}(1 + sin \ 2\gamma \ cos \ \phi + sin \ 2\beta \ cos \ 2\gamma \ sin\phi) \qquad [2.19]$$

When $\gamma = -\pi/4$, we obtain a dark field circular polariscope arrangement; the light intensity is given by:

$$I=\frac{I_0}{2}(1-cos\,\phi)=I_0\,sin^2\frac{\phi}{2} \hspace{2cm} [2.20]$$

We can note that with this arrangement, isoclinics have been eliminated (see Figure 2.6). Outside the model, or when the model is unstressed, the emergent light intensity is zero. Consequently we have a dark field, and the dark fringes are given by an integer number.

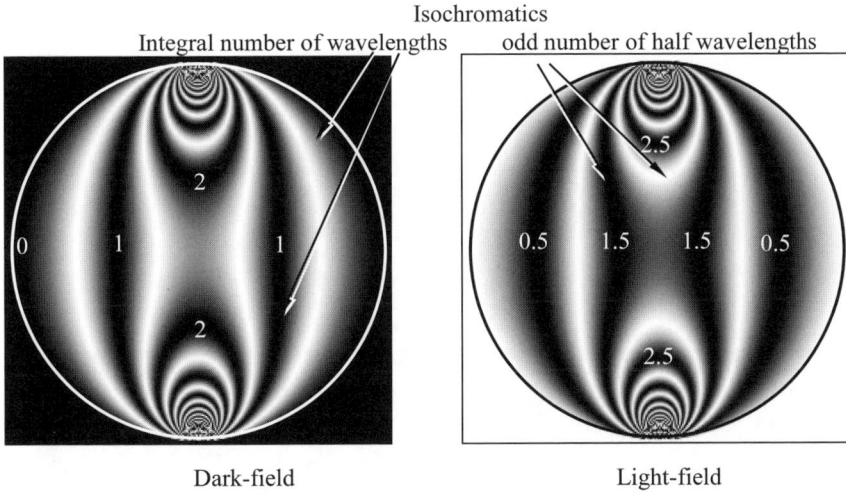

Figure 2.6. *Isochromatic patterns obtained with a circular polariscope*

When $\gamma=\pi/4$, we obtain a light-field circular polariscope arrangement; the emergent light intensity is given by:

$$I=\frac{I_0}{2}(1+cos\,\phi)=I_0\,(cos^2\frac{\phi}{2})=I_0(1-sin^2\frac{\phi}{2}) \hspace{1cm} [2.21]$$

With this configuration, outside the model or when the model is unstressed the emergent light intensity is at a maximum. Consequently we are in the presence of a light field, and the dark fringes are given by an odd number of half wavelengths:

$$I=0 \;\Rightarrow\; cos\frac{\phi}{2}=0 \;\Rightarrow\; \phi=(2k+1)\pi=\frac{2\pi\delta}{\lambda} \;\Rightarrow\; \phi=(k+\frac{1}{2})\lambda$$

$$[2.22]$$

2.5.3. *White light polariscope*

The use of white light has some advantages. First, it facilitates detection of the zero-order fringe isochromatics, since it is the only black fringe in a dark field polariscope (Figure 2.7), while other isochromatics appear in color. Second, with traditional image processing software, we can easily separate a color image into its three main components – blue, green and red (see Figure 2.7). Note that the number of blue fringes is greater because the blue wavelength is smaller. The case is the opposite for a red image. Thus it becomes possible to perform an analysis by multiplying the fringes.

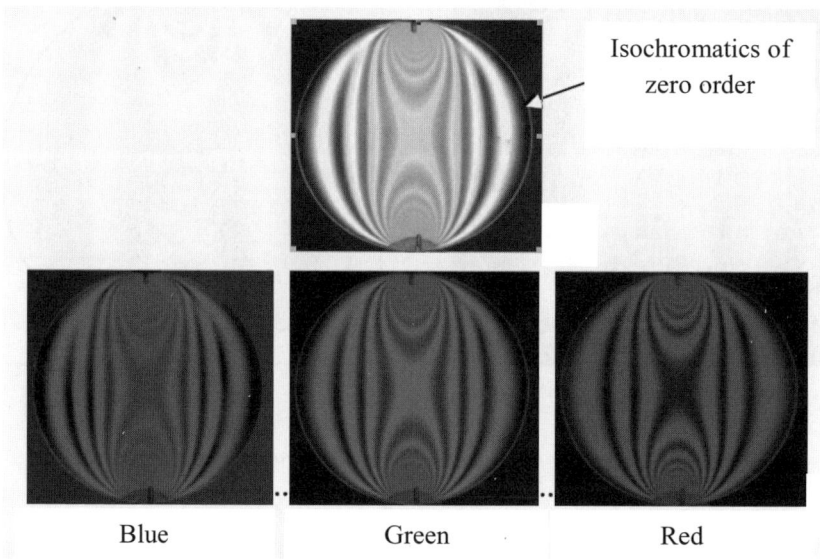

Isochromatics of
zero order

Blue   Green   Red

Figure 2.7. *Isochromatic fringe patterns obtained with a dark-field circular polariscope using white light. For a color version of this figure, see www.iste.co.ut/gh/solidmech.zip*

2.5.4. *Photoelastic coating*

Photoelastic coatings are widely used in industry to determine shear stress at the surface of structural components of a complicated shape (Figure 2.8b) when the material is opaque. A photoelastic (Figure 2.8a) plate is bonded with a reflective adhesive. If this coating is thin compared to the size of the specimen, the surfacic deformations are fully transmitted. Since the material is birefringent, we can see fringes as in conventional photoelasticity (Figure 2.8) [KOB 87].

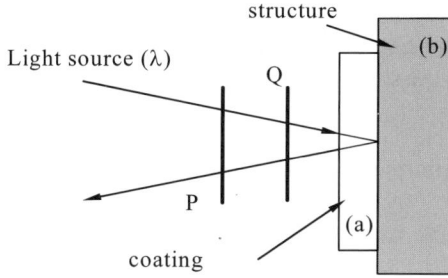

Figure 2.8. *Photoelasticimetry by photoelastic coating*

Moreover, it is reasonable to assume that the normal stress is zero, which means that the photoelastic layer does not disturb the deformation of the structure. This leads to the stress state in the photoelastic material being a mean plane stress state. Thus:

$$\sigma_{3a} = 0, \ \sigma_1 \text{ and } \sigma_2 \text{ are uniform} \qquad [2.23]$$

This situation is valid for plane structures or structures with a large curvature radius. In these conditions, the photoelastic material and the surface to be analyzed have the same principal directions of strain. Thus we can write:

$$(\varepsilon_1 - \varepsilon_2)_a = (\varepsilon_1 - \varepsilon_2)_b \qquad [2.24]$$

Since for each material (a) and (b), we have:

$$\varepsilon_1 = \frac{1}{E}(\sigma_1 - \nu\sigma_2) \ and \ \varepsilon_2 = \frac{1}{E}(\sigma_2 - \nu\sigma_1) \qquad [2.25]$$

in normal incidence with $\delta = 2 \ C \ e \ (\sigma_1 - \sigma_2)_a$ we obtain:

$$(\sigma_1 - \sigma_2)_b = \frac{1 + \nu_a}{E_a}\frac{E_b}{1 + \nu_b}(\sigma_1 - \sigma_2)_a = \frac{1 + \nu_a}{E_a}\frac{E_b}{1 + \nu_b}\frac{\delta}{2Ce} \qquad [2.26]$$

Relationship [2.26] shows that the difference of principal stress in the structure can be determined by measuring the birefringence of the photoelastic coating in normal incidence, when the mechanical characteristics of each material and the photoelastic constant are known. Furthermore, the isoclinics for material (a) are still valid for material (b).

Additional measurements could be made in oblique incidence to determine local stresses. Nevertheless, this approach is not generally sufficiently accurate, particularly because of the stress gradient.

2.6. Measurement of photoelastic constant *C*

Using the previous experimental setup, we visualized the isoclinic and isochromatic fringe patterns. Thus it was possible to determine the orientation of the principal stresses (isoclinics) and the maximum shear stress (isochromatics) at each point of the model when the photoelastic constant *C* is known.

In a plane polariscope, a uniaxial tension test is carried out using a polarizer that is set at $\pi/4$ relative to the direction of loading (which is the direction of principal stress), so no isoclinics are visible. With relationship [2.18], when $\alpha = \pi/4$ we can write:

$$I = I_0 \sin^2 \frac{\phi}{2} = \frac{I_0}{2}(1 - \cos \phi)$$

[2.27]

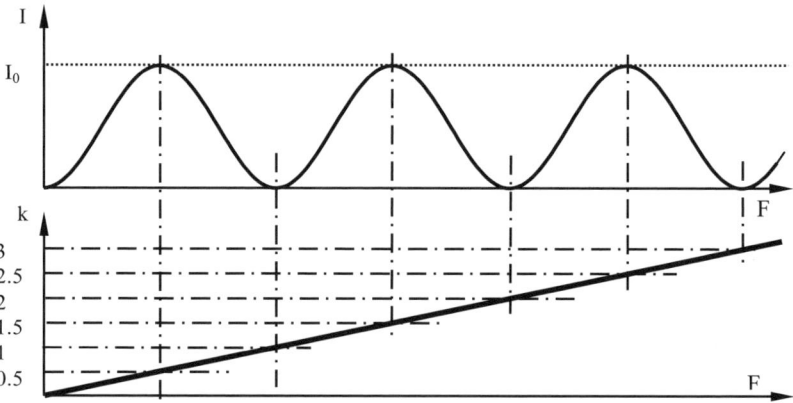

Figure 2.9. *Measurement of the photoelastic constant*

The output light intensity *I* is recorded against the load *F* (see Figure 2.9), which allows the extraction of the fringe order *k* with respect to *F*. In a uniaxial tension test ($\sigma_2 = 0$), if *w* is the width of the specimen, we have:

$$\delta = k\lambda = (n_1 - n_2)t = Ct(\sigma_1 - \sigma_2) = Ct\frac{F}{t\,w}$$

[2.28]

Finally, the slope of the linear response k against F gives C by:

$$C = \frac{k}{F} \, w \, \lambda \qquad\qquad [2.29]$$

2.7. Analysis by image processing

The automatic analysis of photoelastic fringes has been investigated by many authors. Several decades ago, this analysis was generally made point-by-point (Tardy method, Senarmon, etc. [KOB 87]). With the advent of digital acquisition and image processing systems, new techniques using various polariscope arrangements have been developed ([ROB 83], [PAT 91], [ASU 93], [MOR 94], [FOL 96], [LAO 98]). Another approach is based on the use of several wavelengths ([KIH 94], [UME 89]). These procedures, developed to study flat samples in transmission, can be adapted to almost any photoelastic coatings [PAT 98].

Such an analysis is difficult, because two fringe patterns are present in the image. Therefore they must be separated while minimizing the effect of indeterminate values, which can appear when one of these patterns has an amplitude of zero. These two sinusoidal fringe patterns thus have to be analyzed separately. The most common method is to use several polariscope configurations to obtain a demodulated signal, but the quality of the optical elements (polarizers, quarter-wave plate and light source) is important. Since the signal is obtained from trigonometric functions, with discontinuities we then have to apply a phase unwrapping algorithm.

2.7.1. *Using a plane polariscope*

2.7.1.1. *Method with four images*

This is the simplest experimental setup. It consists of using a plane polariscope (see section 2.5.1) with a monochromatic light source and two polarizers [DUP 93].

Let us recall the expression of the output light intensity recorded by a digital camera:

$$I = I_0 \left[cos^2 \, \beta - sin \, 2\alpha \, sin \, 2(\alpha - \beta) \, sin^2 \, \frac{\phi}{2} \right] \qquad\qquad [2.30]$$

From four configurations of the polariscope, we obtain the following four expressions for I:

Angle	Light intensity I
$\beta = 0$	$I_1 = I_0 \left[1 - \sin^2 2\alpha \sin^2 \dfrac{\phi}{2} \right]$
$\beta = \pi/4$	$I_2 = I_0 \left[\dfrac{1}{2} - \sin 2\alpha \cos 2\alpha \sin^2 \dfrac{\phi}{2} \right]$
$\beta = 0$ and rotation of the analyzer $\alpha \to \alpha + \pi/4$	$I_3 = I_0 \left[1 - \cos^2 2\alpha \sin^2 \dfrac{\phi}{2} \right]$
$\beta = \pi/2$	$I_4 = I_0 \left[\sin^2 2\alpha \sin^2 \dfrac{\phi}{2} \right]$

Table 2.2. *Different polariscope arrangements*

From these four arrangements, it is possible to separate both parameters and to calculate the wrapped values. Three steps are required:

1) Calculation of the intensity $I_0 = I_1 + I_4$.

2) Calculation of α:

$$\alpha = \frac{1}{2} tan^{-1} \left[\frac{2I_2 - I_0}{2I_0 - 2I_3} \right] = \frac{1}{2} tan^{-1} \left[\frac{2I_0 - 2I_1}{2I_2 - I_0} \right]$$ [2.31]

3) Calculation of ϕ:

$$\phi = cos^{-1} \left[\frac{2I_1 + 2I_3 - 3I_0}{I_0} \right].$$ [2.32]

The isoclinic parameter α is obtained from modulo $\pi/2$. An unwrapping technique is then performed [BRE 93] to find a continuous value (see Figure 2.10). For the isochromatic parameter ϕ, which is between 0 and π, the unwrapping technique is more complicated because there is no discontinuity as in the previous case (see Figure 2.11) [FOL 96]. One solution is to record a fifth image with a circular polariscope [DUP 93], which enables the isochromatic parameter ϕ to be found from an inverse tangent function.

The isoclinic parameter α is disrupted close to an isochromatic where expressions [2.31] are indeterminate. The accuracy achieved is a function of the setup and the quality of the polarizers, and therefore varies from one setup to another. It is, however, in the order of one degree when no isochromatics disturb the calculation, but it falls to several tens of degrees when $\phi = 2k\pi$. The isochromatic parameter is obtained with an accuracy of about 15 degrees.

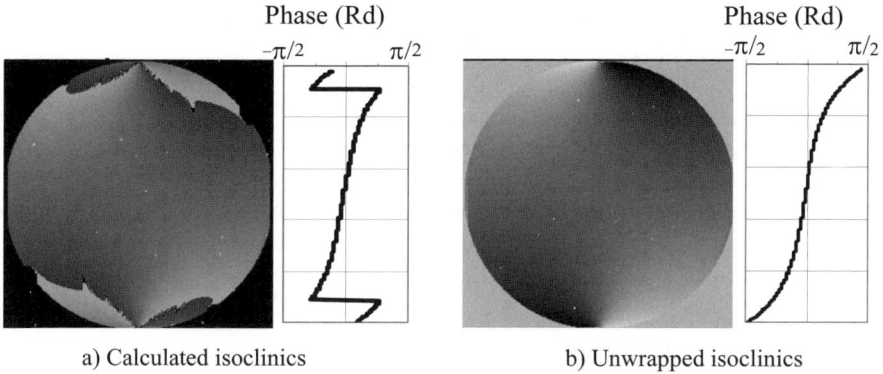

a) Calculated isoclinics   b) Unwrapped isoclinics

Figure 2.10. *Isoclinic parameter mapping and the profile of one column*

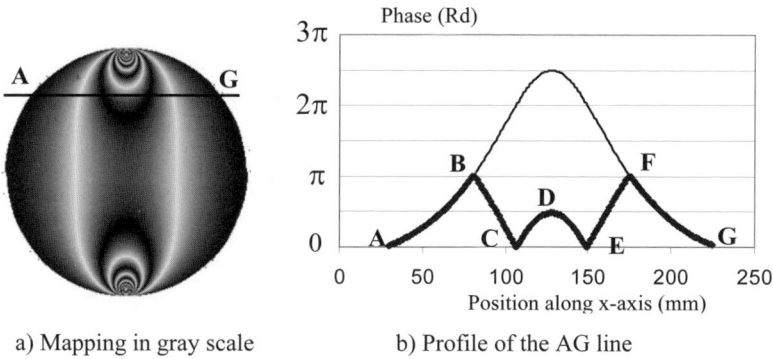

a) Mapping in gray scale   b) Profile of the AG line

Figure 2.11. *Wrapped isochromatic parameter from an inverse cosine function [0, π]*

2.7.1.2. *Analysis using a Fourier transform*

This method again uses the experimental device shown in the previous section. With $\beta = 0$, the technique consists of simultaneously rotating the polarizer and the analyzer [MOR 94] by an angle of θ. Expression [2.17] becomes:

$$I(\theta) = I_0[1 - sin^2 2(\alpha - \theta)sin^2 \frac{\phi}{2}] = I_0[1 - \frac{A}{2}(1 - cos\ 4(\alpha - \theta))] \quad [2.33]$$

For each point of the image, A is a constant ($A = sin^2 \frac{\phi}{2}$).

The signal [2.33] is then a periodical function (with a period of $\pi/2$) with respect to θ. The method consists of calculating the Fourier transform for each pixel of the image as a function of θ.

In zero-order, we obtain the constant part of the signal $I_0(1 - A/2)$, which represents the isochromatic fringe pattern.

When θ varies over $(\pi/2)$, the first order of the spectrum yields the periodic part of signal 4α, and thus the phase at the origin, which is the isoclinic parameter:

$$\alpha = \frac{1}{4} \tan^{-1} [\frac{-\mathrm{Im}(FT_{+1})}{\mathrm{Re}(FT_{+1})}] \text{ , where } \alpha \in [-\pi/4, \pi/4] \tag{2.34}$$

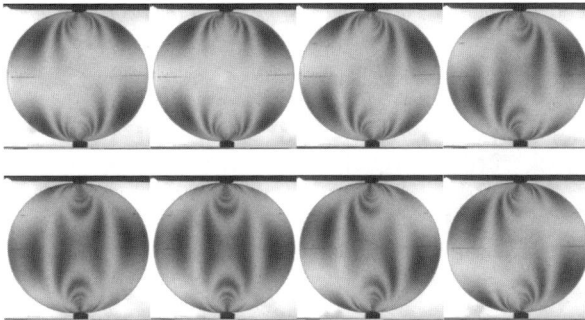

Figure 2.12. *The eight images used for calculation of the isoclinic parameters (plane polariscope)*

The advantage of this method is that we obtain a phase in the Fourier space regardless of variations in light intensity. This technique is less sensitive to the defects of the polariscope and therefore its accuracy is greater. However, it only gives the isoclinic parameter and the isochromatic fringe pattern. Additional procedures must be implemented to extract parameter ϕ:

– by using a circular polariscope in order to find a new combination of images;
or

– by a numerical analysis applied to the fringe pattern [ROB 05].

2.7.2. Using a circular polariscope

The photoelastic model is placed in a circular polariscope. This method consists of rotating the direction of the quarter-wave plate of the analyzer [PAT 91].

Let us recall the relationship giving the output light intensity:

$$I = \frac{I_0}{2}\left(1 + \sin 2\gamma \, \cos \phi + \sin 2\beta \, \cos 2\gamma \, \sin\phi\right) \qquad [2.35]$$

Six arrangements of optical elements are required. They are shown in Table 2.3.

Arrangement	Angle values	Output light intensity
Circular polariscope	$\gamma = \pi/4$, $\alpha+\beta = 0$	$I_1 = I_0/2 \,(1 + \cos \phi)$
	$\gamma = -\pi/4$, $\alpha+\beta = 0$	$I_2 = I_0/2 \,(1 - \cos \phi)$
Output quarter-wave plate and analyzer remaining parallel, with rotation of the polariscope	$\gamma = 0$, $\alpha+\beta = 0$	$I_3 = I_0/2 \,(1 - \sin 2\alpha \sin \phi)$
	$\gamma = 0$, $\alpha+\beta = \pi/4$	$I_4 = I_0/2 \,(1 + \cos 2\alpha \sin \phi)$
	$\gamma = 0$, $\alpha+\beta = \pi/2$	$I_5 = I_0/2 \,(1 + \sin 2\alpha \sin \phi)$
	$\gamma = 0$, $\alpha+\beta = 3\pi/4$	$I_6 = I_0/2 \,(1 - \cos 2\alpha \sin \phi)$

Table 2.3. *Different configurations used with a circular polariscope*

The isoclinic α and isochromatic ϕ parameters are obtained by:

$$\alpha = \frac{1}{2} \, tg^{-1}\left(\frac{I_5 - I_3}{I_4 - I_6}\right) \qquad [2.36]$$

$$\phi = tg^{-1}\left(\frac{I_4 - I_6}{(I_1 - I_2)\cos 2\alpha}\right) = tg^{-1}\left(\frac{I_5 - I_3}{(I_1 - I_2)\sin 2\alpha}\right) \qquad [2.37]$$

The accuracy also depends on the quality of the experimental setup. It is equivalent to that obtained using a plane polariscope. Two difficulties appear when using this technique. First, the calculation of the isoclinics [2.36] is disturbed when the isochromatic parameter is $k\pi$ (twice as often as for the first method). Second, the calculation of the isochromatic parameter is based on the values of the isoclinics, so errors are added.

2.7.3. *Using color images*

Another approach consists of using several wavelengths ([KIH 94], [UME 89], [NUR 97], [LAO 07]). Expressions [2.18] and [2.22] from a plane or a circular polariscope can be rewritten in terms of the wavelength λ, knowing that only the isochromatic parameter can be related to these data [2.5].

2.8. Post-processing of photoelastic parameters

2.8.1. *Drawing of isostatics or stress trajectories*

Once the isoclinic values have been obtained for all points of the specimen, it is possible to plot the isostatics, which represent the principal stress trajectories (see Figure 2.13). The construction is simple to perform numerically. By setting a segment length and a starting point M (see Figure 2.13a), a segment is drawn with respect to the isoclinic value at this point. The end of this segment becomes the beginning of a new segment using the new value of the isoclinic parameter. This procedure is repeated up to the boundary of the specimen. The same procedure is performed in the opposite direction. Gradually, it is thus possible to obtain the isostatic pattern (see Figure 2.13b).

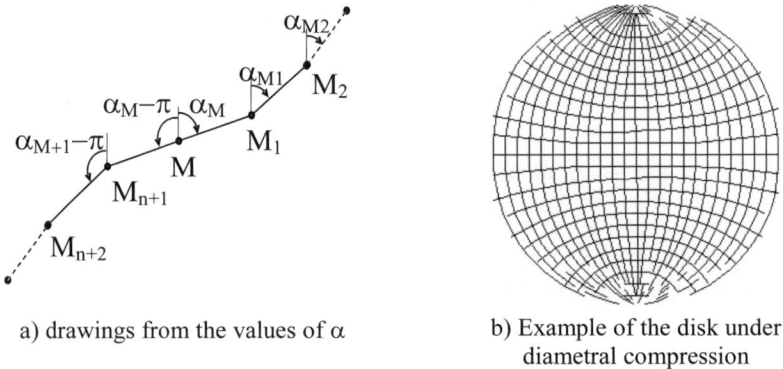

a) drawings from the values of α b) Example of the disk under
   diametral compression

Figure 2.13. *Drawings of stress trajectories*

2.8.2. *Particular points*

Drawing the stress trajectories quickly allows the stress distribution to be observed, thus enabling the optimization of the shape of the work piece. Some particular points appear, called isotropic points. These represent the places where stress is concentrated and thus zones where there is a potential for failure. It is observed that all isoclinics pass through these points. We can deduce that the principal stresses are equal. Two types of isotropic points are possible, depending on the shape of the stress trajectories: negative isotropic points, where non-interlocking stress trajectories occur (Figure 2.14a); or positive isotropic points, where interlocking stress trajectories occur (Figure 2.14b).

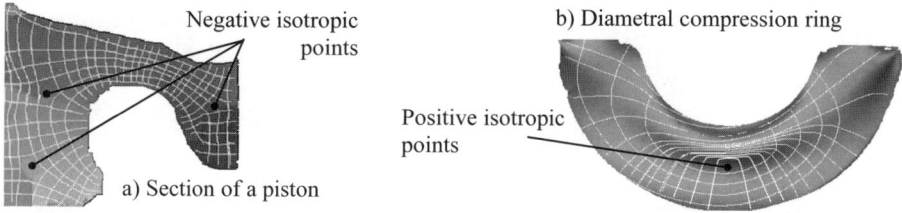

Negative isotropic points

b) Diametral compression ring

Positive isotropic points

a) Section of a piston

Figure 2.14. *Examples of isotropic zones*

2.8.3. *Stress separation and integration of the equilibrium equations*

The equilibrium equation, in the case of plane stress, can be written by neglecting the body forces, as:

$$\begin{cases} \dfrac{\partial \sigma_x}{\partial x} + \dfrac{\partial \sigma_{xy}}{\partial y} = 0 \\ \dfrac{\partial \sigma_y}{\partial y} + \dfrac{\partial \sigma_{xy}}{\partial x} = 0 \end{cases} \text{ with } \begin{cases} \sigma_{xy} = -\dfrac{\sigma_1 - \sigma_2}{2} \sin 2\alpha \\ \sigma_x - \sigma_y = (\sigma_1 - \sigma_2) \cos 2\alpha \end{cases} \quad [2.38]$$

These relationships, written in the form of finite differences, and their integration starting from a given value (in general a boundary conditions $(\sigma_x)_0$ and $(\sigma_y)_0$), lead to σ_x and σ_y [HAA 96].

$$\sigma_x = (\sigma_x)_0 - \sum \frac{\Delta \sigma_{xy}}{\Delta y} \Delta x \text{ and } \sigma_y = (\sigma_y)_0 - \sum \frac{\Delta \sigma_{xy}}{\Delta x} \Delta y \quad [2.39]$$

However, the main drawback of this procedure is that errors are accumulated during the integration.

Another solution can be used to separate stresses: oblique incidence photoelasticity. It consists of analyzing each photoelastic fringe pattern for two orientations of the model (along a vertical axis). This gives us two stress difference equations to solve the problem. Some assumptions are necessary here, however, because we have a three-dimensional medium (see section 2.9)

2.8.4. *Comparison between experimentation and numerical modeling*

It is often possible to model photoelastic fringe patterns (isoclinics and isochromatics) at the end of a finite element procedure in post-processing. In the example below, a beam is subjected to a three-point bending test.

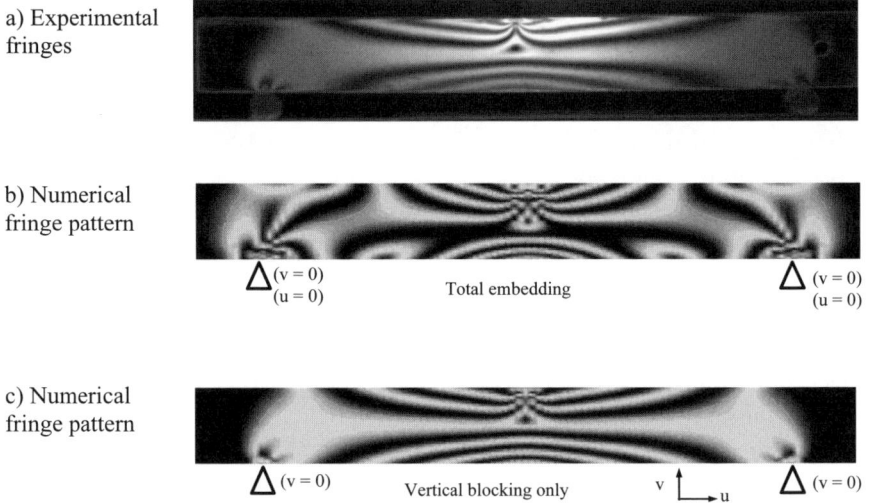

Figure 2.15. *Experimental fringe pattern and numerical simulations (CAST3M) for a three-point bending test (vertical displacement of 0.4 mm)*

The experimental fringe pattern obtained using a circular polariscope with a dark field arrangement is shown in Figure 2.15a. This test was modeled using a mesh of 100 by 40 nodes. Figure 2.15b shows the isochromatics when displacements at the supports are completely blocked, while Figure 2.15c shows isochromatics when displacement is only blocked vertically. We can see a significant difference between the two results. Therefore by comparison with the experimental fringe pattern, we can validate the model and use it to determine the values of the different stress components at any point.

2.9. Three-dimensional photoelasticity

In the study of specimens with complex shapes, a large thickness compared to other dimensions or those subjected to three-dimensional loading, the stress state can no longer be considered as plane. It varies in amplitude and direction. In addition, we can no longer speak of the difference of principal stresses $(\sigma_1 - \sigma_2)$, but of the difference in the secondary principal stresses ($\sigma' - \sigma''$), because the analysis of the thin slice will not necessarily be in the direction of the principal stress axis.

The most commonly-used solution considers an equivalent photoelastic medium ([ABE 66], [ABE 97]). We assume that the difference in secondary principal stresses ($\sigma' - \sigma''$) and the variation in their directions ($d\alpha/dz$) are constant throughout the width of the slice. This provides the two parameters α and ϕ from the knowledge of three physical parameters, R, α^* and ϕ^*. The slice studied is represented by a birefringent medium (α^*, ϕ^*) followed or preceded by an optical element (rotator) that rotates the plane of polarization with an angle R. (Δ'_e, Δ'_s) are the characteristic directions at the input and output, while ($\sigma'e$, $\sigma's$) represent the secondary principal stresses. By introducing parameter α_0, representing the variation in the direction of stress, the terms R, α^*, ϕ^*, α and ϕ are nonlinearly related [DES 81].

Another solution is to consider that the three-dimensional photoelastic medium can be regarded as optically equivalent to an infinite number of birefringent slices with different orientations of their principal axes and different values of optical retardation. This approach is used to calculate the photoelastic fringes of a three-dimensional structure [ZEN 98] and enables the comparison between the fringe patterns observed and those calculated. This method thus makes it possible to validate a mechanical model used in finite element modeling [PRE 08].

2.9.1. *The method of stress freezing and mechanical slicing*

In 1936, Oppel discovered an interesting property of certain polymers, such as epoxy. This phenomenon, called "stress freezing" by Oppel, is based on the biphasic behavior of some polymer materials when heated. These materials are composed of two types of chains whose molecules are linked together: principal chains and secondary chains. At room temperature, these two families of molecular chains resist deformation due to an applied load. However, when the temperature of the polymer increases to the critical temperature about 130°C for an epoxy, the secondary chains break and the principal chains bear the entire load.

If the temperature of the polymer slowly decreases to room temperature while the load is maintained, secondary molecular chains re-form between the strongly deformed principal chains. After unloading, the principal chains recover a little of their original shape, but a considerable proportion of the deformation remains. The elastic deformation of the principal chains is permanently fixed into the material by the re-formed secondary chains. When the model has cooled, it is mechanically cut into slices of appropriate thickness so that the change in the direction of principal stress across the thickness of each slice can be considered negligible. Each separate slice is then analyzed with a circular or plane polariscope. By cutting the photoelastic model into slices for three orthogonal directions, the differences in principal stresses in each direction can be determined.

This method has the following disadvantages: the material has a Poisson ratio of close to 0.5 in the rubbery phase; this value is very different from that of common materials, and can cause problems of similarity; the study of a general case requires two to three experimental models; and mechanical cutting can also disturb the measurements.

2.9.2. Optical slicing

Knowing that the material is transparent, it is possible to analyze either the light that is transmitted (integrated photoelasticity [ABE 79]) or that which is scattered. The latter approach uses the polarization properties of radiated light scattered by a photoelastic material. Indeed, scattered intensity is rectilinearly polarized along direction $\vec{u} \wedge \vec{v}$ (Rayleigh's law) if we make an observation along direction u, which is perpendicular to the direction v of beam propagation.

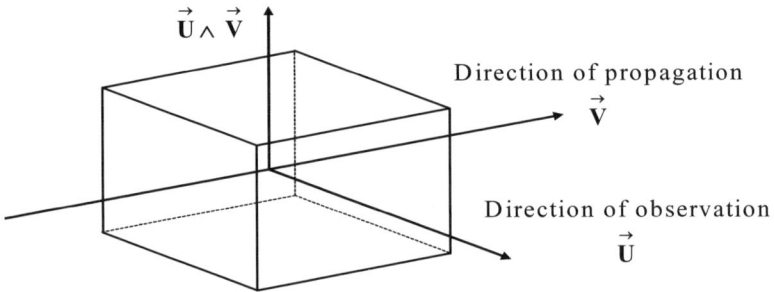

Figure 2.16. *Polarization of scattered light*

In photoelasticity, methods using scattered light phenomenon have been established for cases where the principal directions remain constant along the optical path or where the principal axis rotates, i.e. depending on whether or not a rotator is involved.

This property has been used by R. Weller [WEL 41] and many other authors for studies at a point ([ROB 63], [CHE 67], [BRI 82]) or for entire field analyses ([OF 80], [KIH 06]). These methods have the advantage of being non-destructive.

2.9.2.1. *Scattering used as an analyzer*

Weller's technique consists of illuminating a plane section of the photoelastic model using a plane polarized laser beam and observing the scattered light in various directions normal to the incident beam. This method uses the scattering phenomenon as an analyzer (see Figure 2.17) and assumes that the principal directions do not change along the incident laser beam.

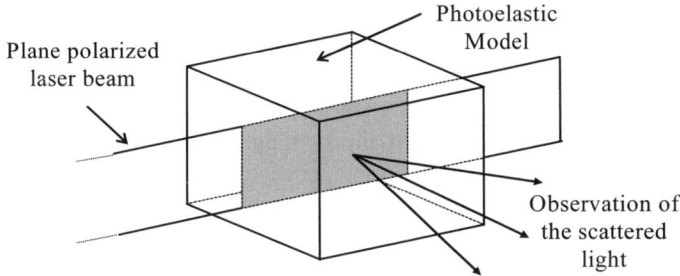

Figure 2.17. *Weller's method*

The fringes observed across the field do not have the same contrast everywhere. This contrast is at a maximum when the polarization directions of incidence and observation are at $\pi/4$ from the secondary principal directions. Contrarily, the fringes disappear when incident polarization is parallel to the input axes of the model. They also disappear in areas where the observation is made along one of the secondary main axes. The intensity of the scattered light is uniform in the case of circularly polarized incident light. The disappearance of the fringes enables the determination of the orientation of the secondary main axes. Measuring the fringe spacing, d, provides the difference between secondary principal stresses (principal directions in a plane perpendicular to a given direction, e.g. the direction of the input light beam) using the following equation:

$$\sigma' - \sigma'' = \frac{\lambda}{Cd}$$

[2.40]

where (σ', σ'') are the secondary principal stresses, C represents the photoelastic constant of the material and λ is the wavelength. This method is simple, but we must verify that the principal axis maintains the same direction at all points of the illuminated section.

From this, Oi and Takashi [OI 98] developed a technique using several images obtained by a combination of incident and observation angles. This full-field technique allows the rotator to be taken into account, but its implementation (set up and analysis) is complex.

2.9.2.2. *Scattering used as a polarizer (optical slicing)*

When a plane coherent light beam passes through a scattering medium, each particle behaves as a light source. These sources being coherent they interfere, creating a speckle field that we can observe in the direction orthogonal to the plane.

When considering a slice of the photoelastic model defined by two parallel plane laser beams (see Figure 2.18), we observe, the superimposition of two speckle fields in the orthogonal direction, each coming from one of the laser beams. The possibility of interference between the two light beams, defined by a correlation factor called γ, depends on the birefringence of the isolated slice [DES 80]. The resulting speckle field intensity I can be expressed as [DUP 97]:

$$\begin{cases} I_1 = I_{1F} + I_{1g} \\ I_2 = I_{2F} + I_{2g} \\ I = I_1 + I_2 + 2\sqrt{I_{1g}}\sqrt{I_{2g}}\,\gamma\,cos(\psi_1 + \psi_2 + \eta) \end{cases} \qquad [2.41]$$

where I_1 and I_2 are the intensities of the plane laser beams, I_{1F} and I_{2F} are the background intensities, ψ_1 and ψ_2 are the random phases of the speckle fields and I_{1g} and I_{2g} are their intensities. η is a function of the optical characteristics of the slice. The correlation factor γ is given by:

$$\gamma^2 = 1 - sin^2\,2\alpha\,sin^2\frac{\phi}{2} \qquad [2.42]$$

where α is the angle formed by one of the principal directions of the slice and the polarization direction of the scattered light, and ϕ is the angular birefringence of the slice.

Expression [2.42] is similar to that commonly used for determining the light intensity obtained when analyzing a similar slice (made, for example, by stress freezing and mechanical slicing) between two parallel polarizers (plane polariscope; relation [2.17]).

In the perceived image corresponding to expression [2.42] consisting of a speckle field, intensities I_1 and I_2 and phases ϕ_1 and ϕ_2 are random variables. Consequently, it is not possible to directly obtain factor γ. Thus the procedure consists of separately analyzing the light intensity of both scattered planes I, then plane 1 alone (I_1) and finally plane 2 alone (I_2). The statistical analysis of these three images, by local calculation of the spatial average (noted $\langle\rangle$) and of the spatial variance (*Var*):

$$Var^2 = \left\langle \left[I - \langle I \rangle \right]^2 \right\rangle = Var_1^2 + Var_2^2 + 2\gamma^2 \left\langle I_{1g} \right\rangle \left\langle I_{2g} \right\rangle \qquad [2.43]$$

where $\langle I \rangle$, $\langle I_{1g} \rangle$ and $\langle I_{2g} \rangle$ are the mean light intensity, gives the value of the correlation factor:

$$k^2 \gamma^2 = \frac{Var^2 - Var_1^2 - Var_2^2}{2 \langle I_1 \rangle \langle I_2 \rangle} \qquad [2.44]$$

where k is a factor related to the material's scattered light properties. Its value is experimentally determined and is approximately equal to 1/10.

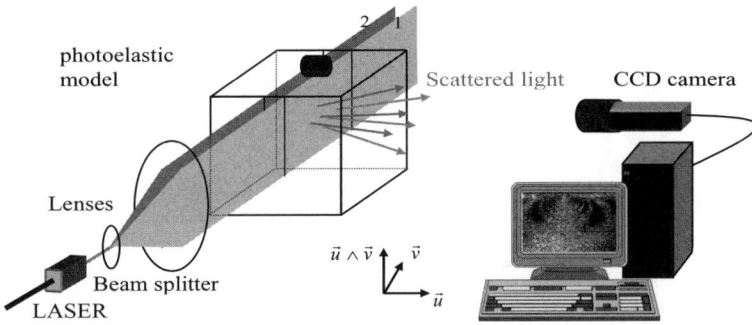

Figure 2.18. *Optical slicing of a photoelastic model*

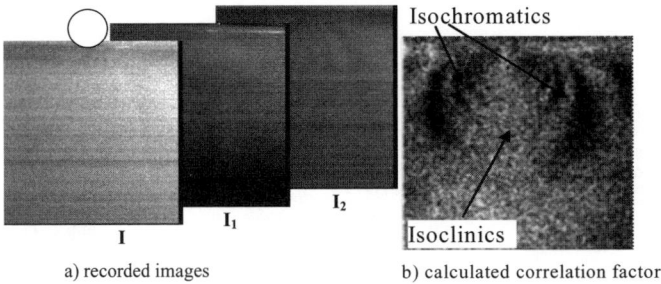

a) recorded images   b) calculated correlation factor

Figure 2.19. *Example of punctual loading*

2.9.3. *Application example*

This example demonstrates the coupling between the numerical and experimental approaches to studying the structure of a complex shape with three-dimensional stress distribution. It concerns the study of spherical plain bearings used in aeronautics for the best mechanical behavior of a pylon-to-engine link [GER 08]. Problems relating to premature scuffing were observed at the surfaces of the spherical ball joints. In addition to tribological aspects, the problem comes from high stress levels. Therefore a numerical tool was developed using finite element simulation in order to approach the pressure distribution in the bearings in a realistic way, depending on their particular geometry. The boundary conditions and the mesh had to be clearly validated. As in section 2.8.4, the experimental fringe patterns were compared to numerically calculated fringe patterns from the three-dimensional stress fields (see Figure 2.20). The model boundary conditions were corrected in the second step (see Figure 2.20c). Contact pressures were then calculated in a more realistic way, according to the damage observed.

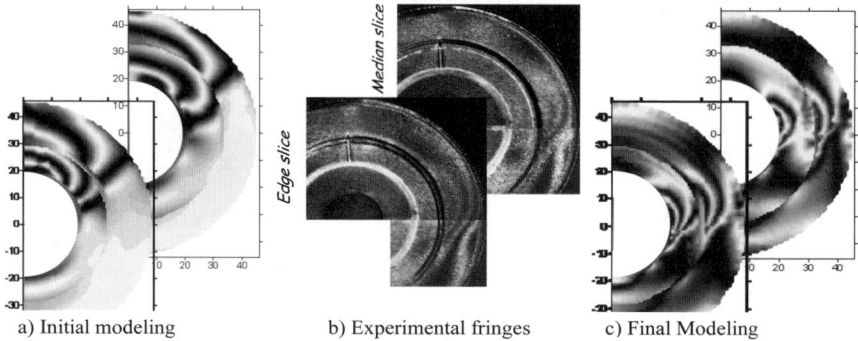

a) Initial modeling   b) Experimental fringes   c) Final Modeling

Figure 2.20. *Comparison between computed and experimental patterns [GER 08]*

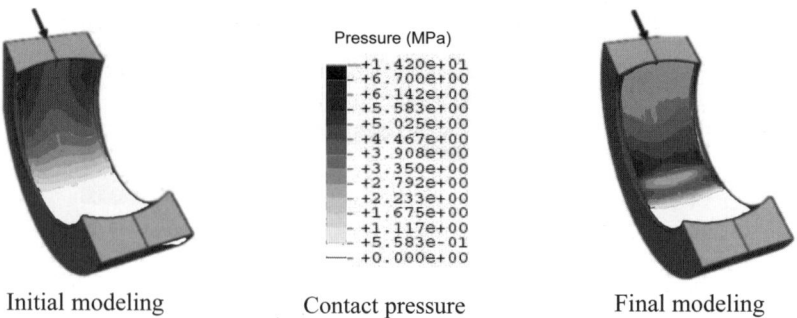

Initial modeling   Contact pressure   Final modeling

Figure 2.21. *Contact pressure on the inner surface of the ring [GER 08]*

2.10. Conclusion

Through recent technological progress (digital image acquisition and processing, numerical modeling and new light sources [LASER]), photoelasticity retains its use in the experimental analysis of structures, and is a current research topic. It is not possible in this chapter to develop all of its aspects; however, we have attempted to show that it can be both an aid for modeling and a simple tool for students, as well as a very powerful approach for engineers.

2.11. Bibliography

[ABE 66] ABEN H., "Optical phenomena in photoelastic models by the rotation of principal axes", *Experimental Mechanics*, vol. 6, no. 1, 1966.

[ABE 79] ABEN H., *Integrated Photoelasticity*, McGraw-Hill, 1979.

[ABE 97] ABEN H., JOSEPSON J., "Strange interference blots in the interferometry of inhomogeneous birefringent objects", *Applied Optics*, vol. 36, no. 28, October 1997.

[AJO 98] AJOVALASIT A., BARONE S., PETRUCCI G., "A review of automated methods for the collection and analysis of photoelastic data", *Journal of Strain Analysis*, vol. 33, no. 2, pp. 75–99, 1998.

[AJO 07] AJOVALASIT A., PETRUCCI G., SCAFIDI M., "Phase shifting photoelasticity in white light", *Optical and Laser in Engineering*, vol. 45, pp. 596–611, 2007.

[ASU 93] ASUNDI A, "Phase shifting in photoelasticity", *Experimental Techniques*, vol. 17, no. 1, pp. 19–23, 1993.

[BRE 93] BREMAND, F., "A phase unwrapping technique for any object relief determination", *International Symposium on Optical tools for manufacturing and Advanced Automation, SPIE*, Boston, September 1993.

[BRI 82] BRILLAUD J., LAGARDE A., "Méthode ponctuelle de photoélasticité tridimensionnelle, Application", *Revue Française de Mécanique*, vol. 84, pp. 41–49, 1982.

[CHE 67] CHENG Y.F., "A dual observation method for determining photoelastic parameters in scattered light", *Experimental Mechanics*, vol. 7, no. 3, pp. 140–144, 1967.

[DES 80] DESAILLY R., LAGARDE A., "Sur une méthode de photoélasticimétrie tridimensionnelle non destructive à champ complet", *Journal de Mécanique Appliquée*, vol. 4, no. 1, pp. 3, 1980.

[DES 81] DESAILLY R., Méthode non-destructive de découpage optique en photoélasticimétrie tridimensionnelle- Application à la mécanique de la rupture, Doctoral Thesis, University of Poitiers, June 1981.

[DUP 93] DUPRÉ J.C., BREMAND F., LAGARDE A., "Photoelastic data processing through digital image processing: isostatics and isochromatics recontruction", *International Conference on Photoelasticity: New Instrumentation and Data Processing Technique*, London, November 1993.

[DUP 97] DUPRÉ J.C., LAGARDE A., "Photoelastic analysis of a three-dimensional specimen by optical slicing and digital image processing", *Experimental Mechanics*, vol. 37, no. 4, pp. 393–397, December 1997.

[GER 08] GERMANEAU A., PEYRUSEIGT F., MISTOU S., DALVERNY O., DOUMALIN P., DUPRÉ J.C., "Verification of a spherical plain bearing FE model using scattered light photoelasticity tests" *Journal of Engineering Tribology*, vol. 222, no. J5, pp. 647–656, 2008.

[HAA 96] HAAKE S.J., PATTERSON E.A., WANG Z.F., "2D and 3D separation of stresses using automated photoelasticity", *Experimental Mechanics*, vol. 36, pp. 269–276, 1996.

[KIH 06] KIHARA T., "Phase unwrapping method for three-dimensional stress analysis by scattered-light photoelasticity with unpolarized light", *Applied Optics*, vol. 45, no. 35, pp. 8848–8854, 2006.

[KIH 94] KIHARA T., "Automatic whole field measurement of principal stress directions using three wavelengths", in SILVA GOMES J.F., *et al.* (eds), *Recent Advances in Experimental Mechanics*, pp. 95–99, A.A. Balkema, Rotterdam, 1994.

[KOB 87] KOBAYASHI A.S., *Handbook on Experimental Mechanics*, Prentice-Hall, 1987.

[MOR 94] MORIMOTO Y., MORIMOTO Y. Jr, HAYASHI T., "Separation of isochromatics and isoclinics using Fourier transform", *Experimental Techniques*, vol. 18, no. 5, pp. 13, September, 1994.

[NUR 97] NURSE A.D., "Full-field automated photoelasticity by use of a three-wavelength approach to phase stepping", *Applied optics*, vol. 36, no. 23, pp. 5781–5786, 1997.

[OI 98] OI T., TAKASHI, M., "An approach to general 3-D stress analysis by multidirectional scattered light technique", *IUTAM Symposium, Advanced Optical Methods and Application in Solid Mechanics*, Poitiers, September 1998.

[PAT 91] PATERSON E.A., WANG Z.F., "Towards full field automated photoelastic analysis of complex components", *Strain*, vol. 27, pp. 49–56, May 1991.

[PAT 98] PATERSON E.A., WANG Z.F., "Simultaneous observation of phase-stepped images for automated photoelasticity", *Journal of Strain Analysis*, vol. 33, no. 33, pp.1–15, 1998.

[PLO 96] PLOUZENNEC N., DUPRÉ J.C., LAGARDE A, "Determination of isoclinic and isochromatic parameters by photoelastic and numerical analysis", *3rd Biennal European Joint Conference on Engineering Systems Design and Analysis*, Montpellier, July 1996.

[ROB 63] ROBERT A., GUILLEMET E., "Nouvelle méthode d'utilisation de la lumière diffusée en photoélasticimétrie à trois dimensions", *Revue Française de Mécanique*, vol. 5 and 6, pp. 147–157, 1963.

[ROB 05] ROBIN E., VALLE V., BRÉMAND F., "Phase demodulation method from a single fringe pattern based on correlation technique with a polynomial form", *Applied Optics*, vol. 44, no. 34, pp. 7261–7269, 2005.

[ROB 83] ROBINSON D.W., "Automatic fringe analysis with a computer image processing system", *Applied Optics*, vol. 22, no. 14, pp. 2168–2176, 1983.

[UME 89] UMEZAKI E., TAMAKI T., TAKAHASHI S., "Automated stress analysis of photoelastic experiment by use of image processing", *Experimental Techniques*, vol. 13, pp. 22–27, 1989.

[WEL 41] WELLER R., "Three dimensional photoelasticity using scattered light", *Journal of Applied Physics*, vol. 12, no. 8, pp. 610–616, 1941.

[ZEN 98] ZENINA A., DUPRÉ J.C., LAGARDE A., "Theoretical and experimental approaches of a three-dimensional photoelastic medium", *11th International Conference on Experimental Mechanics*, Oxford, August 1998.

Chapter 3

Grid Method, Moiré and Deflectometry

3.1. Introduction

The grid method has a very simple basic principle [HUN 98, PAR 90, DAL 91, SEV 93]. Even if it is independent of image formation and acquisition, the grid method is usually classified among optical techniques. The measurement is based on a purely geometrical phenomenon: a geometrical grid is affixed to a substrate. If the substrate deforms, the grid follows any local movement. The distorted grid will carry the displacement information, and its image can be analyzed using a classic modulation/demodulation approach. This will be discussed hereafter.

In recent times, acquisition and image-processing capabilities, as well as advances in image sensor quality, have led to a renewed interest in this particular technique because it is relatively easy to implement. The quality of phase detection algorithms provides an excellent measurement resolution. From a metrological point of view [ZHA 01, SUR 05], this technique is easier to characterize than the digital image correlation.

3.2. Principle

The basic principle of the grid method or geometric moiré is to analyze the displacements of a grid that is glued, deposited or even etched onto a specimen surface. Thus, the grid is supposed to follow the deformation of the substrate. The grid is a spatial carrier with the frequency vector $\mathbf{f} = \mathbf{n}/p$, where $\mathbf{n}$ is the unity vector perpendicular to the grid lines and p is the grid pitch. The intensity $I(\mathbf{R})$ at point

---

Chapter written by Jérôme MOLIMARD and Yves SURREL.

$\mathbf{R} = (X, Y)$ of the grid in its initial (undeformed) state is represented by the following equation:

$$I_i(\mathbf{R}) = A\left[1 + \mathsf{frng}(2\pi\mathbf{f}.\mathbf{R})\right] \tag{3.1}$$

The intensity profile of each grid line is described by the periodic function denoted by frng. Using appropriate mathematical processing for phase detection, harmonics present in the frng function can be filtered out. Thus, it is possible to consider only a sinewave profile:

$$I_i(\mathbf{R}) = A\left[1 + \gamma \, \cos(2\pi\mathbf{f}.\mathbf{R})\right] = A[1 + \gamma \cos \varphi(\mathbf{R})] \tag{3.2}$$

with $\varphi(\mathbf{R}) = 2\pi\mathbf{f}.\mathbf{R}$ and where γ is the *visibility* (or *contrast*) of the carrier signal. It is worth noting that slightly defocusing the lens of the camera acquiring the grid image will smoothen the grid profile. As a result, harmonics are significantly decreased before any numerical processing.

Substrate deformation is described mathematically by the displacement field $\mathbf{U}(\mathbf{R})$. A material particle at a given geometrical point $\mathbf{R}$ will be translated at point $\mathbf{r} = \mathbf{R} + \mathbf{U}(\mathbf{R})$ by the transformation. We can consider either the direct displacement field $\mathbf{U}(\mathbf{R})$ or the inverse displacement field $\mathbf{u}(\mathbf{r})$ defined on the deformed configuration, which will bring the material points back to their initial positions. The relation between these two fields is:

$$\mathbf{U}(\mathbf{R}) = -\mathbf{u}(\mathbf{r}) = -\mathbf{u}[\mathbf{R} + \mathbf{U}(\mathbf{R})] \tag{3.3}$$

and is shown in Figure 3.1. The intensity of the signal observed in the final state at point $\mathbf{r}$ is the same as that observed at point $\mathbf{R}$ in the initial state because the material particle is the same, so the grid morphology around it has not changed. Thus, the intensity $I_f(\mathbf{r})$ observed on the deformed grid in the final state is:

$$I_f(\mathbf{r}) = I_i(\mathbf{R}) = I_i[\mathbf{r} + \mathbf{u}(\mathbf{r})] \tag{3.4}$$

Using [3.2], equation [3.4] can be rewritten as:

$$I_f(\mathbf{r}) = A\left\{1 + \gamma \, \cos[2\pi\mathbf{f}.\mathbf{r} + \varphi_n(\mathbf{r})]\right\} \tag{3.5}$$

with:

$$\varphi_i(\mathbf{r}) = 2\pi\mathbf{f}.\mathbf{u}(\mathbf{r}) = 2\pi\frac{u_i}{p} \tag{3.6}$$

where u_i, $i = x, y$ is the component of the inverse displacement along direction x or y, depending on whether the grid lines are aligned along y or x. Finally, the inverse displacement field appears to be a signal *modulating the phase of the spatial carrier.*

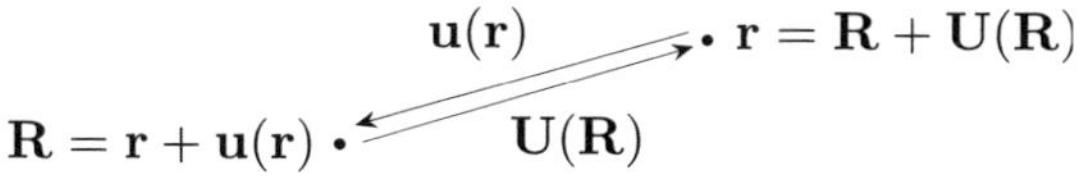

$$\mathbf{u(r)} \qquad \mathbf{\cdot\, r = R + U(R)}$$
$$\mathbf{R = r + u(r)\,\cdot} \qquad \mathbf{U(R)}$$

Figure 3.1. *Direct or inverse displacement fields*

Equation [3.6] describes the linear relationship between the measurand u_i and the phase φ. The *sensitivity* of the phase measurement, which depicts how a small displacement variation is translated into a small phase variation, is the coefficient:

$$s = \frac{\partial \varphi}{\partial u} = \frac{2\pi}{p} \qquad\qquad [3.7]$$

3.3. Surface encoding

One of the trickiest points in setting up the grid technique is the process of attaching the grid to the studied object. There is no standard method, and the experimenter has to choose the most appropriate deposition technique depending on the grid pitch and the substrate material. Common ways are to print the grid directly on a photosensitive film and process it, to glue or transfer the grid and even to engrave the specimen. If the grid pitch lies within the range from 0.1 to a few millimeters, very good results can be obtained by transferring a grid produced by a professional phototypesetter using offset printing[1] (a "film") onto the specimen surface [PIR 04]. The acetate film substrate can then be carefully peeled off, leaving the very thin layer of photosensitive film (where the grid has been printed) on the surface. In most cases, this grid deposition is the most restrictive step, leading to implementation difficulties.

We have to assume a good kinematic transfer between the substrate and the grid through the glue, that is perfect adherence and negligible shear within the glue. In the same way, the grid should not change the global mechanical behavior of the substrate by local stiffening. Note that the same hypothesis has to be made for strain gauge technologies. All these hypotheses should be verified, for example by measuring the thickness or the stiffness of the glue film and/or by developing a numerical model including the grid itself. Currently, the measurement of thin film stiffness is a complicated matter and usually it is necessary to extrapolate results obtained from the bulk material, with a potential lack of representativity.

---

1 Unfortunately, the evolution of the printing industry is tending to reduce the number of phototypesetters, making them less and less available.

3.4. Moiré

Moiré[2] is actually an effect superimposed over the grid technique, modifying the spectral information delivered by the technique. It is the nonlinear effect of beating between two patterns whose spatial frequencies are close to each other (Figure 3.2). This nonlinear effect can be obtained in many ways (multiplication, XOR operation, etc.).

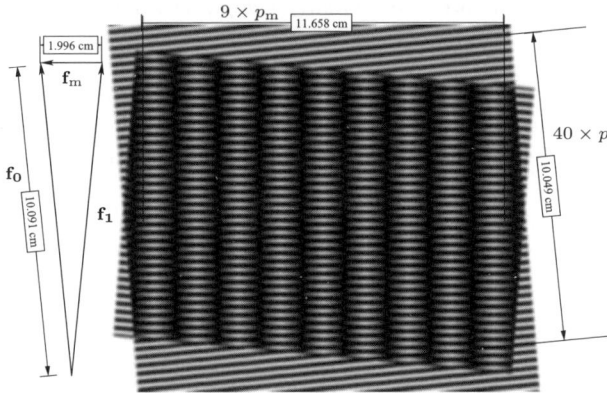

Figure 3.2. *Moiré example and the corresponding vectorial diagram of spatial frequencies*

Let us take as an example the superposition of two transparent sheets with transparencies $t_0(\mathbf{r})$ and $t_1(\mathbf{r})$ described by:

$$t_i(\mathbf{r}) = \frac{1}{2}[1 + \cos(2\pi \mathbf{f}_i . \mathbf{r})] \quad i = 0, 1 \tag{3.8}$$

where the spatial frequency vectors $\mathbf{f}_0$ and $\mathbf{f}_1$ are close to each other. The resulting transparency function will be:

$$t(\mathbf{r}) = t_0(\mathbf{r})t_1(\mathbf{r})$$

$$= \frac{1}{4} + \frac{1}{2}[\cos(2\pi \mathbf{f}_0 . \mathbf{r}) + \cos(2\pi \mathbf{f}_1 . \mathbf{r})] \tag{3.9}$$

$$+ \frac{1}{8}\{\cos[2\pi(\mathbf{f}_0 + \mathbf{f}_1).\mathbf{r}] + \cos[2\pi(\mathbf{f}_0 - \mathbf{f}_1).\mathbf{r}]\}$$

Moiré corresponds to the low-frequency component: the beat; thus the local frequency of the moiré fringes is:

$$\mathbf{f}_m = \mathbf{f}_0 - \mathbf{f}_1 \tag{3.10}$$

---

2 It is a common mistake to write moiré with an uppercase "M". Actually it is a French common noun, not the name of an inventor.

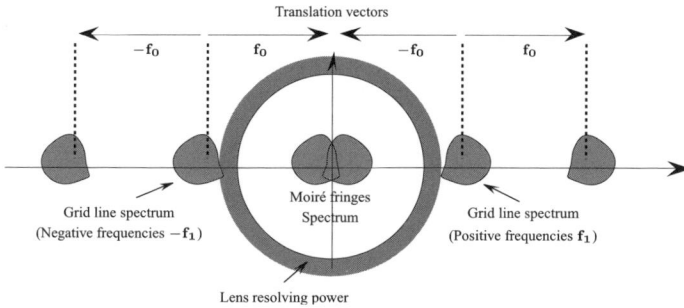

Figure 3.3. *Spectrum shift due to moiré effect*

The main interest of moiré is, precisely, the translation in the frequency domain. Considering that grid 1 is deformed (in the frequency domain, this results in a spreading of the frequencies around the nominal frequency f_1) and grid 0 is the reference, the beat phenomenon results in a translation of the information contained around f_1 and $-f_1$ toward the origin. This effect is shown in Figure 3.3. A circle representing the resolving power of the optical instrument used for image acquisition is shown; it represents the fact that image forming cannot be perfect in the sense that all spatial frequencies present in an object cannot be surely transferred to the image; there is always a "cutoff frequency" that prevents high spatial frequencies from passing (low-pass filtering). In the case depicted in Figure 3.3 where the grid frequency falls beyond the optical resolving power of the image-forming lens, the frequency shift performed by moiré allows information corresponding to basically unreachable frequencies to be obtained. Thus, the moiré fringes are recorded while individual grid lines are not resolved. It is therefore possible to use grids with a very small pitch (down to 40 lines per millimeter) in order to increase sensitivity to displacements. But the implementation is complicated by the use of a reference grid: this must be as close as possible to the deformed grid in order to increase fringe visibility and avoid parasitic fringes due to parallelism discrepancies (e.g. air wedges) between the two grids.

One of the more interesting features of the moiré effect is the "amplification" of deformations. The deformation of moiré fringes is definitely more visible than the deformation of the active grid. This effect is explained in Figure 3.4, where the spatial period p of the grid lines or moiré fringes is represented as a function of the spatial frequency f, following the simple relationship $p = 1/f$. The variation in the grid spatial frequency caused by substrate deformation leads to a variation in the moiré fringe pitch that is much higher (and thus more visible) than the grid pitch variation itself.

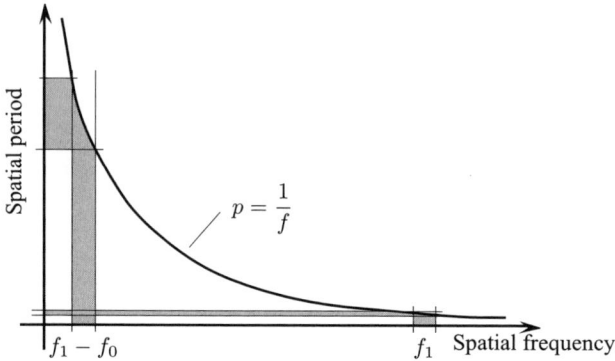

Figure 3.4. *Amplification of visible deformation by moiré: a small variation*
δf in the spatial frequency around f_1 (spatial frequency of the grid)
corresponds to a pitch (period) variation much lower than the corresponding
moiré fringe pitch variation around the spatial frequency $f_1 - f_0$

3.5. Phase detection

The phase of the modulating grid can be extracted in two different ways: a global phase detection, based on the theory of analytical signal processing, and a local phase detection, based on spatial phase shifting.

3.5.1. *Global extraction procedure*

This method is known as global Fourier transform. More precisely, it comprises the extraction of the *analytical signal* [ROD 87, TAK 82, WOM 84, TAK 90].

The intensity of a sinusoidal grid subjected to a deformation is modulated and is written as:

$$I(x) = A\left\{1 + \gamma \cos\left[2\pi f_0 x + \varphi(x)\right]\right\}$$

$$= A + \frac{A\gamma}{2}\exp(\mathrm{j}\,2\pi f_0 x)\exp[\mathrm{j}\,\varphi(x)] + \frac{A\gamma}{2}\exp(-\mathrm{j}\,2\pi f_0 x)\exp[-\mathrm{j}\,\varphi(x)]$$

$$= A + C(x)\exp(\mathrm{j}\,2\pi f_0 x) + C^\star(x)\exp(-\mathrm{j}\,2\pi f_0 x)$$

[3.11]

where the star denotes complex conjugation and:

$$C(x) = \frac{A\gamma}{2}\exp[\mathrm{j}\,\varphi(x)]$$

[3.12]

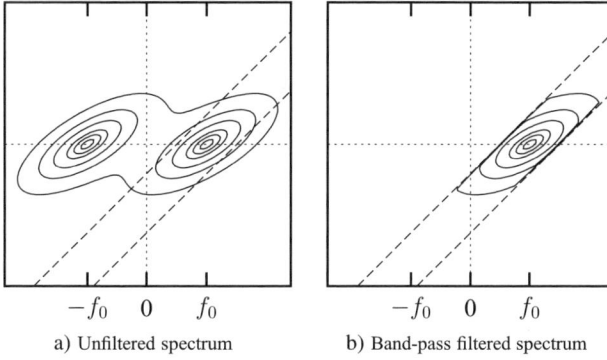

a) Unfiltered spectrum   b) Band-pass filtered spectrum

Figure 3.5. *Frequency spectrum of a distorted vertical grid, showing two lateral lobes centered in $\mathbf{f_0}$ and $-\mathbf{f_0}$ (The Dirac function at 0 is not represented for clarity)*

The global Fourier transform [3.11] gives:

$$\widetilde{I}(f) = A\delta(f) + \widetilde{C}(f - f_0) + \widetilde{C^\star}(f + f_0)$$

$$= A\delta(f) + \widetilde{C}(f - f_0) + \widetilde{C}^{\,\star}(-f - f_0)$$

[3.13]

where ~ denotes the Fourier transform. Suppose that the function $C(x)$ contains only low frequencies, then $\widetilde{C}(f - f_0)$ and $\widetilde{C}^{\,\star}(-f - f_0)$ spread slightly around f_0 and $-f_0$, respectively. Thus, a filter in the Fourier domain can be used to isolate the frequency area $\widetilde{C}(f - f_0)$ (Figures 3.5(a) and (b)). Finally, using [3.12], the argument of this term gives $\varphi(\mathbf{r}) + 2\pi\mathbf{f_0}.\mathbf{r}$; then it is easy to subtract the linear carrier component $2\pi\mathbf{f_0}.\mathbf{r}$.

The second method is local; we recommend this approach because this local aspect leads to a better evaluation of the detected phase uncertainty. It is based on *phase shifting*.

3.5.2. *Local phase detection: phase shifting*

3.5.2.1. *Number of unknowns*

In equation [3.1], the number of unknowns is infinite, because the shape of the fringe function frng, described as a Fourier series, is *a priori* unknown. The mean intensity A and the visibility γ are unknown, as well as the sought phase φ. In common illumination cases, A and γ present low-frequency variations over the map.

In an ideal case, the real intensity field represented by equation [3.1] reduces to the sinusoidal shape described by equation [3.2]. The number of unknowns is then reduced to three. At the sensor level, that is for one pixel of the CCD camera, a single intensity value is insufficient to recover phase information, because of these three unknowns. It is therefore necessary to add other information sources. These are to be found either over time, using other intensity map recordings, or over space, using neighboring pixels. One of the most efficient ways used in most modern measuring systems is the *phase-shifting* method.

Before going further into its principle, let us recall that the user will always have to make some assumption about the simplification of the function frng, to reduce the number of unknowns. In other words, he/she has to decide how many parasitic harmonics (i.e. frequencies that are integer multiples of the fundamental) in the signal have to be rejected.

3.5.2.2. *Principle*

Phase shifting consists of using many samples I_k, $k = 0, 1,\ldots, M - 1$, with a constant phase shift δ:

$$I_k = I(\varphi + k\delta) \tag{3.14}$$

The sampled data points are best fitted to a sine wave. The phase lag of this sine is the sought phase φ (Figure 3.6). In the grid technique, spatial phase shifting is used and the sampled points correspond to the neighboring pixels of the CCD sensor.

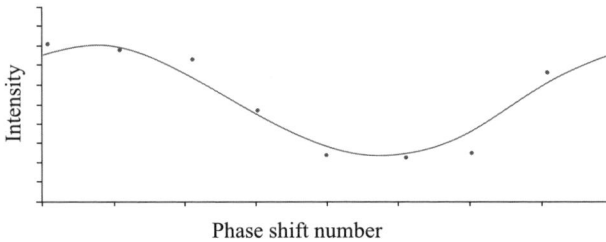

Figure 3.6. *Sampled intensity versus phase shift*

In the literature, the formula used to extract the phase φ from the intensity samples $\{I_k\}$ is often called the *algorithm*. Phase detection algorithms can be written in the general form:

$$[\varphi] = \arctan_{2\pi} \left[\frac{\sum\limits_{k=0}^{M-1} b_k I_k}{\sum\limits_{k=0}^{M-1} a_k I_k} \right] \tag{3.15}$$

where $[\varphi]$ is the measured phase. Brackets are used to recall that the phase is obtained modulo 2π, that is in the $[-\pi, \pi]$ range. Equation [3.15] can be seen as the argument of the complex linear combination[3]:

$$S(\varphi) = \sum_{k=0}^{M-1} c_k I_k \qquad [3.16]$$

with $c_k = a_k + \mathrm{j}\, b_k$

3.5.2.3. *N-bucket or DFT algorithm*

In general, to eliminate harmonics up to order j, it is necessary to choose $\delta = 2\pi/N$ with $N = j + 2$. The resulting algorithm is called the *N-bucket algorithm*. With this algorithm, $c_k = \exp(-\mathrm{j}\,2k\pi/N)$ and the algorithm is written as:

$$[\varphi] = -\arctan_{2\pi} \left[\frac{\displaystyle\sum_{k=0}^{N-1} I_k \sin(2k\pi/N)}{\displaystyle\sum_{k=0}^{N-1} I_k \cos(2k\pi/N)} \right] \qquad [3.17]$$

It is easy to see that this algorithm corresponds to the calculation of the argument of the second coefficient of the discrete Fourier transform (DFT) of the data set $\{I_k\}$. The total number of necessary samples for this algorithm is $M = N$ with a phase shift equal to $2\pi/N$.

This algorithm correctly detects the phase if the sampling data are exactly phase shifted by $2\pi/N$. But this is generally not the case in the grid technique: it is in practice very difficult to adjust (or *tune*, like a radio that has to be tuned to the emission carrier frequency in order to demodulate the signal) the lens magnification so that N pixels exactly sample a grid period. Moreover, the magnification is not constant due to lens distortions. Finally, the deformation applied to the sample gives some local changes of the grid pitch. Thus, it is impossible to have a perfect sampling in the initial and in the final state. A robust algorithm with respect to phase-shifting imperfections is necessary.

3.5.2.4. *Windowed discrete Fourier transform (WDFT) algorithm*

In [SUR 96a], the robustness of phase-shifting algorithms was thoroughly studied using the characteristic polynomial theory. The best algorithm in the case of detuning is the windowed discrete Fourier transform (WDFT) algorithm. This corresponds to

---

3 Sometimes called a *phasor*.

sampling over two periods, followed by triangular windowing (Figure 3.7), and is written as:

$$\varphi = \arctan_{2\pi} \left[-\frac{\sum\limits_{k=1}^{N-1} k(I_{k-1} - I_{2N-k-1})\sin(2k\pi/N)}{NI_{N-1} + \sum\limits_{k=1}^{N-1} k(I_{k-1} + I_{2N-k-1})\cos(2k\pi/N)} \right] \qquad [3.18]$$

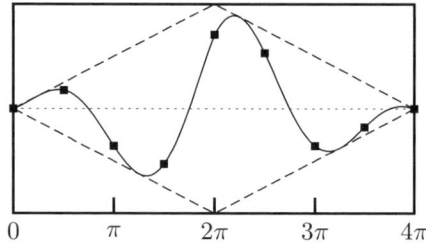

Figure 3.7. *Triangular window used in the window Fourier transform algorithm. This signal can be seen as a* wavelet *adapted to phase detection*

It can easily be seen that this algorithm corresponds to the estimation of the third coefficient of the DFT of the data set $\{I_k\}$. The total number of samples for this algorithm is $M = 2N - 1$ with a $2\pi/N$ phase shift.

3.5.3. *Measuring both components of the displacement*

With one set of parallel lines, the grid technique can measure only one displacement component, perpendicular to the grid line direction. It is possible, however, to use a crossed grid to measure simultaneously both in-plane components. A simple image process can be written to separate information along x or y axes.

Let us suppose that the grid line sampling is done with N points per grid period. The basic idea is to perform a moving average over a length corresponding to the period of the signal to be eliminated (Figure 3.8). Mathematically, consider the following signal:

$$f(x) = A + B\cos(2\pi f x) \qquad [3.19]$$

It can easily be seen that averaging this function over a width $p = 1/f$ will provide the constant A, wherever this averaging is performed along the x axis. We can say that the periodicity is "eradicated" by the moving average.

So, to eliminate horizontal (or vertical) lines in the grid image, a horizontal (or vertical, respectively) moving average has to be made (Figure 3.8).

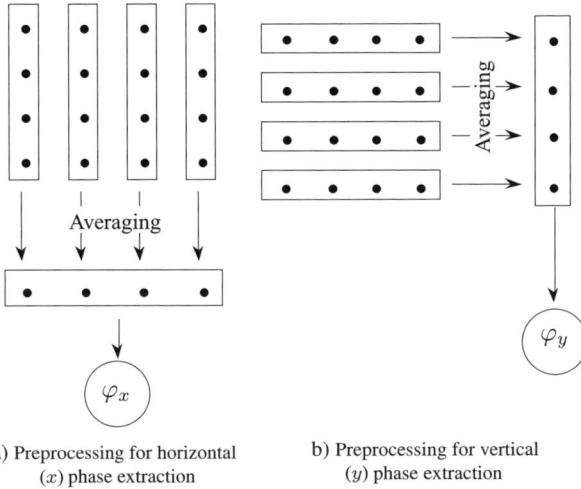

Figure 3.8. *Filtering out horizontal or vertical lines by averaging spatial frequencies, on a set of 4 × 4 pixels*

3.6. Sensitivity to out-of-plane displacements

Sensitivity to out-of-plane displacements should not be neglected. If d and d' are respectively the distances from the camera lens optical center to the object and image, the lens magnification G is given by:

$$G = -\frac{d'}{d} \qquad\qquad [3.20]$$

and the period p' corresponding to the image of an object grid with a period p is:

$$p' = Gp \qquad\qquad [3.21]$$

If the distance d varies by δd, the period p of the object grid being fixed, there will be an apparent strain ϵ of the grid image equal to:

$$\epsilon = \frac{\delta p'}{p'} = \frac{\delta G}{G} = -\frac{\delta d}{d} \qquad\qquad [3.22]$$

This apparent strain will bias the mechanical strain measurement.

For example, a 1 mm out-of-plane movement observed at a distance $d = 1$ m induces a 0.1 % apparent strain, which cannot be neglected in most situations.

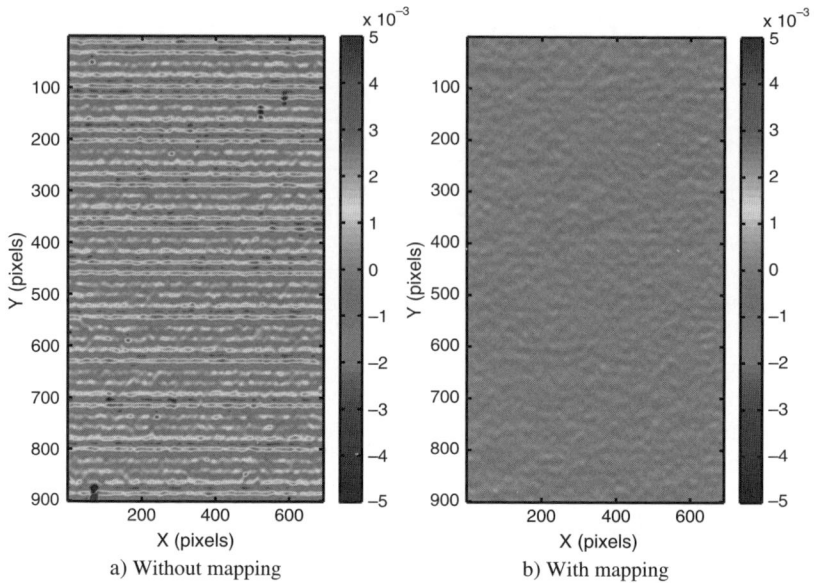

Figure 3.9. *Apparent strain measurement in the case of rigid body movement. Effect of mapping the final configuration onto the initial configuration by using the unwrapped phase information [BAD 09]*

3.7. Grid defects

It is necessary to take into account grid defects in order to obtain the best results. The (unwrapped!) phase of the grid lines can be used as Lagrangian coordinates of the material points of the surface, even if the grid mesh is not perfect in the initial state. The most important point is that there is an unequivocal correspondence between any point M of the surface and the grid phases $(\varphi_x(M), \varphi_y(M))$ at that point. Thus, the spatial coordinates of any point can be identified in the initial and final images, and displacement vectors can be calculated[4]. It is also possible to take into account grid defects by mapping the final phase map onto the initial configuration [BAD 09]. This is shown in Figure 3.9. A large vertical rigid body movement is applied to a specimen, and the apparent strain is calculated. Without mapping (Figure 3.9(a)), the grid defects appear clearly, as the grid phase derivatives are subtracted at the same pixels in the final and initial states, thus at different material points. On the other hand, if the final state is mapped onto the initial state (Figure 3.9(b)), only the measurement noise can be observed around the nominal zero strain.

---

4 The phase is only used here as a definite spatial marker of the surface points; no assumption about spatial regularity is required here.

3.8. Large deformation/large strain

In its simplest implementation, the grid method is based on the assumption that the reference and distorted configurations coincide geometrically, that is the grid line phase variation at a given pixel is supposed to describe the corresponding point displacement. If the displacement/deformation between the reference and the distorted state is significant (e.g. during a bending test), physical points have to be matched together. Consequently, the displacement can be described either in the *reference state*, or in the *distorted state*. Two strategies, comparable to those used in computational mechanics, are applicable in order to evaluate the displacement field in one of these two states. The first, hereafter, referred to as *explicit* consists of an accumulation of small variations. The second, called *implicit*, establishes the displacement field between two mechanical situations by an iterative process without knowledge of the kinematic history. In addition, large strains induce not only grid phase variations but also large grid pitch variations. If this variation is too large, using a single phase detection algorithm (which has to be tuned to the grid pitch) is no longer possible, since it will detune. An estimation of the grid pitch for each loading situation is an elegant way to directly estimate the strain field.

3.8.1. *Explicit method*

The first method consists of calculating as many intermediate states as necessary to keep the small perturbation hypothesis valid; an incremental strain lower than one pixel per grid period is a good practical value. The incremental deformation U_k is given for the geometry of state k. It is necessary to map this deformation in the first (initial) state, and then add it to the total deformation map. This operation is tricky because the interpolation of the current displacement map propagates the random or systematic errors present in the total displacement field U_{tot}. A regularization, based on low-pass filtering or a kinematic hypothesis, might become necessary [AVR 08].

This approach is very efficient for displacement-controlled mechanical tests, but it is time consuming. Its main interest is to address the large displacement problem and to allow the use of *temporal phase unwrapping*. If the image acquisition is run from the very beginning of the test, and if this test is performed at a constant speed, the result is an *absolute* displacement, and not a relative displacement with an unknown number (integer) of grid pitches p, as is the case with the classical method. Finally, the accumulation of infinitesimal deformations avoids grid defect decorrelation. The grid defect errors are compensated for from one step to the next. This eliminates the bias error related to the grid signature, and increases the quality of the strain map.

The algorithm described here is based on the processing of a large number of images. The easiest input file format is a video file; the output file can be a video file too, even if this kind of file is not as easy to manipulate as an image file.

3.8.2. *Implicit method*

If no intermediate state is known, the displacement fields connecting the reference and the current state are obtained using an iterative procedure [MOL 09]. The first estimate can be based on characteristic points of the images, or on the displacements obtained under the small perturbation hypothesis. The reference image is then deformed according to the approximate displacement field U^1_{tot}. This operation implies an interpolation of gray-level images, which adds some errors to the final displacement estimate (see the image correlation chapter). The interpolated reference image is compared to the current image using classic phase analysis, giving a corrective displacement field denoted by U^k for each iteration k, along the x and y directions. The correction fields are mapped into the current reference frame; they must be added to the previous displacement estimate U^k_{tot}, giving a new displacement estimate to be tested. This procedure is repeated until convergence is attained, that is when the corrective displacement field U^k is negligible:

$$\sigma \left(\left\| U^k \right\| \right) \approx \frac{P}{2\pi} \sigma_\varphi \qquad\qquad [3.23]$$

where σ_φ is the standard deviation of the phase.

Convergence is achieved with a small number of iterations, three being a typical value. Comparing with the previous method, note that the final result does not depend on any intermediate state. One of the difficulties of this approach is the phase unwrapping necessary for the iterative displacement field U^k; a robust spatial phase unwrapping algorithm is required [VIO 06]. The absolute displacement of at least one point of the specimen is also necessary to obtain the absolute displacement map. Because only two mechanical states are processed, this method is convenient when the surface is not visible during the deformation process, for example during metal forming (Figures 3.10(a) and (b)).

3.8.3. *Large strain*

As stated earlier, when the strain increases greatly, the grid frequency varies too much to avoid detuning the phase detection algorithm. A solution is to re-estimate this frequency. It is feasible to retrieve the phase and grid pitch for a set of regions of interest, which are as small as possible, to obtain a local information. In [DUP 93], it is proposed to couple the 2D Fourier transform of a $N \times N$ region of interest to a 2D interpolation of the spectrum frequency. The period p of the signal, its initial phase φ and the complex amplitude A are functions of the complex values of the frequency peak found at coordinate K and the surrounding peaks. Denoting $\tilde{I}_k$ as the complex amplitudes of the Fourier transform of the signal, the quantity β is then written as:

$$\beta = \Re \left(\frac{\tilde{I}_{K-1} - \tilde{I}_{K+1}}{2\tilde{I}_K - \tilde{I}_{K+1} - \tilde{I}_{K-1}} \right) \qquad\qquad [3.24]$$

Then the signal period, complex amplitude and phase are [DUP 93]:

$$p = N/(K + \beta) \text{ with } -0.5 < \beta < 0.5 \tag{3.25}$$

$$A = \frac{j2\pi\beta(\beta^2 - 1)}{1 - e^{j2\pi\beta}}(2\tilde{I}_K - \tilde{I}_{K-1} - \tilde{I}_K) \tag{3.26}$$

$$\varphi = \arctan_{2\pi} \frac{\Im(A)}{\Re(A)} \tag{3.27}$$

The typical lateral extension of the zone of interest in the spectral analysis method is 32 × 32 pixels.

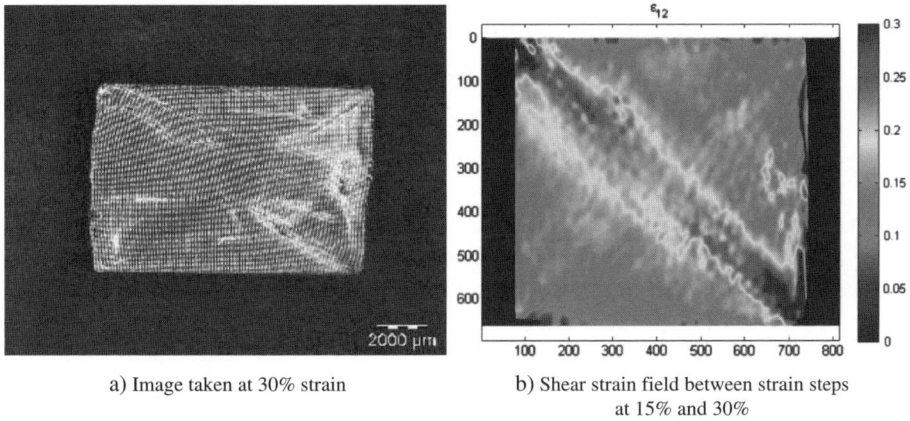

a) Image taken at 30% strain

b) Shear strain field between strain steps at 15% and 30%

Figure 3.10. *Implicit method applied to plane compression [MOL 09]*

3.9. Fringe projection

Whereas the grid technique uses a grid *affixed* to the specimen, fringe projection uses a regular grid pattern *projected* onto the surface. The technique allows the *shape* of an object to be measured. The basic principle of fringe projection is presented in Figure 3.11. A regular pattern of bright lines is projected onto the object, for example, using a video projector. The surface image is acquired by a CCD camera from a different point of view characterized by an angle θ. The video projector and the camera form a stereoscopic configuration. The consequence is that the camera will see the bright line pattern distorted by the object shape. The fringe projection technique uses the light diffused by the object, therefore the object must be diffusive enough, without too much specular reflection. The problem is that direct reflected light will, in general, create very bright zones in the image (because these zones correspond to locations on the object where the light source can be seen by reflection), preventing the homogeneous visibility of the projected lines. Note, however, that little diffused light is necessary if the object slopes are such that no reflected rays within the field of view (FOV) are captured by the camera lens.

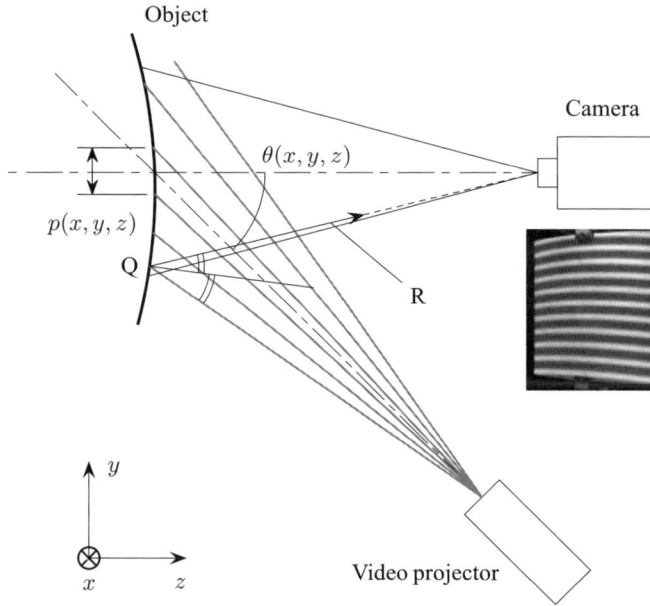

Figure 3.11. *Basic principle of fringe projection: the image of the object is formed with diffused light. Direct reflected light (ray R) would create a shiny spot around point Q. In the small image shown, the light incidence angles on the object are nowhere such that the reflected light directly enters the camera lens: there is no shiny spot*

As described earlier, the temporal phase-shifting technique can be used to record a phase map: a set of M fringe patterns with a known phase shift $2\pi/N$ are successively projected onto the surface. If the measurement has to be dynamic, that is the object shape changes in time, temporal phase shifting cannot be used, but spatial phase shifting may be possible if the surface is regular enough [WAN 10]. In the first case, the spatial resolution is one pixel[5], whereas it is equal to the grid period $p(x, y, z)$ in the second case.

A first basic relationship concerns a small altitude variation $\delta z(x, y)$, and the corresponding small phase variation $\delta\varphi(x, y)$ at each point of the field is given as:

$$\delta\varphi(x, y) = \frac{2\pi \tan \theta(x, y, z)}{p(x, y, z)} \delta z(x, y) \qquad [3.28]$$

---

5 More precisely, the length over the object that corresponds to one sensor pixel.

where $p(x, y, z)$ is the local grid pitch and $\theta(x, y, z)$ is the angle between the illumination and observation systems. Thus, the *sensitivity* is

$$s(x, y, z) = \frac{\delta\varphi(x, y)}{\delta z(x, y)} = \frac{2\pi \tan \theta(x, y, z)}{p(x, y, z)} \qquad [3.29]$$

As can be seen, this sensitivity varies all over the field[6].

To convert the measured phase into an altitude, a detailed geometrical modeling and/or calibration procedure is/are necessary [TAK 83, BRE 04].

As usual, phase extraction algorithms provide a wrapped phase, and thus it *must* be unwrapped. If the surface shows large stepwise altitude changes (introducing phase changes of the order of magnitude of or larger than 2π), or if the measurement cannot provide a simply connected measurement field (due to edges, holes, shadowing, etc.), spatial phase-unwrapping algorithms will fail. In that case, other unwrapping approaches can be used. Gray codes [SAN 99] is the most common method in commercial systems, but others exist, such as temporal phase unwrapping [SAL 97], spatiotemporal phase unwrapping [ZHA 99] and synthetic wavelength [TOW 05]. Color-based multiplexing even allows the use of one single frame [LIU 00]. An up-to-date review of the fringe projection technique can be found in [GOR 10].

Typical metrological performances of the shape measurement setup are a spatial resolution of one pixel in the case of temporal phase-shifting implementation, with a resolution that depends on the phase measurement resolution, expressed as the fraction $1/F$ of fringe[7] (i.e. of 2π rad) that corresponds to the noise level (as explained at length in section 1.2.2.2 of Chapter 1). It can be seen from equation [3.28] that with a typical angle θ of $45°$ ($\tan \theta = 1$), this corresponds to a z measurement resolution equal to[8] p/F. With a high fringe density, for example 4 pixels per fringe at the object level, the total number of fringes in the FOV will be $P/4$, where P is the number of sensor pixels along the direction orthogonal to the fringes, so the length corresponding to one fringe is FOV $\times 4/P$. Thus, the resolution will be FOV $\times 4/FP$. When using a $1,000 \times 1,000$ pixel sensor, an easy-to-achieve phase resolution of 1/200 fringe (i.e. $F = 200$) will provide a z resolution of FOV/50,000, for example 4 μm over a 200 mm FOV.

---

6 Unless the fringe projection is collimated and the camera is equipped with a telecentric lens, conditions that correspond to a very restricted practical use, since the attainable field of view cannot in that case exceed some tens of millimeters.

7 F is sometimes called the *finesse* by analogy with the Perot–Fabry interferometer.

8 Here, we drop the dependence of p upon (x, y, z) and consider only an average value of the fringe pitch.

The fringe projection method has been described and used in experimental mechanics by many authors, among them [LAG 02, HUA 03, SCI 05]. Here, a simple example is given in Figure 3.12 [GIG 06]: a square [0, 90°] composite layup (carbon fiber-reinforced polymer) has been subjected to a temperature variation, from its curing temperature (130 °C) down to ambient temperature (25 °C). Experimental shapes are presented using an adjustable color scale. Global curvatures are calculated using a least squares algorithm, and compared to a nonlinear Rayleigh–Ritz algorithm: at high temperatures (from 110 to 130 °C), two opposite curvatures exist, corresponding to a saddle shape; at lower temperatures (from 25 to 110 °C), one of the two curvatures almost disappears while the other increases greatly, which corresponds to a cylindrical shape. The transition point (around 110 °C) corresponds to a bifurcation in the mathematical model.

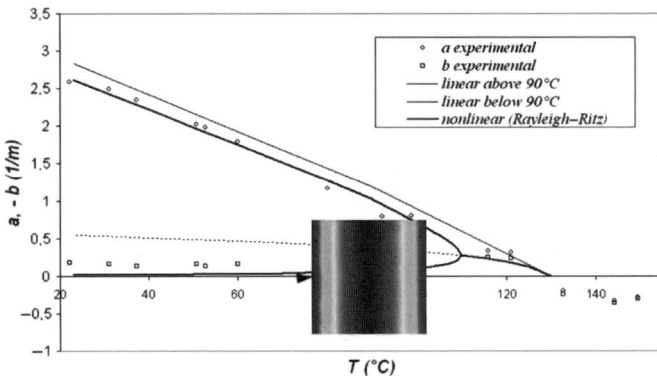

Figure 3.12. *Application of fringe projection to composite dimensional stability monitoring [GIG 06]. For a color version of this figure, see www.iste.co.uk/gh/solidmech.zip*

3.10. Deflectometry

Deflectometry is a technique that is directly sensitive to the surface *slope*. It is applied in experimental mechanics to plate structures, but also in industry for *aspect quality control.*

The principle of deflectometry is presented in Figure 3.13. As in previous sections, it is based on the measurement of displacements, but in this case, these displacements are *apparent* within an image. A camera is used to record the *reflected* image of a black and white grid.

Even if it is apparently close to the fringe projection technique, deflectometry is different both in its physical principle and in its measuring characteristics: fringe projection is based on *diffused* light, and specular reflection is an adverse feature; on the contrary, deflectometry is based on *reflected* light. Fringe projection basically measures altitudes, and deflectometry measures slopes. In other words, fringe projection is appropriate for measuring an absolute shape, perhaps missing very small defects, and deflectometry is effective for measuring shape defects, because slopes carry information about small altitude variations over small distances, a situation often corresponding to shape defects.

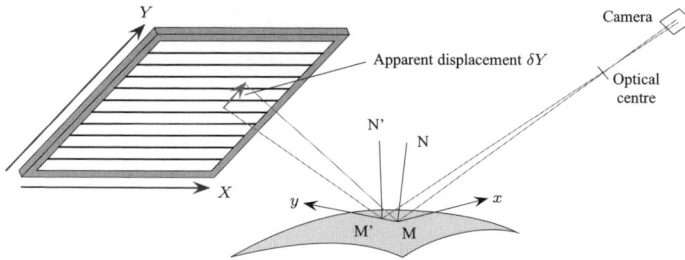

Figure 3.13. *At each pixel location, the measured phase is related to the corresponding source point, and consequently to the normal and the slope of the surface*

The phase of the reflected grid is obtained either using temporal phase stepping (the displayed grid is shifted by an integer fraction of its period), or spatial phase stepping (a single image is necessary, if the surface is smooth enough and has a high spatial frequency grid). At each image pixel location, the measured phase is related to the corresponding source point. A model of beam propagation in space gives the local slope at each point.

Local slopes of a surface defined as $z = z(x, y)$ are the quantities $\partial z / \partial x$ and $\partial z / \partial y$. Under classic plate theory assumptions, the first derivative of the slopes are the local curvatures $k_{xx} = \partial^2 z / \partial x^2$ and $k_{yy} = \partial^2 z / \partial y^2$. The cross-derivative is the torsion $k_{xy} = \partial^2 z / \partial x \partial y = k_{yx}$.

In normal incidence, the implementation is simple. Figure 3.14 shows a setup using a stroboscopic lamp to study a vibrating plate.

In this example, an artificial defect has been embedded in the composite plate. Experimental curvatures are compared with those obtained using a finite element model (Figure 3.15).

Figure 3.16 shows the spectacular performance this technique can have in the domain of aspect quality control. Nanometric shape defects can be detected if they extend over a short distance, because the resulting slopes will be detectable. To outline this point, consider a surface described by the following x profile:

$$z(x) = a \cos \left(\frac{2\pi x}{\Lambda} \right)$$

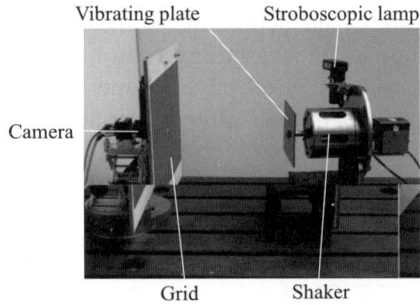

Figure 3.14. *Deflectometry measurement of a vibrating plate [GIR 06]; the sample surface is processed to enhance reflection and the camera, located behind a hole in the center of the grid, records the image of the reflected grid*

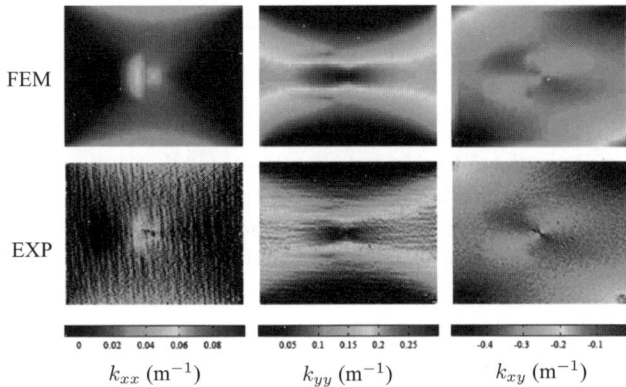

Figure 3.15. *Experimental curvatures (EXP) compared with finite element model (FEM) [GIR 06]; the experimental setup is presented in Figure 3.14*

The x slope is:

$$\alpha(x) = -\frac{2\pi}{\Lambda} \sin\left(\frac{2\pi x}{\Lambda}\right)$$

and the curvature:

$$k(x) = \frac{4\pi^2}{\Lambda^2} \cos\left(\frac{2\pi x}{\Lambda}\right)$$

Thus, the curvature varies as the inverse of the *square* of the defect wavelength, and consequently increases very rapidly when the wavelength decreases.

Local defects correspond to curvatures that are generally easily detectable using deflectometry. In practice, the main problem is not the sensitivity but the illumination source size: on highly curved convex objects, the measurable field, that is the area where the reflected grid can be observed, can become dramatically small.

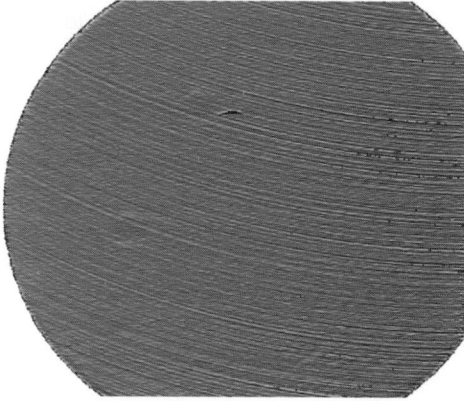

Figure 3.16. *Curvatures on a flatness standard (certified shape deviation <400 nm, roughness <6 nm). Part diameter 40 mm*

3.11. Examples

3.11.1. *Off-axis tensile test of a unidirectional composite coupon*

Figure 3.17 shows an example of a tensile test on a unidirectional (UD) composite specimen.

Figure 3.18 presents the boundary conditions for an off-axis tensile test on a UD carbon fiber-reinforced composite specimen. This test is used in order to characterize the shear modulus G of an orthotropic material. Let us recall here that for such a material, the shear modulus is independent of the Young's modulus E and Poisson's ratio ν, and that the relationship $G = E/2(1 + \nu)$ does not hold, as it is only valid for isotropic materials. This is why specific tests have to be designed for the characterization of the shear modulus, among which is the off-axis test.

The simplest kinematic configuration is one where a *uniform* stress is imposed. The specimen then deforms as a parallelogram, and the shear strain is uniform. Isodisplacement lines are parallel straight lines with an oblique orientation[9]. The measurement of shear strain using an electrical rosette strain gauge in the middle of the specimen is sufficient to obtain the shear modulus G. However, in practice, a load

---

9 In the case of an isotropic material, isodisplacement lines are obviously parallel and normal to the specimen axis.

frame really *applies a displacement*. The ends of the specimen being clamped, any rotation of the specimen ends is restricted, so the real boundary conditions are closer to a uniformly applied displacement than to a uniformly applied stress. As a result, an in-plane torsion torque appears, and the strain field is no longer homogeneous.

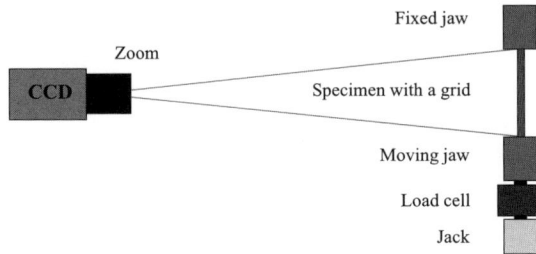

Figure 3.17. *Displacement measurement using spatial phase shifting*

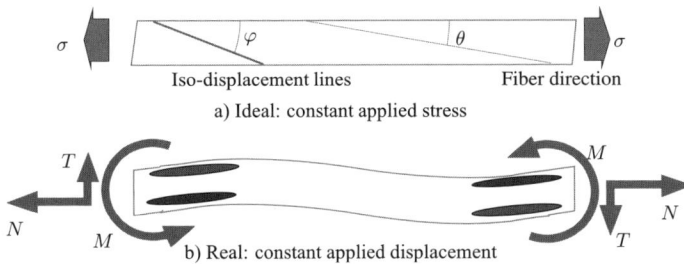

Figure 3.18. *Off-axis tensile test on a UD composite specimen*

Many different solutions have been proposed, such as correction factors on the strain gauge measurement (not very reliable, as they attempt to model an unknown kinematic state), rotating clamps (expensive), etc. A very elegant solution [SUN 93] is to use *oblique tabs*, that is tabs with a trapezoidal shape at the ends of the specimen, with one edge aligned with the theoretical isodisplacement lines, in order to impose the displacement along this line. We present different boundary conditions that have been tested, with a displacement field measurement using the grid method (Figure 3.19). The chosen grid pitch was $p = 0.61$ mm, and the studied field length was 120 mm.

The boundary conditions were:

– without tabs;

– with rectangular $[\pm 45]_s$ composite tabs;

– with aluminum oblique tabs.

When the latter are used, the homogeneity of the strain field is dramatically improved, and corresponds to the almost perfectly straight and equidistant isodisplacement lines that can be observed in that case. It is worth noting that the $[\pm 45]_s$ composite tabs enable partial rotations of the ends, but the effect is not sufficient by itself to ensure a homogeneous strain field.

During this test, the *maximum displacement* of the movable jaw was 58 μm, which corresponds to only 1/10 of grid pitch. In other words, the entire measured phenomenon takes place "within 1/10 of fringe". The measured resolution, obtained from the noise evaluation, was 1.10 μm, or $p/500$.

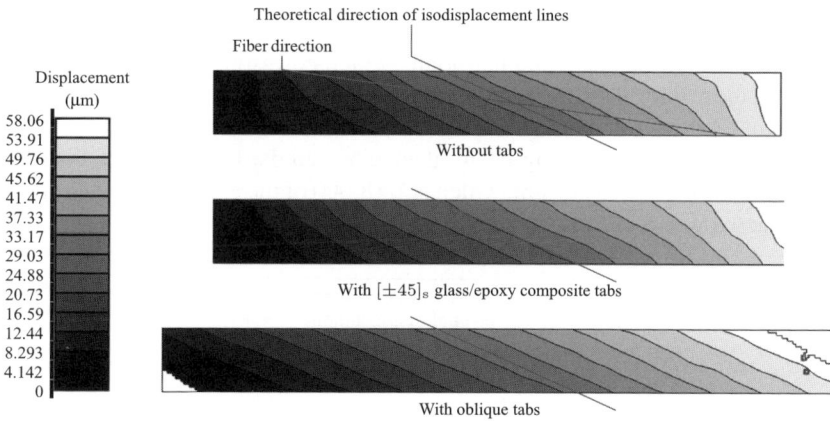

Figure 3.19. *Axial displacements*

3.11.2. *Rigid body displacement*

In the case of a rigid body displacement, the interest of using an optical full-field technique is the large number of independent measurements it allows. Thus, noise can be dramatically reduced by statistical averaging, with the consequence of greatly increasing the *resolution*. The example presented in [SUR 96b] concerns the study of the displacement of a motorized translation stage, actuated by a stepper motor. The nominal displacement step is 1 μm, but it can be controlled with a resolution of 1/10 step, that is 100 nm. A grid with a pitch $p = 125$ μm placed on the stage was observed with an 8-bit 768 × 574 pixel CCD camera. The lens magnification was chosen such that spatial phase stepping could be used with $N = 4$ pixels per period, which corresponds to $167 \times 122 = 20{,}374$ independent measurements. The resolution was estimated from the noise level measurement at 1.31 μm, or 1/100 of the grid pitch. The resolution on the average displacement obtained from the whole field is evaluated to be $1.31/\sqrt{20{,}374}$ μm $\approx$ 9 nm. With more recent equipment (1,000 × 1,000 or 2,000 × 2,000 pixels, 10- or 12-bit cameras), it is clear that a significantly lower resolution value can be achieved.

Figure 3.20 shows the results for 100 successive measurements, corresponding to a set of 10 nominal displacements of 100 nm, each of them successively measured 10 times (the two curves are slightly shifted horizontally for sake of legibility). Here, for a given position, the repeatability fluctuations are significantly lower than the nominal displacement. Using two series of 500 displacements (10,000 measurements), a systematic cyclic displacement error, with a period of 1 μm was characterized. This seems to have been caused by the stepper motor controller in its special 1/10 step control mode.

This systematic error is clearly visible in Figure 3.20, the measurement being adjusted to the nominal displacement for a value close to 500 nm. This systematic error was identified by a Fourier analysis of the deviation between the measured and nominal values. The fundamental harmonic, which corresponds to a period of 1 μm, was subtracted from this difference. A low-frequency variation was also subtracted in order to take into account the possible thermal drift, no specific precautions against this phenomenon having been taken. The standard deviation of the residual fluctuations was measured at 7 nm, which is 1/18,000 of the grid period, corresponding to the expected value, or even better.

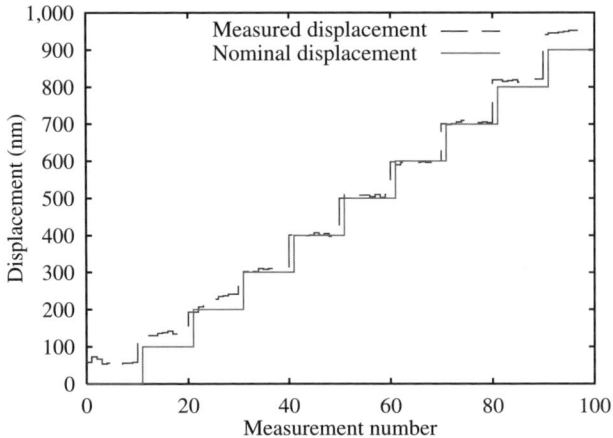

Figure 3.20. *Ten successive displacements with increments of 1/10 of the nominal step of 1 μm, each displacement being measured 10 times*

3.11.3. *SEM measurement*

This example shows, on the one hand, the possibility of using the grid technique with any kind of imaging system and, on the other hand, the efficiency of signal harmonics rejection if a large pixel number per period is chosen when the physical grid has a cyclic ratio [10] far from unity.

The example shown in Figure 3.21(a) is an image obtained using a scanning electron microscope (SEM) on a "grade 702" zirconium alloy. The material was rolled, and thus is textured. This alloy is used for tanks in nuclear waste reprocessing plants. The lines have a very small width in order not to hide the substrate. The mechanical test is a simple tensile test along the transverse rolling direction. The grain size is around 15 μm. The grid was deposited using SEM micro-electro-lithography (PMMA resin directly insulated by the SEM beam, partial dissolution, gold sputtering, total dissolution of the resin). The grid pitch is $p = 2$ μm. Phase extraction was performed by one of the authors[11].

3.11.4. *Characterization of lens distortion*

Most full-field optical methods use a camera and a lens for image acquisition. Optical lenses are affected by *image distortion*. This aberration transforms the geometry of the object plane. It is classically illustrated using a cartesian grid. The two classic types are *barrel* (Figure 3.22(a)) and *pincushion* (Figure 3.22(b)) distortions. The distortion is described in the image plane by a displacement vector field connecting the image points without distortion to effective image points.

The grid technique coupled with spatial phase stepping can be used to characterize distortion. This approach takes advantage of the excellent resolution of phase detection techniques and of the high density of information, because the grid used as the object can be chosen with a small pitch.

By analogy with a displacement field in mechanics, the phase of the distorted lines is represented by:

$$\varphi(\mathbf{r}) = 2\pi \mathbf{f}.\mathbf{u}^{-1}(\mathbf{r}) \qquad [3.30]$$

where $\mathbf{f}$ is the spatial frequency vector of a non-distorted image and $\mathbf{u}^{-1}(\mathbf{r})$ is the field of inverse displacements that convert the distorted grid back to the perfect grid. With a crossed-lines grid, each line direction allows us to measure one component of the inverse displacement.

An estimate of the spatial frequency $\mathbf{f}$ is given by the phase gradient at the center of the image, where distortion is taken as null:

$$\mathbf{f} = -\frac{1}{2\pi} \left.\nabla\varphi\right|_{\mathbf{r}=0} \qquad [3.31]$$

---

10 Ratio between the widths of the black and white lines of the grid.

11 Initially, the grid line pattern was created to be used at LMS-X for digital image correlation.

a) SEM image   b) Zoom

Displacement
(% pitch)

105.0
90.00
75.00
60.00
45.00
30.00
15.00
.0000
−15.00
−30.00
−45.00
−60.00
−75.00
−90.00
−105.0

c) u_x (left) and u_y (right) displacements

Strain
(%)

14
12
10
8
6
4
2
0
−2
−4
−6
−8
−10
−12
−14

d) Tensile strain ϵ_{xx} (left) and shear strain ϵ_{xy} (right)

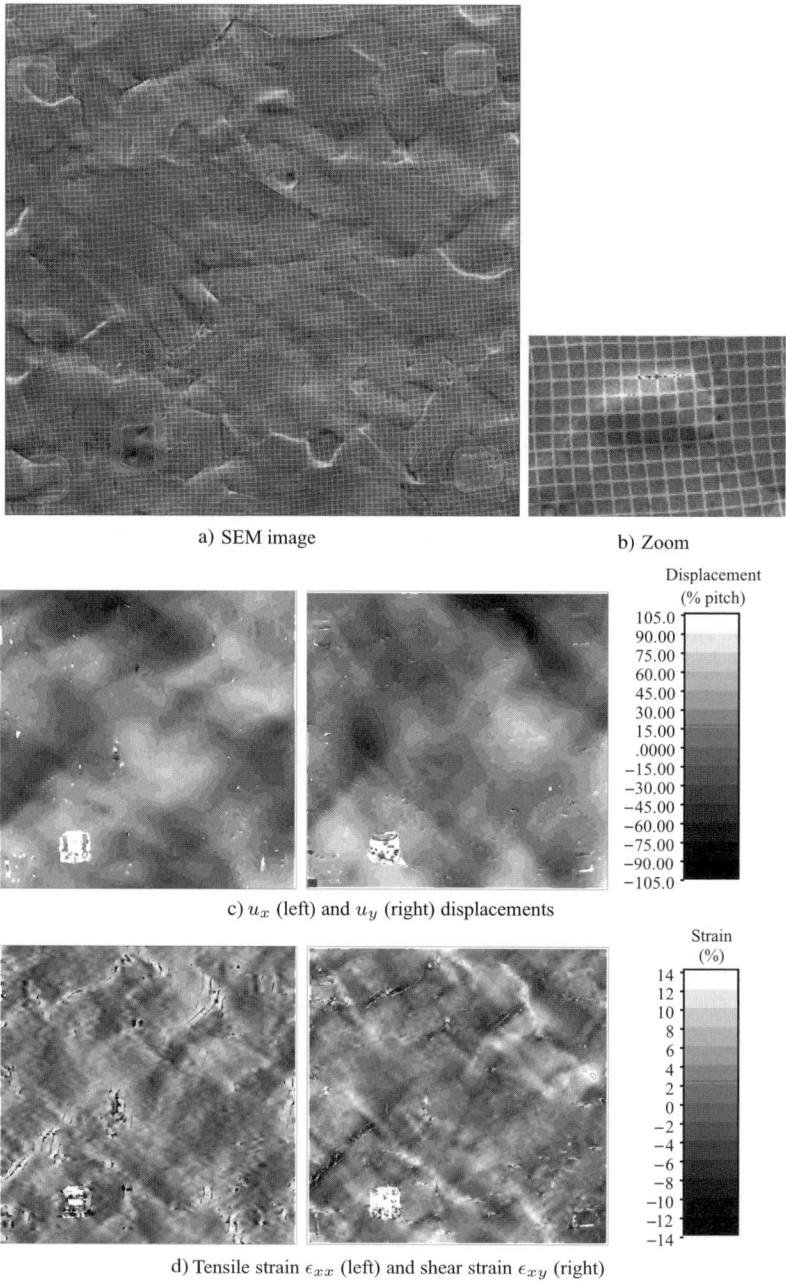

Figure 3.21. *Tensile test of a zirconium alloy under SEM (courtesy of LMS, École Polytechnique, M. Dexet and J. Crépin)*

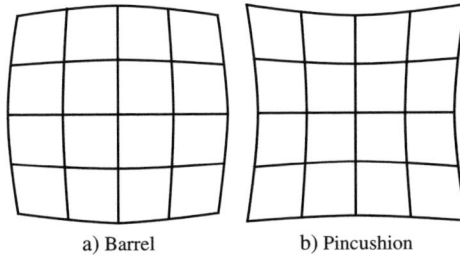

a) Barrel   b) Pincushion

Figure 3.22. *Distortion*

It is not necessary to know the exact grid pitch, but if it is known, the spatial frequency as defined above, combined with the pixel size, gives the magnification. The measured spatial period can be expressed in pixels per grid pitch, and corresponds to a 2π rad phase variation. In a similar way, the phase related to the distortion can be converted into pixels of apparent displacement. It is really the *inverse* displacement that can be readily used afterward for image correction.

The distortion of a 16 mm focal length lens, with $f/4$ aperture, is presented in Figure 3.23 [COU 04]. The grid used had a 4 mm pitch and was printed by an A0 plotter. The grid defects can be considered as negligible compared to the lens distortion to be evaluated. The grid was positioned 1.35 m away from the lens and the image acquisition was performed using an 8-bit $1,296 \times 1,030$ pixel digital camera. Thus, one grid period was sampled by seven contiguous pixels. The distortion can be observed by downsampling the image by a factor of 7 (i.e. by keeping only one of every seven pixels in the image), which creates a moiré effect, amplifying grid deformation (Figure 3.23(a)). There is no interest in showing here the grid without this moiré effect because the deformation is scarcely visible to the naked eye. Let us mention however that the distortion here is a barrel- and not a pincushion-type, as the moiré image falsely suggests.

Displacement fields along x and y are reproduced in Figures 3.23(c) and (d). It is worth noting that the displacements induced by the distortion are noticeable: approximately 25 pixels at the boundaries. After measurement of the inverse displacement fields by the grid method, it is possible to undistort the original image corresponding to Figure 3.23(a) to obtain the corrected image shown in Figure 3.23(b). To test the quality of this correction, it is possible to characterize in the same way the residual distortion of the corrected image. Residual displacement fields are presented in Figures 3.23(e) and (f). The standard deviation of the residual distortion was found to be of 0.007 pixel.

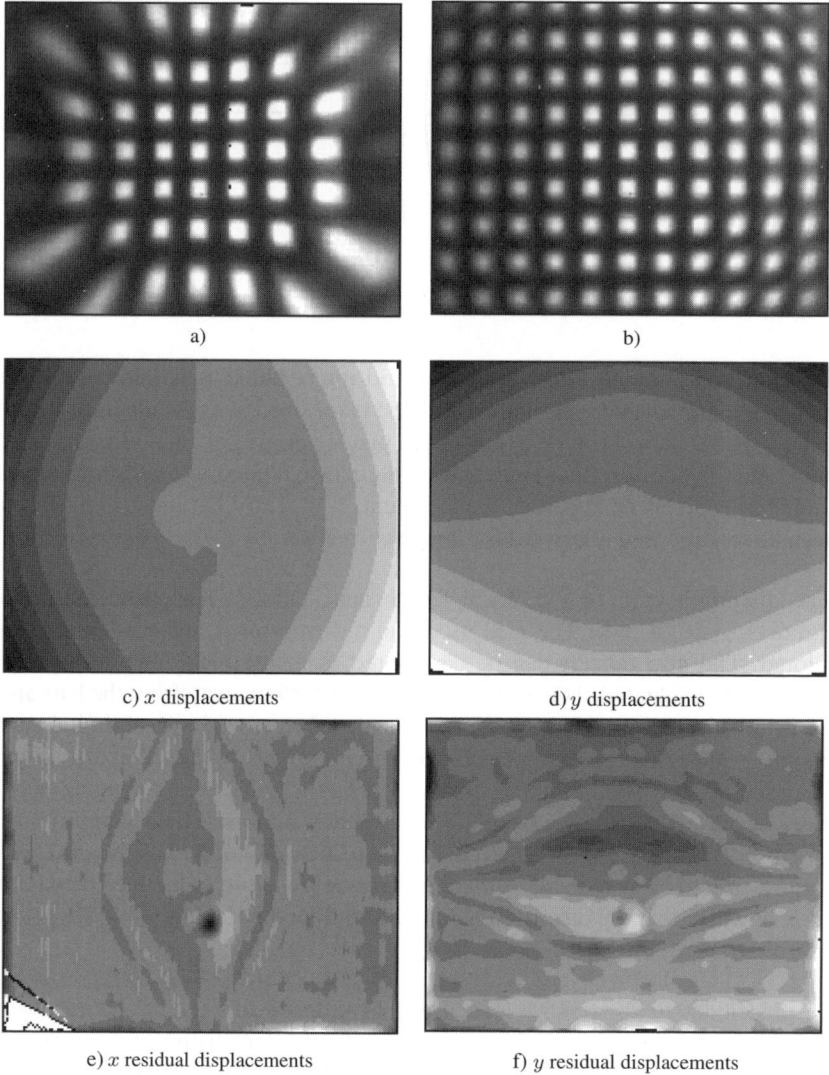

a)

b)

c) x displacements

d) y displacements

e) x residual displacements

f) y residual displacements

Figure 3.23. *a) Moiré effect due to downsampling a grid image observed with a short focal lens, and affected by significant distortion. b) Idem with the corrected image; c) and d) inverse displacement fields due to lens distortion (scaling: 3.5 pixels per color); e) and f) residual displacement fields calculated from the corrected image (scaling: 0.013 pixels per color)*

3.12. Conclusion

In this chapter, the grid method has been described as a simple method, relatively easy to implement and having a good resolution. Because it is not based on interferometry, it is independent of the light wavelength and it can be used at different scales: macroscopic structure, lab specimen or micro specimen imaged using SEM or atomic force microscope (AFM). The resolution of the displacement is calculated using the grid pitch and the phase noise level, and can be estimated *in situ* with a repeatability test.

If the resolution of the phase measurement is equal to $2\pi/F$ rad, then the resolution of the displacement measurement is p/F. Typical values of F range from 200 to 500 or even 1,000 with recent equipment: 10- or 12-bit digital cameras, high quality illumination systems (LED), good quality lenses, etc. When using a grid period of 100 or 200 μm (5–10 lines per mm), displacement resolution can be less than 1 μm, which is of great interest in experimental mechanics.

Furthermore, it was shown that such a full-field method can even be used on a homogeneous field (i.e. a rigid body motion), taking advantage of the large number of independent measurements to reduce noise and increase resolution. An example was given from the field of nanometric displacements.

Finally, an application beyond the field of experimental mechanics was presented: the grid method can be efficiently used to characterize lens distortion.

Finally, it is important to outline the main drawbacks of the grid method. In the case of in-plane displacement measurements, the first disadvantage is the necessity to prepare the surface by printing, gluing, engraving or depositing a grid in some way (e.g. microlithography for microgrids). This task should not be underestimated and can be problematic. The second limitation is that only plane surfaces subjected to in-plane displacements can be investigated.

3.13. Bibliography

[AVR 08] Avril S., Feissel P., Pierron F., Villon P., "Estimation of the strain field from full-field displacement noisy data. Comparing finite elements global least squares and polynomial diffuse approximation", *European Journal of Computational Mechanics*, vol. 17, no. 5–7, pp. 857–868, 2008.

[BAD 09] Badulescu C., Grédiac M., Mathias J.-D., Roux D., "A procedure for accurate one-dimensional strain measurement using the grid method", *Experimental Mechanics*, vol. 49, no. 6, pp. 841–854, 2009.

[BRE 04] Breque C., Dupré J., Brémand F., "Calibration of a system of projection moiré for relief measuring: biomechanical applications", *Optics and Lasers in Engineering*, vol. 41, no. 2, pp. 241–260, 2004.

[COU 04] Coudert S., Triconnet K., Zebiri A., Surrel Y., "Étalonnage transverse d'un objectif de caméra par la méthode de la grille", *Photomécanique - Étude du comportement des matériaux et des structures*, École des mines d'Albi Carmaux, Albi, France, 04 May 2004.

[DAL 91] DALLY J., RILEY W., *Experimental Stress Analysis*, 3rd ed., McGraw-Hill, 1991.

[DUP 93] DUPRÉ J.C., BRÉMAND F., LAGARDE A., "Numerical spectral analysis of a grid: application to strain measurement", *Optics and Lasers in Engineering*, vol. 18, pp. 159–172, 1993.

[GIG 06] GIGLIOTTI M., MOLIMARD J., JACQUEMIN F., VAUTRIN A., "On the nonlinear deformations of thin unsymmetric 0/90 composite plates under hygrothermal loads", *Composites Part A: Applied Science and Manufacturing*, vol. 37, no. 4, pp. 624–629, 2006.

[GIR 06] GIRAUDEAU A., GUO B., PIERRON F., "Stiffness and damping identification from full-field measurements on vibrating plates", *Experimental Mechanics*, vol. 46, pp. 777–787, 2006.

[GOR 10] GORTHI S.S., RASTOGI P., "Fringe projection techniques: whither we are?", *Optics and Lasers in Engineering*, vol. 48, no. 2, pp. 133–140, 2010.

[HUA 03] HUANG P.S., ZHANG C., CHIANG F.-P., "High-speed 3-D shape measurement based on digital fringe projection", *Optical Engineering*, vol. 42, pp. 163–168, 2003.

[HUN 98] HUNTLEY J., "Automated fringe pattern analysis in experimental mechanics: a review", *Journal of Strain Analysis*, vol. 33, no. 2, pp. 105–125, 1998.

[LAG 02] LAGARDE J., ROUVRAIS C., BLACK D., DIRIDOLLOU S., GALL Y., "Skin topography measurement by interference fringe projection: a technical validation", *Skin Research and Technology*, vol. 7, no. 2, pp. 112–121, 2002.

[LIU 00] LIU W., WANG Z., MU G., FANG Z., "Color-coded projection grating method for shape measurement with a single exposure", *Applied Optics*, vol. 39, no. 20, pp. 3504–3508, 2000.

[MOL 09] MOLIMARD J., DARRIEULAT M., "Mesoscopic investigation of the heterogeneities caused by channel-die compression", *Journals of Engineering Materials and Technology*, vol. 54, pp. 7–17, 2009.

[PAR 90] PARKS V., "Strain measurement using grids", *Optical Engineering*, vol. 21, no. 4, pp. 663–669, 1990.

[PIR 04] PIRO J., GRÉDIAC M., "Producing and transferring low-spatial-frequency grids for measuring displacement fields with moiré and grid methods", *Experimental Techniques*, vol. 28, pp. 23–26, 2004.

[ROD 87] RODDIER C., RODDIER F., "Interferogram analysis using Fourier transform techniques", *Applied Optics*, vol. 26, pp. 1668–1673, 1987.

[SAL 97] SALDNER H., HUNTLEY J., "Temporal phase unwrapping: application to surface profiling of discontinuous objects", *Applied Optics*, vol. 36, no. 13, pp. 2770–2775, 1997.

[SAN 99] SANSONI G., CAROCCI M., RODELLA R., "Three-dimensional vision based on a combination of gray-code and phase-shift light projection: analysis and compensation of the systematic errors", *Applied Optics*, vol. 38, no. 31, pp. 6565–6573, 1999.

[SCI 05] SCIAMMARELLA C., LAMBERTI L., SCIAMMARELLA F., "High-accuracy contouring using projection moiré", *Optical Engineering*, vol. 44, no. 9, pp. 093605.1–093605.12, 2005.

[SEV 93] SEVENHUIJSEN P., SIRKIS J., BRÉMAND F., "Current trends in obtaining deformation data from grids", *Experimental Techniques*, vol. 17, no. 3, pp. 22–26, 1993.

[SUN 93] SUN C.T., CHUNG I., "An oblique end-tab design for testing off-axis composite specimens", *Composites Part A*, vol. 24, no. 8, pp. 619–623, 1993.

[SUR 96a] SURREL Y., "Design of algorithms for phase measurements by the use of phase stepping", *Applied Optics*, vol. 35, no. 1, pp. 51–60, 1996.

[SUR 96b] SURREL Y., FOURNIER N., "Displacement field measurement in the nanometer range", in GÓRECKI C. (ed.), *Optical Inspection and Micromeasurements*, Besançon, vol. SPIE 2782, pp. 233–242, 10–14 June, 1996.

[SUR 05] SURREL Y., "Les techniques optiques de mesure de champ: essai de classification", *Instr. Mes. Metrol.*, vol. 5/1–2, 2005.

[TAK 82] TAKEDA M., INA H., KOBAYASHI S., "Fourier-transform method of fringe-pattern analysis for computer-based topography and interferometry", *Journal of the Optical Society of America*, vol. 72, pp. 156–160, 1982.

[TAK 83] TAKEDA M., MUTOH K., "Fourier transform profilometry for the automatic measurement of 3-D object shapes", *Applied Optics*, vol. 22, no. 24, pp. 3977–3982, 1983.

[TAK 90] TAKEDA M., "Spatial carrier fringe pattern analysis and its applications to precision interferometry and profilometry: an overview", *Industrial Metrology*, vol. 1, pp. 79–99, 1990.

[TOW 05] TOWERS C., TOWERS D., JONES J., "Absolute fringe order calculation using optimised multi-frequency selection in full-field profilometry", *Optics and Lasers in Engineering*, vol. 43, no. 7, pp. 788–800, 2005.

[VIO 06] VIOTTI M., KAUFMANN G., "Measurement of elastic moduli using spherical indentation and digital speckle pattern interferometry with automated data processing", *Optics and Lasers in Engineering*, vol. 44, pp. 495–508, 2006.

[WAN 10] WANG P., DRAPIER S., MOLIMARD J., VAUTRIN A., MINNI J., "Characterization of liquid resin infusion (LRI) filling by fringe pattern projection and in situ thermocouples", *Composites Part A: Applied Science and Manufacturing*, vol. 41, no. 1, pp. 36–44, 2010.

[WOM 84] WOMACK K.H., "Frequency domain description of interferogram analysis", *Optical Engineering*, vol. 23, pp. 396–400, 1984.

[ZHA 99] ZHANG H., LALOR M.J., BURTON D.R., "Spatiotemporal phase unwrapping for the measurement of discontinuous objects in dynamic fringe-projection phase-shifting profilometry", *Applied Optics*, vol. 38, no. 16, pp. 3534–3541, 1999.

[ZHA 01] ZHAO B., ASUNDI A., "Microscopic grid methods – resolution and sensitivity", *Optics and Lasers in Engineering*, vol. 36, pp. 437–450, 2001.

Chapter 4

Digital Holography Methods

4.1. Introduction

Holography was invented in 1947 by the Hungarian physicist Dennis Gabor during his research into electronic microscopy [GAB 48]. This invention won him the Nobel Prize for physics in 1971. It took until 1962 for the first lasers using this technique to find concrete applications [LEI 61, DEN 62, POW 65, COL 71]. In the 1960s, 1970s and 1980s, cineholography was successfully applied to non-destructive testing and three-dimensional industrial applications [PAQ 65, ALB 76, FAG 83, FAG 86, SMI 88]).

Holography is a productive mix of interference and diffraction. Interference encodes the amplitude and relief of a three-dimensional object on a photographic silver plate, and diffraction functions as a decoder, reconstructing a wave that seems to come from the object that was originally illuminated. This encoding contains all the information about a given scene: amplitude and phase – and thus relief. Considering the constraints involved in the treatment of classical holograms, which make their industrial use difficult (for quality control in production lines, for example), the replacement of the silver support by a matrix of the discrete values of the hologram was envisaged in 1967 [GOO 67]. The idea was to replace the analog recording/decoding with digital recording/decoding, simulating diffraction by a digital network [HUA 71, KRO 72].

Even though the concept of digital holography had been known for some time, it took until the 1990s for "digital" holography based on array detectors to emerge

Chapter written by Pascal PICART and Paul SMIGIELSKI.

[SCH 94, SCH 05]. At the end of the 1980s there were important developments in two sectors of technology: micro-technological procedures allowed the creation of image sensors with numerous miniaturized pixels; and the rapid computational processing of images became accessible with the appearance of powerful processors and an increase in storage capacities. In recent years, digital Fresnel holography has been greatly developed and applied with success to a large number of domains, and fascinating possibilities have been demonstrated:

– phase contrast digital holographic microscopy [ZHA 98, CUC 99a, CUC 99b, PED 02, FER 06, CHA 07, MAN 08, KUH 09];

– imaging across diffusive media [LEC 00];

– digital color holography [YAM 02, PIC 08b];

– measurement of surface profiles [YAM 01b];

– characterization of micro-components [SEE 01, CHA 06a];

– synthetic aperture imaging [LEC 00, JAC 01, BIN 02];

– compensation for lens aberrations [STA 00];

– mechanical measurement by spatial multiplexing [PIC 03a, PIC 04]; and

– phasing of laser beams [BEL 09, PAU 09].

Furthermore, the rise of robust and reliable high-energy dual-pulse lasers has enabled applications in vibration mechanics [PED 95]. Digital holography is also a promising technology for the recognition and comparison of three-dimensional objects, as evidenced by recent works on the subject [JAV 00, OST 02].

This chapter presents methods for digital holographic interferometry. Digital holographic interferometry exploits not only the amplitude of the object, but also its phase. The review of the applications of digital Fresnel holography that is proposed is not intended to be exhaustive, as it will focus on the methods that exploit the phase of the hologram for quantitative measurement in the field of solid mechanics. After a presentation of the basics of digital holography, we will discuss methods based on the spatial multiplexing of digital holograms and on digital color holography.

4.2. Basics of wave optics

Holography is closely related to the wave description of light, and particularly to diffraction theory and light interference [COL 71, BOR 99, HAR 02, HAR 06]. The hologram includes full information on any object, in terms of the complex amplitude of the object being studied: thus it is possible to reconstruct the object field in its entirety (amplitude and phase) from the digitally-encoded hologram.

4.2.1. Light diffraction

Figure 4.1 illustrates the geometry of diffraction and the variables used in this chapter. Consider an extended object illuminated with a monochromatic wave. This object diffracts a wave to the observation plane located at distance $d_0 = |z_0|$. The surface of the object generates a wave front that will be noted A:

$$A(x, y) = A_0(x, y) \exp(j\psi_0(x, y))$$

[4.1]

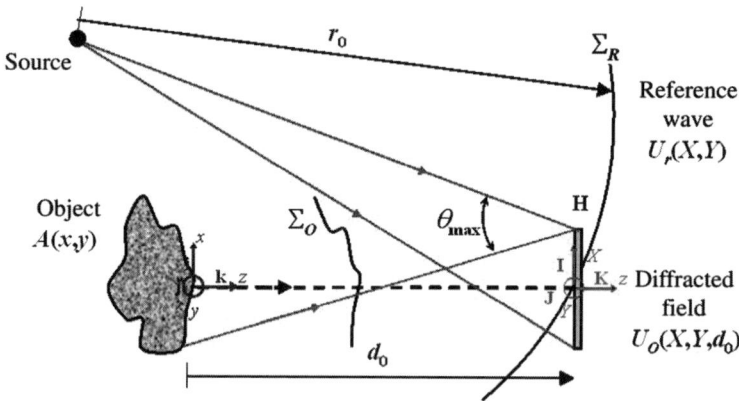

Figure 4.1. Free space diffraction, interferences and notations

Amplitude A_0 describes the reflectivity of the object, and phase ψ_0 is related to its surface and shape. (x, y, z) is a set of reference axes attached to the object's surface. The natural roughness of the object means that ψ_0 is a random variable uniformly distributed over $]-\pi, +\pi]$. Considering the diffraction theory under the Fresnel approximations [GOO 05], the diffracted field U_O at distance d_0 and at spatial coordinates (X, Y) of the observation plane is given by the two-dimensional Fresnel transform, according to:

$$U_O(X, Y, d_0) = -\frac{j}{\lambda d_0} \exp\left(\frac{2j\pi d_0}{\lambda}\right) \exp\left(\frac{j\pi}{\lambda d_0}(X^2 + Y^2)\right)$$

$$\times \iint A(x, y) \exp\left(\frac{j\pi}{\lambda d_0}(x^2 + y^2)\right) \exp\left(-\frac{2j\pi}{\lambda d_0}(xX + yY)\right) dx dy$$

[4.2]

In the observation plane, this wave can also be written as:

$$U_o(X,Y,d_o) = a_o(X,Y)\exp(j\varphi_o(X,Y)),$$

[4.3]

with a_O being the modulus of the complex amplitude and φ_O its optical phase. Since the object is naturally rough, the diffracted field at distance d_0 has a speckle pattern [DAI 84, GOO 07].

4.2.2. Interference

Figure 4.1 illustrates the geometry of the interference. Let us consider U_r, which is the complex amplitude of the reference wave front, at the recording plane. Here we have:

$$U_r(X,Y) = a_r(X,Y)\exp(j\varphi_r(X,Y)),$$

[4.4]

with a_r being the modulus and φ_r the optical phase. The reference wave front usually comes from a small pinhole; thus it is a spherical divergent wave, impacting the plane with a non-zero incident angle. Considering (x_s, y_s, z_s) to be the coordinates of the source point in the hologram reference frame $(z_s < 0)$, the optical phase of the reference wave can be written in the Fresnel approximations by [GOO 05]:

$$\varphi_r(X,Y) \cong \frac{\pi}{\lambda z_s}\left((X - x_s)^2 + (Y - y_s)^2\right)$$

[4.5]

This optical phase can also be written in the form:

$$\varphi_r(X,Y) = 2\pi\left(f_{xr}X + f_{yr}Y\right) + \frac{\pi}{\lambda r_0}\left(X^2 + Y^2\right) + \varphi_s$$

[4.6]

where (f_{xr}, f_{yr}) are the carrier spatial frequencies, $r_0 \approx |z_s|$ is the curvature radius of the wave and φ_s is a constant that can be omitted. As a general rule, we are interested in adjusting the reference wave so that it has uniform amplitude, i.e. $a_r(X,Y) = C^{te}$. The total illumination, noted H, is then written [KRE 96, HAR 02, KRE 04]:

$$H = |U_r + U_o|^2 = |U_r|^2 + |U_o|^2 + U_r^* U_o + U_r U_o^*$$

[4.7]

This equation can also be written:

$$H = a_r^2 + a_0 + 2a_r a_0 \cos(\varphi_r - \varphi_0)$$

[4.8]

Equations [4.7] and [4.8] constitute what is classically called the *digital hologram*.

4.3. Basics of digital holography

4.3.1. *Recording the hologram*

The energy received by the sensor depends on the exposure time Δt and on the illumination H, such that [KRE 96]:

$$W = \int_{t_1}^{t_1+\Delta t} H dt = \Delta t |U_r|^2 + \Delta t |U_o|^2 + \Delta t U_r^* U_o + \Delta t U_r U_o^* \qquad [4.9]$$

The main physical parameter that characterizes the sensor is the mean sensitivity, noted W_0 (J/cm^2) here. The characteristic curve between the energy and the digital transmittance of the sensor exhibits a linear part, centered at mean sensitivity W_0. Figure 4.2 illustrates the typical digital transmittance obtained with a charge-coupled device (CCD) sensor having *nbits* digitization. The number of gray levels varies from 0 to $2^{nbits}-1$, thus the digital transmittance is proportional to the energy received. The essence of digital holography relies on this fact: if the sensor illumination is sinusoidal as $W = W_0 + \Delta W \cos(\varphi)$ and included in the linear part of the characteristic curve, then the transmittance of the digital hologram will be proportional to the illumination received. Thus, we will have $t = t_0 + \beta \Delta W \cos(\varphi)$, as shown in Figure 4.2. Coefficient β represents the slope of the transmittance with regard to the illumination at coordinates (W_0, t_0) for the mean illumination.

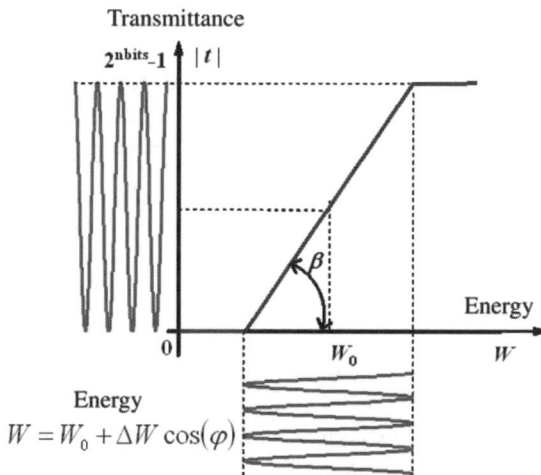

Figure 4.2. *Transmittance curve and illumination received*

In the linearity zone, the complex transmittance, t, is related to the energy received according to this relationship:

$$t = t_0 + \beta(W - W_0) \qquad\qquad [4.10]$$

The mean transmittance is noted t_0, and W_0 represents the mean energy. According to Figure 4.2, the value of W must not be very far from W_0, so the sinusoidal variation remains in the linear part. This means that the sinusoidal fringes must be weakly modulated. From the previous equations, we have:

$$W_0 = \Delta t |U_o|^2 + \Delta t |U_r|^2 \qquad\qquad [4.11]$$

It follows that:

$$t = t_0 + \beta \Delta t \left(U_r^* U_O + U_r U_O^* \right) \qquad\qquad [4.12]$$

These equations show that the phase and amplitude of the object wave front are encoded in the digital hologram. Two types of digital sensors can be distinguished: CCD sensors and complementary metal-oxide semi-conductor (CMOS) sensors. Pixels have a pitch varying between 2 μm and 20 μm. The spatial resolution provided by such devices is between 50 mm^{-1} and 500 mm^{-1}. The mean sensitivity of these sensors is about $W_0 = 5 \times 10^{-4}$J/m^2, but it depends on quantum efficiency, pixel surface and the number of electrons at pixel saturation.

A digital hologram is a speckle pattern [DAI 84]. Locally, the micro-fringes generated by the two monochromatic waves with an angle θ (see Figure 4.1) impose Shannon conditions for recording. The Shannon theorem applied to digital holography leads to the optimal recording distance between the object and sensor. It is given by [PIC 08a]:

$$d_0 = \frac{Ca}{2\lambda} \max(p_x, p_y) \qquad\qquad [4.13]$$

where $C = 2 + 3\sqrt{2}$ for a circular object with a diameter of d, $C = 8$ for a square object with a side of $\Delta A_x = \Delta A_y = a$, and where p_x and p_y are the pixel pitches.

Ideally, the optimal spatial frequencies of the reference wave are found to be $(f_{xr}, f_{yr}) = (\pm(1/2 - 1/(2 + 3\sqrt{2}))/p_x, \pm(1/2 - 1/(2 + 3\sqrt{2}))/p_y)$ for the circular object and $(f_{xr}, f_{yr}) = (\pm 3/8p_x, \pm 3/8p_y)$ for the square object. For the square object, we may also choose $f_{xr} = 0$ or $f_{yr} = 0$ [PIC 08a].

4.3.2. *Numerical reconstruction with the discrete Fresnel transform*

Numerical reconstruction is based on the principle of diffraction. Complex transmittance at the digital hologram plane is then given by equation [4.12]. The three orders of diffraction orders zero order, +1 order and −1 order. The +1 order is useful for measuring displacement fields. The phase and amplitude of the object wave front are reconstructed by the numerical simulation of the diffraction of a reference wave impacting the digital hologram constituted by the pixel matrix ($M \times N$ pixels). The reference can be chosen to have a unitary amplitude and null phase [KRE 97]. We have to take into account the spatial sampling in the hologram plane $(X, Y) = (np_x, mp_y)$ where $(m;n) \in (-M/2, + M/2-1; -N/2, + N/2-1)$, and since the processor cannot calculate indefinitely, we also have to take into account the spatial sampling in the reconstructed plane. For a numerical distance equal to $-d_0$, the discrete version of the Fresnel transform is given by the following, where pixels are considered to be non-extended [SCH 94, KRE 97, PIC08c]:

$$A_R(x,y,d_0) = \frac{j}{\lambda d_0} \exp\left(-\frac{2j\pi d_0}{\lambda}\right) \exp\left(-\frac{j\pi}{\lambda d_0}(x^2 + y^2)\right)$$

$$\times \sum_{k=-K/2}^{k=K/2-1} \sum_{l=-L/2}^{l=L/2-1} w(lp_x, kp_y) r(lp_x, kp_y, d_0) \exp\left(-\frac{j\pi}{\lambda d_0}(l^2 p_x^2 + k^2 p_y^2)\right)$$

$$\times \exp\left(+\frac{2j\pi}{\lambda d_0}(lp_x x + kp_y y)\right) \tag{4.14}$$

where $w(lp_x, kp_y)$ is the numerical reference wave considered to be the wave impacting at the digital hologram plane.

From the discrete Fresnel transform, and in the case where $w(lp_x, kp_y) = 1$, the field diffracted in +1 order is found to be [PIC 03a, PIC 08a]:

$$A_R^{+1}(x,y,-d_0) = \beta \Delta t \lambda^2 d_0^2 \exp\left(-j\pi \lambda d_0 (f_{xr}^2 + f_{yr}^2)\right)$$

$$\times U_r^*(x,y) A(x - \lambda f_{xr} d_0, y - \lambda f_{yr} d_0) \tag{4.15}$$

The +1 order is then localized at spatial coordinates $(\lambda d_0 f_{xr}, \lambda d_0 f_{yr})$. Spatial resolution in the reconstructed plane is found to be $\rho_x = \lambda d_0/Np_x$ and $\rho_y = \lambda d_0/Mp_y$ [KRE 97, YAM 01a, PIC 08a]. These two quantities impose an ultimate limit on the details that can be recovered with the digital holographic method. Note that the spatial resolution is proportional to the wavelength and the recording distance, and inversely proportional to the sensor width. Application with $\lambda = 0.6328$ μm, $d_0 = 500$ mm, $p_x = p_y = 5$ μm and $N = M = 1{,}024$ leads to $\rho_x = \rho_y = 61.79$ μm.

It should be pointed out that the consideration of the reference wave in digital computing is not mandatory if the physical wave is uniform and plane, i.e. $|r_0| = \infty$ and $a_r(X,Y) = C^{te}$ [KRE 97]. In the case where the physical reference wave is spherical, $|r_0| \neq \infty$, its curvature must be taken into account, and $t(l, k, d_0)$ is multiplied by the numerical complex conjugate of equation [4.4].

From the expression of the discrete Fresnel transform, if the reconstructed field is calculated with $(K, L) \geq (M, N)$ data points, then the sampling pitch is equal to $\Delta\eta = \lambda d_0/Lp_x$ and $\Delta\xi = \lambda d_0/Kp_y$ [KRE 97, YAM 01a]. The spatial sampling in the image plane is $x = l\Delta\eta$ and $y = k\Delta\xi$, with l and k varying from $-L/2$ to $L/2-1$ and from $-K/2$ to $K/2-1$. The schematic diagram of the algorithm is given in Figure 4.3. This algorithm is known as the S-FFT algorithm, since it uses only a single FFT computation.

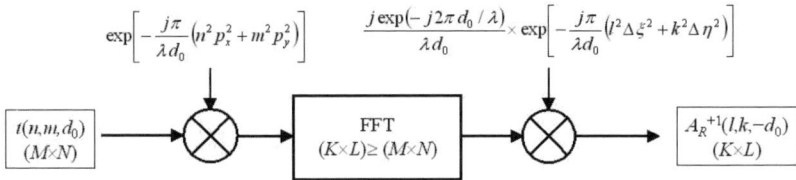

Figure 4.3. *Reconstruction algorithm with the discrete Fresnel transform*

Numerically computing the object field enables the estimation of the object's amplitude $|A_R^{+1}|$ and phase $\psi_r = \arg(A_R^{+1})$, calculated by an arctangent formula [KRE 97].

4.3.3. Numerical reconstruction using convolution with adjustable magnification

The discrete Fresnel transform is adapted to a large range of object sizes and shapes. In such computations, the sampling pitch in the reconstructed plane depends on physical parameters such as the wavelength. This is not suitable for digital color holography. The second possibility for reconstruction of the object is based on the diffraction convolution formulae [KRE 97]. This means that the reconstructed field is obtained by this convolution equation ($\otimes$ means convolution), at any distance d_r:

$$A_R^{+1}(x,y,d_r) = \left(w(x,y)t(x,y,d_0)\right) \otimes h(x,y,d_r)$$ [4.16]

where $h(x, y, d_r)$ is the kernel associated with diffraction along distance d_r, and $w(x, y) = \exp(j\pi(x^2 + y^2)/\lambda R_c)$ is a numerical spherical wave front with a curvature radius of R_c. The reconstruction parameters are linked by the magnification of the reconstructed image, γ, such that $d_r = -\gamma d_0$, $R_c = \gamma d_0/(\gamma - 1)$. The magnification can be chosen as $\gamma = \min(Lp_x/\Delta A_x, Kp_y/\Delta A_y)$, meaning that the reconstructed object will lie fully in the reconstructed horizon sized $Lp_x \times Kp_y$ [LI 09, LI 10]. The angular spectrum transfer function can be used as the transfer function of the reconstruction process [GOO 05]. In this case, the mathematical expression has to be adapted and is given by [LI 09]:

$$G(u,v,d_r) = \begin{cases} \exp\left(2j\pi d_r/\lambda\sqrt{1 - \lambda^2(u - f_{xr})^2 - \lambda^2(v - f_{yr})^2}\right) \\ \quad \text{if } |u - f_{xr}| \le Lp_x/2\lambda d_r \text{ and } |v - f_{yr}| \le Kp_y/2\lambda d_r \\ 0 \quad \text{elsewhere} \end{cases} \quad [4.17]$$

The reader may consult references [LI 09, PIC 09, TAN 10b] for further details regarding the reconstruction process. The practical computation of such an equation can be performed according to the properties of the Fourier transform, thus leading to a *double-Fourier transform algorithm* (D-FFT):

$$A_R^{+1} = FT^{-1}\left[FT[w \times t] \times G\right] \quad [4.18]$$

In D-FFT algorithms, the reconstructed object is sampled by a number of data points that can be chosen freely with $(K, L) \ge (M, N)$, whereas with the S-FFT algorithm, the number of useful data points from the reconstructed object is given by the ratio $(\Delta A_x/\Delta\eta; \Delta A_y/\Delta\xi)$. The synoptic of the convolution algorithm with the D-FFT algorithm is given in Figure 4.4.

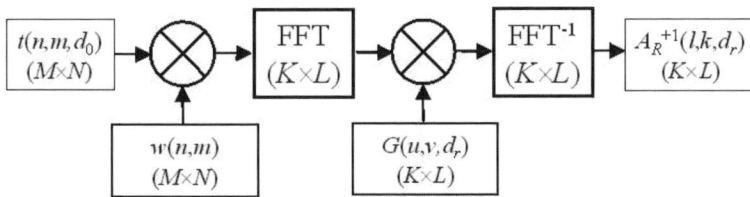

Figure 4.4. *Reconstruction algorithm using convolution with adjustable magnification*

4.3.4. Sensitivity vector

Non-contact measurements using digital holographic methods are based on the variation in the optical phase of the reconstructed object when it is subjected to a stress. This stress may be of a biological, chemical or physical nature. Let us imagine a point A at the light source and a point B attached to the object. When the object is slightly deformed by a stress, point B undergoes a three-dimensional change, represented by displacement vector $\mathbf{U}(U_x, U_y, U_z)$, which generates variations in the optical path from A to B and from B to C (see Figure 4.5). These variations are much smaller than the absolute values of these path lengths, and have modules in the order of tens or hundreds of wavelengths of the used light. The variation in the optical path length is [KRE 04]:

$$\delta_{opt}(ABC) = n\mathbf{K}_e.\mathbf{U} - n\mathbf{K}_o.\mathbf{U} = n\mathbf{U}.(\mathbf{K}_e - \mathbf{K}_o) = n\mathbf{U}.\mathbf{S} \qquad [4.19]$$

where the "observation" vector $\mathbf{K}_o$ is related to the direction of observation, the "illumination" vector $\mathbf{K}_e$ represents the direction of illumination, n is the refractive index around the object, which is often equal to one in solid mechanics, and $\mathbf{S} = \mathbf{K}_e - \mathbf{K}_o$ is the "sensitivity vector".

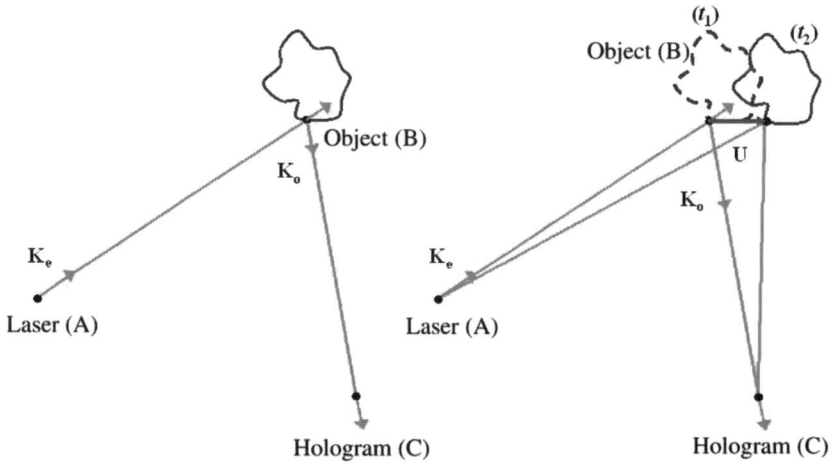

Figure 4.5. *Variation in the source–object–hologram path length*

The sensitivity vector corresponds to the difference between the illumination and observation vectors. It indicates the direction of displacement in which sensitivity is optimal. Knowledge of the coordinates of this vector is essential for the precise analysis of the amplitudes of displacement. The relation between the displacement of the object and variation in the optical phase is given by:

$$\Delta\varphi = \frac{2\pi}{\lambda}\delta_{opt}(ABC) = \frac{2n\pi}{\lambda}\mathbf{U}.\mathbf{S} \qquad [4.20]$$

4.4. Basics of digital holographic interferometry

4.4.1. *Phase difference*

The measurement of optical phase variations generated by the object requires the recording and reconstruction of at least two holograms (the principle of double exposure). The first corresponds to a reference hologram, and the second corresponds to a hologram of the stressed object. Consequently, the phase variation of equation [4.20] may be evaluated by calculating the difference in the optical phase between the two holograms. Let ψ_{r1} and ψ_{r2} be the optical phases of the first and second hologram respectively. We then have:

$$\Delta\varphi = \psi_{r2} - \psi_{r1} \quad \mathrm{mod}(2\pi) \tag{4.21}$$

This phase variation will produce digital interference fringes, modulo 2π, which enable us to quantify the displacement of the object between the two states. The optical path variation seen in digital holographic interferometry therefore corresponds to the variation in the position of the object projected onto the sensitivity vector. As an illustration, Figure 4.6 shows two phases, ψ_{r1} and ψ_{r2}, of the first and second hologram, respectively, as well as the phase-difference calculated modulo 2π. The two phases are random and uniformly distributed across $[-\pi, +\pi]$. The phase difference is also obtained in $[-\pi, +\pi]$. We observe digital interference fringes that represent phase jumps each time that $\Delta\varphi$ passes $-\pi$ or $+\pi$. We also observe that the result is noisy, which is translated into the appearance of a "salt-and-pepper" texture in the image. This noise is due to the decorrelation of the speckle pattern that exists for more or less each movement of the object [DAI 84, SLA 11, KAR 12a].

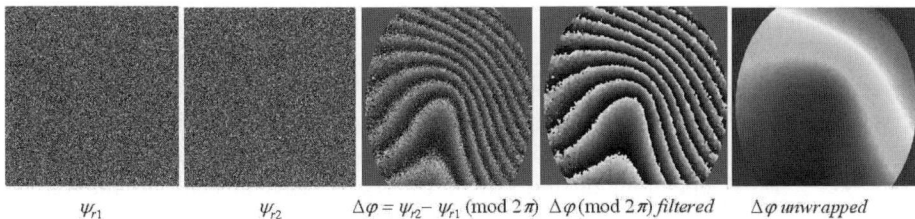

ψ_{r1} ψ_{r2} $\Delta\varphi = \psi_{r2} - \psi_{r1}$ (mod 2π) $\Delta\varphi$ (mod 2π) filtered $\Delta\varphi$ unwrapped

Figure 4.6. *Illustration of the phase-subtraction method and unwrapping procedure. For a color version of this figure, see www.iste.co.uk/gh/solidmech.zip*

Such results lead to two conclusions: it is necessary to apply a spatial low-pass filter to the result to reduce the level of noise; and to reconstruct the continuity of the phase variation, which is destroyed by calculation using the arctangent function.

4.4.2. *Spatial filtering of the phase and phase unwrapping*

The filtering procedure must delete the noise without degrading the phase jumps. The most widely-used method is sine/cosine phase filtering (simple or iterative) [AEB 99]. First we calculate $\cos(\Delta\varphi)$ and $\sin(\Delta\varphi)$ and then, from a filter kernel of $n \times n$ points, apply a linear "moving average" or "moving median" filter to the cosine and sine. The results obtained are generally very close together for a given value of n. The choice of n is imposed by the spatial frequency of the digital fringes. The closer together they are, the smaller n must be. Generally, we treat the image with odd values of n: 3×3, 5×5, 7×7 or 9×9 (or greater). As a consequence of the filtering, there is degradation in the spatial resolution of $2n$ pixels.

Phase unwrapping consists of reconstructing the physical continuity of the phase map, which is calculated modulo 2π. Of course, this continuity may only be reconstructed to within a phase constant, unless we know at which points of mapping the phase is strictly zero. The unwrapping algorithm searches for the phase jumps; at each one detected, -2π or $+2\pi$, it adds $+2\pi$, or -2π so that the phase of the two neighboring points is continuous. An example of phase unwrapping is given in Figure 4.6 with an illustration of the result. Phase unwrapping techniques have become more sophisticated in recent years with the appearance of powerful algorithms whose implementation is not always straightforward. For a complete description, the reader should refer to reference [GHI 98].

4.5. Digital holographic interferometry with spatial multiplexing

4.5.1. *Principle*

The most interesting feature of off-axis digital holography is that it enables the optical phase to be directly computed from a single recording, whereas this is not the case for techniques using phase shifting [YAM 97]. Thus, comparison by subtraction leads to the phase change evaluation $\Delta\varphi$. We only need two holograms recorded at two different instants to determine phases ψ_{r1} and ψ_{r2} and then calculate $\Delta\varphi = \psi_{r2} - \psi_{r1}$. Depending on the sensitivity vector, it is possible to measure a weighted mixing of the displacement components, the out-of-plane component or one of the in-plane components. In order to simultaneously measure two or three components of the displacement field, we need two or three illumination directions combined with a direction of observation.

It is possible to simultaneously measure two components by spatially multiplexing digital holograms corresponding to two directions of illumination. In such a set-up, the two holograms have different spatial reference frequencies [PIC 03a, PIC 04, PIC 05]. Spatial multiplexing is based on the incoherent mixing of

2×2 coherent waves. Each coherent duplet is composed of a reference wave (smooth and plane) and the diffracted object wave. The incoherent mixing is performed by crossing the polarizations of the optical waves. If the object depolarizes light, a delay line can be used to destroy the coherence of the laser. The two reference waves must be adjusted so that there is no overlapping between the useful orders.

4.5.2. Theory

Here U_{ri} is the complex amplitude of the ith reference wave and U_{Oi} is that of the object under the ith illumination. The digital hologram is:

$$H = |U_{r1}|^2 + |U_{o1}|^2 + |U_{r2}|^2 + |U_{o2}|^2$$
$$+ U_{r1}U_{o1}^* + U_{r2}U_{o2}^* + U_{r2}^*U_{o2} + U_{r1}^*U_{o1}$$

[4.22]

The last two terms constitute the +1 orders. According to equation [4.15], the +1 orders are then localized at spatial coordinates $(\lambda d_0 f_{xr1}, \lambda d_0 f_{yr1})$ and $(\lambda d_0 f_{xr2}, \lambda d_0 f_{yr2})$. The reconstructed field exhibits two images of the object, which do not overlap, that can be extracted to give individual optical phases along each direction of illumination. The simultaneous measurement of two displacement components is related to the sensitivity vectors produced by the directions of illumination. If the illuminations are symmetrical with respect to the direction normal to the object's surface, then the in-plane component and the out-of-plane component are calculated by the difference and the sum of the optical phase changes $\Delta\varphi_1$ and $\Delta\varphi_2$, given by each duplet respectively [PIC 03a]. According to the geometry defined in Figure 4.7, we have:

$$\Delta\varphi_{OP} = \Delta\varphi_1 + \Delta\varphi_2 = \frac{2\pi}{\lambda} \mathbf{U}.(\mathbf{K}_{e1} - \mathbf{K}_{e2} - 2\mathbf{K}_o) = \frac{4\pi}{\lambda}(1 + \cos\theta)U_z$$

[4.23]

$$\Delta\varphi_{IP} = \Delta\varphi_1 - \Delta\varphi_2 = \frac{2\pi}{\lambda} \mathbf{U}.(\mathbf{K}_{e1} - \mathbf{K}_{e2}) = \frac{4\pi}{\lambda}\sin\theta U_x$$

[4.24]

4.5.3. Experimental set-up

Figure 4.7 illustrates the experimental set-up. Spatial multiplexing is carried out using a twin polarizing Mach-Zehnder interferometer. A linearly polarized non-expanded helium neon laser beam ($\lambda = 632.2$ nm) passes first through a 50% transmission/reflection beam-splitter that splits the light into the two holographic interferometers. Then, along each interferometer, the laser beam meets a half-wave plate with adjustable directions of polarization in order to adjust the amount of light in the object and reference beams. The polarizations are then crossed between the interferometers. In the first interferometer the polarization is oriented along s and in the second interferometer the polarization is oriented following p. The extinction ratio between s and p is better in reflection than in transmission so a linear polarizer is placed after transmission by the cube.

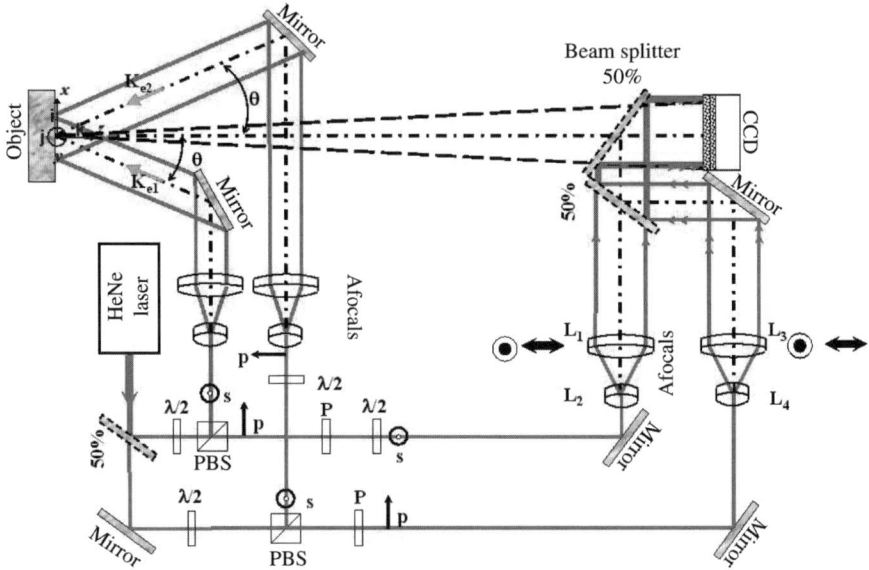

Figure 4.7. *Experimental set-up for spatial multiplexing of digital holograms (PBS: polarizing beam splitter cube)*

The adjustment of the paths at the same polarization is performed using a fixed half-wave plate that turns the polarization in a suitable direction. In this way, there is no parasitic interference between the cross paths of the twin interferometric setup that could produce unwanted coherent noise. The laser beams along the object paths are expanded by means of the two afocal systems and they illuminate the object following vectors $\mathbf{k}_{e1}$ and $\mathbf{k}_{e2}$. Smooth reference beams are produced using pinholes (not represented in Figure 4.7) and afocal systems, and they are recombined onto the CCD area with a beam splitter placed in front of the detector. In this example, the off-axis holographic recording is performed using lenses L_1 and L_3, which are displaced outside the afocal axis by means of two micrometric transducers [PIC 03a].

As an illustration of this technique, Figure 4.8a shows the reconstructed field of an aluminum wafer 24 mm in diameter placed $d_0 = 995$ mm from the sensor (S-FFT reconstruction). Figure 4.8b shows out-of-plane and in-plane phase changes (modulo 2π) extracted from the spatially-multiplexed digital holograms.

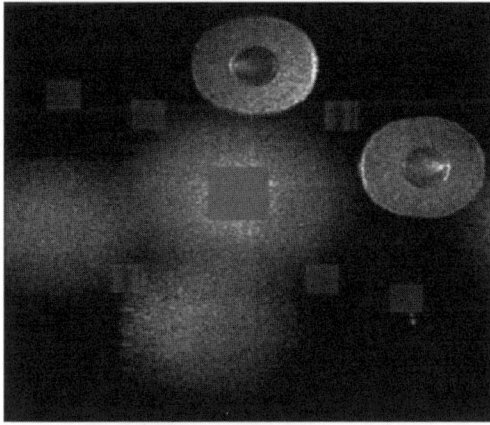

a) Spatially multiplexed holograms

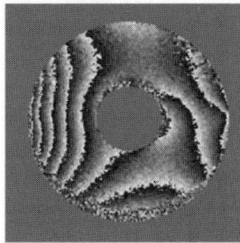

$\Delta\varphi_{OP}$

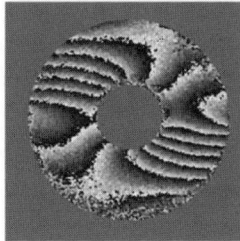

$\Delta\varphi_{IP}$

b) Phase changes

Figure 4.8. *Spatial multiplexing and two-dimensional deformation measurement*

4.5.4. *Application to synthetic concrete subjected to three-point bending*

In this study, a concrete test specimen was subjected to three-point bending for the evaluation of its flexural behavior [PIC 04]. The test specimens were constituted of epoxy resin concrete, which has a resin matrix and sand 4 mm in diameter as the aggregates. The mass percentage of resin in the concrete mix was 15%, with 85%

sand. Figure 4.9 shows the mechanical configuration. The test specimens were 95 mm long and 15 mm wide, with a thickness of 8 mm. The distance between the external supports was 75 mm. During the test, the span supports were mobile and the central one was fixed. The mechanical set-up was located at $d_0 = 1,045$ mm from the sensor and the illumination angle was $\theta = 45°$. Numerical computing was performed with the S-FFT algorithm and $K = L = 1,024$. The spatial resolution in the reconstructed plane was then 138.8 μm.

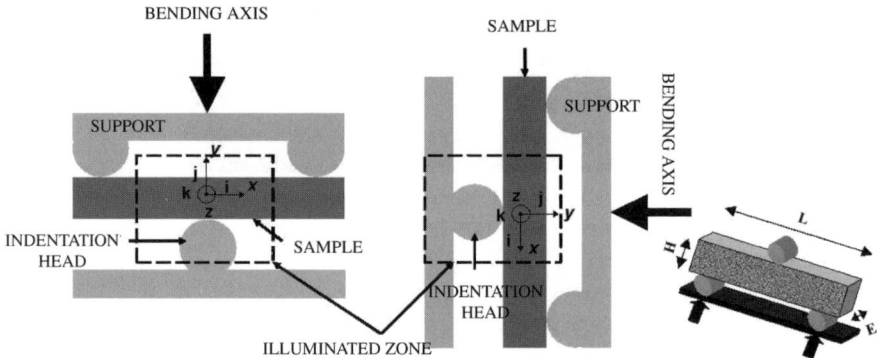

Figure 4.9. *Mechanical set-up for three-point bending*

The observed part of the sample is indicated by the dashed line in Figure 4.9. At each step of the test, the deformation was limited to a low fringe number in order to retain compatibility with the spatial resolution of the sample image. With each increment in displacement, the hologram of the current state was recorded and this hologram was used as a reference for the following state. The test ran until the sample fractured; at that time, 123 holograms had been recorded.

Fracture occurred at a deflection of approximately 400 μm for the sample described above. In order to obtain three displacement components, we performed the test twice with two samples coming from the same mold and with 90° rotation of the indentation head between the two tests. The **k** component was systematically measured and we noted that sample behavior was similar between the tests and that the cracks appeared at the same loadings. We can thus report results on estimated three-dimensional components with two different samples.

The displacement fields are represented in Figure 4.10 with wrapped fringes modulo an adapted spatial wavelength. For component **i**, it is noted that the part located between the central support and the neutral axis of the sample was in compression and that the other part was in tension.

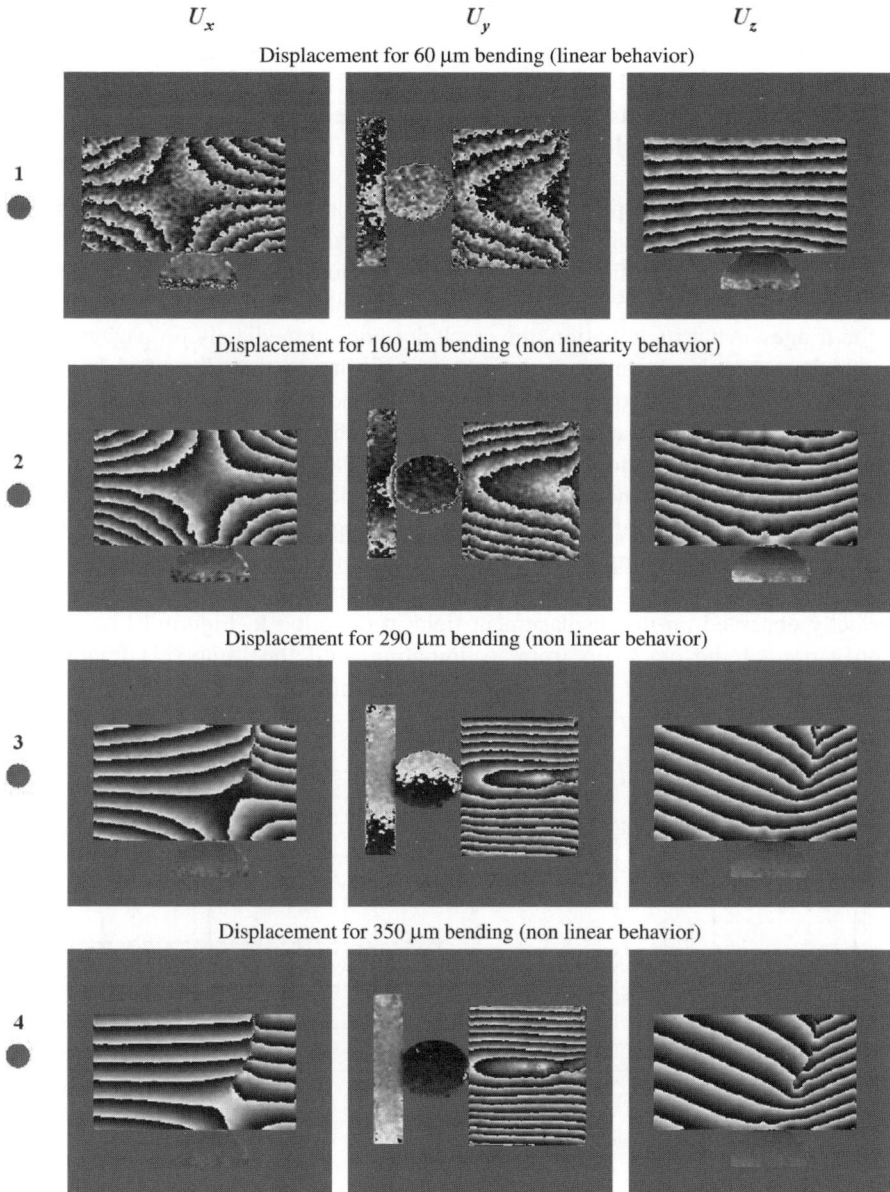

Figure 4.10. *Displacement fields measured with spatially-multiplexed holograms*

Furthermore, these results show a perfect linearity of behavior from the first to the 65th hologram, after which we locally observe a break in the fringes. The small dissymmetry of the displacement field along the direction of **i** is due to the adhesion between the span supports and the specimen, which does not impose exactly the same displacement on the left and the right.

Observation of the displacement field in direction **k** makes it possible to see the Poisson effect, which results in a rotation of the face around the **i** axis. The contour lines of the displacement field confirm the observations made for direction **i**. In fact, the crack appears in the 66th hologram, and we can observe a change in the direction of the fringes. After its initiation, the crack does not immediately propagate through the thickness, and we can observe behavior similar to that described before the initiation of this crack over the course of 20 holograms (point No. 2). From hologram No. 85, the crack propagated through the thickness until the specimen fractured (points No. 3 and 4). During this stage, the evolution of the displacement field in direction **j** does not present any modification. However, for the other two directions, the initiation of the crack causes a strong gradient in component **k**. This is explained by the opening of the crack in direction **i**, causing a relaxation of stress along this axis. Furthermore, this locally weakens the Poisson effect, which is perfectly observed on the displacement field in direction **k**. Figure 4.11 shows the propagation of the crack tip from its initiation until the sample is fractured, as deduced from the set of holograms. This evolution can be associated with the nonlinear behavior of the resin concrete [PIC 04].

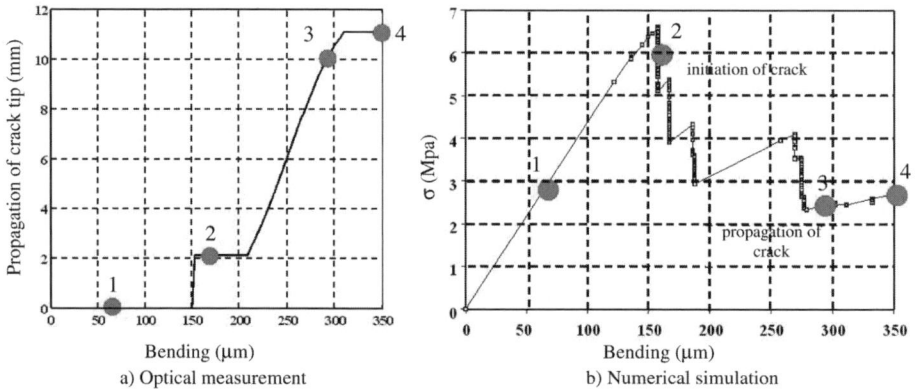

a) Optical measurement   b) Numerical simulation

Figure 4.11. *Crack tip propagation during test and load-bending curve obtained using finite element method*

Mechanisms of fracture in synthetic concretes are three-dimensional phenomena and three-dimensional modeling is necessary to study crack propagation. Liliu et al. [LIL 03]

developed a three-dimensional model in which the failure constraint of elements representing the interface is chosen to be the same as that of the matrix, the behavior of the composite structure being unaffected by the proportion of aggregate. Heterogeneity is introduced through mechanical characteristics using a statistical distribution (for example Gaussian distribution). Berthelot *et al.* [BER 04] studied the influence of the width of the distribution of failure stresses on damage in heterogeneous materials. These results were adapted to the three-dimensional analysis of synthetic concretes [PIC 04]. All of the meshing elements have the same dimensions, as can be seen in Figure 4.12, and this study supposes that there is a characteristic dimension for the evolution of damage. During the simulation performed using the Permeas code, damage is created by changing the stiffness matrix of the damaged element, and an element is only supposed to fail in mode I. At each step of the computation, the location of the element under greater stress is determined by the search for the maximum of ratios $\sigma_j(i)/\sigma_{ju}(i)_{i=1,...,n}$ and thus of the direction of failure j for element (i), $\sigma_j(i)$ being the average stress of element (i).

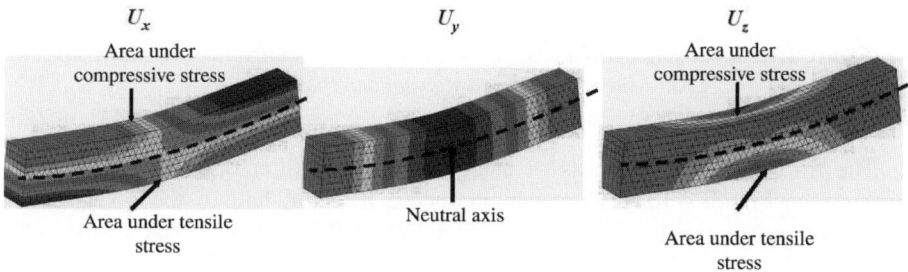

U_x U_y U_z

Area under compressive stress

Area under tensile stress

Neutral axis

Area under compressive stress

Area under tensile stress

Figure 4.12. *Three-dimensional displacement fields obtained with the finite element method. For a color version of this figure, see www.iste.co.uk/gh/solidmech.zip*

In the load–displacement curve of Figure 4.11b, two tendencies are observed before the maximum load: before point No. 1, the structure exhibits linear behavior and the number of damaged elements is low; between points No. 1 and 2, we can observe nonlinear behavior associated with a decrease in the stiffness of the structure – this is the cause of the appearance and propagation of microcracks. The results presented in Figure 4.10 are consistent with the simulation: the first microcracks appear at around 150 μm in the two curves. Beyond point No. 2, the coalescence of microcracks generates a macrocrack that leads to destruction of the sample. Figure 4.12 shows the three-dimensional sample that was simulated. Due to the gradient of the constraint field in the sample, the crack has initiated in the opposite zone from the load. This is consistent with the optical measurements provided by the spatial multiplexing of digital holograms.

4.6. Digital color holography applied to three-dimensional measurements

4.6.1. *Recording digital color holograms*

Figure 4.13 illustrates different approaches for *simultaneously* recording digital color holograms.

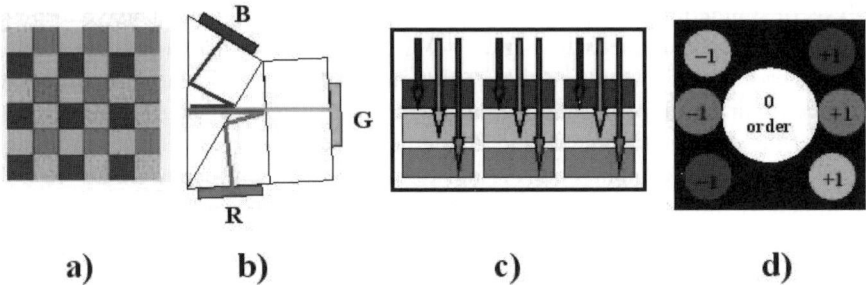

Figure 4.13. *Recording of digital color holograms. For a color version of this figure, see www.iste.co.uk/gh/solidmech.zip*

The first possibility consists of using a chromatic filter organized in a Bayer mosaic (see Figure 4.13a). However, in such a detector, half of the pixels detect green, and only a quarter detect red or blue [YAM 02, DES 11]. The spatial color filter creates holes in the mesh and therefore a loss of resolution.

The second possibility consists of using three detectors organized as a "triple-CCD", the spectral selection being carried out by a prism with dichroic treatments (see Figure 4.13b). Such a detector guarantees a high spatial resolution and a spectral selectivity compatible with the constraints of digital color holography for non-contact metrology [DES 12].

A third possibility consists of using a color detector based on a stack of photodiodes (see Figure 4.13c) [DES 08, TAN 10a, TAN 10b, DES 11]. Spectral selectivity is related to the mean penetration depth of the photons in the silicon: blue photons at 425 nm penetrate to around 0.2 μm, green photons at 532 nm to around 2 μm and red photons at 630 nm to around 3 μm. Thus the construction of junctions at depths at around 0.2 μm, 0.8 μm and 3.0 μm give the correct spectral selectivity for color imaging. The architecture guarantees maximum spectral resolution, since the number of effective pixels for each wavelength is that of the entire matrix. However, spectral selectivity is not perfect, since green photons may be detected in the blue and red bands.

A final possibility consists of using a monochromatic detector combined with spatial chromatic multiplexing (see Figure 4.13d). Each reference wave must have different separately-adjusted spatial frequencies according to its wavelength (see Figure 4.7). The complexity of the experimental apparatus increases with the number of colors. For two-color digital holography, this is acceptable; for three colors, it becomes prohibitive. A demonstration of this approach is given in references [KUH 07, MAN 08, PIC 09, TAN 11].

Multidimensional mechanical deformation measurements have stimulated a lot of research [PED 97, SCH 99, PIC 03a, TAN 10a]. The use of digital color holography for multidimensional measurement was proposed in 2008 using a two-color digital Fresnel hologram [PIC 08b]. The advantage of digital color holography is that the simultaneous recording of colors is possible with a high resolution. The reference beams can be superposed so that the set-up can be considerably simplified compared to a system including a spatial multiplexing scheme. The use of three colors provides three directions of illumination, and thus three sensitivity vectors. Three-dimensional deformation measurements can therefore be achieved directly with single-shot recording and without any sequential recording.

4.6.2. *Application to composite material subjected to a short beam test*

During the late 1990s, composite materials conquered high-technology domains, such as aeronautics, sports, automobiles, etc. The reasons for such success are attributed to their superior mechanical characteristics, which depend on the reinforcement and on the matrix of the material. However, the mechanical properties of composite structures deteriorate over time, leading to cracks and delamination between two consecutive plies. Various experimental methods are generally used to measure displacements or crack propagation, or more simply to detect defects [REI 82, BOR 98, BOS 02]. Nevertheless, no experimental method can provide high-sensitivity full-field three-dimensional information simultaneously on both mechanical displacement and crack initiation. The use of digital three-color holography is a pertinent choice for the detection of premature cracks in composite materials.

The experimental set-up adapted to the investigation of the short beam test is described in Figure 4.14. It uses three wavelengths produced by three DPSS lasers (red 671 nm, green 532 nm and blue 471 nm). Each laser beam is separated into a probe beam and a reference beam by three polarizing beam splitters. The three reference beams are combined thanks to dichroic plates and impact the sensor at a constant incident angle (off-axis holography). Thus, the particularity of this set-up is that it uses a single reference beam including the three wavelengths. Polarizations are tuned so that they are parallel in each arm so interference can occur between each reference and probe beam. The three probe beams illuminate the object under different incident angles. Thus, the setup provides a simultaneous triple sensitivity to the displacement field at the surface of the structure. The color sensor uses three CCDs and includes $N \times M = 1344 \times 1024$ pixels with pitches $p_x = p_y = 6.45$ μm. It allows the three holograms to be recorded simultaneously.

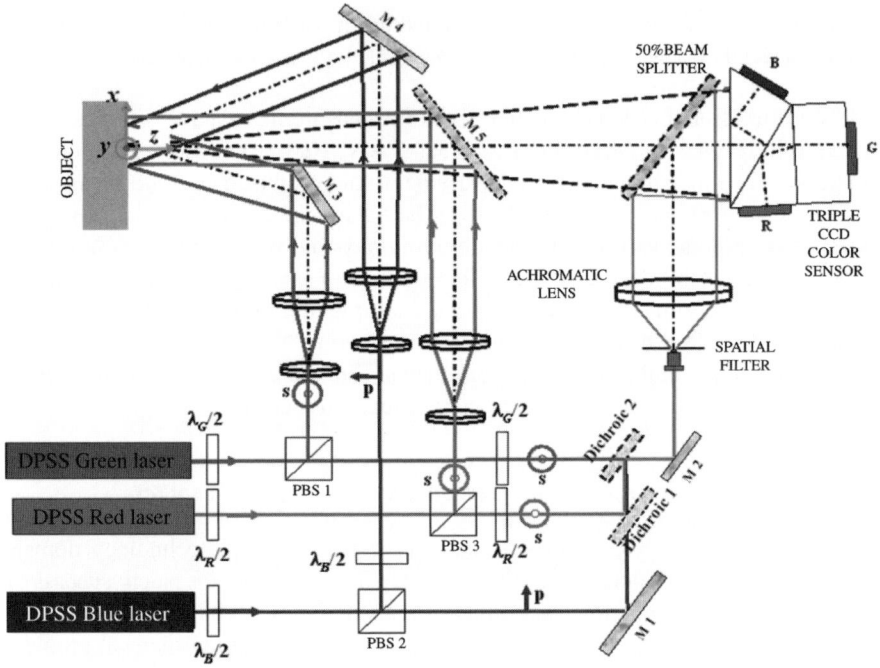

Figure 4.14. *Digital three-color holographic interferometer. For a color version of this figure, see www.iste.co.uk/gh/solidmech.zip*

The three directions of illumination produce the three sensitivity vectors [TAN 11]. Figure 4.17a illustrates the illuminating geometry. The illumination angles are $\theta^B_{xz} = 35.67$, $\theta^G_{xz} = -42.70$, $\theta^R_{yz} = 29.94$ and $\theta^R_{xz} = -2.75$, and we have $\theta^B_{yz} = \theta^G_{yz} = 0$. The relation between the displacement vector $\mathbf{U}$ and the illumination geometry is $\Delta\varphi_\lambda = 2\pi\mathbf{U}.(\mathbf{K}^\lambda_e - \mathbf{K}_o)/\lambda$ for each wavelength (λ refers to R, G or B), where $\mathbf{K}^\lambda_e = -\cos\theta^\lambda_{yz}\sin\theta^\lambda_{xz}\mathbf{i} - \sin\theta^\lambda_{yz}\mathbf{j} - \cos\theta^\lambda_{yz}\cos\theta^\lambda_{xz}\mathbf{k}$ is the illumination vector and $\mathbf{K}_o \cong \mathbf{k}$ is the observation vector. The three-dimensional displacement field (U_x, U_y, U_z) at the surface of the sample can thus be extracted from the set of equations given by the measurement of $\Delta\varphi_R$, $\Delta\varphi_G$ and $\Delta\varphi_B$. Note that the sampling of the reconstructed field calculated with the S-FFT algorithm depends on wavelength. Thus, S-FFT algorithms are not suitable for digital color holography. For this reason, digital color holograms are reconstructed using the D-FFT algorithms with adjustable magnification. With such an approach, the physical scale of the reconstructed field is kept constant for each wavelength, and the monochromatic image can be superposed to calculate the three-dimensional displacement fields.

To illustrate the method, a test object is subjected to mechanical loading along the y direction of the reference axis. Recording a three-color digital hologram and computing it directly with the discrete Fresnel transform along each individual monochromatic image (R, G, B) leads to the numerical reconstructions of Figures 4.15a, 14.15b and 14.15c for the R, G and B channels, respectively. The wavelength-dependant size can be observed. Figures 4.15d, 4.15e and 4.15f show amplitude images computed with the D-FFT algorithm given in Figure 4.4, exhibiting a constant object size. Therefore, unlike the discrete Fresnel transform, the D-FFT algorithm adapts the size of the objects so that superimposition of the three holograms may be obtained. After recording three-color digital holograms for different mechanical loadings, numerical reconstruction yields the optical phases along each RGB channel. Furthermore, three phase changes $\Delta\varphi_R$, $\Delta\varphi_G$ and $\Delta\varphi_B$ along the R, G and B channels were computed and unwrapped. Figure 4.16 shows the three-dimensional plots of the (U_x, U_y, U_z) components of the displacement field obtained from the data set.

Figure 4.15. *Numerical reconstructions of the object using the discrete Fresnel transform (a, b, c) and the convolution algorithm (d, e, f)*

Figure 4.17b gives the scheme of the short beam shear test [KAR 12b]. Figure 4.17c is a photograph illustrating the "white" illumination of the sample by the three laser wavelengths. The sample (2 mm × 10 mm × 20 mm) is subjected to three-point bending with progressive loading. The curvature radius is 5 mm for the indentation head and 2 mm for the supports (see Figure 4.17b). The distance between the supports is 10 mm and a force sensor records the effort applied to the sample. The materials tested contain 40% fiber, and are composed of glass/epoxy (20 plies) or linen/epoxy (12 plies, results not presented here).

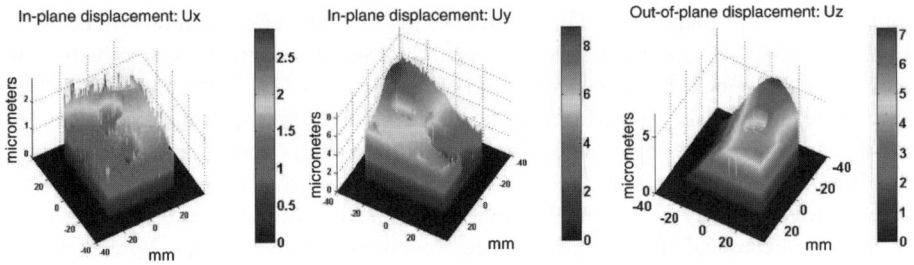

Figure 4.16. *Three-dimensional plots of the displacement field obtained from the digital color holograms. For a color version of this figure, see www.iste.co.uk/gh/solidmech.zip*

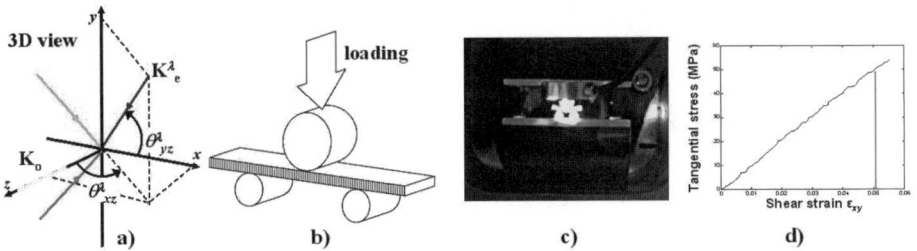

Figure 4.17. *Mechanical configuration of short beam test*

The distance between the sensor and the sample is about $d_0 = 650$ mm. The D-FFT algorithm (see Figure 4.4) is applied with the following parameters: magnification is $\gamma = 0.66$ and reconstruction distance is calculated to be $d_r = -429.3$ mm. The phase differences between two deformation states are calculated for each wavelength and then unwrapped, and the three components of the displacement field are computed. The shear strains ε_x, ε_y and ε_{xy} are estimated by calculating the derivatives of the displacement fields. A maximum force of 1,200 N was applied, corresponding to the recording of 230 digital color holograms. Figure 4.17d shows the stress/strain curve. In a short beam shear test, the focus is on the inter-laminar shear stresses – due to tangential efforts – which are not negligible compared to normal stresses, due to the flexure of the sample.

Unfortunately, normal stresses, as well as compression due to the indentation head, can generate premature failure of the sample. However, it is not possible to find the first type of failure simply by analyzing the stress/strain curve (see Figure 4.17d) or by performing a post-mortem inspection. The high sensitivity of digital color holography enables the investigation to be carried out *in situ*.

Figure 4.18 shows a comparison between the experimental measurement of the three-dimensional displacement field using the holographic set-up, and finite element method modeling performed with CATIA software, for an applied force of 50 N. We assume the load to be inside the elastic part of the loading curve. The black line indicates the coincidence zone for both the real sample and the numerical sample, which is performed in two dimensions. Experimental and numerical results are found to be in a very good agreement along this line.

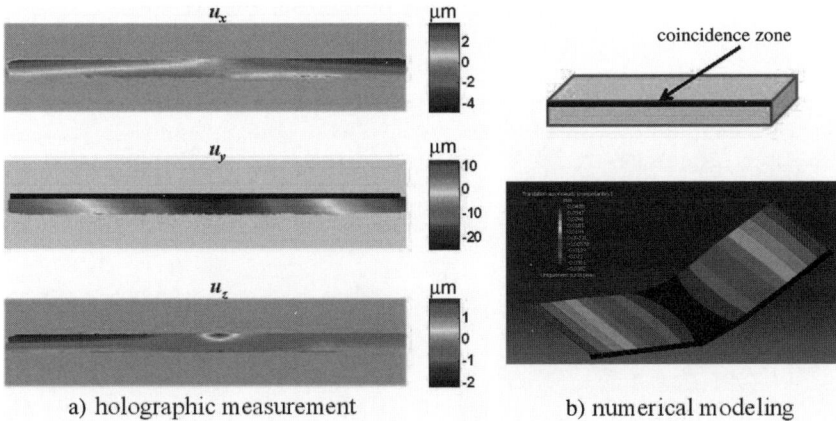

a) holographic measurement   b) numerical modeling

Figure 4.18. *a) Experimental measurement of the three-dimensional displacement field; and b) bottom right: numerical modeling of U_y; top right: coincidence line for experiment/modeling comparison. For a color version of this figure, see www.iste.co.uk/gh/solidmech.zip*

From these results, the shear strains are calculated and compared to the numerical model. Figure 4.19 shows the good agreement between the experimental and numerical results, thus confirming the suitability of the three-color holographic set-up to provide precise, full-field and high-resolution measurements of the three-dimensional displacement fields. It is then possible to perform an inverse identification of the G13 shear modulus.

Moreover, the high sensitivity of the set-up enables premature cracks to be detected. Crack initiation can be tracked by the phase changes between two loading states and can be clearly exhibited with the evaluation of the shear strains. This gives pertinent information regarding the initiation of the crack. In the current example, the premature crack is slightly visible at the middle-left of the sample (see Figure 4.20).

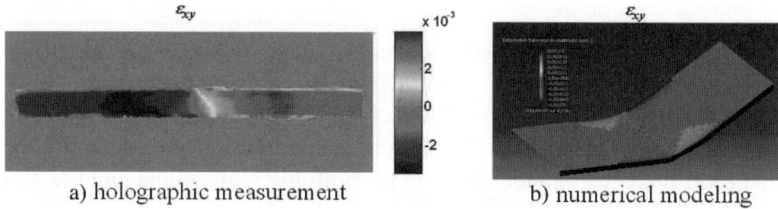

a) holographic measurement   b) numerical modeling

Figure 4.19. *Shear strain: a) experimental results; and b) numerical modeling. For a color version of this figure, see www.iste.co.uk/gh/solidmech.zip*

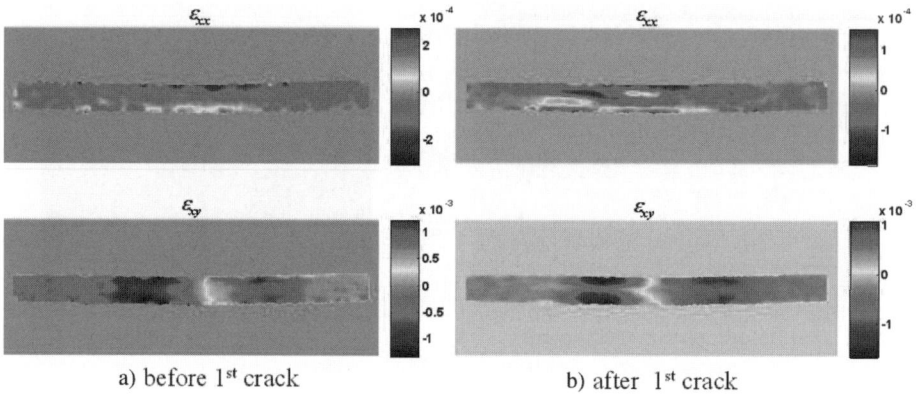

a) before 1st crack   b) after 1st crack

Figure 4.20. *Shear strains before and after the first crack. For a color version of this figure, see www.iste.co.uk/gh/solidmech.zip*

Crack initiation clearly appears between holograms 180 and 182, since it corresponds to shear strain values of almost zero (see Figure 4.20b). The red line in the curve of Figure 4.17d indicates the first premature crack detected with the proposed method.

4.7. Conclusion

This chapter has described methods of digital holography that can be used in solid mechanics. We have focused on the methods that are the most appropriate for the mechanical engineer and that enables the simultaneous measurement of two or three components of the displacement field. Other methods exist in the literature, and for an exhaustive presentation the reader is invited to consult references [KRE 05, PIC 12] and an online course [PIC 08c, PIC 08d] which is free to access via the Internet.

4.8. Acknowledgment

Some of the experimental results presented in this chapter were obtained through a research program funded by the French National Agency for Research (ANR) under grant agreement n°ANR 2010 BLAN 0302.

4.9. Bibliography

[AEB 99] AEBISCHER H.A., WALDNER S., "A simple and effective method for filtering speckle-interferometric phase fringe patterns", *Optics Communications*, vol. 162, pp. 205–210, 1999.

[ALB 76] ALBE F., SMIGIELSKI P., FAGOT H, "Use of holographic interferometry for quantitative investigations of deformation of surface of a material impacted by a projectile", *SMPTE Journal*, vol. 85, pp. 18–25, 1976.

[BEL 09] BELLANGER C., BRIGNON A., COLINEAU J., HUIGNARD J.P, "Coherent fiber combining by digital holography", *Proceedings SPIE*, vol. 7195, 2009.

[BER 04] BERTHELOT J.-M., FATMI L., "Statistical investigation of the fracture behaviour of inhomogeneous materials in tension and three-point bending", *Engineering Fracture Mechanics*, vol. 71, pp. 1535–1556, 2004.

[BIN 02] BINET R., COLINEAU J., LEHUREAU J.C., "Short-range synthetic aperture imaging at 633nm by digital holography", *Applied Optics*, vol. 41, pp. 4775–4782, 2002.

[BOR 98] BORZA D.N., "Specialized techniques in holographic non-destructive testing of composite", *Composites Part B: Engineering*, vol. 29, pp. 497–504, 1998.

[BOR 99] BORN M., WOLF E., *Principles of Optics*, 7th Edition, Cambridge University Press, 1999.

[BOS 02] BOSIA F., BOTSIS J., FACCHINI M., GIACCARI P., "Deformation characteristics of composite laminates", *Composites Science and Technology*, vol. 62, pp. 41–54, 2002.

[CHA 06a] CHARRIÈRE F., KÜHN J., COLOMB T., MONTFORT F., CUCHE E., EMERY Y., WEIBLE K., MARQUET P., DEPEURSINGE C., "Characterisation of micro lens by digital holographic microscopy", *Applied Optics*, vol. 45, pp. 829–835, 2006.

[CHA 06b] CHARRIÈRE F., MARIAN A., MONTFORT F., KUHN J., COLOMB T., CUCHE E., MARQUET P., DEPEURSINGE C., "Cell refractive index tomography by digital holographic microscopy", *Optics Letters*, vol. 31, pp. 178–180, 2006.

[CHA 07] CHALUT K., BROWN W., WAX A., "Quantitative phase microscopy with asynchronous digital holography", *Optics Express*, vol. 15, pp. 3047–3052, 2007.

[COL 71] COLLIER R.J., BURCKHARDT C.B., LIN L.H., *Optical Holography*, Academic Press, New York, NY, 1971.

[CUC 99a] CUCHE E., MARQUET P., DEPEURSINGE C., "Simultaneous amplitude-contrast and quantitative phase-contrast microscopy by numerical reconstruction of Fresnel off-axis holograms", *Applied Optics*, vol. 38, pp. 6994–7001, 1999.

[CUC 99b] CUCHE E., BEVILACQUA F., DEPEURSINGE C., "Digital holography for quantitative phase contrast imaging", *Optics Letters*, vol. 24, pp. 291–293, 1999.

[DAI 84] DAINTY J.C., *Laser Speckle and Related Phenomena*, Berlin, Springer-Verlag, 1984.

[DEN 62] DENISYUK Y.N., "Manifestation of optical properties of an object in wave field of radiation in scatters", *Doll. Akad. N. SSSR*, vol. 144, pp. 1275, 1962.

[DES 08] DESSE J.M., PICART P., TANKAM P., "Digital three-color holographic interferometry for flow analysis", *Optics Express*, vol. 16, pp. 5471–5480, 2008.

[DES 11] DESSE J.M., PICART P., TANKAM P., "Sensor influence in digital 3λ holographic interferometry", *Measurement Science & Technology*, vol. 22, pp. 064005, 2011.

[DES 12] DESSE J.M., PICART P., TANKAM P., "Digital color holography applied to fluid mechanics and structure mechanics", *Optics and Lasers in Engineering*, vol. 50, pp. 18–28, 2012.

[FAG 83] FAGOT H, SMIGIELSKI P., ALBE F., ARNAUD J.L., "Pulsed holographic non-destructive testing on aircraft", *Proceedings SPIE*, vol. 369, pp. 493–496, 1983.

[FAG 86] FAGOT H, SMIGIELSKI P., "Cineholography and interferometry", *Comptes Rendus de l'Académie des Sciences*, vol. 302, pp. 157–163, 1986.

[FER 06] FERRARO P., ALFERI D., DE NICOLA S., DE PETROCELLIS L., FINIZIO A., PIERATTINI G., "Quantitative phase-contrast microscopy by a lateral shear approach to digital holographic image reconstruction", *Optics Letters*, vol. 31, pp. 1405–1407, 2006.

[GAB 48] GABOR D., "A new microscopic principle", *Nature*, vol. 161, pp. 777–778, 1948.

[GHI 98] GHIGLIA D.C., PRITT M.D., *Two-dimensional Phase Unwrapping: Theory, Algorithms and Software*, Wiley & Sons Ltd., New York, NY, 1998.

[GOO 67] GOODMAN J.W., LAWRENCE R.W., "Digital image formation from electronically detected holograms", *Applied Physics Letters*, vol. 11, pp. 77–79, 1967.

[GOO 05] GOODMAN J.W., *Introduction to Fourier Optics*, 3rd ed., Roberts and Company Publishers, Green Wood Village, CO, 2005.

[GOO 07] GOODMAN J.W., *Speckle Phenomena in Optics*, Ben Roberts and Co, Swansea, 2007.

[HAR 02] HARIHARAN P., *Basics of Holography*, Cambridge University Press, New York, NY, 2002.

[HAR 06] HARIHARAN P., *Basics of Interferomerty*, Academic Press, New York, NY, 2006.

[HUA 71] HUANG T.S., "Digital holography", *Proceedings of the IEEE*, vol. 159, pp. 1335–1346, 1971.

[JAC 01] JACQUOT M., SANDOZ P., TRIBILLON G., "High resolution digital holography", *Optics Communications*, vol. 190, pp. 87–94, 2001.

[JAV 00] JAVIDI B., TAJAHUERCE E., "Three-dimensional object recognition by use of digital holography", *Optics Letters*, vol. 25, pp. 610–612, 2000.

[KAR 12a] KARRAY M., SLANGEN P., PICART P., "Comparison between digital Fresnel holography and digital image-plane holography: the role of the imaging aperture", *Experimental Mechanics*, 2012.

[KAR 12b] KARRAY M., POILANE C., MOUNIER D., GARGOURY M., PICART P., "Investigation of crack initiation with a three-color digital holographic interferometer", *Proceedings SPIE of 5th International Conference on Speckle Metrology*, Vigo, Spain, 10–12 September 2012.

[KRE 96] KREIS T.H., *Holographic Interferometry – Principles and Methods*, Berlin, Akademie Verlag Gmbh, 1996.

[KRE 97] KREIS T.H., ADAMS M., JÜPTNER W., "Methods of digital holography: a comparison", *Proceedings of SPIE*, vol. 3098, pp. 224–233, 1997.

[KRE 04] KREIS T.H., *Handbook of Holographic Interferometry Optical and Digital Methods*, Weinheim, Wiley-VCH, 2004.

[KRO 72] KRONROD M.A., MERZLYAKOV N.S., YAROSLAVSKII L.P., "Reconstruction of a hologram with a computer", *Soviet Physics Technical Physics*, vol. 17, pp. 333–334, 1972.

[KUH 07] KUHN J., COLOMB T., MONTFORT F., CHARRIERE F., EMERY Y., CUCHE E., MARQUET P., DEPEURSINGE C., "Real-time dual-wavelength digital holographic microscopy with a single hologram acquisition", *Optics Express*, vol. 15, pp. 7231–7242, 2007.

[LEC 00] LE CLERC F., COLLOT L., GROSS M., "Numerical heterodyne holography with two-dimensional photodetector arrays", *Optics Letters*, vol. 25, pp. 716–718, 2000.

[LEI 61] LEITH E., UPATNIEKS J., "New technique in wavefront reconstruction", *J. Opt. Soc. Am. A*, vol. 51, pp. 1469, 1961.

[LI 09] LI J.C., TANKAM P., PENG Z., PICART P., "Digital holographic reconstruction of large objects using a convolution approach and adjustable magnification", *Optics Letters*, vol. 34, pp. 572–574, 2009.

[LI 11] LI J.C., PENG Z., TANKAM P., SONG Q., PICART P., "Digital holographic reconstruction of local object field using an adjustable magnification", *J. Opt. Soc. Am. A*, vol. 28, pp. 1291–1296, 2011.

[MAN 08] MANN C.J., BINGHAM P.R., PAQUIT V.C., TOBIN K.W., "Quantitative phase imaging by three wavelength digital holography", *Optics Express*, vol. 16, pp. 9753–9764, 2008.

[OST 02] OSTEN W., BAUMBACH T., JUPTNER W., "Comparative digital holography", *Optics Letters*, vol. 27, pp. 1764–1766, 2002.

[PAQ 65] PAQUES H., SMIGIELSKI P., "Cineholography", *Comptes Rendus de l'Académie des Sciences*, vol. 260, pp. 6562, 1965.

[PAU 09] PAURISSE M., HANNA M., DRUON F., GEORGES P., BELLANGER C, BRIGNON A., HUIGNARD J.P., "Phase and amplitude control of a multimode LMA fiber beam by use of digital holography", *Optics Express*, vol. 17, pp. 13000–13008, 2009.

[PED 95] PEDRINI G., TIZIANI H.J., "Digital double pulse holographic interferometry using Fresnel and image plane holograms", *Measurement*, vol. 18, pp. 251–260, 1995.

[PED 97] PEDRINI G., TIZIANI H., "Quantitative evaluation of two-dimensional dynamic deformations using digital holography", *Optics & Laser Technology*, vol. 29, pp. 249–256, 1997.

[PED 02] PEDRINI G., TIZIANI H.J., "Short-coherence digital microscopy by use of a lens less holographic imaging system", *Applied Optics*, vol. 41, pp. 4489–4496, 2002.

[PIC 03a] PICART P., MOISSON E., MOUNIER D., "Twin sensitivity measurement by spatial multiplexing of digitally recorded holograms", *Applied Optics*, vol. 42, pp. 1947–1957, 2003.

[PIC 03b] PICART P., LEVAL J., MOUNIER D., GOUGEON S., "Time-averaged digital holography", *Optics Letters*, vol. 28, pp.1900–1902, 2003.

[PIC 04] PICART P., DIOUF B., BERTHELOT J.-M., "Investigation of fracture mechanisms in resin concrete using spatially multiplexed digital Fresnel holograms", *Optical Engineering*, vol. 43, pp. 1169–1176, 2004.

[PIC 05] PICART P., LEVAL J., GRILL M., BOILEAU J.P., PASCAL J.C., BRETEAU J.M., GAUTIER B., GILLET S., "2D full field vibration analysis with multiplexed digital holograms", *Optics Express*, vol. 13, pp. 8882–8892, 2005.

[PIC 08a] PICART P., LEVAL J., "General theoretical formulation of image formation in digital Fresnel holography", *J. Opt. Soc. Am. A*, vol. 25, pp. 1744–1761, 2008.

[PIC 08b] PICART P., MOUNIER D., DESSE J.M., "High resolution digital two-color holographic metrology", *Optics Letters*, vol. 33, pp. 276–278, 2008.

[PIC 08c] PICART P., *Optics for Engineers*, available at: http://www.optique-ingenieur.org/en/courses/–OPI_ang_M02_C10/co/OPI_ang_M02_C10_web.html, 2008.

[PIC 08d] PICART P., *Optics for Engineers*, available at: http://www.optique-ingenieur.org/en/courses/–OPI_ang_M02_C11/co/OPI_ang_M02_C11_web.html, 2008.

[PIC 09] PICART P., TANKAM P., MOUNIER D., PENG Z., LI J.C., "Spatial bandwidth extended reconstruction for digital color Fresnel holograms", *Optics Express*, vol. 17, pp. 9145–9156, 2009.

[PIC 12] PICART P., LI J.C, *Digital Holography*, ISTE Ltd., London and John Wiley & Sons, New York, 2012.

[POW 65] POWELL R.L., STETSON K.A., "Interferometric analysis by wavefront reconstruction", *J. Opt. Soc. Am.*, vol. 12, pp. 1593–1598, 1965.

[REI 82] REIFSNIDER K.L., "Damage in composite materials", *American Society for Testing and Materials*, vol. 775, 1982.

[SCH 94] SCHNARS U., JÜPTNER W., "Direct recording of holograms by a CCD target and numerical reconstruction", *Applied Optics*, vol. 33, pp. 179–181, 1994.

[SCH 99] SCHEDIN S., PEDRINI G., TIZIANI H.J., SANTOYO F.M., "Simultaneous three-dimensional dynamic deformation measurements with pulsed digital holography", *Applied Optics*, vol. 38, pp. 7056–7062, 1999.

[SCH 05] SCHNARS U., JUEPTNER W., *Digital Holography – Digital Hologram Recording, Numerical Reconstruction, and Related Techniques*, Berlin, Springer, 2005.

[SEE 01] SEEBACHER S., OSTEN W., BAUMBACH T., JUPTNER W., "The determination of material parameters of microcomponents using digital holography", *Optics and Lasers in Engineering*, vol. 36, pp. 103–126, 2001.

[SLA 11] SLANGEN P.R.L., KARRAY M., PICART P., "Some figures of merit so as to compare digital Fresnel holography and speckle interferometry", *Proceedings of SPIE*, vol. 8082, pp. 808205–1–808205–13, 2011.

[SMI 88] SMIGIELSKI P., FAGOT H, ALBE F., "Holographic cinematography and its applications", *Proceedings of SPIE*, vol. 673, 1988.

[STA 00] STADELMAIER A., MASSIG J.H., "Compensation of lens aberration in digital holography", *Optics Letters*, vol. 25, pp. 1630–1632, 2000.

[TAN 10a] TANKAM P., SONG Q., KARRAY M., LI J.C., DESSE J.M., PICART P., "Real-time three-sensitivity measurements based on three-color digital Fresnel holographic interferometry", *Optics Letters*, vol. 35, pp. 2055–2057, 2010.

[TAN 10b] TANKAM P., PICART P., MOUNIER D., DESSE J.M., LI J.C., "Method of digital holographic recording and reconstruction using a stacked color image sensor", *Applied Optics*, vol. 49, pp. 320–328, 2010.

[TAN 11] TANKAM P., PICART P., "Use of digital color holography for crack investigation in electronic components", *Optics and Lasers in Engineering*, 2011.

[YAM 97] YAMAGUCHI I., ZHANG T., "Phase-shifting digital holography", *Optics Letters*, vol. 22, pp. 1268–1270, 1997

[YAM 01a] YAMAGUCHI I., KATO J., OHTA S., MIZUNO J., "Image formation in phase shifting digital holography and application to microscopy", *Applied Optics*, vol. 40, pp. 6177–6186, 2001.

[YAM 01b] YAMAGUCHI I., KATO J., OHTA S., "Surface shape measurement by phase shifting digital holography", *Optical Review*, vol. 8, pp. 85–89, 2001.

[YAM 02] YAMAGUCHI I., MATSUMURA T., KATO J., "Phase shifting color digital holography", *Optics Letters*, vol. 27, pp. 1108–1110, 2002.

[ZHA 98] ZHANG T., YAMAGUCHI I., "Three-dimensional microscopy with phase shifting digital holography", *Optics Letters*, vol. 23, pp. 1221–1223, 1998.

Chapter 5

Elementary Speckle Interferometry

5.1. Introduction

One of the key roles of interferometry is the measurement of optical path lengths. The wavelength is the natural measurement unit. In the center of the visible electromagnetic spectrum, the wavelength is equal to 0.5 μm. The visible part of the spectrum is therefore ideally adapted to the measurement of micro-displacements. However, due to the extremely high oscillation frequency of the electromagnetic field (0.6 PHz) in practice optical path lengths cannot be directly measured. Interferometry specifically solves the problem by comparing the lengths of two paths – the reference and the "object" paths. By using highly coherent sources, paths exhibiting a difference of an integer multiple of the wavelength interfere constructively, giving a strong optical signal. Path differences with an excess of a half-wavelength are destructive, resulting in a null signal.

Classical interferometry deals with optically smooth and polished surfaces. The geometry of the optical paths and their lengths are perfectly defined. The creation and the significance of the interference signals are therefore conceptually straightforward. Nothing is as simple as this in speckle interferometry. Speckle interferometry examines the case of objects with ordinary diffusion, characterized by a RMS roughness state greater than the wavelength. Illuminated with coherent light, in reflection or transmission, these objects give rise to speckle fields – i.e. strongly fluctuating distributions of light intensities and phases. This chapter focuses on how to deal with this new speckle phenomenon associated with any rough surface in the field of interferometry (SI).

Chapter written by Pierre JACQUOT, Pierre SLANGEN and Dan BORZA.

5.2. What is speckle interferometry?

Speckle interferometry (SI) can be defined as the set of techniques that aim to create, record and take advantage of a two-beam interference pattern involving at least one speckle wave ([JON 89, JAC 08]). As with any monochromatic, linearly polarized wave, a speckle wave can be represented by a generic expression of scalar complex amplitude, $U(\mathbf{r})$, whose main property is that it is a solution to the Helmholtz equation [HEC 02]:

$$U(\mathbf{r}) = A(\mathbf{r})\exp\mathbf{j}\varphi(\mathbf{r});\, \mathbf{j}^2 = -1; U^*(\mathbf{r}) = A(\mathbf{r})\exp-\mathbf{j}\varphi(\mathbf{r}) \qquad [5.1]$$

A is the amplitude of the electrical field vector $\mathbf{E}$ defining the direction of polarization, φ the phase of the wave, and $\mathbf{r}$ the spatial position vector where the field is observed. The complex amplitude representation ignores the fast temporal variation of the electric field oscillating at the optical frequency ν. The phase φ, however, shows modulo 2π in which the value of its cycle $\mathbf{E}$ is found, with respect to a time origin decided by convention. Photodetectors are not fast enough to follow the rapid oscillations of the electric field and are therefore basically sensitive to the average time of its square modulus, which is called light intensity, I. With complex notation, the light intensity is simply given by:

$$I(\mathbf{r}) = 2\left\langle E^2(\mathbf{r},t)\right\rangle_\tau = U(\mathbf{r})U^*(\mathbf{r}) = \left|U(\mathbf{r})\right|^2 \qquad [5.2]$$

Speckle waves exhibit rapid spatial intensity and phase fluctuations. Such waves are spontaneously reflected or transmitted by diffusing surfaces that are uneven on the wavelength scale, being opaque or transparent respectively, when illuminated with coherent light. Figure 5.1 represents the typical granular appearance of the intensity and phase of a speckle pattern. Further explanations of the phenomenon will be given in section 5.2.2.

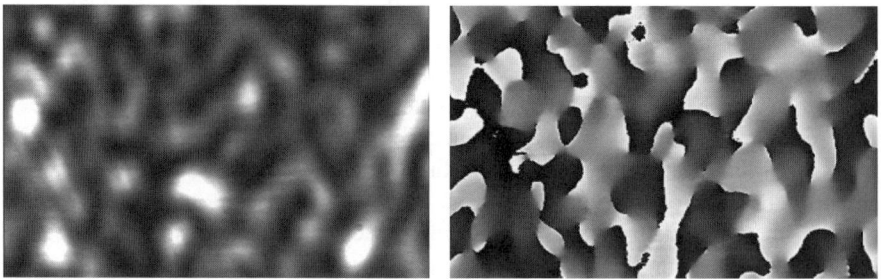

Figure 5.1. *Intensity (left), and phase (right) distributions of a speckle pattern; the average size of the random spots is typically in the order of a few microns*

According to the above definition, SI is governed by the two-beam interference formula giving the resultant intensity distribution obtained in air, known as the aerial intensity distribution:

$$I(\mathbf{r}) = |U_1(\mathbf{r}) + U_s(\mathbf{r})|^2 = I_1(\mathbf{r}) + I_s(\mathbf{r}) + 2\sqrt{I_1(\mathbf{r})I_s(\mathbf{r})}\cos\tilde{\varphi}(\mathbf{r}) \qquad [5.3]$$

where subscript s stands for the speckle wave, while subscript 1 characterizes either a simple smooth wave or a second speckle wave. $\tilde{\varphi}$ is the phase difference between the two waves, which is often referred to as the phase of the interference pattern.

We have a *specklegram* when this interference pattern is recorded linearly. The two-beam interference formula is a direct consequence of the Young's principle of superposition stating that the effect of the "interference" of two waves is simply described by the sum of their complex amplitudes taken separately. The two waves play a fully symmetrical role. The spatial periodicity of the waves along the propagation direction – by definition the wavelength λ is reminiscent of the divisions on a ruler – offers a remarkable means of measuring the phase shift between the two waves as a function of the geometrical path difference δ travelled:

$$\tilde{\varphi}(\mathbf{r}) = \frac{2\pi}{\lambda}n\delta(\mathbf{r}) \qquad [5.4]$$

Here, λ is the vacuum wavelength of the light source and n the refractive index of the propagation medium.

At least one of the components is a speckle wave, so the resultant interference pattern has a random aspect similar to the one observed in the intensity distribution of a single speckle wave. This randomness enables us to distinguish between SI and classical interferometry. At this point it is worth noting that the hologram of a diffusing object is the recording of an intensity distribution obeying equation [5.3].

However, holographic techniques, whether analogue or digital, are all composed of a three-dimensional step reconstructing the object wave and its propagation, which is unknown in SI. This reconstruction step is sufficient to make a clear distinction between SI and holographic techniques, even though some holographic and speckle reconstructions may appear to be similar. To sum up, SI gathers together the set of techniques that aim to create an intensity distribution of the type given in equation [5.3], with the purpose of directly exploiting the two-dimensional phase information contained in the recording plane.

5.2.1. *Simplified principle – correlation fringes*

The unpredictable nature of the specklegram makes its direct use problematic. SI basically operates in differential mode. Two correlated evolutions of a specific interference state are compared. An initial state, usually related to a fixed configuration of the interferometric system, is chosen as a reference state. The corresponding specklegram is captured by a digital acquisition unit. The final state results from a small perturbation of the interferometric system (e.g. micrometric displacements, deformations of the object under examination, deformation of the components of the optical setup, or small changes in the wavelength or refractive index of the surrounding medium). By assuming that the phases of the two waves change between the initial and final states by φ_{1P} and φ_{SP}, respectively, the two states being compared, I_R and I_P, read:

$$I_{R \, or \, P} = \left| U_{1R \, or \, 1P} + U_{SR \, or \, SP} \right|^2 \; ; U_{1P \, or \, SP} = U_{1R \, or \, SR} \, \exp \mathrm{j} \varphi_{1P \, or \, SP}$$

$$I_R = I_{1R} + I_{SR} + 2\sqrt{I_{1R} I_{SR}} \, \cos \tilde{\varphi}$$

$$I_P = I_{1R} + I_{SR} + 2\sqrt{I_{1R} I_{SR}} \, \cos \left(\tilde{\varphi} + \varphi_{SP} - \varphi_{1P} \right)$$

[5.5]

Subscripts $_R$ and $_P$ stand for the "reference" and "perturbed" states, and the argument $\mathbf{r}$ is omitted for simplicity. In the regions where the difference $\varphi_{SP}(\mathbf{r}) - \varphi_{1P}(\mathbf{r})$ is an integer multiple of 2π, the initial and perturbed specklegrams are identical. For odd multiples of π, the sum of the two intensities is twice the incoherent addition of the two waves, i.e. $2(I_{1R} + I_{SR})$.

Thus, simple arithmetic operations enable us to disclose the regions of the two specklegrams that are either fully correlated or have no correlation at all. Though multiplication, addition and squared difference are sometimes used, absolute value subtraction, $I_- = |I_P - I_R|$, best reveals the correlation fringes:

$$I_- = 4\sqrt{I_{1R} I_{SR}} \left| \sin \left[\tilde{\varphi} + \frac{\varphi_{SP} - \varphi_{1P}}{2} \right] \right| \left| \sin \left[\frac{\varphi_{SP} - \varphi_{1P}}{2} \right] \right|$$

[5.6]

The square root and the first sine term are fluctuating terms; the second sine is a deterministic modulation. A one-dimensional cut of I_- is shown in Figure 5.2.

Quite naturally, the spatial fluctuations are very rapid so the envelope of the absolute subtraction represents the deterministic modulation or, in other words, the photometric profile of the correlation fringes. The correlation fringes are the loci of equal values to the difference in the phase changes of the two waves,

$\varphi_{SP}(\mathbf{r}) - \varphi_{1P}(\mathbf{r})$. In particular, black fringes, centered on the zeros of intensity, are given by the following expression:

$$\varphi_{SP}(\mathbf{r}) - \varphi_{1P}(\mathbf{r}) = \pm 2\pi n;\ n: \text{integer} \qquad [5.7]$$

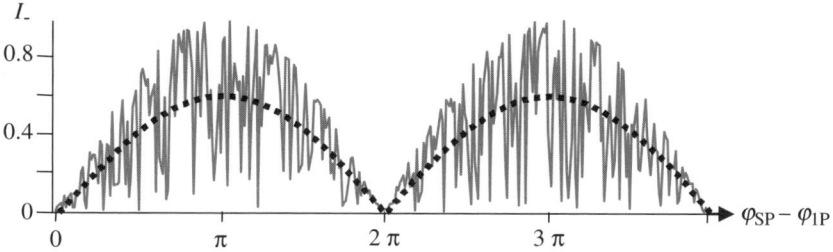

Figure 5.2. *Photometric profile of correlation fringes (numerical simulation)*

Fluctuations with high spatial frequency can be smoothed using a low-pass filter. Today acquisition systems and numerical image processing enable us to display the absolute subtraction I_- and the low pass filtering at a temporal frequency in the kHz range "in real time"; see Figure 5.3.

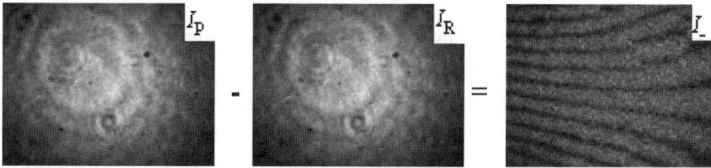

Figure 5.3. *Absolute subtraction of the reference and perturbed specklegrams; the residual grainy appearance is due to an incomplete filtering of the spatial fluctuations*

Leendertz and Butters are given credit for being the first to realize that the process described above gives access to the phase change of the speckle wave without the need to fully reconstruct it, as is the case in holography ([LEE 70, BUT 71]). Precursory work by Archbold, Burch, Ennos and Taylor also deserves a mention here [ARC 69].

5.2.2. *Speckle field and specklegram statistics in a nutshell*

Equations [5.3] and [5.6] highlight the necessity of comprehending the statistical properties of the intensity and phase of the speckle field, as well as those of related random variables. A thorough analysis of the calculation can be found in [GOO 07] and [LEH 01]. Here we limit ourselves to a brief reminder of the so-called "Gaussian" model. The speckle amplitude, U, shown in Figure 5.4, is the result of a random walk in the complex plane. With some reasonable statistical assumptions, such as a uniform distribution of the orientation of the elementary contributions and the application of the central limit theorem, it turns out that the speckle aerial amplitude and intensity, respectively, obey a centered Gaussian and a negative exponential probability density. Unsurprisingly, the phase φ is uniformly distributed over an interval of 2π.

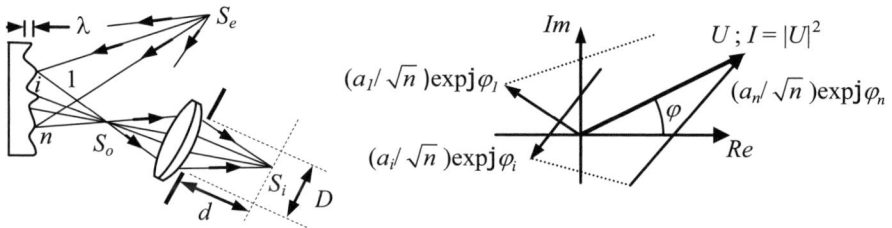

Figure 5.4. *Speckle pattern formation in a defocused imaging geometry as the addition of n contributions, randomly distributed in amplitude and phase, arising from the roughness of the surface*

In practice, the recording of an aerial speckle pattern goes hand-in-hand with spatially integrating it to some extent. In this case, the intensity statistics degenerate into a gamma distribution, which then develops into a very good approximation of a non-centered Gaussian as soon as the independent elementary contributions are in excess of 10. Moreover, the process of spatial integration leaves equation [5.6] formally unchanged. The speckle amplitude and intensity autocorrelations are peak-shaped functions, whose first zeros define the mean size $\langle\Delta X\rangle$, $\langle\Delta Y\rangle$ and $\langle\Delta Z\rangle$ of the small spots composing the patterns:

$$\langle\Delta X\rangle = \langle\Delta Y\rangle = \frac{\lambda d}{D} ; \langle\Delta Z\rangle = \frac{8\lambda d^2}{D^2} \qquad [5.8]$$

It is worth noting that these dimensions only depend on the wavelength and the beam aperture D/d; see Figure 5.4. For standard apertures ($d/D>1$), speckle grains tend to be longer than they are wide and thus allude to the shape of a cigar. The same analysis can be made and the same results obtained for the processes of focused imaging and free space propagation. The random walk mechanism also emphasizes another key property of speckle fields: the existence of singularities. Singularities are points of zero intensity arising if and only if both real and imaginary parts of the complex amplitude are simultaneously null. The speckle phase is then indeterminate. Singularities still exist in the spatial integration process. Interferometric measurements are meaningless at the location of singularities. In an observation plane, there are as many singularities as there are speckle grains defined by equation [5.8]. Although they have the topological nature of a point, singularities cover much less surface area.

Random variables that are more specific to SI are the arithmetic and geometric means of determining the intensity of waves, I_F and I_M respectively. The interference formula, equation [5.3], can in fact be rewritten as:

$$I = 2\left(I_F + I_M \cos \tilde{\varphi}\right); I_F = \frac{I_1 + I_S}{2}; I_M = \sqrt{I_1 I_S}; I_M \leq I_F \qquad [5.9]$$

All possible determinations of the interferometric signal are bound to stay within the interval $[2(I_F - I_M), 2(I_F + I_M)]$, where I_F and I_M are the "background" and "modulation". Both obey four types of behavior, depending on whether the two interfering waves are one smooth and one speckle wave or two speckle waves, and whether the recording is spatially resolved or not. The corresponding eight probability densities for I_F and I_M are all known [LEH 01]. Together with the characteristic response curves of the photodetectors, these densities enable us to optimize the recording step.

5.2.3. Speckle field transformation – small perturbation theory

In Chapter 4 of [SCH 79], the transformation of the speckle complex amplitude is thoroughly analyzed when the perturbation consists of a micro-deformation of the opaque diffusing object. In the presence of small displacements, **U(r)**, of its surface and as a first approximation, the complex amplitude of the perturbed state is deduced from the reference amplitude by the following relation:

$$U_{SP}\left(\mathbf{r}\right) = U_{SR}\left(\mathbf{r} + \mathbf{\Omega}(\mathbf{r})\right) \exp j\varphi_{SP}\left(\mathbf{r}\right) \qquad [5.10]$$

Here, $\mathbf{\Omega}$ is a small vector that is linearly related to the displacement vector **U** and its six first surface partial derivatives. The 9×3 coefficients of the linear relationship depend on the geometrical parameters of the illumination and observation of the object. The phase behavior was partially anticipated in section 5.2.1 (the perturbed state is deduced from the reference state purely and simply by a phase change).

Equation [5.10] provides more information: the speckle wave undergoes a local translation equal to the vector $\mathbf{\Omega}$ in response to surface deformation. In the general case, the relationship between $\mathbf{\Omega}$ and **U** is quite complicated. In contrast to this, when the imaging system is sharply focused at the surface of the object, this relation takes the very simple form of equality, $\Omega = U$. Consequently, from now on we shall only consider focused imaging processes. When the perturbation is not a micro-displacement field, but consists of small changes in the wavelength or refractive index of the propagation medium or a small modification of the geometry of the illumination–observation system for instance, it is easy to transpose the results presented in this section.

5.2.4. *Phase change–deformation law – sensitivity vector*

The expression of the phase change φ_{SP} as a function of vector U is explicitly derived in [SCH 79]. Instead of this, a simple geometrical argument is used here with respect to Figure 5.5. To ensure that the two states compared are correlated. The displacement $\Omega = U$ must not be spatially resolved in the image plane. In other words, no two image points M_i and M'_i, where M'_i is the displaced position of M_i, may be optically separated.

From the point of view of image correlation methods, this condition is paradoxical, since resolution of the displacement field is at the very root of these methods. In interferometry, the displacement is encoded by the associated phase change or optical path change. Assuming small displacements, the optical path difference, δ_{opt}, can be safely approximated by the projection of the displacement vector in the directions of illumination and observation, namely:

$$\delta_{opt}(M) = U(M) \cdot (K_o(M) - K_e(M)) = U(M) \cdot S(M)$$
$$S(M) = K_o(M) - K_e(M) \qquad\qquad [5.11]$$

Here, S is the *sensitivity vector*, which is aligned with the bisector of the angle between the directions of illumination and observation defined by the unit vectors K_e and K_o. Depending on the relative orientation of these two directions, S gives a weight between 0 and 2 to the components of the displacement vector U. For the sake of simplicity, the second wave – a simple or speckle wave, arising from the same object for the former or a different object for the latter – is not displayed in Figure 5.5.

The corresponding phase change, φ_{SP}, is immediately deduced from the path difference:

$$\varphi_{SP}(M) = \frac{2\pi}{\lambda} S(M) \cdot U(M) = k\, S(M) \cdot U(M) \qquad\qquad [5.12]$$

where $k = 2\pi / \lambda$ is the wave number. Equation [5.12] is the fundamental formula of SI. The formula shows that the phase change is linearly related to the components of the displacement field, with the geometrical parameters of the optical setup playing the role of weighting factors. We shall see that the formula can be applied to all SI arrangements. Lastly, the formula highlights one of the main properties of SI: total freedom in the choice of the sensitivity vector, since for diffusing objects the directions of illumination and observation are arbitrary and independent, as opposed to classical or grating interferometry where these directions are bound by reflection or diffraction laws.

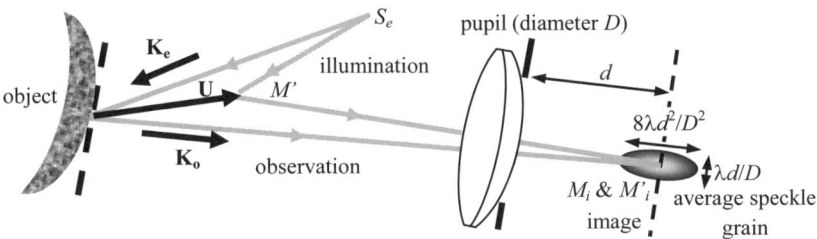

Figure 5.5. *Phase change associated with the object displacement field **U***

5.2.5. *Success or failure of experiments – central role of decorrelation*

Formula [5.10] is only valid in the hypothesis of micro-deformations. Sizable perturbations, which are "large" in terms of the wavelength, and are related to the illumination–observation geometry, the surface roughness or the wavelength itself, would create a final state that was completely decorrelated from the reference state. This would lead to the disappearance of the correlation fringes. Even within the framework of the assumption of micro-perturbation, however, a partial loss of correlation will always occur: the imperfect superimposition represented by the wave shift Ω, taking place inevitably in the pupil plane as well as in the image plane, will reduce the degree of correlation between the two states being compared. A comprehensive analysis of the effects of decorrelations in the pupil and image planes is given in [LEH 01]. Only the outer limits are briefly discussed here. Total decorrelation, meaning there is no interferometric signal, arises when Ω is either greater than the pupil diameter, D, or the mean speckle size; see Figure 5.5 and equation [5.8]. Due to the elongated shape of the speckle grains, the correlation condition is easier to fulfill for the longitudinal component of Ω. On the one hand, the observation system must be sufficiently stopped down, so that the object displacement field is not optically resolved. On the other hand, the pupil diameter should be large enough to ensure an image formation with, as far as possible, the same contributing wave fronts. These competing demands lead to a compromise.

Even with quite a small aperture, the diameter of the pupil is generally large compared to Ω, even when there is quite a small aperture that in principle has the same order of magnitude as U. A noticeable exception to this is when the object surface undergoes micro-rotations. In this case, as with mirrors or gratings, the amplifying lever effect may result in large displacements in the pupil plane. Finally, the optimal aperture is found experimentally using a trial and error procedure backed up by theoretical safeguards [LEH 01]. It is also often pertinent to rely on numerical simulations using adequate speckle field modeling [EQU 06].

5.3. Optical point of view

A large number of SI arrangements have been proposed [SIR 99]. From an optical point of view, much of the process involves looking for various means by which to divide and recombine two waves, at least one of them being a speckle wave. Three main categories of optical setups are represented in Figure 5.6.

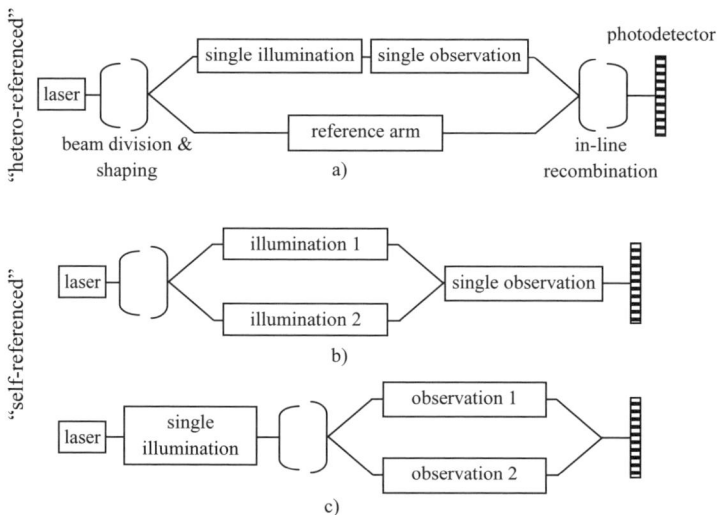

Figure 5.6. *Three main categories of SI setup: a) in-line reference; b) two illuminations; and c) two observations*

For the so-called "hetero-referenced" setups, the laser beam is divided to create the reference beam on one arm of the interferometer, and the illumination and observation beams of the object on the other arm. Object and reference beams are recombined in-line in front of the detector. A photographic lens usually serves to focus the object onto the photodetector. The angle between object and reference beams, θ, must be small enough for the carrier spatial frequency f of the recorded interference pattern to be resolved by the detector (θ is smaller than 3° at the center wavelength of the visible spectrum):

$$\sin\theta < \lambda f \qquad\qquad\qquad [5.13]$$

The reference wave can be a simple wave, an arbitrary speckle wave, or a speckle wave generated by a comparison object that is macroscopically identical to the object being investigated. Hetero-referenced SI setups are quite close to the configuration of standard digital holography. However, contrary to the holographic scheme, the speckle method forms an image of the object on the photodetector and does not aim to reconstruct the object wave in three-dimensional space. The intensity and phase of the specklegrams are simply and directly exploited in the recording plane. Despite geometrical similarities, hetero-referenced SI and digital holography are thus clearly separate techniques.

The laser beam can be divided to create two illumination beams for the object. Each of these beams generates a speckle field of its own, see Figure 5.6b. The two speckles are phase-modified by object deformation. According to equation [5.5], the phase change of the new interference pattern, which is related to the deformed state, is given by the difference between the phase increments of the two speckles. Thanks to the diffusion properties of rough objects, the recombination of the two speckle waves is a fully natural process, rendering additional optical components unnecessary.

Division and recombination can be carried out on the observation side only, using a single-beam illumination, as shown in Figure 5.6c. Together, Figures 5.6b and 5.6c represent the so-called "self-referenced" interferometers. In practice, double-object observation can be achieved in different ways: either by observing the object from separate directions or at unequal distances, or by optically transversally shifting the duplicated images. This last possibility opens the door to a vast subset of methods that are often referred to as "shearing SI" or more simply "shearography" [STE 03]. It is also possible that the two superimposed images come from two separate objects or from two distinct parts of the same object.

Figure 5.6 suggests that the practical creation of speckle interferometers is much more open than that of their classical or grating counterparts. As mentioned above, this is mostly due to the diffusion properties of uneven surfaces, which re-emit light in all directions, while continuous or periodic smooth surfaces reflect and diffract light in well-defined, discrete directions. Nonetheless, SI also makes use of classical interferometric schemes, typically of amplitude or wavefront division types. These are represented by empty parentheses in Figure 5.6. "Amplitude division" signifies that the splitting and recombination of the two waves are achieved by means of semi-transparent polished glass plates. "Wavefront division" consists of spatially individuating two portions of a wave using simple masks or any reflective, refractive or diffractive optical component, as done by bi-mirrors, bi-prisms, bi-lenses. The rectangles and parentheses in Figure 5.6 are thus the elementary building blocks of countless optical systems.

If, instead of a deformation of the surface of the object the perturbation is a small change in wavelength, the refractive index of the surrounding medium or geometrical modification of the setup, the changes in the corresponding optical path depend on the shape of the object. In these cases, the setups of Figure 5.6 can therefore all serve to measure the shape of rough objects.

5.4. Mechanical point of view: specific displacement field components

Advances in SI have been fostered by applications of experimental mechanics. Fundamentally, SI techniques are sensitive to changes in the optical paths associated with object deformation. These changes in turn depend on the illumination and observation geometry according to relations [5.7] and [5.12]. Plenty of geometrical configurations have been devised in order to make the variations in the optical path proportional to specific Cartesian components of the displacement fields. Only three of the most popular arrangements, depicted in Figure 5.7, are discussed here. They provide the measurements of the out-of-plane and in-plane components and their first-order partial derivatives, respectively.

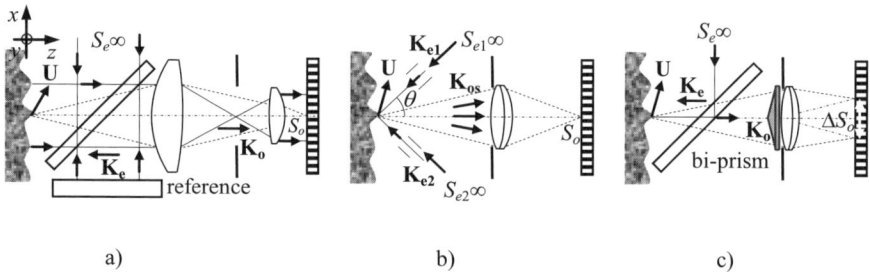

a)   b)   c)

Figure 5.7. *SI schematics for displacement component measurements: a) out-of-plane; b) in-plane; and c) partial derivatives (shearography)*

5.4.1. *Measurement of the out-of-plane component*

As shown in Figure 5.7a, the reference wave is a simple or speckle wave propagating along the optical axis of the observation system. The object wave is created by collimated illumination of the object under normal incidence, followed by a telecentric observation, again in a direction normal to the surface of the object. Applying equation [5.12], it can immediately be seen that the sensitivity reduces to the sole out-of-plane component of displacement u_z. Section 5.5 will propose several techniques enabling the computation of the resulting phase change φ_m, corresponding to u_z, and therefore the computation of this out-of-plane component:

$$u_z = \frac{\lambda}{4\pi}\varphi_m \qquad\qquad\qquad [5.14]$$

Due care is not always exercised in the choice of a telecentric observation system. On the contrary, a system ruled by the central projection principle, as in Figure 5.7b, introduces small spurious additional sensitivities to the in-plane components (u_x, u_y). These components actually contribute to the optical path changes or the phase changes with a weight that is equal to the tangent of the field angles of the object points observed. By illuminating and observing the object symmetrically with respect to its surface normal, under an angle θ, the setup is desensitized by a factor of $1/\cos\theta$.

The setup presented in Figure 5.7a is flexible enough to give rise to a variety of modes of measurement where, locally or as a whole, object displacement can be optically compensated for or directly compared to that of a reference object.

5.4.2. Measurement of the in-plane component [LEE 70]

Measurement of the in-plane displacement components u_x or u_y is carried out using the setup shown in Figure 5.7b. A double symmetrical illumination of the object, under incidences θ_x in the (x,z) plane or θ_y in the (y,z) plane, is performed. As a result, two uncorrelated speckle fields are created in any common direction of observation. Both interfering speckle fields are influenced by object deformation. Once again application the sensitivity formula, equations [5.7] and [5.12], immediately gives the expected result in each of the incidence planes, (x,z) or (y,z):

$$u_{x,y} = \frac{\lambda}{4\pi\sin\theta_{x,y}}\varphi_{mx,y} \qquad\qquad\qquad [5.15]$$

Point-by-point laser velocimetry makes use of the same optical setup. As in this case the illuminated area is point-like, no imaging lens is necessary. The interfering speckle fields are observed in free space propagation and the recording device is a simple photodiode.

5.4.3. *3C–3D: three components attached to three-dimensional objects*

The needs expressed by the experimental mechanics community prompted engineers to develop optical systems that are able to provide the three components of the displacement field attached to arbitrary three-dimensional skew surfaces. In principle, a multitude of arrangements combining the two generic setups presented above can solve the problem. In practice, two main categories of solutions have emerged:

– the multiple illuminations scheme, creating speckles, each one interfering with a single in-line reference wave [GAU 05]; and

– the multiple illuminations scheme, but where the speckles created interfere two-by-two, without a reference wave [GOU 05].

In both cases, at least three illuminations are needed – and a fourth is needed when the absolute fringe order cannot be determined by *a priori* or auxiliary knowledge, such as for example the displacement vector of some known points [SIE 04]. Assuming no uncertainty in the absolute phase, three couples of illuminating beams, namely three independent sensitivity vectors, give rise to three independent phase maps $\varphi_m^{(1),(2),(3)}$ through the procedures described below in section 5.5. These phase maps verify:

$$\begin{bmatrix} \varphi_m^{(1)} \\ \varphi_m^{(2)} \\ \varphi_m^{(3)} \end{bmatrix} = k\left[\mathbf{M}\right] \begin{bmatrix} u_x \\ u_y \\ u_z \end{bmatrix}; \left[\mathbf{M}\right] = \begin{bmatrix} \mathrm{K_{ox}}\text{-}\mathrm{K_{ex}^{(1)}} & \cdot & \cdot \\ \cdot & \mathrm{K_{oy}}\text{-}\mathrm{K_{ey}^{(2)}} & \cdot \\ \cdot & \cdot & \cdot \end{bmatrix} \text{or} \begin{bmatrix} \mathrm{K_{ex}^{(2)}}\text{-}\mathrm{K_{ex}^{(1)}} & \cdot & \cdot \\ \cdot & \mathrm{K_{ey}^{(3)}}\text{-}\mathrm{K_{ey}^{(2)}} & \cdot \\ \cdot & \cdot & \cdot \end{bmatrix} \qquad [5.16]$$

Once more, matrix coefficients of [**M**] are obtained by applying the sensitivity law [5.12]. Two examples of these coefficients are given for the two categories of solutions, clarifying their structure. Superscripts [1], [2] and [3] refer to the three illumination geometries considered. The inversion of the linear system [5.16] returns the three sought-after displacement components. Setup designs that as far as possible involve orthogonal sensitivity vectors should be promoted [OST 96]. If high accuracy is an issue, it is useful to increase the number of independent acquisitions and solve the over-determined linear system [5.16] by a least-square method.

There are advantages and drawbacks associated with each type of solution. Multiple illuminations are easy to execute and transform, particularly if optical fibers are used. Intrinsically, they exhibit a predominant sensitivity to the in-plane components (u_x, u_y). Setups with an in-line reference wave are a little more difficult to adjust, but the interference patterns are generally less noisy and the laser power that is available is more efficiently exploited. These setups are predominantly sensitive to the out-of-plane component u_z. For mechanical reasons, it may be convenient to build in sensitivity to a particular projection of the displacement vector. In this case, it is possible to generate synthetic sensitivity vectors $\mathbf{S}_{syn}$ by using a digital procedure consisting of computing suitable linear combinations of the experimental sensitivity vectors of leading term $\kappa_i \mathbf{S}_i$, and deriving the corresponding synthetic phase maps:

$$\varphi_{syn} = \sum_i \kappa_i \varphi_{mi} = \frac{2\pi}{\lambda} \mathbf{U} \cdot \sum_i \kappa_i \mathbf{S}_i = \frac{2\pi}{\lambda} \mathbf{U} \cdot \mathbf{S}_{syn} \qquad [5.17]$$

Measurement of the three components of the displacement vector is very involved and calls for the acquisition of a great number of specklegrams: a minimum of 18 for a three-image phase-shifting algorithm without shape measurement, as we shall see below. A general approach to the solution is to resort to multiplexing techniques.

5.4.4. Partial derivatives of the displacement – shearography

The two interfering speckle fields arise from image doubling by means of a bi-prism or any interferometric device mounted in front of the observation system, see Figure 5.7c. The two images are transversally shifted, one with respect to the other, by an amount of $\pm\Delta S_o / 2$. The two speckles are affected by object deformation and, to the first order, increment their phase by:

$$\varphi_{\pm}\left(S_o \pm \frac{\Delta S_0}{2}\right) = \varphi(S_o) \pm \frac{\Delta S_0}{2} \cdot \mathbf{grad}\,\varphi(S_o) \qquad [5.18]$$

The phase change measured is therefore given by:

$$\varphi_m(S_o) = \Delta S_0 \cdot \mathbf{grad}\,\varphi(S_o) = \frac{2\pi}{\lambda} \Delta S_0 \cdot \mathbf{grad}\left[\mathbf{S}(S_o) \cdot \mathbf{U}(S_o)\right] \qquad [5.19]$$

As anticipated, shearography actually purports to be a method devoted to the measurement of the gradient of the speckle phase change. Developing equation [5.19] in the (x,y,z) Cartesian coordinate system, it can be seen in the general case that shearography is sensitive to a linear combination of the six first-order surface partial derivatives of the displacement vector. This number is reduced to three when the image shift ΔS_o is chosen parallel to the x or y axis, and only one at the center of the field for the geometric configuration represented in Figure 5.7c, since the components of the sensitivity vector $\mathbf{S}(S_0)$ are $(0, 0, 2)$. In this example, which frequently occurs in practice, the slopes of the deformed surfaces are measured:

$$\frac{\partial u_z}{\partial x} = \frac{\lambda}{4\pi\Delta x}\varphi_{mx} \text{ or } \frac{\partial u_z}{\partial y} = \frac{\lambda}{4\pi\Delta y}\varphi_{my} \qquad [5.20]$$

A shearographic system is basically a common path interferometer, the two paths of the interfering waves spatially being very close to each other: the optical heads can be built following a very compact and stable design; the image doubling device is easily adjustable to zero optical path difference, thereby considerably reducing the coherence requirement of the source; the image shift can be of a transversal, radial or rotational nature, with many possible implementations in each case [SIR 99]; as with derivation operators, shearographic setups are largely unaffected by rigid body motion of the object; multiple directions of illumination and multiple wavelengths can be quite simply introduced in these setups; the sensitivity can be adapted by two independent means: through the choices of the illumination/observation geometry and the amplitude of the shift between the twin images. From the end-user's point of view, shearography thus has many clear advantages. By superimposing two spatially shifted wavefronts, shearography is nevertheless sensitive to air turbulence in front of the object surface, like any shearing interferometer. As we shall see in section 5.5, phase extraction may require the addition of carrier fringes to the correlation fringes. In shearography, this is accomplished by creating an external quadratic phase change between the two states being compared, namely a change in the curvature of the illumination or observation waves in practice.

5.4.5. *Shape measurement and other considerations*

The phase changes of a specklegram can be attributed to causes other than object deformation. Small changes in wavelength, direction of illumination, object orientation or refractive index induce phase changes proportional to the height of the object, $z(x, y)$:

$$\varphi_m = \frac{2\pi}{\lambda} \alpha z \qquad [5.21]$$

These intentional perturbations must remain small enough to prevent the correlation between the compared states being compromised. It is thus convenient to proceed by concatenating the series of n results obtained by dividing the total perturbation into n equal small fractions.

Many setups are candidates for measuring z. With reference to Figure 5.7b, a rotation $\Delta\theta$ of the two illumination beams in the same direction amounts to expressing the proportionality constant α of equation [5.21] as $2\,\Delta\theta\sin\theta$; for an in-line reference setup and normal illumination and observation, constant α is either $2\Delta\lambda / \overline{\lambda}$ or $2\Delta n$, depending on whether the perturbation is a change in wavelength $\Delta\lambda$ for a mean wavelength $\overline{\lambda}$ or a change in refractive index Δn. It is a decisive advantage of SI to enable both shape and displacement field measurements using the same setup. Shape measurement is also indirectly useful in determination of the sensitivity vector, which usually varies over the surface of the object.

A plethora of articles describe particular SI setups with interesting properties or that are devoted to specific applications [MEI 96]. For illustrative purposes only, two such examples will be mentioned:

– based on the scheme in Figure 5.7b, Albertazzi has developed a "radial speckle metrology" that is adapted to residual stress analyses [ALB 06]; and

– Bruno has just proposed a very compact and stable type of shearography relying on radial shifts [BRU 07].

5.5. Phase extraction

The emergence of new image acquisition and digitization systems starting in the 1980s, as well as a considerable increase in computing power, has led to the development of effective methods of automatic phase extraction. From being qualitative, interferometry became quantitative. Automatic processing is now a routine task in SI. The task is, however, more difficult for SI than for classical interferometry due to the random nature of the speckle waves, giving rise to strong and rapid spatial fluctuations of terms I_F and I_M in equation [5.9]. Needless to say, a variety of noises also corrupt the signals.

In spite of the numerous challenges, many methods of phase extraction have been successfully developed for SI and provide suitable solutions to the problems. A review is presented in [EQU 09a]. Only the most well-known methods are considered in the following.

5.5.1. *One-image methods*

The so-called One-image methods are particularly well suited to the study of dynamic deformations. A single short-exposure specklegram is required. Single pulse or repetition rate Q-switched lasers usually provide these very short exposures, acting like a flash in photography. The most common method is the one- or two-dimensional, temporal or spatial, Fourier transform method [TAK 82]. A carrier frequency needs to be introduced into the specklegram. This is simply achieved by the deliberate introduction of an external linear phase variation between the interfering waves. In the two-dimensional spatial case, the "carrier" takes the form of equidistant parallel fringes with a period p, or (p_x, p_y), along the directions of the co-ordinate system. The problem is then reduced to extracting the phase φ from an interferogram $I(\mathbf{r})$ obeying:

$$I(\mathbf{r}) = 2I_F(\mathbf{r}) + 2I_M(\mathbf{r})\cos\left(\varphi(\mathbf{r}) + 2\pi\frac{\mathbf{s}\cdot\mathbf{r}}{p}\right) \tag{5.22}$$

where $\mathbf{s}$ is the unit vector orthogonal to the direction of the carrier fringe. Using complex notation, equation [5.22] can be rewritten:

$$I(\mathbf{r}) = 2I_F(\mathbf{r}) + I_{MC}(\mathbf{r})\exp\left(j2\pi\frac{\mathbf{s}\cdot\mathbf{r}}{p}\right) + I_{MC}^*(\mathbf{r})\exp\left(-j2\pi\frac{\mathbf{s}\cdot\mathbf{r}}{p}\right)$$
$$I_{MC}(\mathbf{r}) = I_M(\mathbf{r})\exp(j\varphi(\mathbf{r})) \tag{5.23}$$

The Fourier transform of equation [5.23] is made up of a central term, or zero order, centered on the zero spatial frequency, and two symmetric off-set terms centered on the carrier spatial frequency ($\pm 1/p_x$, $\pm 1/p_y$). If these three orders are well separated, it is possible to filter out the zero and -1 orders and to translate back to the origin the +1 order. The Fourier transform $\mathcal{I}_{MC}(\rho)$ of $I_{MC}(\mathbf{r})$ is thus obtained. A second inverse Fourier transform gives $I_{MC}(\mathbf{r})$ itself. The phase sought is the argument of $I_{MC}(\mathbf{r})$:

$$\varphi(\mathbf{r}) = \mathrm{arctg}\left[\frac{Im\{I_{MC}(\mathbf{r})\}}{Re\{I_{MC}(\mathbf{r})\}}\right]; \varphi \in [0, 2\pi] \tag{5.24}$$

The carrier fringes can be introduced as a slight modification of the angle between the two interfering waves, either at each acquisition (for the in-line reference setups) or between the two states being compared (for all setups).

The wavelet transform is another tool that is well suited to phase extraction in dynamic deformation processes. The tool is thoroughly presented in [COL 97]. Spatial and temporal versions are feasible. The Morlet wavelet is to lead to the best compromise with respect to time–frequency resolutions.

5.5.2. Phase-shifting methods

Phase-shifting methods are undoubtedly the best studied and the most efficient phase extraction procedures [CRE 88]. Phase changes ψ_i are deliberately introduced into one arm of the interferometer. The unknown phase is solved modulo 2π by an expression involving linear combinations of the n intensities, I_i corresponding to each phase increment:

$$I_i = 2\left(I_F + I_M\cos(\varphi + \psi_i)\right)$$
$$\varphi = \mathrm{arctg}\left[\frac{\sum_{i=1}^n \alpha_i I_i}{\sum_{i=1}^n \beta_i I_i}\right]; \varphi \in [0, 2\pi] \tag{5.25}$$

where α_i and β_i are coefficients particular to the n-image algorithm selected. I_F and I_M can also be computed using a similar formula. Phase shifts ψ_i are performed by changing the optical paths or refractive index. In practice, a piezoactuator-mounted mirror located in one arm of the interferometer is generally moved stepwise, each step being equal to a fraction of the wavelength; other means include phase modulators, such as nematic liquid crystals [SLA 07] or polarization plates.

Phase shifts can be achieved in spatial or temporal domains. In the temporal domain, the best-known algorithms rely on three 120° or four 90° phase-shifted specklegrams. The information redundancy of the 5-, 6-, 7-, ..., n-image algorithms make it possible to minimize several detrimental effects, such as the nonlinearities of the recording device and phase-modulator, the uncertainty of the phase shifts, and the sources of various noises [SUR 00]. The time taken to acquire the series of specklegrams and the necessity to keep the interferometric setup perfectly stable during this time are limiting factors in the acquisition of an adequate number of images.

Several possible approaches for data processing in time-resolved speckle interferometry are presented by [BOR 12]. Numerical processing concerns classical bi-dimensional processing as well as the one-dimensional temporal histories of individual pixels at acquisition rates of up to 26,000 fps. The least-squares approach is generalized, and different processing techniques are compared in terms of variance after phase unwrapping.

Spatial phase-shifting is feasible with the in-line reference setups, provided that a small angle exists between object and the directions of the reference beam. Parallel equidistant micro-fringes are thus created that must be resolved by the photodetector: their period must be greater than the width of several pixels. Large speckle grains are necessary order to ensure negligible phase variations of $\tilde{\varphi}$ over the fringe period. These conditions are highly restricting, requiring the imager to have a very small aperture. In compensation, the definite advantage is that we are able to extract the phase in one step.

The modulo 2π phase maps obtained by phase-shifting techniques exhibit a much lower noise level than the correlation fringes, and can be further processed to eliminate the 2π jumps.

5.5.3. *Advanced methods*

Interferometric signal processing and phase extraction are currently among the most active research areas in SI. The same is true for digital holography, from which some methods can be borrowed, [PIC 09]. In SI, the most significant developments are for dynamic processes in the subfield of processing temporal pixel signals. From these raw signals, zero-mean centered signals of the same phase can be built by means of simple and fast algorithms using empirical mode decomposition. Such pre-processed signals can then be efficiently processed by Hilbert transform to recover the phase [EQU 09b]. Another treatment, derived from the Hilbert transform and based on fitting circles with pixel signals in the complex plane, also appears to be very promising [EQU 11]. The fact that SI signals are intrinsically devoid of information in certain random places calls for non-standard interpolation procedures, such as Delaunay triangulation [EQU 10].

5.5.4. *Phase unwrapping*

Whatever the phase extraction method, the periodicity of the cosine function imposes a phase determination modulo 2π: while the phase sought is generally continuous, the computed phase maps display artificial discontinuities everywhere integer multiples of 2π are reached. One- or two-dimensional phase unwrapping methods aim to eliminate these artificial jumps, see Figure 5.8. They can be grouped into two main categories:

– *Filtering methods*, suppressing the fluctuations of the raw maps by as much as possible, yet always to the detriment of the spatial or temporal resolutions. For instance, wrapped phase maps are efficiently smoothed by linearly filtering the sine and cosine of these maps; a simple test identifying the neighboring pixels showing a phase jump of close to 2π solves the problem; adding or subtracting 2π at these points eliminates the discontinuity; the algorithm is quite simple and fast, but prone to the propagation of error.

– *Two-dimensional global methods*, based in general on the properties of phase maps; along any closed path encircling any point, the phase integral after unwrapping must be zero; a different result indicates an incorrect point, which is eliminated and the phase is replaced at these points by neighboring values. The algorithms are powerful, and preserve spatial resolution, though they require a much longer computation time. In both cases, a prerequisite is to be certain that the real phase variation is spatially resolved according to the Nyquist criterion.

Although simpler to process, one-dimensional unwrapping of temporal pixel signals from the dynamic processes must also distinguish the intrinsic 2π jumps from unexpected and sudden fluctuations. The first temporal unwrapping algorithm was published by Huntley [HUN 93].

Temporal algorithms provide the possibility of bypassing some of the difficulties encountered in spatial unwrapping. Errors remain attached to faulty pixels and do not spread. However the signal must be sampled at least twice in its shortest period in order to be safely processed. In SI such a sampling frequency is often high, requiring the recording of a great number of specklegrams. This is achieved by high-speed cameras linked to large memory-acquisition systems.

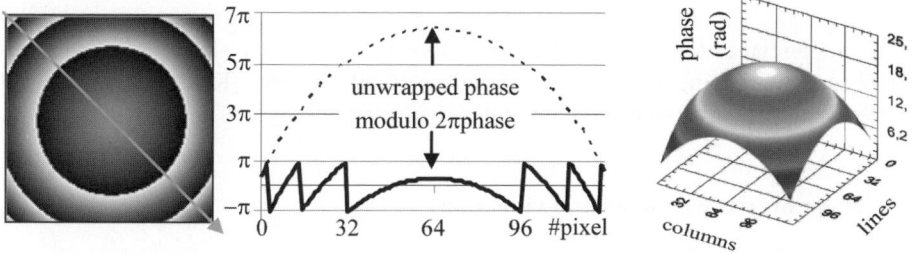

wrapped phase map   diagonal profile   unwrapped phase map

a) spatial unwrapping

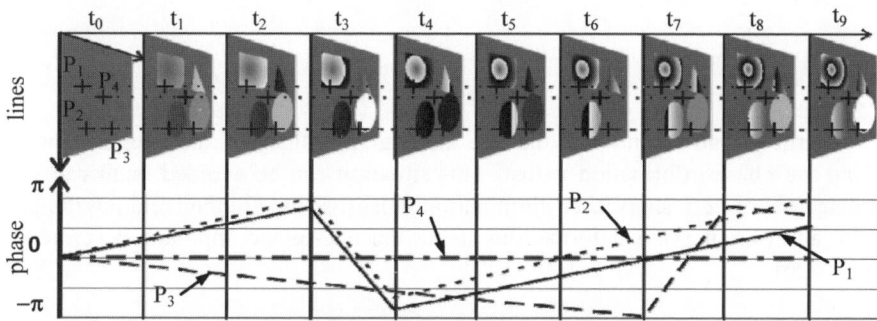

Computed temporal phase evolution

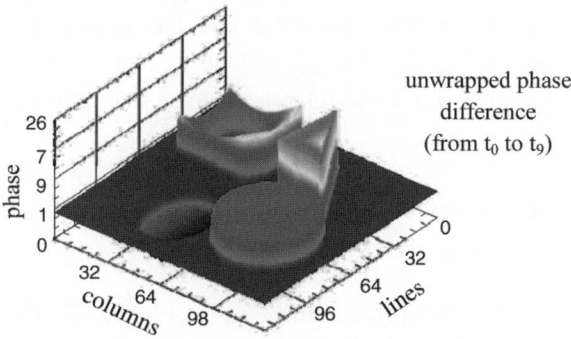

b) temporal unwrapping

Figure 5.8. *a) Spatial and b) temporal phase unwrapping. For a color version of this figure, see www.iste.co.uk/gh/solidmech.zip*

The problem of phase unwrapping disappears when the quantity sought is the spatial derivative of the phase maps $\partial\varphi/\partial\xi$, with ξ representing either x or y, since the calculation comes down to deriving the sine and cosine of the wrapped maps, thereby removing the discontinuities from these maps.

Commercial and free-access phase unwrapping software is available. A full analysis of phase unwrapping is given in references [GHI 98] and [KIM 02]. Figure 5.8 shows examples of spatial and temporal unwrapping.

5.6. Dynamic deformations and vibrations

In dynamic processes, the measurement acquired is the time-integrated intensity. Taking τ as the time constant of the photodetector or the duration of the laser pulse, the integrated intensity is given by:

$$I_\tau = \frac{2}{\tau}\int_0^\tau \left(I_F + I_M \cos\varphi(t)\right)dt \qquad [5.26]$$

For large variations in phase that are much greater than 2π, the integral tends to $2I_F$ and the phase information is lost. This situation can be avoided thanks to very fast detectors or very short laser illumination pulses. There are then real possibilities for the analysis of rapid deformations using the one-image approach described in section 5.5.1.

Vibrations represent another interesting case of particular temporal behavior. In response to a forced sinusoidal excitation of the object at frequency v_p, the temporal phase behavior is also sinusoidal and has amplitude V at the same frequency ([VIK 89, STE 99]). With φ_0 accounting for the difference in static random phase between the object and reference waves, and provided that $\tau \gg 1/v_p$, the intensity recorded becomes:

$$\begin{aligned}
&\varphi(t) = V\sin\left(2\pi v_p t\right) + \varphi_0 \\
&I_\tau = 2\left[I_F + I_M \cos\varphi_0 J_0(V)\right]; \tau \gg 1/v_p
\end{aligned} \qquad [5.27]$$

J_0 is the zero-order first kind Bessel function. V is related to the object vibration amplitude by relation [5.12]. It is worth noting that no synchronization is required between excitation and acquisition. This is called the "time average" technique and it provides vibration contours of equal amplitude. It has no limit with respect to high frequencies. The brightest fringes represent the nodal lines of the vibration, and fringe contrast decreases with fringe orders. Time average fringe patterns can be improved by the absolute value subtraction of a static reference state or by subtracting two consecutive frames, n and $n + 1$, while introducing an external π phase change in the interferometer:

$$\left| I_r - I_i \right| = 2I_M \cos\varphi_0 \left(1 - J_0 \left(V \right) \right)$$
$$\left| I_{n+1} - I_n \right| = 4 \left| I_M \cos\varphi_0 \right| \left| J_0 \left(V \right) \right|$$

[5.28]

An even better improvement is attained by applying the standard temporal phase-shifting techniques presented in section 5.5.3 [BOR 04]. By subtracting vibration and static phase maps, the following phase map is obtained:

$$\varphi = \beta\pi; \beta = 0 \text{ for } J_0 \left(V \right) \geq 0; \beta = 1 \text{ for } J_0 \left(V \right) < 0$$

[5.29]

This is a binary image, containing half as many fringes as the Bessel function, as can be seen in Figure 5.9.

The same technique eliminates the phase noise term $\cos\varphi_0$ perturbing signal $J_0(V)$ in the two orthogonal components of the time average specklegram. The two components are filtered before recombination, giving rise to fringe patterns with high spatial resolution [BOR 06]. Figure 5.9 shows a typical result.

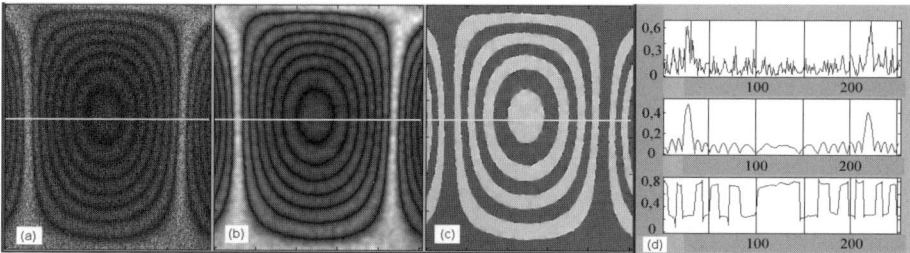

Figure 5.9. *a) Time average specklegrams; b) high spatial resolution; c) binary; and d) profiles along the central horizontal line*

Stroboscopic techniques are another way to cope with vibratory deformation processes [DOV 00]. The vibration is frozen by a series of short acquisitions that are equidistant in time, see Figure 5.10. By adjusting the delay between the excitation and shutter signals, it is possible to decompose large object vibrations into smaller intervals that are compatible with the limited dynamic range of SI.

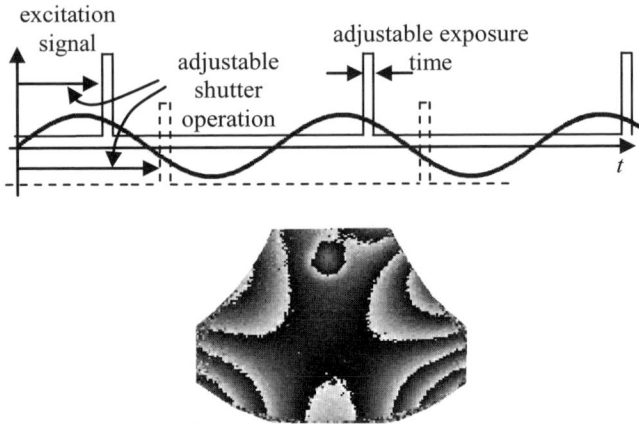

Figure 5.10. *Stroboscopic recording and example of a standard phase-shifting result: vibration of a fan blade*

5.7. Setup calibration

Experimental mechanics applications require knowledge of the displacement field components in a definite Cartesian coordinate system, while phase maps obtained in SI are generally related to linear combinations of these components. Therefore these maps should be reduced to pure single-component maps, see Figure 5.11 which illustrates such a reduction to a pure (u_x, u_y, u_z) sensitivity in the case of deformations undergone by a stacked piezotransducer [BOR 08].

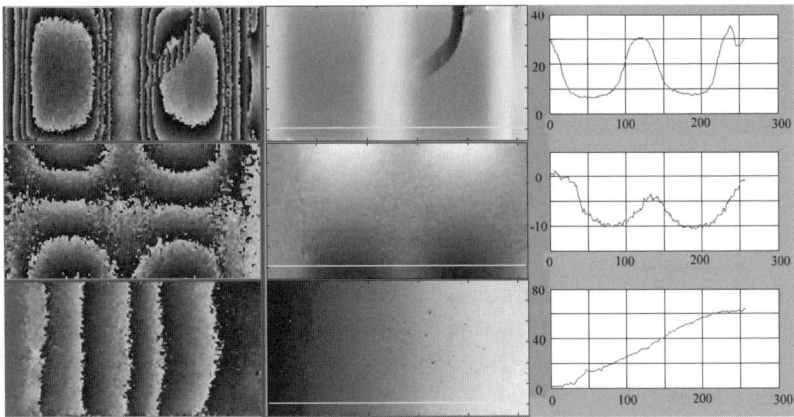

Figure 5.11. *Separation of the phase maps representing u_z (top), u_y (middle) and u_x (bottom). For each of the three components, and from left to right we have the modulo 2π phase map, unwrapped phase map, and phase profiles along the selected horizontal lines*

Digital images are subjected to a variety of processing techniques, including arithmetic operations and nonlinear transformations such as inverse tangent or even complex treatments like filtering or unwrapping. All of these processes start in the $(X, Y, 0)$ pixel rectangular coordinate system of the digital images. In the end, it is more convenient to express the displacement maps in a Cartesian coordinate system that is defined with respect to the object surface for each pure component. Such transformations, or mappings, depend on the object and geometry of the setup, the sensitivity vector and the imager characteristics.

5.7.1. *Specifying the material point in object coordinates*

In order to express the results in the object's Cartesian coordinate system (x, y, z), possibly as column vectors $[x, \ y, \ z, \ u_x, \ u_y, \ u_z]$, two main factors must be taken into account: the perspective projection related to the object–image geometry and the distortions suffered by the image due to aberrations in the imaging system. The perspective projection is relatively simple, since it is based on known analytical bi-linear models ([KRE 05, BOU 99]). To establish the transformation from the two-dimensional camera coordinates to the three-dimensional object coordinate system, at least two independent perspective views of the object are necessary. With more views, least-square methods and nonlinear optimization procedures are exploited. Lens distortions are rarely taken into account in SI, since they only slightly affect the positions of the points but not their displacement. In the context of coordinate mappings, some research works propose to affix the displacement components to the nodes of finite element meshes [MOR 08].

5.7.2. *Determination of the sensitivity vector*

The setup geometry, as well as the position and shape of the object, must be known in order to determine the sensitivity vector at each point of the object surface. As a matter of good practice, this requires the application of a series of qualifying procedures to the entire interferometric system, including the imager itself, the optical setup and the phase modulator. The main procedures are described in [GOU 05].

At each object point, and for each source of illumination and observation point in the setup, the three components of the unit vector from the directions of illumination and observation must be determined. The coordinates of the source points and the projection centers must thus be known. To do this, any standard method of three-dimensional metrology can be used. Conversely, the SI method itself can contribute to solving the problem. By subjecting the object to a series of known rigid body micro-displacements (u_{xi}, u_{yi}, u_{zi}) and by measuring the corresponding phase φ_{mi}, a set of linear equations is obtained:

$$
\begin{bmatrix} \cdots \\ \cdots \\ \varphi_{mi} \\ \cdots \end{bmatrix} = \frac{2\pi}{\lambda} \begin{bmatrix} \cdots & \cdots & \cdots \\ \cdots & & \\ u_{xi} & u_{yi} & u_{zi} \\ \cdots & & \end{bmatrix} \begin{bmatrix} S_x \\ S_y \\ S_z \end{bmatrix}
$$ [5.30]

The system is solved by standard least-square techniques, yielding the three components of the sensitivity vector (S_x, S_y, S_z) for each object point.

5.8. Specifications and limits

It can be difficult to globally compare two SI systems, insofar as they are often devoted to quite different applications. The specifications of systems measuring the in-plane, out-of-plane or derivative components are strongly connected to the characteristics of the principal constituents of these systems: laser, camera, optical elements and software. For instance, the dimensions of the object surface and the spatial resolution can be clarified from knowledge of the nature of the imaging lens (aperture, field of view, etc.), laser specifications (type, polarization, energy/power, coherence length, etc.) and camera properties (number of pixels, acquisition rate, etc.). The resolution of the phase measurement is highly dependent on the choice of particular algorithms and processing software. A rather large discrepancy can be observed in the specifications given by the suppliers of SI systems.

One possible definition of the signal-to-noise ratio is the ratio between $2I_M$, the modulation of the specklegram and σ_I, the standard deviation of the fluctuations in intensity:

$$
R_{SB} = \langle 2I_M \rangle / \sigma_I
$$ [5.31]

where the average modulation of all pixels is considered.

For continuous wave lasers, the temporal resolution is determined by the acquisition rate of the camera and choice of the number of images for the phase-shifting algorithm. Common acquisition rates are 25–50 images/s. In the case of temporal phase-shifting, current processors enable modulo 2π phase maps to be displayed at standard video rates [STE 99].

The reference [MOO 99] describes a particular system where the acquisition rate is increased to 4,500 images/s for an image format of 256 × 256 pixels. The phase-shifted correlation fringes are displayed afterwards. For pulsed laser systems, the temporal resolution is given by the minimum time taken to digitize a full frame of the camera, presently in the order of 0.1 μs.

Spatial resolution is essentially governed by the number of pixels and size of the imager, the optical performance and magnification of the imaging lens, and the speckle size. For some suppliers, spatial resolution is only characterized by the quantity of pixel and the dimensions of the field of view. As a rule, it is generally widely accepted that the spatial fringe density is limited by the necessity to count at least six to eight pixels per fringe period. A smaller number would prevent these fringes from being correctly resolved and would make either fringe order numbering or phase unwrapping problematic.

Uncertainties in phase measurements are related to multiple causes: optoelectronic noise in the interferometric system, bit-limited digitization, mechanical perturbations and air turbulence in the optical setup, speckle phase singularities, and above all decorrelations. When the latter are under control, displacement resolutions associated with phase resolutions can reach a value as small as one tenth of a micron [LEH 01]. Phase resolution is generally better in the case of temporal unwrapping than in the case of spatial unwrapping.

5.9. Final remarks, outlook and trends

Figure 5.12 summarizes the multiple measurement approaches that can be implemented using speckle fields. In this chapter, these measurement possibilities are sometimes presented superficially, sometimes in greater depth and with emphasis on SI. It must be recognized that the rapid growth and the favorable impact on experimental mechanics of SI techniques is largely due to the progress of CCD and CMOS photoelectric imagers. Moreover, ultrafast cameras (10 kHz, 1 megapixel) can now be used for time-resolved measurements in SI.

SI techniques continue to flourish in different directions: low-coherence processes, new spectral domains in the infrared or ultraviolet regions and even in the visible part of the spectrum with multiple wavelength sources for multiplexing purposes. An atypical choice of SI for studying the temporal behavior of phase objects has recently been revisited [SLA 10].

Following the mainstream point of view, the phase singularities here have been considered to be noise. New approaches, however, consider that singularities can be exploited in a metrology context [WAN 06]. In the same way that speckle fields themselves were previously sometimes perceived as noise and sometimes as signals or information carriers, it could be that enhancing the properties of singularities will open the door to new developments in SI.

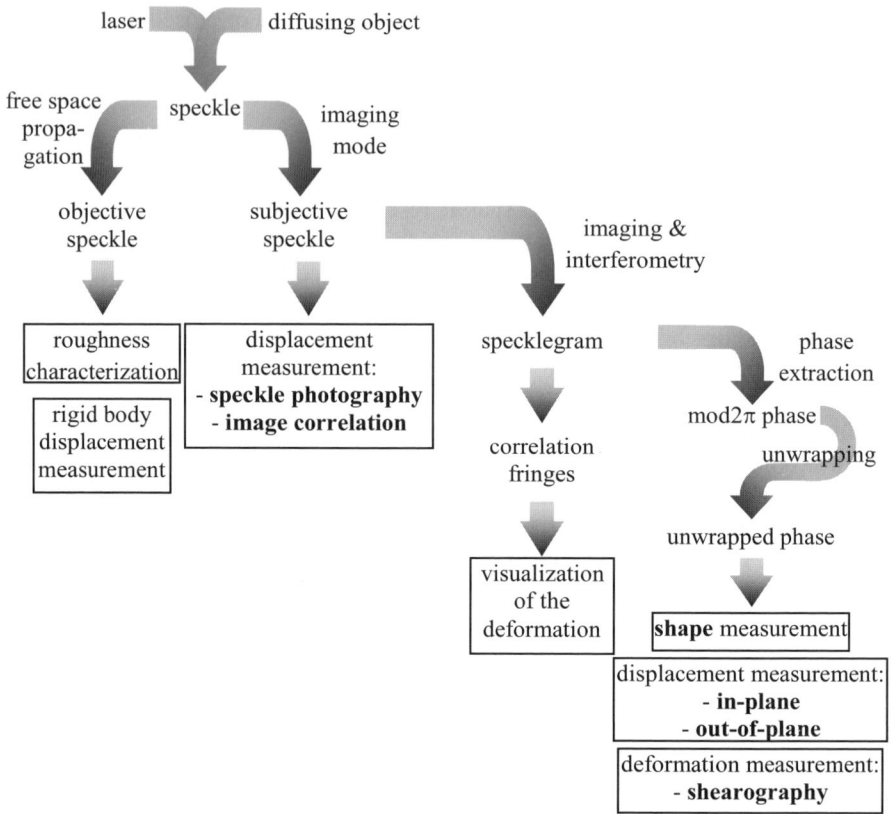

Figure 5.12. *Summary of the measurement possibilities offered by speckle fields*

Finally, recorrelation techniques are continuously progressing. Excessively large speckle grain displacements, preventing the formation of contrasted fringe patterns, can be measured and compensated for. These large speckle displacements can be obtained, for example, by applying image correlation techniques to the random phase maps of the two states being compared. Once correctly re-superimposed, the subtraction of the two phase maps gives the usual displacement phase map [HIN 09].

It has recently been demonstrated that the complex amplitude of a speckle field can be reconstructed from single wave intensity recordings [ALM 09]; this again could bring noteworthy simplification and extension to SI.

5.10. Bibliography

[ALB 06] ALBERTAZZI A.G. JR., "Radial metrology with electronic speckle pattern interferometry", *Journal of Holography and Speckle*, vol. 3, pp. 117–124, 2006.

[ALM 09] ALMORO P.F., MAALLO A.M.S., HANSON S.G., "Fast-convergent algorithm for speckle-based phase retrieval and a design for dynamic wavefront sensing", *Appl. Opt.*, vol. 48, pp. 1485–1493, 2009.

[ARC 69] ARCHBOLD E., BURCH J.M., ENNOS A.E., TAYLOR P.A., "Visual observation of surface vibration nodal patterns", *Nature*, vol. 222, pp. 263–265, 1969.

[BOR 04] BORZA D.N., "High-resolution time-average electronic holography for vibration measurement", *Opt. Lasers Eng.*, vol. 41, pp. 415–527, 2004.

[BOR 06] BORZA D.N., "Full-field vibration amplitude recovery from high-resolution time-averaged speckle interferograms and digital holograms by regional inverting of the Bessel function", *Opt. Lasers Eng.*, vol. 44, pp. 747–770, 2006.

[BOR 08] BORZA D.N., LEMOSSE D., PAGNACCO E., "Full-field experimental–numerical study of mechanical static strain and stress in piezoelectric multilayer compression-type actuators", *Comp. Struct.*, vol. 82, pp. 36–49, 2008.

[BOR 12] BORZA D.N., NISTEA I., "High temporal and spatial resolution in time resolved speckle interferometry", *Opt. Lasers Eng.*, vol. 50, 2012.

[BOU 99] BOUGUET J.-Y., Visual methods for three-dimensional modeling, Thesis, California Institute of Technology, 1999.

[BRU 07] BRUNO L., POGGIALINI A., "A novel operating principle in speckle interferometry: the double-focusing", *Opt. Express*, vol. 15, pp. 8787–8796, 2007.

[BUT 71] BUTTERS J.N., LEENDERTZ J.A., "Speckle pattern and holographic techniques in engineering metrology", *Opt. Laser Technol.*, vol. 3, pp. 26–30, 1971.

[COL 97] COLONNA DE LEGA X., Processing of non-stationary interference patterns: adapted phase-shifting algorithms and wavelet analysis. Application to dynamic deformation measurements by holographic and speckle interferometry, EPFL thesis no. 1666, Lausanne, 1997.

[CRE 88] CREATH K., "Phase-measurement interferometry techniques", in: *Progress in Optics XXVI*, Elsevier, pp. 349–393, 1988.

[DOV 00] DOVAL A.F., "A systematic approach to TV holography", *Meas. Sci. Tech.*, vol. 11, pp. R1–R36, 2000.

[EQU 06] EQUIS S., JACQUOT P., "Simulation of speckle complex amplitude: advocating the linear model", *Speckle06: From grains to flowers*, SPIE vol. 6341, 634138-1-6, 2006.

[EQU 09a] EQUIS S., Phase extraction of non-stationary signals produced in dynamic interferometry involving speckle waves, EPFL thesis no. 4514, Lausanne, 2009.

[EQU 09b] EQUIS S., JACQUOT P., "The empirical mode decomposition: a must-have tool in speckle interferometry?", *Opt. Express*, vol. 17, pp. 611–623, 2009.

[EQU 10] EQUIS S., JACQUOT P., "Coping with low modulation in speckle interferometry: a novel approach based on the Delaunay triangulation", *Speckle 2010: Optical Metrology*, SPIE, vol. 7387, 738709-1-10, 2010.

[EQU 11] EQUIS S., FLANDRIN P., JACQUOT P., "Phase extraction in speckle interferometry by a circle fitting procedure in the complex plane", *Optics Letters*, vol. 36, pp. 4617–4619, 2011.

[GAU 05] GAUTIER B., Etudes et réalisation d'un interféromètre de speckle à mesure de forme intégrée, Thesis, Ecole des Mines de Paris, Alès, 2005.

[GHI 98] GHIGLIA D.C., PRITT M.D., *Two-dimensional Phase Unwrapping: Theory, Algorithms, and Software*, John Wiley & Sons, New York, NY, 1998.

[GOO 07] GOODMAN J.W., *Speckle Phenomena in Optics,* Roberts & Company Publishers, 2007.

[GOU 05] GOUDEMAND N., 3D-3C speckle interferometry: optical device for measuring complex structures, ETHZ thesis no. 15961, Swiss Federal Institute of Technology, Zürich, 2005.

[HEC 02] HECHT E., *Optics*, 4th edition, San Francisco, Addison Wesley, 2002.

[HIN 09] HINSCH K.D., ZEHNDER K., JOOST H., GÜLKER G., "Monitoring detaching murals in the Convent of Müstair (Switzerland) by optical metrology", *Journal of Cultural Heritage*, vol. 10, pp. 94–105, 2009.

[HUN 93] HUNTLEY J.M., SALDNER H., "Temporal phase unwrapping algorithm for automated interferogram analysis", *Appl. Opt.*, vol. 32, pp. 3047–3052, 1993.

[JAC 08] JACQUOT P., "Speckle interferometry: A review of the principal methods in use for experimental mechanics applications", *Strain*, vol. 44, pp. 57–69, 2008.

[JON 89] JONES R., WYKES C., *Holographic and Speckle Interferometry,* Cambridge University Press, 1989.

[KIM 02] KIM S., KIM Y.S., "Two-dimensional phase unwrapping using wavelet transform", *Electron. Lett.*, vol. 38, pp. 19–20, 2002.

[KRE 05] KREIS T., *Handbook of Holographic Interferometry, Optical and Digital Methods*, Weinheim, Wiley-VCH Verlag & Co., 2005.

[LEE 70] LEENDERTZ J.A., "Interferometric displacement measurement on scattering surfaces utilizing speckle effect", *J. Phys. E: Sci. Instrum*, vol. 3, pp. 214–218, 1970.

[LEH 01] LEHMANN M., "Speckle statistics in the context of digital speckle interferometry", in *Digital Speckle Pattern Interferometry and Related Techniques*, John Wiley & Sons, pp. 1–58, 2001.

[MEI 96] MEINLSCHMIDT P., HINSCH K.D., SIROHI R.S. (Eds), *Selected Papers on Electronic Speckle Pattern Interferometry: Principles and Practice*, SPIE Milestone Series Vol. MS 132, 1996.

[MOO 99] MOORE A., HAND D., BARTON J., JONES J., "Transient deformation measurement with electronic speckle pattern interferometry and a high-speed camera", *Appl. Opt.*, vol. 38, pp. 1159–1162, 1999.

[MOR 08] MOREAU A., BORZA D.N., NISTEA I., "Full-field vibration measurement by time-average speckle interferometry and by Doppler vibrometry – a comparison", *Strain*, vol. 44, pp. 386–397, 2008.

[OST 96] OSTEN W., JÜPTNER W., "Measurement of displacement vector fields of extended objects", *Opt. Lasers Eng.*, vol. 24, pp. 261–285, 1996.

[PIC 09] PICART P., TANKAM P., MOUNIER D., PENG Z., LI J.C., "Spatial bandwidth extended reconstruction for digital color Fresnel holograms", *Opt. Express*, vol. 17, pp. 9145–9156, 2009.

[SCH 79] SCHUMANN W., DUBAS M., *Holographic Interferometry*, Springer Verlag, 1979.

[SIE 04] SIEBERT T., SPLITTHOF K., ETTEMEYER A., "A practical approach to the problem of the absolute phase in speckle interferometry", *Journal of Holography and Speckle*, vol. 1, pp. 32–38, 2004.

[SIR 99] SIROHI R.S., CHAU F.S., *Optical Methods of Measurement Wholefield Techniques*, Marcel Dekker, Inc., 1999.

[SLA 07] SLANGEN P., "Phase shifting speckle interferometry with nematic liquid crystals light valve", *SPIE*, vol. 7008, 7008-OW-32, 2007.

[SLA 10] SLANGEN P., APRIN L., HEYMES F., EQUIS S., JACQUOT P., "Liquid blending: an investigation using dynamic speckle interferometry", *Speckle 2010: Optical Metrology*, SPIE, vol. 7387, 738719-1-8, 2010.

[STE 03] STEINCHEN W., YANG L., *Digital Shearography: Theory and Applications of Digital Speckle Pattern Shearing Interferometry*, SPIE Press, Bellingham, WA, 2003.

[STE 99] STETSON K.A., "1999 William M. Murray Lecture, The problems of holographic interferometry", *Exp. Mech.*, vol. 39, pp. 249–255, 1999.

[SUR 00] SURREL Y., "Fringe analysis, photomechanics, topics" *Appl. Phys.*, vol. 77, pp. 55–102, 2000.

[TAK 82] TAKEDA M., INA H., KOBAYASHI S., "Fourier-transform method of fringe-pattern analysis for computer-based topography and interferometry", *J. Opt. Soc. Am.*, vol. 72, pp. 156–160, 1982.

[VIK 89] VIKHAGEN E., "Vibration measurement using phase shifting TV-holography and digital image processing", *Opt. Commun.*, vol. 69, pp. 214–218, 1989.

[WAN 06] WANG W., YOKOZEKI T., ISHIJIMA R., WADA A., MIYAMOTO Y., TAKEDA M., HANSON S.G., "Optical vortex metrology for nanometric speckle displacement measurement", *Opt. Express*, vol. 14, pp. 120–127, 2006.

Chapter 6

Digital Image Correlation

6.1. Background

Digital image correlation (DIC), which appeared in the early 1980s [LUC 81, BUR 82, SUT 83, SUT 86], has had a major impact in the field of mechanics of solids and structures. Today, it is still undergoing very spectacular developments.

The challenge is to *measure* the displacement *fields* of surfaces (or in volumes) of stressed specimens and structures from images acquired at different stages of loading. A specific advantage of this tool is that it exploits numerical images that are usually acquired by optical means. Imaging devices have made significant progress, in terms of not only quality and definition, but also (lower) cost. These imaging means are inherently contactless, non-intrusive, tolerant to aggressive conditions (e.g. temperature and chemical environment), easy to use, efficient and cheap, many of which features cannot fail to be appealing in the context of mechanical tests.

DIC can easily be used at different scales of space and time to the extent that it relies on principles applicable to pictures obtained by very different imaging systems. It is nowadays possible to use images shot by fast and ultra-fast cameras at time scales down to the microsecond or less [SCH 03a, SCH 03b, SIE 07, TIW 07, BES 08, BES 10], acquired by a scanning electron microscope (SEM) [DOU 00a, DOU 00b, SOP 01, DOU 03, TAT 03, SUT 06, SUT 07a] or an atomic force microscope (AFM) [CHA 02, CHO 05, CHO 07a, CHO 07b, HAN 10] at nanometric scales, but also satellite images at geophysical scales [SCA 92, LEP 07]. Multicamera systems give access to three-dimensional (3D) shapes and displacement fields of the surfaces of an observed object (see section 6.4). 3D images obtained by computed

---

Chapter written by Michel Bornert, François Hild, Jean-José Orteu and Stéphane Roux.

(micro)tomography [BAY 99, BOR 04, LEN 07, BAY 08, RAN 10] or magnetic resonance imaging [NEU 08, BEN 09] can also be used to measure 3D displacement fields in the bulk of various (optically opaque) materials.

The wealth of kinematic data obtained allows not only for a quantitative exploitation through identification techniques (see the following chapters), but also for the validation of this identification, to enrich or degrade it according to the needs. By specifying appropriate forms of the displacement fields to be measured, image correlation can directly address this identification step (see section 6.2.8).

6.2. Surface and volume digital image correlation

DIC techniques can be applied indifferently to classical 2D or volumetric images to which recent imaging tools give access. As a consequence, the term "pixel" will refer in the following to the elementary discrete datum of a digital image defined in a 2D or 3D space. We detail in this section the algorithms providing evaluations of the apparent mechanical transformation, Φ_a, that link two images of the same mechanical system under two different configurations.

After a first section devoted to the guiding principles and ingredients common to all the presented algorithms, the discussion then will focus on the so-called "local" image correlation techniques, which evaluate the Φ_a transformation piecewise, through a large number of independent analyses on subimages, called "correlation windows" or "domains". Several variants of the more recent "global" approaches, which may be enriched by some *a priori* mechanical information suited to each problem dealt with, will subsequently be described. The main sources of uncertainties and their quantification will be discussed in section 6.3.

6.2.1. *Images*

The input data of the analysis are positive integers, called *gray levels*, of the image of the first configuration, known as the "reference image", denoted f_I, and those of the second configuration, called the "deformed image", denoted g_I, where the subscript I refers to a pair of integers (column, line) in the case of 2D images or to a triplet (column, line, plane) in the case of 3D images. These integers vary between 0 and some maximum values related to the *digitization* or *encoding depth* (or *dynamic range*) of the image sensor (e.g. 256 levels for 8-bit images and 4,096 for 12-bit images). Indices I vary in planar or volumetric domains, whose extension characterizes the *image definition* (e.g. $[0; 1,023] \times [0; 1,023]$ for a one-megapixel 2D image), and in which it is possible to define continuous positions, with real-valued coordinates x^1. The two images can be extracted from a temporal sequence

---

1 In the following, $x(I)$ is the point with integer coordinates I.

comprising a large number of images, of which any image can be selected as the reference image or the deformed image. However, we do not address in the following the algorithms that process such a sequence globally.

Images f_I and g_I result from a complex acquisition chain, which is today ever more various. It is essential to consider this acquisition process in the final interpretation of the estimated transformation, and to take into account the presence of biases that could result from it. In particular, the analysis of this chain would make it possible to specify the link between some physical quantity, continuously varying in space, denoted by $\tilde{f}$, and the discrete gray level f_I that this quantity induces through the imaging system at position I in the image. Because of space limitations, this aspect will not be described here.

DIC principles are based on the essential assumption that the physical quantity that leads to the image is associated with some physical property of the matter that constitutes the analyzed system, and, as a consequence, is transported by the apparent sought mechanical transformation Φ_a. Denoting by $\tilde{g}$ the continuous physical quantity associated with the deformed image, this assumption is written as

$$\tilde{g}(\Phi_a(\boldsymbol{x})) = \tilde{f}(\boldsymbol{x}) \tag{6.1}$$

6.2.2. Texture of images

As will be made more specific later on, DIC exploits a texture that must be the signature of each surface (or volume) element, simply transported, according to equation [6.1], by the displacement field, without any other deterioration. Moreover, the gray levels representing this texture in the images need to exhibit a broad dynamic range that covers as much as possible the available encoding depth of the images, that is from 8 to 16 bits, without however exhibiting saturation. Finally, it is desirable to have strong contrasts from one pixel to the next in order to be sensitive to small displacement amplitudes.

The simplicity of implementation of the method would suggest that we might rely on the natural texture of the studied material. This is sometimes possible, as shown in Figure 6.1. However, the correlation function of the texture will dictate the performances of the analysis. Thus, for the example of steel shown in Figure 6.1(a), the various phases revealed by selective chemical etching have an orientation that will allow a good estimate of the displacement along the horizontal direction but a worse one in the perpendicular direction. In this context, the surface topography of silica observed by AFM (Figure 6.1(c)) does not lead to a strong sensitivity. Conversely, the phase image of the same zone (Figure 6.1(d)) can lead to a much better resolution. The selection of an appropriate mode of observation, associated when required with an adequate surface preparation, is thus an essential element for the use of natural contrast in DIC.

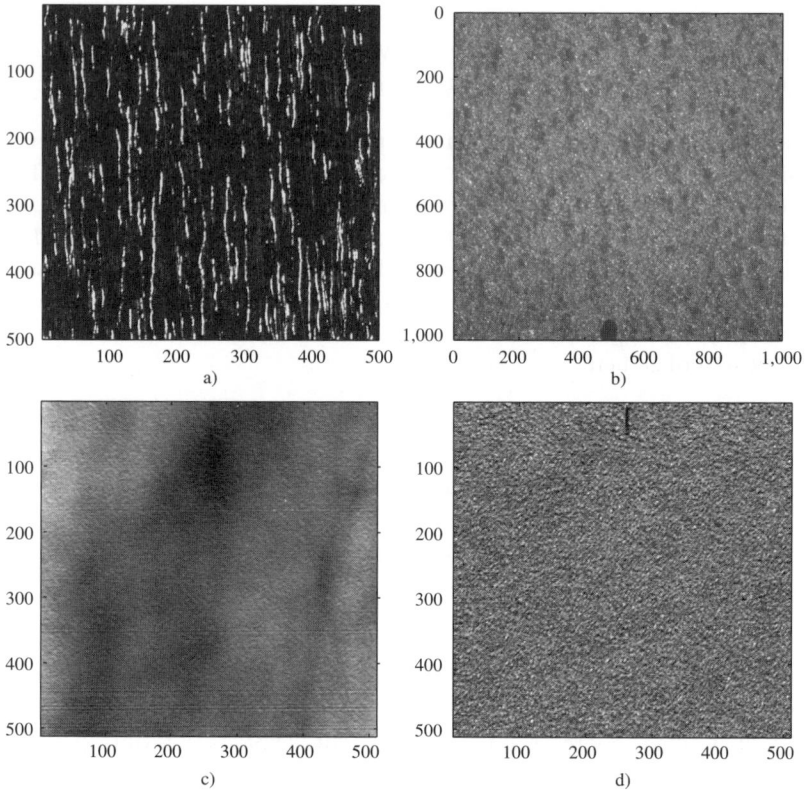

Figure 6.1. *Examples of natural image textures suitable for digital image correlation. a) 304 L stainless steel (1 pix. = 3.2 μm); b) silicon carbide (1 pix. = 1.85 μm); c) topographic AFM image of silica glass (1 pix. = 2 nm); d) phase AFM image (same area as in c)*

The existence of such a contrast is not guaranteed. A possible way to circumvent this difficulty is to deposit an artificial texture on the surface of the studied samples, as shown in Figure 6.2. The pulverization of fine droplets of black paint on a white background (or conversely) is a process called "speckle painting" that can be used on many materials. An airbrush enables the formation of an aerosol whose droplet size can be adjusted by an adapted nozzle. For the largest scales (e.g. for civil engineering structures), marking can be done with a stencil set. A limitation can emerge from the strong deformability of materials, as in the example of Figure 6.2(c), which is an elastomer. It is difficult for a paint deposit to follow deformations of large amplitude. In this case, a powder deposit (talc in this example) allows for a good texture stability of up to several hundred percent of strain.

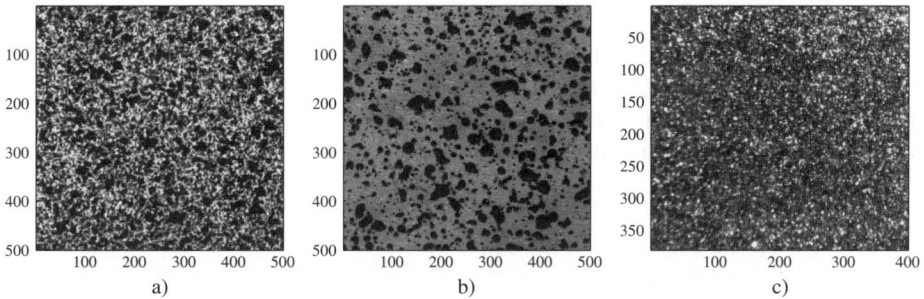

Figure 6.2. *Examples of artificial image textures suitable for digital image correlation*

The deposition of particles on the surface can also be used in the context of SEM imaging [SCR 07]. Electron lithography techniques, in particular, in their simplified version, which can be implemented without expensive equipment other than the SEM itself [ALL 94], make it possible to mark surfaces by means of microgrids with a step varying from a few tens of micrometers to less than 1 μm, by metal deposition or chemical engraving. The periodicity of the obtained local contrast requires, however, some minor adaptations of the correlation algorithms to avoid offsets of one or several grid steps in the evaluation of the displacement field. An advantage of this type of marking is that it does not hide the microstructure and thus facilitates the micromechanical analysis of heterogeneous materials [DOU 00a, HER 07]. However, the analysis cannot be as dense as with fine speckle painting.

Finally, when 3D images of computed tomography are to be exploited, texture in the bulk of the studied material must be relied on (Figure 6.3). Some materials are thus privileged, such as most rocks [LEN 07], stone wool [HIL 09b], nodular graphite cast iron [RAN 10] and foams [ROU 08], whose micropores, inclusions or heterogeneous microstructures behave like speckle paint. In the absence of such a contrast, or within the framework of an analysis at a finer scale, it is sometimes possible to reinforce it artificially (e.g. by the diffusion of a heavy element to mark grain boundaries, or by the addition of fine markers during the processing of the material [BOR 04]), but the difficulty then is to make sure that this marking does not introduce any modification of the mechanical properties, and to take into account the modified microstructure of the material in the interpretation of the results.

6.2.3. *Guiding principles*

The aim of the exposed methods is to evaluate the apparent transformation Φ_a on a region of interest (ROI) R of the reference image starting from the knowledge of the gray levels f_I and g_I. This is essentially an *ill-posed* problem insofar as the available information (i.e. the gray levels of the pixels) is insufficient to uniquely determine a vectorial displacement at each pixel. It is thus imperative to regularize the problem by restricting this determination to a particular family of transformations $\Phi_0(\alpha, \cdot)$, parameterized by N scalars, α, written here in vector form. Image correlation methods can then be given the following general formulation

$$\alpha_{\min} = \underset{\alpha \in V_\alpha}{\text{Argmin}} \, C(\Phi_0(\alpha, \cdot), R, [f], [g]) \qquad [6.2]$$

where the *correlation coefficient* C is a scalar that measures the similarity[2] between the ROI R of the image $[f]$ (i.e. the set of pixels f_I) transformed by $\Phi_0(\alpha, \cdot)$ and the corresponding region of the image $[g]$. The parameters α belong to a certain domain V_α of $\mathbb{R}^N$. The numerical resolution of this minimization problem leads to an evaluation $\Phi_0(\alpha_{\min}, \cdot)$ of the sought transformation $\Phi_a(\cdot)$.

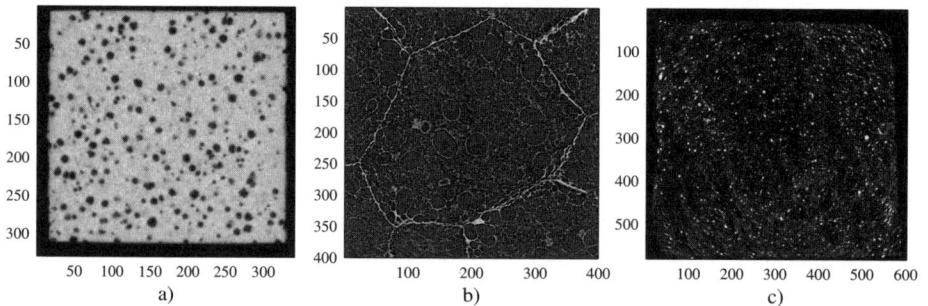

Figure 6.3. *Examples of sections through 3D x-ray computed tomography images: a) cast iron with spheroidal graphite nodules; b) polymeric foam; and c) stonewool*

The various available image correlation algorithms derive from specific choices of the expressions of the correlation coefficient C (section 6.2.4), of the considered family of transformations and its parameter settings, which distinguish in particular the local methods (section 6.2.6) from the global methods (section 6.2.8), as well as of the minimization algorithms in use. Moreover, the evaluation of a continuous transformation from the knowledge of discrete values of gray levels provided by the input images requires recourse to an interpolation method at some stage of the minimization problem. The various ways to do so define additional alternatives when specifying image correlation algorithms (section 6.2.5). These interpolation algorithms make it possible to evaluate the local displacement field with a resolution that is notably lower than one pixel, known as *subpixel* accuracy. For the sake of simplicity, the discrete nature of the input data will be ignored in the following

---

2 By convention, we consider here that the similarity is larger when C is smaller, although the reverse could be adopted (by changing the sign of C, for example).

presentation, except in section 6.2.5 where this question is explicitly addressed. This simplification consists of assuming the pixels to be infinitely small with respect to the spatial gray level fluctuations, so that discrete gray levels f_I and g_I can be represented by continuous functions $f(\boldsymbol{x})$ and $g(\boldsymbol{x})$.

6.2.4. *Correlation coefficients*

The simplest and most classic similarity criterion is the sum of squared differences between the images[3]

$$C_1 = \int_R [f(\boldsymbol{x}) - g(\Phi_0(\boldsymbol{x}))]^2 \ \mathrm{d}\boldsymbol{x} = \int_R [f(\boldsymbol{x}) - g(\boldsymbol{x} + \boldsymbol{u}_0(\boldsymbol{x}))]^2 \ \mathrm{d}\boldsymbol{x} \qquad [6.3]$$

This quantity is clearly positive. It disappears when the transformation Φ_0 coincides with the apparent transformation Φ_a and when the conservation condition [6.1] is exactly satisfied, not only for the physical quantities $\tilde{f}$ and $\tilde{g}$ at the origin of the gray levels in the images, but also for the continuous gray levels f and g. The choice of a squared difference induces some operational advantages but may be replaced by an absolute value or any other measure of discrepancy.

Some imaging systems (e.g. SEM) do not make it possible to guarantee the stability of the conversion into gray levels f_I of the physical quantity $\tilde{f}$ at the origin of the image contrast and convected by motion according to equation [6.1]. We can then choose a less demanding similarity measure such as

$$C_2 = \min_{a,b} \int_R [f(\boldsymbol{x}) - (a \ g(\Phi_0(\boldsymbol{x})) + b)]^2 \ \mathrm{d}\boldsymbol{x} \qquad [6.4]$$

which consists of seeking the best linear regression between the gray levels of the two paired images. The optimized coefficients a and b can be interpreted as changes in contrast and brightness, respectively, of the image between the two configurations[4]. Optimization in equation [6.4] is straightforward and leads to $a = \int_R (f - \bar{f})(g - \bar{g}) \ \mathrm{d}\boldsymbol{x} / \int_R (g - \bar{g})^2 \ \mathrm{d}\boldsymbol{x}$ and $b = \bar{f} - a\bar{g}$, where $\bar{f}$ and $\bar{g}$ are the averages of f and $g \circ \Phi_0$ over R. As $\int_R (f - \bar{f})^2 \ \mathrm{d}\boldsymbol{x}$ does not depend on Φ_0, optimizing C_2 is then equivalent to minimizing

$$C_3 = 1 - \frac{\left| \int_R (f - \bar{f})(g - \bar{g}) \ \mathrm{d}\boldsymbol{x} \right|}{\sqrt{\int_R (g - \bar{g})^2 \ \mathrm{d}\boldsymbol{x} \int_R (f - \bar{f})^2 \ \mathrm{d}\boldsymbol{x}}} \qquad [6.5]$$

---

3 For ease of notation, the various arguments of C are omitted in the following definitions.

4 If these changes can be quantified independently, it is possible not to carry out optimization, and to apply the criterion C_1 to the image f and the rescaled image $a \ g + b$.

which can be interpreted as the correlation coefficient of a linear regression and varies from 0 (perfect similarity) to 1 (no link between images)[5].

The definition of these criteria leads immediately to the definition of the image of the residuals, whose gray levels defined on R are given by $f(x) - g(\Phi_0(x))$ or $f(x) - (a\,g(\Phi_0(x)) + b)$, depending on which correlation criterion is considered. The qualitative or quantitative analysis of these residuals is a means of evaluating the relevance of the optimal transformation with respect to the actual unknown transformation. If the registration were perfect, the residuals would contain nothing but image noise. It is worth noting that, in this case and under the assumption of a moderate noise level, the criterion C_3 is equal at its optimum to $(1/2)(\sigma(f')/\sigma(f))^2$, where $\sigma(f)$ and $\sigma(f')$ are the standard deviations of the noiseless image f and its noise f' on R. With an optimal contrast occupying the full dynamic range of the gray levels and a noise level of about 1% of the latter, as is often observed with images from conventional optical cameras, the minimum of the C_3 coefficient might become as small as 10^{-3}. This concept of residuals will be discussed again in section 6.2.8.

The above list of criteria is not exhaustive [CHA 11] but gathers those most commonly used for quantitative analyses in mechanics. Other similarity measures might be used within other image registration applications (e.g. satellite or medical imaging), in particular when the condition of convection of the gray levels is not well satisfied, because, for instance, of high noise levels [BOR 04] or of too strong a change in image contrast between the various configurations.

6.2.5. Subpixel interpolation

All DIC algorithms need to specify the interpolation method required to access a resolution in displacement of notably less than one pixel. We should be aware of the fact that the associated technical choice might have considerable consequences on measurement accuracy.

The computation of the correlation coefficient, whatever the formulation retained among those presented in section 6.2.4, needs to address the discrete nature of the images. The integration in equations [6.3]–[6.5] is, in practice, replaced by a summation over a set of pixels in the reference image and requires that we specify how the gray level in the deformed image at point $\Phi_0(x(I))$ is calculated. This point has, in general, non-integer coordinates and its gray level can only be evaluated by means of some interpolation scheme making use of the gray levels of surrounding pixels. There are numerous possible choices to do so, such as evaluation by the gray level of the nearest neighbor, bilinear interpolation making use of the first four neighbors (eight

---

5 When suppressing the absolute value of the numerator of this expression, the criterion varies between 0 and 2, the maximum value corresponding to a "perfect inversion" of the contrast.

neighbors for a trilinear interpolation in 3D), bicubic interpolation (various possible formulations) based on the first 16 neighbors (64 in 3D), spline functions of various orders, Fourier transforms on windows of adjustable size or even wavelets. A high-order interpolation will ensure the continuity of the correlation coefficient and its derivative to various orders with respect to the kinematic parameters α, required by some optimization techniques, but will be greedier in computation time.

The errors induced by this interpolation take the form of an over- or under estimation of the displacement components, which depends, on average, on its fractional part, expressed in pixels, as a consequence of the periodicity of the sensor behavior. The average of these errors is thus a function of this fractional part, which exhibits a symmetry with respect to the half-pixel displacement, and is often referred to as the "S-shaped" systematic error curve [CHO 97, SCH 00]. The precise shape and amplitude of this curve express the level of efficiency of the interpolation scheme that is used to restore the evolution of gray levels in an image induced by a real subpixel translation. This error must be quantified and if possible controlled when small strains are to be analyzed, by means of the selection of an interpolation scheme adapted to the actual texture of the image, or conversely by means of a modification of the latter [YAN 10], which depends on the marking of the sample and the parameters of the optical system.

Various procedures have been proposed to determine this S-shaped curve. It is possible, for instance, to generate an artificial translated image from some real reference image and compare the results of the DIC analysis to the prescribed translation. However, the representativity of the virtually translated image with respect to a real image itself depends on the numerical translation procedure used. A Fourier-based translation [SCH 00, HIL 06], consisting of a multiplication in Fourier space by a phase term $e^{ik\delta}$ for each wave number k, is a particular procedure that leads to a C_∞ interpolation of the reference image, which exhibits some satisfactory properties with respect to this question. Another possibility is to numerically simulate the integration of the optical signal by the sensor [BOR 09, DUP 10] with the difficulty in that case of defining a good numerical model of the sample texture. A last option consists of analyzing a pair or a sequence of real images corresponding to a well-known transformation. This can be done but is delicate in practice for a pure translation [WAN 09], although it can easily be achieved with out-of-plane motions that induce homogeneous apparent deformation gradients [YAN 10, DAU 11]. Recent analyses have also shown that this systematic error is strongly sensitive to image noise [WAN 09], at least for some algorithms [DUP 10].

It would ideally be advisable to reduce the amplitude of the systematic errors to a level lower than that of the random errors, which correspond to the *fluctuations* of the measured displacements. The amplitude of the latter may also depend on the fractional part of the displacement field. When systematic errors are appropriately monitored, random errors constitute the actual limiting factor. This question will be addressed again in section 6.3.2 to discuss measurement uncertainty.

6.2.6. *Local approaches*

In traditional approaches, the transformation Φ_0 over R is decomposed into a multitude of *independent and local transformations*, or *shape functions*, parameterized by the coefficients of their local expansion near centers x^0 and used in the neighborhood D_{x^0} of these centers

$$\forall \boldsymbol{x} \in D_{x^0}, \quad \Phi_0(\boldsymbol{\alpha}, \boldsymbol{x}) - \boldsymbol{x} = \boldsymbol{u}_0(\boldsymbol{x}) = \sum_i \left[\alpha_i^0 + \sum_j \alpha_{ij}^1 (x_j - x_j^0) \right.$$

$$\left. \cdots + \sum_{jk} \alpha_{ijk}^2 (x_j - x_j^0)(x_k - x_k^0) + \ldots \right] \boldsymbol{e}_i \qquad [6.6]$$

where $\boldsymbol{e}_i$ is the unit vector along direction i in the image and x_i the coordinates of point $\boldsymbol{x}$ ($i \in \{1,2\}$ in 2D and $i \in \{1,2,3\}$ in 3D). Zeroth-order methods are limited to the translation vector $\boldsymbol{\alpha}^0$ and were historically the first to be used. First-order methods are the most common and take into account the components of the first deformation gradient α_{ij}^1. There are thus six parameters to locally describe Φ_0 in 2D and 12 in 3D. Some variants also include the second-order cross-terms (α_{112}^2 and α_{212}^2 in 2D, leading to a total of eight coefficients). We may also choose to restrict the local transformation to rigid motions (i.e. two translation components and one rotation angle in 2D, three components and three angles in 3D). Higher order expansions are seldom used.

With such a choice of local parametric descriptions, the standard processing procedure consists of defining in the ROI R a set of points x^k, often organized into rows and columns (and planes in 3D) and usually regularly spaced, and associating with each of them a small local domain D_{x^k}, in general square (or cubic in 3D), called *correlation window*, *subset* or *domain*. Problem [6.2] is then decomposed into as many independent optimizations leading to local parameters $\boldsymbol{\alpha}^k$ for each point x^k, which define local estimations of the transformation Φ_a over the domain D_{x^k}, from which in general only the value $\Phi_0(\alpha^k, \boldsymbol{x}^k)$ at the center is retained. The final result is then a discrete collection of displacement vectors on a set of points $x^k \in R$.

Such a description of the mechanical transformation by means of a set of local shape functions requires, thus, essentially three choices: the order of the local transformation, the size of the correlation window and the sampling step of the ROI. These choices are usually left to the user in classic image correlation software applications, but they have some implications with respect to the relevance of the kinematic measurements (displacement or strain). While questions relative to experimental errors will be addressed in detail in section 6.3, the following aspects are briefly emphasized:

– The selection of the order of the local transformation is strongly linked to the window size. The systematic analysis carried out in the work of [BOR 09] on virtual images exhibiting non-uniform deformation gradients with variable characteristic lengths demonstrates the existence of two error regimes. The first error is linked to the inability of the shape function to describe the real transformation over the whole correlation window. This error is all the more important if the window is large and the order of the transformation is low. The second error regime, referred to as the *ultimate error*, coincides with the error encountered when the real transformation is a pure translation. It is, on the contrary, characterized by a decrease in window size and an increase in the order of the shape function. This latter phenomenon is discussed in detail in section 6.3.2.

– Sampling step and window size can be chosen independently. We often select a step that is less than or equal to the window size, so as to use the whole of the information contained in the image. An arbitrarily fine sampling of the image can even be chosen, at the price of increased computation time. However, it is important to be aware of the difference between this measurement step and the spatial resolution of the measurement of the displacement field, which remains controlled by the correlation window size. The displacement measurement associated with each position of the ROI sampling is the result of an averaging process over the whole correlation window.

– The sampling step may be a parameter entering implicitly or explicitly in the procedures used to compute strain components from the discrete values of the displacement components evaluated at each measurement point. We refer the reader to Chapter 7 for a deeper discussion of this question. Note, however, that measurement accuracy for the components of the deformation gradient relative to a certain gauge length is inversely proportional to the latter. When low strain levels are to be quantified, it is necessary to privilege larger sampling steps or make use of differentiation procedures based on displacements evaluated over several sampling steps [ALL 94]. Moreover, in case of overlapping correlation windows, displacement measurements at neighboring positions are no longer independent, which may induce some bias in the computation of the gradient components.

– DIC analysis results in a discrete sampling of the displacement vectors (and their gradients) over a regular mesh of the ROI, the value at each sampling point being itself an average of the local displacement field over the correlation window. To define a displacement field at each position x in the ROI, these values need somehow to be interpolated. There is, however, no systematic way to evaluate the distance between the interpolated displacement field and the experimentally investigated field.

To conclude this short description of local formulations, let us note that they can also deal with an arbitrary selection of measurement points, adapted to the problem under consideration. These points may, for instance, be defined as the nodes of the finite element mesh of a structural calculation modeling the experiment [HER 07], as some isolated points of interest when the analysis does not require a dense evaluation of the kinematics – in such a case, the DIC techniques would operate as marker tracking techniques – or as the only points with an appropriate local image contrast, in the case of a DIC analysis of a system with a non-uniformly distributed natural local contrast. Local formulations can also make use of arbitrarily shaped correlation windows, for instance, associated with some discrete entities of the considered system (such as the grains in a sand sample [HAL 10]), or delimited by characteristic boundaries of the system (such as, the interface between constitutive phases of a heterogeneous material [RUP 07]). The use of image masks is useful for the practical implementation of such alternatives to the general algorithm.

6.2.7. *Optimization algorithms*

At this stage, it is possible to compute C and its gradients for any set of parameters α. The optimization, equation [6.2], can then in principle be performed with any iterative numerical algorithm. A minimization algorithm, making use of gradients (e.g. gradient descent, Newton–Raphson and Levenberg–Marquardt) or not (e.g. simplex and Powell), will enable a more or less rapid convergence toward a local minimum close to the initializing set of parameters. To avoid convergence to an erroneous local minimum, various strategies might be used, such as the preliminary systematic scanning of a set of displacement fields sufficiently large to contain the actual field, the use of known displacement fields relative to a previous stage of the mechanical test and the displacement of already analyzed neighboring positions, in the context of a "propagating front" type algorithm, or even, in the most difficult cases, the assistance of an operator. In this context, preliminary systematic scanning is often restricted to integer components of the translation vector α_i^0, with other possible parameters set to zero or frozen to some value determined by other means. To speed up this calculation stage, the gray-level interpolation scheme may be set to a simple nearest neighbor algorithm; an alternative is to compute correlation coefficients in Fourier space by means of efficient fast Fourier transform algorithms.

It is also possible to make use of these discrete evaluations of the correlation coefficient to interpolate this quantity by a biquadratic (or triquadratic in 3D) polynomial function near its discrete optimum, making use of the nine (or 27) discrete values surrounding this optimum, and to analytically seek the continuous optimum of this interpolating function. Such an algorithm allows subpixel accuracy to be reached at a particularly low computation cost [WAT 01], but is restricted to the optimization of a limited number of components of the vector α.

Another proposed strategy to avoid local minima consists of a multiscale algorithm [GAR 01a, HIL 02]. The idea is to start with strongly low-pass filtered images in order to artificially enlarge the well associated with the global minimum and to avoid local trappings far away from the solution. However, the obtained estimate is not accurate, as details of the image have been filtered out. The strategy consists then of gradually restoring these details and performing the optimization again with these richer images, starting the algorithm with the displacement obtained from the coarser images, until the original image is recovered. This procedure (which also applies to the global approaches described in the following section) turns out to be extremely robust and makes it possible to benefit from an optimization by descent, while limiting the effects of local minima.

This wide variety of options, which can be combined with each other, leads to a very large range of DIC algorithms, which furthermore will be enriched by the global approaches described in the following section. The selection of a particular combination of options is a compromise, whose definition depends on the available local image contrast, the expected strain level, the required measurement resolution and its spatial resolution, as well as the numerical cost and the robustness of the algorithms.

6.2.8. *Global approaches*

The main difference between the local and global approaches is the choice of the displacement basis used to account for the transformation Φ_0 and its spatial definition. The local approach results in an independent calculation of the transformation on each correlation (small) window. However, no specific regularity of the displacement field is exploited. Conversely, it is possible to choose a continuous transformation basis in a global approach. The displacement field is expressed as

$$\boldsymbol{u}_0(\boldsymbol{x}) = \Phi_0(\boldsymbol{\alpha}, \boldsymbol{x}) - \boldsymbol{x} = \sum_i \alpha_i \boldsymbol{\psi}_i(\boldsymbol{x}) \qquad [6.7]$$

where the fields $\boldsymbol{\psi}_i$ form the chosen kinematic basis that can be defined over the whole ROI. There is no restriction at this stage for the definition of this basis. The superposition of different displacement fields throughout domain R makes the sought amplitudes $\boldsymbol{\alpha}$ interdependent, hence the term "global" to address this problem. The determination of $\boldsymbol{\alpha}$ is performed as above, via the minimization of the sum of squared differences, $C_1(\Phi_0(\boldsymbol{\alpha}, \cdot), R, [f], [g])$. A Newton–Raphson scheme, which resorts to successive linearizations, is implemented to determine the solution, provided the amplitudes of the displacement fields are sufficiently small. As mentioned above, a multiscale strategy can significantly minimize this requirement [BES 06], and large displacements can be measured.

Unlike interpolated fields obtained with a local approach, each pixel is directly confronted with the value of the displacement field at this point and no interpolating step is required. The correlation residual normalized by the dynamic range of gray levels is comparable to that encountered in local approaches, that is of the order of 1%. As mentioned in section 6.2.4, the residual field allows correlation errors to be spatially located. This property enables the user to quickly check for the good convergence of the algorithm, and especially to possibly correct/extend the chosen basis of fields. For example, if a fractured medium is observed and the chosen displacement basis is continuous, then the residual error will focus on the support of the kinematic discontinuity. This is a very effective way of precisely locating the crack front in all its geometric complexity [RAN 10] as shown in Figure 6.4.

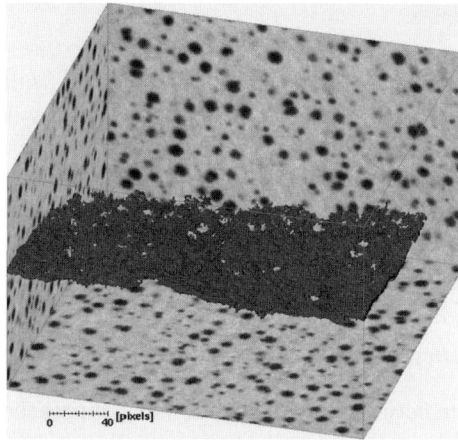

Figure 6.4. *Three-dimensional residual field showing the concentration of errors on the support of the crack (after [RAN 10])*

The nature of the kinematic basis was not specified beyond its overall character, and great latitude exists for this choice. Sometimes some specific informations (or assumptions) on the expected displacement field are available, and this is a valuable guide in the choice of this basis. Let us cite some examples:

– Without any specific information on the kinematics, the representation of the displacement field by finite element shape functions is a convenient possibility that will enable an easy and direct interface with numerical simulations. The so-called Q4-elements in 2D [BES 06], or C8-elements in 3D [ROU 08] (i.e. four-noded quadrilaterals with bilinear interpolation (2D) or eight-noded cubes (3D) with trilinear interpolations), are very well adapted to the discretization of the image. An example of a displacement field obtained with Q4 elements is shown in Figure 6.5(a). It is also possible to work with an unstructured mesh [LEC 09] or to use different shape functions [RÉT 09a]. Specific finite elements can then be considered with enriched kinematics, as in the extended finite element method (X-FEM) framework [RÉT 08] to add discontinuities to a given mesh (see Chapter 14). Finally, the fact that discretizations of the displacement field are shared with numerical modeling makes the coupling direct and *seamless* between image correlation on the one hand and, for example, an elastic calculation on the other hand (see [RÉT 09b] for such an example).

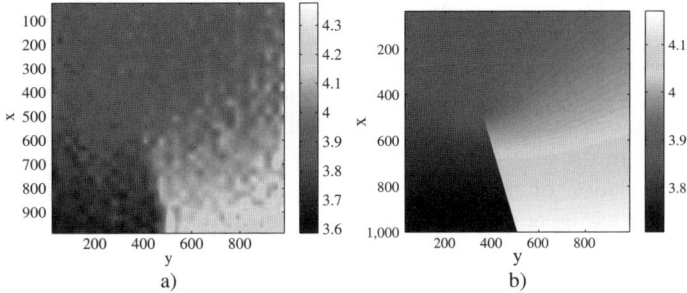

Figure 6.5. *Horizontal displacement component obtained on a silicon carbide specimen with a) Q4 finite elements (of size 32 pixels) or an integrated approach with a few closed-form solutions b). The physical size of one pixel is 1.85 μm. The displacements are expressed in pixels (after [ROU 09])*

– The test can be analyzed with an analytical expression for the displacement field in the context of elastic behavior. This solution can be an element of the selected kinematic basis. It is generally supplemented by additional degrees of freedom (such as rigid body motions). An example of this approach is the analysis of diametral compression (i.e. "Brazilian test" [HIL 06]) or a point force loading (i.e. Flamant's problem [ROU 05]). Note, however, that the solution involves the Poisson's ratio ν of the material. If it is not known, the general solution can be split into two terms: one dependent on ν and the other not. Similarly, the amplitude of the displacement field (except rigid body motions) is proportional to the ratio of the mechanical loading level to the shear modulus of the material. Thus, measuring the amplitudes of the basic fields, when the load level is known, enables the direct assessment of the elastic properties of the considered medium [ROU 05]. Beyond the pure kinematic measurement, we begin to see the identification of mechanical properties (elastic in this case) without any additional work via global DIC.

– A particularly important case concerns cracks [ROU 06, ROU 09]. A family of analytical solutions exists, namely Williams' series [WIL 57], which satisfies the cancelation of the tractions on the crack mouth. An appropriate selection of some terms of this infinite series can account for the sought displacement field in the vicinity of the crack tip. Among the selected fields, it is natural to include those associated with a $1/\sqrt{r}$ singularity for the stresses and strains. The amplitudes of these fields are related (via the physical size of the pixel and the elastic properties of the medium) to the stress intensity factors in modes I and II. An example of such a determination is shown in Figure 6.5(b) for the same pair of images as that used in the left part of the figure. However, this family of fields requires *a priori* knowledge of the geometry of the crack (here assumed to be straight). The support of the crack is generally easy to appreciate since it is apparent in the displacement field or in the residual field (see Figure 6.4). The position of the crack tip is yet another issue. One solution is to use specific terms of Williams' series (i.e. strain and stress fields with $r^{-3/2}$ singularity [ROU 09]). The amplitudes of these fields provide information about the actual position of the crack tip.

– When considering slender objects, beam-type models are often used. This description assumes particular (e.g. Euler–Bernoulli) kinematics that can be taken into account as in the displacement basis. This approach then allows the assessment by image correlation of the motion of a beam or a frame in a language directly suitable for modeling purposes [HIL 09a]. In this context, it is possible to finely characterize nonlinear modes (e.g. plastic hinge formation) in the framework of the strength of materials (i.e. constitutive law of a localized mode, equivalent position of a point corresponding to a localized failure mechanism, see Chapter 12). Moreover, the measurement with elastic kinematics of a given part of the beam gives access to the actual loading conditions.

– Beyond elasticity, it is sometimes possible to have *a priori* information to define a suitable displacement basis. The formation of shear bands in a uniaxial tensile test results in specific fields obeying Hadamard's conditions and that can be accounted for (see Chapter 14).

– Finally, the basis fields can be computed numerically. If we wish to consider several unknowns (e.g. boundary conditions and material parameters), it suffices to calculate the various fields associated with changes in the parameters associated with the unknowns of the problem and consider them as a kinematics basis. The values of the amplitudes of these fields correspond to the best evaluation reflecting the observed situation [LEC 09].

6.3. Errors and uncertainties

Image correlation gives access to quantitative measurements of displacement fields. It is therefore crucial to assess the sources and levels of error and uncertainty associated with its implementation.

6.3.1. *Main error sources*

The first source of errors is extrinsic to DIC because they result from image acquisition. Among them, we have cited the following:

– Those related to the 3D–2D projection, which will be described in more detail in section 6.4. In particular, out-of-plane motions can induce artificial yet quantifiable displacements and strains [SUT 08b]. The use of telecentric lenses largely restricts some of these artifacts. It is also possible to correct for variations in optical magnification induced by global motions of an object with respect to the imaging system if one has additional means for quantifying them [YAN 10].

– Those related to the imperfect positioning of pixel coordinates. For optical images, the distortions induced by objective lenses fall within this category. For images obtained by scanning (e.g. SEM and AFM), the imperfect positioning of the locally probed area (by the electronic spot or cantilever tip) may also introduce significant spurious strains.

– Those due to intrinsic noise in images. Two images acquired in identical conditions have gray levels that differ by a random amount whose standard deviation varies typically between a few thousandths (very good camera) and a few tenths (SEM image acquired very quickly) of the dynamic range of the gray levels. The impact of noise on image correlation [ROU 06] is detailed in section 6.3.3.

A second category of errors is due to DIC:

– The selected basis of displacement fields is unable to describe the apparent transformation Φ_a. This error regime has already been mentioned in section 6.2.5 and has been investigated in detail in the work of [BOR 09] for various software applications based on local approaches.

– The second type of error is related to the way the gray levels (or correlation coefficient) are interpolated in the correlation procedure when moving images for non-integer values (section 6.2.5).

We could also mention the errors generated by possible changes in local contrast and brightness, invalidating the basic assumption of gray-level conservation [6.1]. If it is possible to model them, they can be taken into account [HIL 12b] in the correlation procedure (e.g. for the analysis of AFM images [HAN 10]). Finally, when talking about errors or uncertainties, it is important to remember that the correlation residuals are very good indicators. They provide information on the local quality of the registration between images. They also allow for the modification or enrichment of the chosen kinematic basis [RÉT 08, RAN 10].

6.3.2. *Uncertainty and spatial resolution*

The choice of discretization method for the displacement field is crucial. Systematic studies on virtual images show that two different regimes of error occur [BOR 09]. The first is related to the inability of the shape function to describe the actual transformation of the entire correlation window (for a local approach) or the discretized kinematics (for a finite element-based global approach). The second type of error, referred to as ultimate, is similar to that encountered for pure translation transformations. It is characterized by a decrease in the error with decreased window size and increased order of the function shapes. It is a direct consequence of the ill-posedness of the correlation problem [6.2]. The information contained in a small correlation window will only identify a limited number of parameters and an increase in their number will be accompanied by increasing uncertainty about the value of each of them. The user is faced with a dilemma, which is comparable in some ways to Heisenberg's uncertainty principle, where position and displacement cannot be fully resolved simultaneously. Good spatial resolution (i.e. small window or element size) will be accompanied by a high displacement uncertainty and vice versa. A way of breaking this limit is to resort to regularized DIC [ROU 12] or Digital Volume Correlation [LEC 11].

To quantify *a priori* measurement uncertainty, we restrict ourselves to the second case by choosing a simple kinematics (i.e. uniform translation) that belongs to the space of shape functions. We proceed as in the evaluation of the interpolation subpixel error (section 6.2.5). From the displacement field evaluated for an image pair comprising the reference picture and its translated copy, the standard deviation of the displacement field gives an *a priori* estimate of the measurement uncertainty. This uncertainty is usually much higher than the systematic error. It is customary to observe a dependence of the standard displacement uncertainty, σ_u, on the size, ℓ, in pixels, of the interrogation window (local approach) or elements (global approach using finite elements) as a power law

$$\sigma_u = A\ell^{-\eta} \tag{6.8}$$

where A is of the order of unity and η varies between 1 and 2 as appropriate. This allows the evaluation of measurement uncertainties (excluding other factors related to noise or the complexity of the observed displacement field) that commonly reach a hundredth or even a thousandth of one pixel in the most favorable cases.

From this estimate it is also possible to evaluate strain uncertainty when it is obtained, for instance, by simple finite differences from the displacement field. It should be noted however that the fluctuations that will be observed on strains are strongly correlated (or anticorrelated) at small scales. This is especially important for the subsequent use of these kinematic measurements for the identification of mechanical properties (see the second part of this book) when these correlations are not taken into account. It is therefore advisable to prefer formulations of inverse problems that are based on displacements rather than strains, or to be aware of the implicit filtering often linked to the calculation of strain (see Chapter 7).

6.3.3. *Noise sensitivity*

Another limiting factor is due to the presence of noise in images. This noise can be of different kinds and therefore has very different statistical features. It can sometimes be reduced by image averaging, provided that the sample does not evolve (i.e. move) between acquisitions. A good approximation is often given by white noise (i.e. uncorrelated from pixel-to-pixel). Its effect can be followed up to the estimate of displacement fluctuations. In the case of a local approach, when the correlation windows do not overlap, the displacement fluctuations exhibit no correlation between different discrete estimates. However, when these windows overlap, or in the case of a global approach, the correlation matrix of the kinematic degrees of freedom is no longer diagonal, but its calculation is possible for a low noise level using a linearized operator for the final stage of the displacement estimation [HIL 12a].

In general, the standard displacement uncertainty (i.e. its standard deviation σ_u) is proportional to that of the image noise ϵ (by linearity). The amplitude will be inversely proportional to the mean gradient of the image, and will decrease as a power law with the number of pixels that are used to measure the considered motion. Thus, in d dimensions, for both local and global approaches

$$\sigma_u \propto \frac{\epsilon}{\ell^{d/2} \langle |\nabla f|^2 \rangle^{1/2}} \qquad [6.9]$$

A more detailed discussion of this evaluation is provided in [BES 06, ROU 06, HIL 12a].

It is possible to take this expression as a practical recommendation on image quality and the correlation parameters to use. Acquisition should minimize image noise (i.e. ϵ), and acquire images with a large depth of gray level (as long as it is not effectively truncated by the noise level ϵ). The texture of the analyzed images also affects displacement uncertainty. Dependence on the gray-level gradient calls for textures having a high contrast, and also having a short correlation length. The limit to this suggestion is the ability to interpolate gray levels at a subpixel level, which instead requires a regular variation at the pixel level. A good compromise is to deal with textures whose correlation length (radius) is of the order of two to five pixels. Finally, for the parameters of DIC itself, the size of the correlation window (or element) should increase to reduce its sensitivity to noise, but at the cost of lower spatial resolution. Again, a compromise must be made. We can also note that the sensitivity of different correlation algorithms with respect to this image noise is not identical [DUP 10].

6.4. Stereo-correlation or 3D-DIC

The main advantage of the 2D-DIC techniques described in section 6.2 is their (apparent) simplicity and versatility, such as a single camera is sufficient and the object surface preparation (when required) can easily be performed (spraying paint is an easy way to create an appropriate speckle pattern at the surface of the object). However, some important points need to be borne in mind:

– By using a single camera, only in-plane displacements/strains can be measured on a planar object.

– From a practical point of view, the camera and the object need to be aligned so that the camera image plane and the planar object are parallel and remain parallel throughout the experiment.

– If the objects undergo some out-of-plane displacements, they will not be properly detected and they will produce false apparent strains in the images [SUT 08b].

With the stereo-correlation technique (also called 3D-DIC) that is described hereafter, all these problems can be tackled. This technique enables the measurement of 3D displacements (in-plane and out-of-plane) and 2D strains undergone by any 3D object (not necessarily planar) while the experimental constraints are minimized [ORT 09a].

It should be noted that stereo-correlation procedures combine two techniques:

– *The DIC technique*, which is an image-matching technique to compute the 2D displacement of pixels between two images. With this technique, several images taken at different instants with a single camera can be registered (see section 6.2), or two images (or more) taken at the same instant by two cameras (or more) can be matched.

– *The stereovision technique*, which is a triangulation-based 3D reconstruction technique to compute the 3D position of a scene point from its stereo projections in two images (or more).

6.4.1. *The stereovision technique*

Binocular stereovision is a technique for recovering the 3D structure of a scene from two (or more) different viewpoints (Figure 6.6(a) where P is the 3D point to be measured, p_1 and p_2 are its stereoprojections in the images and C_1 and C_2 are the optical centers of the two cameras). In Figure 6.6(a), we can see that it is possible to compute the 3D coordinates of point P provided that:

1) The two image points p_1 and p_2, which correspond to the projection onto the images of the same physical point P, can be identified. This step is called *stereo image matching* or *search for stereo-correspondents*. It is the critical step of the stereovision technique (see section 6.4.1.3).

2) The two lines $C_1 p_1$ and $C_2 p_2$ that intersect at P can be computed. This step requires that the intrinsic parameters (focal length, pixel size and distortion coefficients) of each camera and the extrinsic parameters of the stereo rig (relative

position and orientation of the two cameras) be known. The intrinsic and extrinsic parameters are obtained by using an offline calibration process (see section 6.4.1.2).

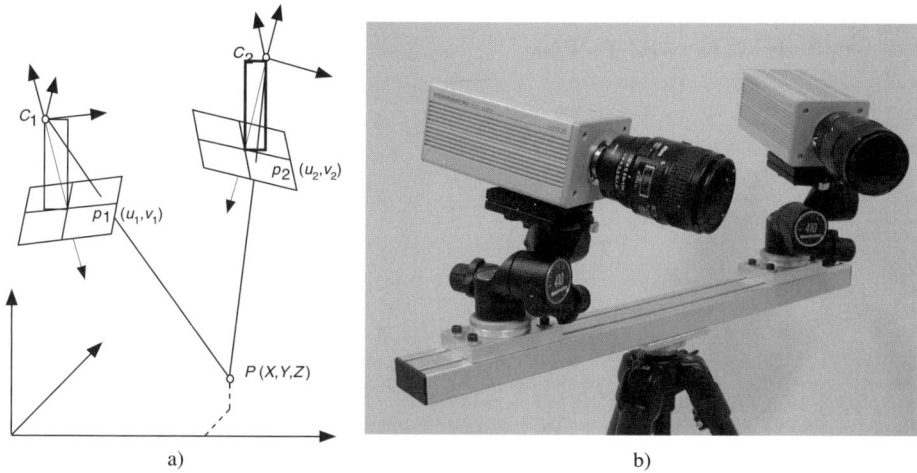

Figure 6.6. *a) The principle of stereovision and b) picture of a stereo rig*

6.4.1.1. *Geometry of a stereovision sensor*

A binocular stereovision sensor is made up of two cameras positioned in such a way that their field of view intersects. Let us call $\mathcal{R}_c$ the reference frame associated with the left camera, $\mathcal{R}'_c$ the reference frame associated with the right camera and $\mathcal{R}_w$ the so-called world reference frame, in which the 3D measurements will be expressed. $\mathbf{T}$, $\mathbf{T}'$ and $\mathbf{T}_s$ are the rigid-body transformations (a rotation matrix and a translation vector) that link these reference frames (Figure 6.7). These three transformations are linked by $\mathbf{T}_s = \mathbf{T}' \, \mathbf{T}^{-1}$. $\mathbf{T}_s$ is often called the *stereoscopic transformation*. It represents the orientation and translation of one camera with respect to the other. It should be noted that the 3D measurements are often expressed in the reference frame associated with the left or the right camera. In that case, $\mathbf{T}$ or $\mathbf{T}'$ is the identity matrix.

By modeling each camera with the classical linear *pinhole model*[6] and by labeling $\mathbf{K}$ and $\mathbf{K}'$ the intrinsic parameters of the left and right camera, respectively, the

---

6 The pinhole model is based on the perspective projection. In such a model, the distortions (which can be more or less important) induced by the lenses are not taken into account. For metrology applications, which require a high degree of measurement accuracy, distortions must be taken into account and more sophisticated camera models should be used [GAR 01a, GAR 01b, COR 05b, ORT 09b, ORT 09a, SUT 09].

positions m and m' of the 2D image points corresponding to a 3D point M are written as[7]

$$\tilde{m} = \mathbf{K\,T\,}\tilde{M} \qquad\qquad\qquad\qquad\qquad\qquad [6.10]$$

$$\tilde{m}' = \mathbf{K'\,T'\,}\tilde{M} = \mathbf{K'\,T_s\,T\,}\tilde{M} \qquad\qquad\qquad\quad [6.11]$$

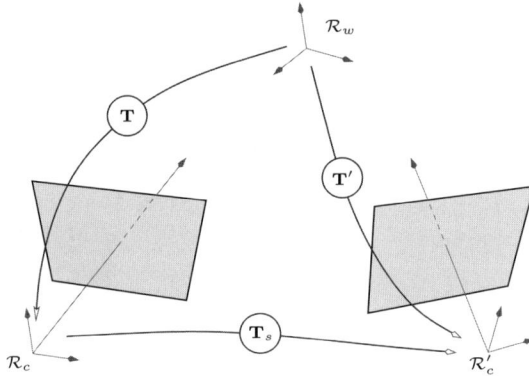

Figure 6.7. *The three reference frames associated with a stereovision sensor*

The image registration problem mentioned in section 6.4.1 consists of finding the image points m and m' that are stereo-correspondent.

In stereovision, the *epipolar geometry* [HOR 95] shown in Figure 6.8 provides the following important geometrical property (also called *epipolar constraint*):

> Given a point m in the left image, its corresponding point m' in the right image always appears to be lying along a line of the right image entirely defined by the coordinates of m. This line is called the *epipolar line* associated with m.

Because of this important geometrical property, which is inherent to any stereo imaging system, the search for the stereo-correspondent of a given point in the left image is simplified from a 2D search across the entire image to a 1D search along its epipolar line. Using this geometrical constraint, image registration can be faster and more robust (since possible false registrations that could be found away from the epipolar line are avoided).

---

7 To be able to write the pinhole model in a matrix form, *homogeneous coordinates* are used [FAU 93, FAU 01]. $\tilde{m}$ are the homogeneous coordinates associated with m.

6.4.1.2. *Calibration of a stereovision sensor*

Camera calibration is an important task in 3D computer vision, in particular when metric data are required for applications involving accurate dimensional measurements. Calibrating a camera involves determining its intrinsic parameters (matrix $\mathbf{K}$, and possibly distortion parameters). Calibrating a stereovision sensor composed of two cameras involves determining the intrinsic parameters of each camera and the relative position and orientation between the two cameras ($\mathbf{T}_s$ transformation). These calibration data are required to compute, by triangulation, the 3D coordinates of a point corresponding to matched pixels on the two images. For more details on the calibration of a stereovision sensor, see [GAR 01a, GAR 01b and ORT 09b].

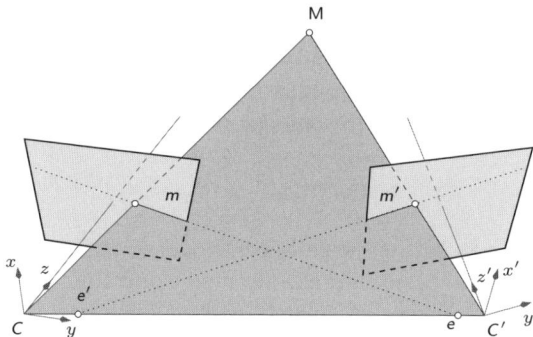

Figure 6.8. *Epipolar geometry: the epipolar lines are defined by the intersection of the image plane of each camera with the plane defined by points* M, C *and* C', *where* C *and* C' *are the optical centers of the left and right camera, respectively*

6.4.1.3. *Matching of stereo images by DIC*

This section is dedicated to the matching of stereo images by DIC (which leads to the so-called *stereo-correlation* technique). It should be noted that stereovision can be used without DIC-based matching. This is, for instance, the case when stereo images are matched using feature matching [ORT 02, ORT 09a].

DIC-based image matching simply consists of using the DIC technique described in section 6.2 in order to find stereo-correspondents. The only significant difference is that stereo-correlation can use (it is not compulsory) the epipolar constraint so that the search for correspondent pixels is restricted to the epipolar line (or to a pixel band around the epipolar line).

6.4.1.4. *3D reconstruction by triangulation*

If the image points m $= (u, v)$ and m' $= (u', v')$ (provided by registration) and matrices $\mathbf{K}$, $\mathbf{K}'$, $\mathbf{T}$ and $\mathbf{T}_s$ (provided by the calibration of the stereovision sensor) are known, equations [6.10] and [6.11] lead to an overdetermined system of four equations and three unknowns that are the three coordinates of the sought 3D point $M = (X, Y, Z)$ [HAR 97]

$$\begin{pmatrix} u \\ v \\ u' \\ v' \end{pmatrix} = \begin{pmatrix} & H_{4 \times 3} & \end{pmatrix} \begin{pmatrix} X \\ Y \\ Z \end{pmatrix} \qquad [6.12]$$

Equation [6.12] can be written as $b = H \, M$ and the 3D coordinates of point M are determined analytically using the pseudo-inverse method

$$M = \left[(H^T H)^{-1} H^T \right] b$$

The triangulation problem can also be solved using nonlinear optimization by minimizing the distance between the measured image points $(u, v)_m$ and the image points $(u, v)_p$ predicted by the camera model [SUT 09]

$$\chi^2 = \sum \left((u_m - u_p)^2 + (v_m - v_p)^2 + (u'_m - u'_p)^2 + (v'_m - v'_p)^2 \right) \qquad [6.13]$$

Note that equation [6.12] implies a linear model of the camera (pinhole model). When lens distortions are taken into account, the camera model is no longer linear. To compute the 3D coordinates of point M using equation [6.12], the image points need to be corrected for their distortion beforehand. If the distortion is not corrected beforehand, then equation [6.13] should be used.

6.4.2. 3D displacement measurement by stereo-correlation

Using the stereovision technique, the shape variation of an object can be measured by analyzing a sequence of pairs of stereo images. However, in experimental mechanics, we are generally interested in the surface strain field, which can be obtained by tracking the displacement of some points at the surface of an object undergoing mechanical or thermal loading, as shown in Figure 6.9.

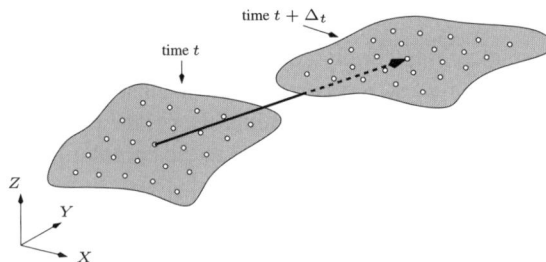

Figure 6.9. *Three-dimensional displacement field corresponding to two deformation states of the object*

The 2D technique described in section 6.2 allows the 2D displacement field to be computed at the surface of a planar object. By combining the stereo-correlation technique (DIC-based spatial registration between two cameras and triangulation, to obtain a cloud of 3D points) and DIC-based temporal matching between the images acquired by the left or the right camera at different instants, the 3D displacement field of any object (with any complex shape) can be computed. This is shown in Figure 6.10.

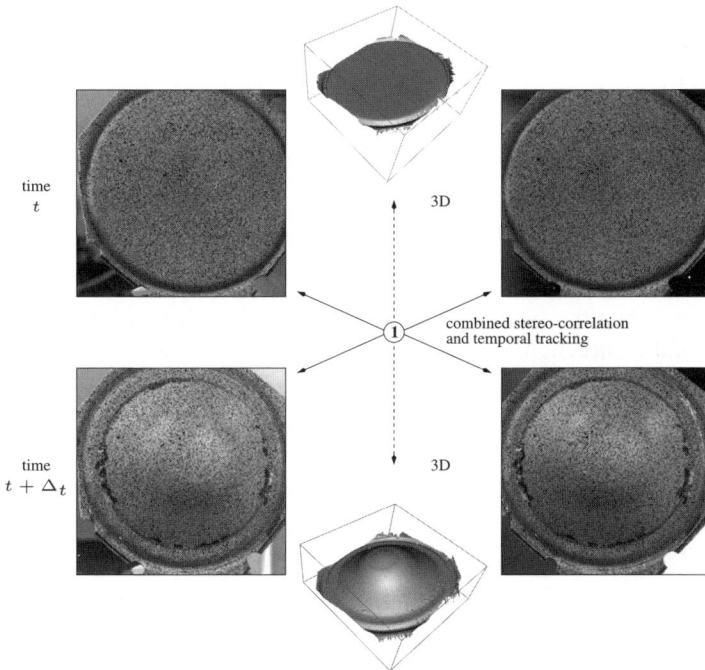

Figure 6.10. *Three-dimensional displacement field computation. Stereo-correlation (DIC-based spatial matching + triangulation) provides the shape of the object at a given instant. DIC-based temporal matching between the images taken by the left or right camera at different instants enables the 3D points of the shapes to be temporally linked*

This measurement technique provides 3D displacements from which strains can be computed as described in the following section.

6.4.3. *Computation of surface strains from 3D displacements*

The stereo-correlation technique combined with temporal image matching provides the 3D displacement field of any object with any complex shape (not necessarily planar). The surface strain field can be computed from the 3D displacement field by differentiation of the 3D displacement at each point of the surface, which can be a complex task.

From a practical point of view, the computation of the strain at a given point P of the surface is performed through several steps [COR 05a, SUT 09]:

– First, a small contiguous collection of 3D surface points (e.g. a 7×7 array with center point P) is selected from the initial surface shape.

– Next, a least squares plane is fitted to this 3D position data set providing the tangent plane.

– A local coordinate system $(\mathbf{n}, \mathbf{b}, \mathbf{t})$ is associated with the tangent plane and centered on P.

– All displacement components for each of the specimen points are converted into the local system. A set of displacement components (d_t, d_b, d_n) is obtained and centered (d_X, d_Y, d_Z).

– The array of data points for each displacement component is then fitted with a least squares function of the form $\mathbf{g}(X, Y)$.

– Differentiating the functional fit $\mathbf{g}(X, Y)$ for each displacement component in both the X and Y directions produces the $\varepsilon_{XX}(\mathrm{P})$, $\varepsilon_{YY}(\mathrm{P})$ and $\varepsilon_{XY}(\mathrm{P})$ components of the strain tensor at point P.

6.4.4. *Applications*

The stereo-correlation technique has proved to be a powerful non-contact technique for measuring 3D displacement/2D strain fields on any 3D object surface and is now widely used for industrial applications in different fields of mechanics and materials [SUT 09], such as fracture mechanics [SUT 92, LUO 94, LUO 00, SUT 07b], biomechanics [SUT 08a, TYS 02]; and for the study of different types of materials, such as composites [MCG 01, LOP 04, MUL 05, KAR 06], ceramics/concrete [ORT 07, ROB 07], foams [GUA 07], elastomers [MIS 05, JON 06], tires [MOS 07] and soft membranes [VIA 05]. The technique is very versatile and can be used for large scale experimental mechanics problems. It has been applied at micro [SCH 04] or macro scales [HEL 03] (small or large structures); it can measure both small (a few hundreds of micro-strains) and large strains (>200%); and it can be applied to high-speed dynamic tests by using high-speed cameras [SCH 03a, SCH 03b, SIE 07, TIW 07, BES 08, BES 10].

6.5. **Conclusions**

DIC, as we have seen through its many variants, in 2D or 3D, surface and even volume, at different scales of space and time, allows for the measurement of displacement fields with a rich description and remarkable accuracy, especially in light of its ease of implementation. Image correlation can exploit with great flexibility pictures acquired by different imaging means, which are every day better and more informative. Many tools are available to incorporate into this measurement procedure any *a priori* knowledge that is available about the kinematics to assess measurement uncertainty, or the covariance matrix of the measured degrees of freedom induced by image noise, and finally to visualize the residual field, often containing very valuable information. Today's experiments provide a quantity of measured data comparable with or sometimes even higher than what was traditionally the privilege of numerical simulations.

These measurements are used, now more than ever before, to monitor mechanical tests, whether in the context of conventional characterization or when used in an industrial context. The latter distinction is increasingly blurred as kinematic measurements, in particular by stereo-correlation, and can account for complex loadings and geometries. This merger opens up outstanding perspectives in terms of the quantitative use of tests, and therefore relevant responses to the use of materials, structures or assemblies under realistic conditions, representative of their in-service life.

Finally, what explains the spectacular development of this technique is probably its coupling with numerical simulations for identification and validation purposes. The end of this book is devoted to this broad issue. Again, the classic boundary between the experimental and numerical fields is gradually vanishing due to measured fields. They thus appear in different dimensions as the focal point of a particularly fruitful dialog between all parties interested in the mechanical behavior of materials and structures, experimentalists and modelers, designers and users, researchers and producers.

6.6. Bibliography

[ALL 94] ALLAIS L., BORNERT M., BRETHEAU T., CALDEMAISON D., "Experimental characterization of the local strain field in a heterogeneous elastoplastic material", *Acta Materialia*, vol. 42, no. 11, pp. 3865–3880, 1994.

[BAY 99] BAY B.K., SMITH T.S., FYHRIE D.P., SAAD M., "Digital volume correlation: three-dimensional strain mapping using X-ray tomography", *Experimental Mechanics*, vol. 39, pp. 217–226, 1999.

[BAY 08] BAY B.K., "Methods and applications of digital volume correlation", *Journal of Strain Analysis for Engineering Design*, vol. 43, no. 8, pp. 745–760, 2008.

[BEN 09] BENOIT A., GUÉRARD S., GILLET B., GUILLOT G., HILD F., MITTON D., PÉRIÉ J.-N., ROUX S., "3D analysis from micro-MRI during *in situ* compression on cancellous bone", *Journal of Biomechanics*, vol. 42, pp. 2381–2386, 2009.

[BES 06] BESNARD G., HILD F., ROUX S., "Finite-element displacement fields analysis from digital images: application to Portevin-Le Chatelier bands", *Experimental Mechanics*, vol. 46, pp. 789–803, 2006.

[BES 08] BESNARD G., ETCHESSAHAR B., LAGRANGE J.-M., VOLTZ C., HILD F., ROUX S., "Metrology and detonics: analysis of necking", *Proceedings of the 28th International Congress on High-Speed Imaging and Photonics (SPIE)*, 71261N, Canberra, Australia, 9–14 November 2008.

[BES 10] BESNARD G., LAGRANGE J.-M., HILD F., ROUX S., VOLTZ C., "Characterization of necking phenomena in high speed experiments by using a single camera", *EURASIP Journal on Image and Video Processing*, vol. 2010, no. 215956, 2010.

[BOR 04] BORNERT M., CHAIX J.-M., DOUMALIN P., DUPRÉ J.-C., FOURNEL T., JEULIN D., MAIRE E., MOREAUD M., MOULINEC H., "Mesure tridimensionnelle de champs cinématiques par imagerie volumique pour l'analyse des matériaux et des structures", *Instrumentation, Mesures, Métrologie*, vol. 4, pp. 43–88, 2004.

[BOR 09] BORNERT M., BRÉMAND F., DOUMALIN P., DUPRÉ J.-C., FAZZINI M., GRÉDIAC M., HILD F., MISTOU S., MOLIMARD J., ORTEU J.-J., ROBERT L., SURREL Y., VACHER P., WATTRISSE B., "Assessment of digital image correlation measurement errors: methodology and results", *Experimental Mechanics*, vol. 49, pp. 353–370, 2009.

[BUR 82] BURT P.J., YEN C., XU X., "Local correlation measures for motion analysis: a comparative study", *Proceedings of the IEEE Conference on Pattern Recognition and Image Processing*, Las Vegas (NV), USA, pp. 269–274, 14–17 June 1982.

[CHA 11] CHAMBON S., CROUZIL A., "Similarity measures for image matching despite occlusions in stereo vision", *Pattern Recognition*, vol. 44, no. 9, pp. 2063–2075, 2011.

[CHA 02] CHASIOTIS I., KNAUSS W.G., "A new microtensile tester for the study of MEMS materials with the aid of atomic force microscopy", *Experimental Mechanics*, vol. 42, pp. 51–57, 2002.

[CHO 97] CHOI S., SHAH S.P., "Measurement of deformations on concrete subjected to compression using image correlation", *Experimental Mechanics*, vol. 37, pp. 307–313, 1997.

[CHO 05] CHO S.W., CHASIOTIS I., FRIEDMAN T.A., SULLIVAN J., "Young's modulus, Poisson's ratio and failure properties of tetrahedral amorphous diamond-like carbon for MEMS devices", *Journal of Micromechanics and Microengineering*, vol. 25, pp. 728–735, 2005.

[CHO 07a] CHO S.W., CHASIOTIS I., "Elastic properties and representative volume element of polycrystalline silicon for MEMS", *Experimental Mechanics*, vol. 47, no. 1, pp. 37–49, 2007.

[CHO 07b] CHO S.W., JONNALAGADDA K., CHASIOTIS I., "Mode I and mixed mode fracture of polysilicon for MEMS", *Fatigue and Fracture of Engineering Materials and Structures*, vol. 30, pp. 21–31, January 2007.

[COR 05a] CORNILLE N., Accurate 3D shape and displacement measurement using a scanning electron microscope, PhD Thesis, INSA Toulouse (France) and University of South Carolina, Columbia (USA), 2005.

[COR 05b] CORNILLE N., GARCIA D., SUTTON M.A., MCNEILL S., ORTEU J.-J., "Calibrage d'imageurs avec prise en compte des distorsions", *Instrumentation, Mesures, Métrologie*, vol. 4, nos. 3–4/2004, pp. 105–124, 2005.

[DAU 11] DAUTRIAT J., BORNERT M., GLAND N., DIMANOV A., RAPHANEL J., "Localized deformation induced by heterogeneities in porous carbonate analysed by multi-scale digital image correlation", *Tectonophysics*, vol. 503, nos. 1–2, pp. 100–116, 2011.

[DOU 00a] DOUMALIN P., BORNERT M., "Micromechanical applications of digital image correlation techniques", in JACQUOT P., FOURNIER J.M. (eds), *Interferometry in Speckle Light: Theory and Applications*, Springer, pp. 67–74, 2000.

[DOU 00b] DOUMALIN P., Microextensométrie locale par corrélation d'images numériques; Application aux études micromécaniques par microscopie électronique à balayage, PhD Thesis, École Polytechnique, France, 2000.

[DOU 03] DOUMALIN P., BORNERT M., CRÉPIN J., "Caractérisation de la répartition de la déformation dans les matériaux hétérogènes", *Mécanique et Industries*, vol. 4, pp. 607–617, 2003.

[DUP 10] DUPRÉ J.C, BORNERT M., ROBERT L., WATTRISSE B., "Digital image correlation: displacement accuracy estimation", *Proceedings of the 14th International Conference on Experimental Mechanics*, EPS Web of Conferences 6, 31006, pp. 8, Poitiers, France, 4–9 July 2010.

[FAU 93] FAUGERAS O., *Three-Dimensional Computer Vision: A Geometric Viewpoint*, The MIT Press, 1993.

[FAU 01] FAUGERAS O., LUONG Q.T., PAPADOPOULOS T., *The Geometry of Multiple Images*, The MIT Press, 2001.

[GAR 01a] GARCIA D., Mesure de formes et de champs de déplacements tridimensionnels par stéréo-corrélation d'images, PhD Thesis, National Polytechnic Institute of Toulouse, France, 2001.

[GAR 01b] GARCIA D., ORTEU J.-J., DEVY M., "Calibrage précis d'une caméra CCD ou d'un capteur de vision stéréoscopique", *Photomécanique 2001*, Futuroscope, Poitiers, France, pp. 24–26, 2001.

[GUA 07] GUASTAVINO R., GÖORANSSON P., "A 3D displacement measurement methodology for anisotropic porous cellular foam materials", *Polymer Testing*, vol. 26, no. 6, pp. 711–719, 2007.

[HAL 10] HALL S., BORNERT M., DESRUES J., PANNIER Y., LENOIR N., VIGGIANI C., BÉSUELLE P., "Discrete and continuum analysis of localised deformation in sand using X-ray micro-CT and volumetric digital image correlation", *Géotechnique*, vol. 60, no. 5, pp. 315–322, 2010.

[HAN 10] HAN K., CICCOTTI M., ROUX S., "Measuring nanoscale stress intensity factors with an atomic force microscope", *Europhysics Letters*, vol. 89, pp. 66003, 2010.

[HAR 97] HARTLEY R.I., STURM P., "Triangulation", *Computer Vision and Image Understanding (CVIU'97)*, vol. 68, no. 2, pp. 146–157, 1997.

[HEL 03] HELM J.D., SUTTON M.A., MCNEILL S.R., "Deformations in wide, center-notched, thin panels, part I: three-dimensional shape and deformation measurements by computer vision", *Optical Engineering*, vol. 42, no. 5, pp. 1293–1305, 2003.

[HER 07] HERIPRE E., DEXET M., CREPIN J., GELEBART L., ROOS A., BORNERT M., CALDEMAISON D., "Coupling between experimental measurements and polycrystal finite element calculations for micromechanical study of metallic materials", *International Journal of Plasticity*, vol. 23, no. 9, pp. 1512–1539, 2007.

[HIL 02] HILD F., RAKA B., BAUDEQUIN M., ROUX S., CANTELAUBE F., "Multiscale displacement field measurements of compressed mineral-wool samples by digital image correlation", *Applied Optics*, vol. 41, no. 32, pp. 6815–6828, 2002.

[HIL 06] HILD F., ROUX S., "Digital image correlation: from displacement measurement to identification of elastic properties – a review", *Strain*, vol. 42, pp. 69–80, 2006.

[HIL 09a] HILD F., ROUX S., GRAS R., MARANTE M.E., GUERRERO N., FLÓREZ-LÓPEZ J., "Displacement measurement technique for beam kinematics", *Optics and Lasers in Engineering*, vol. 47, pp. 495–503, 2009.

[HIL 09b] HILD F., MAIRE E., ROUX S., WITZ J.-F., "Three dimensional analysis of a compression test on stone wool", *Acta Materialia*, vol. 57, pp. 3310–3320, 2009.

[HIL 12a] HILD F., ROUX S., "Comparison of local and global approaches to digital image correlation", *Experimental Mechanics*, 2012.

[HIL 12b] HILD F., ROUX S., "Digital image correlation", in RASTOGI P., HACK E. (eds), *Optical Methods for Solid Mechanics, A Full-Field Approach*, Wiley-VCH, Weinheim, Germany, pp. 183–228.

[HOR 95] HORAUD R., MONGA O., *Vision par ordinateur: outils fondamentaux*, 2nd ed., Hermes, 1995.

[JON 06] JONES A., SHAW J., WINEMAN A., "An experimental facility to measure the chemorheological response of inflated elastomer membranes at high temperature", *Experimental Mechanics*, vol. 46, no. 5, pp. 579–587, 2006.

[KAR 06] KARAMA M., LORRAIN B., "Modélisation numérique et expérimentale du comportement de structures sandwich", *Mécanique & Industries*, vol. 7, pp. 39–48, 2006.

[LEC 09] LECLERC H., PÉRIÉ J.-N., ROUX S., HILD F., "Integrated digital image correlation for the identification of material properties", *Lecture Notes in Computer Science*, vol. 5496, pp. 161–171, 2009.

[LEC 11] LECLERC H., PÉRIÉ J.-N., ROUX S., HILD F., "Voxel-scale digital volume correlation", *Experimental Mechanics*, vol. 51, no. 4, pp. 479–490, 2011.

[LEN 07] LENOIR N., BORNERT M., DESRUES J., BESUELLE P., VIGGIANI G., "Volumetric digital image correlation applied to X-ray microtomography images from triaxial compression tests on argillaceous rock", *Strain*, vol. 43, no. 3, pp. 193–205, 2007.

[LEP 07] LEPRINCE S., BARBOT S., AYOUB F., AVOUAC J.-P., "Automatic and precise orthorectification, coregistration, and subpixel correlation of satellite images, application to ground deformation measurements", *IEEE Transactions on Geoscience and Remote Sensing*, vol. 45, no. 6, Part 1, pp. 1529–1558, 2007.

[LOP 04] LOPEZ-ANIDO R., EL-CHITI F.W., MUSZYÑSKI L., DAGHER H.J., THOMPSON L.D., HESS P.E., "Composite material testing using a 3-D digital image correlation system", *Proceedings of the Composites 2004 Conference*, Tampa, FL, 6–8 October 2004.

[LUC 81] LUCAS B.D., KANADE T., "An iterative image registration technique with an application to stereo vision", *Proceedings of the 7th International Joint Conference on Artificial Intelligence, IJCAI'81*, vol. 2, Morgan Kaufmann Publishers Inc., San Francisco, CA, pp. 674–679, 1981.

[LUO 94] LUO P.F., CHAO Y.J., SUTTON M.A., "Application of stereo vision to three-dimensional deformation analyses in fracture experiments", *Optical Engineering*, vol. 33, no. 3, pp. 981–990, 1994.

[LUO 00] LUO P.F., HUANG F.C., "Application of stereovision to the study of mixed-mode crack-tip deformations", *Optics and Lasers in Engineering*, vol. 33, no. 5, pp. 349–368, 2000.

[MCG 01] MCGOWAN D.M., AMBUR D.R., HANNA T.G., MCNEILL S.R., "Evaluating the compressive response of notched composite panels using full-field displacements", *AIAA Journal of Aircraft*, vol. 38, no. 1, pp. 122–129, 2001.

[MIS 05] MISTOU S., KARAMA M., DESMARS B., PERES P., PIRON E., HEUILLET P., "Application de la méthode de stéréo-corrélation d'images à la caractérisation des élastomères en grandes déformations", *Instrumentation, Mesures, Métrologie*, vol. 4, nos. 3–4, pp. 147–166, 2005.

[MOS 07] MOSER R., LIGHTNER III J. G., "Using three-dimensional digital imaging correlation techniques to validate tire finite-element model", *Experimental Techniques*, vol. 31, no. 4, pp. 29–36, 2007.

[MUL 05] MULLE M., PÉRIÉ J.-N., ROBERT L., COLLOMBET F., GRUNEVALD Y., "Mesures de champs par stéréo-corrélation sur structures composites instrumentées par fibres optiques à réseaux de Bragg", *Instrumentation, Mesures, Métrologie*, vol. 4, nos. 3–4, pp. 167–192, 2005.

[NEU 08] NEU C.P., WALTON J.H., "Displacement encoding for the measurement of cartilage deformation", *Magnetic Resonance in Medicine*, vol. 59, pp. 149–155, 2008.

[ORT 02] ORTEU J.-J., "Mesure 3D de formes et de déformations par stéréovision", Les Techniques de L'Ingénieur, vol. BM 7015 of *Traité Génie Mécanique* – Travail des matériaux, 2002.

[ORT 07] ORTEU J.-J., CUTARD T., GARCIA D., CAILLEUX E., ROBERT L., "Application of stereovision to the mechanical characterisation of ceramic refractories reinforced with metallic fibres", *Strain*, vol. 43, no. 2, pp. 96–108, 2007.

[ORT 09a] ORTEU J.-J., "3-D computer vision in experimental mechanics", *Optics and Lasers in Engineering*, vol. 47, nos. 3–4, pp. 282–291, 2009.

[ORT 09b] ORTEU J.-J., Calibrage géométrique d'une caméra ou d'un capteur de vision stéréoscopique, Ressources en ligne UNIT, Optique pour l'Instrumentation, 2009. Available at http://optique-instrumentation.fr/.

[POL 02] POLLEFEYS M., Visual 3D modeling from images – a tutorial, 2002. Available at http://www.cs.unc.edu/ marc/tutorial/.

[RAN 10] RANNOU J., LIMODIN N., RÉTHORÉ J., GRAVOUIL A., LUDWIG W., BAÏETTO-DUBOURG M.-C., BUFFIÈRE J.-Y., COMBESCURE A., HILD F., ROUX S., "Three dimensional experimental and numerical multiscale analysis of a fatigue crack", *Computer Methods in Applied Mechanics and Engineering*, vol. 199, pp. 1307–1325, 2010.

[RÉT 08] RÉTHORÉ J., HILD F., ROUX S., "Extended digital image correlation with crack shape optimization", *International Journal for Numerical Methods in Engineering*, vol. 73, pp. 248–272, 2008.

[RÉT 09a] RÉTHORÉ J., ELGUEDJ T., SIMON P., CORET M., "On the use of NURBS functions for displacement derivatives measurement by digital image correlation", *Experimental Mechanics*, vol. 50, no. 7, pp. 1099–1116, 2009.

[RÉT 09b] RÉTHORÉ J., HILD F., ROUX S., "An extended and integrated digital image correlation technique applied to the analysis fractured samples", *European Journal of Computational Mechanics*, vol. 18, pp. 285–306, 2009.

[ROB 07] ROBERT L., NAZARET F., CUTARD T., ORTEU J.-J., "Use of 3-D digital image correlation to characterize the mechanical behavior of a fiber reinforced refractory castable", *Experimental Mechanics*, vol. 47, no. 6, pp. 761–773, December 2007.

[ROU 05] ROUX S., HILD F., PAGANO S., "A stress scale in full-field identification procedures: a diffuse stress gauge", *European Journal of Mechanics A/Solids*, vol. 24, pp. 442–451, 2005.

[ROU 06] ROUX S., HILD F., "Stress intensity factor measurements from digital image correlation: post-processing and integrated approaches", *International Journal of Fracture*, vol. 140, pp. 141–157, 2006.

[ROU 08] ROUX S., HILD F., VIOT P., BERNARD, D., "Three dimensional image correlation from X-ray computed tomography of solid foam", *Composites A: Applied Science and Manufacturing*, vol. 39, pp. 1253–1265, 2008.

[ROU 09] ROUX S., RÉTHORÉ J., HILD F., "Digital image correlation and fracture: an advanced technique for estimating stress intensity factors of 2D and 3D cracks", *Journal of Physics D: Applied Physics*, vol. 42, pp. 214004, 2009.

[ROU 12] ROUX S., HILD F., LECLERC H., "Mechanical assistance to DIC", *Proceedings on Full Field Measurements and Identification in Solid Mechanics*, Elsevier, IUTAM Procedia, Cachan, France, vol. 4, pp. 159–168, 4–8 July 2012.

[RUP 07] RUPIN N., Déformation à chaud de métaux biphasés: modélisations théoriques et confrontations expérimentales, PhD Thesis, École Polytechnique, France, 2007.

[SCA 92] SCAMBOS T.A., DUTKIEWICZ M.J., WILSON J.C., BINDSCHADLER, R.A., "Application of image cross-correlation to the measurement of glacier velocity using satellite image data ", *Remote Sensing of Environment*, vol. 42, no. 3, pp. 177–186, 1992.

[SCH 03a] SCHMIDT T.E., TYSON J., GALANULIS K., "Full-field dynamic displacement and strain measurement – specific examples using advanced 3D image correlation photogrammetry: part II", *Experimental Techniques*, vol. 27, no. 4, pp. 22–26, 2003.

[SCH 03b] SCHMIDT T.E., TYSON J., GALANULIS K., "Full-field dynamic displacement and strain measurement using advanced 3D image correlation photogrammetry: part I", *Experimental Techniques*, vol. 27, no. 3, pp. 47–50, 2003.

[SCH 00] SCHREIER H.W., BRAASCH J.R., SUTTON M.A., "Systematic errors in digital image correlation caused by intensity interpolation", *Optical Engineering*, vol. 39, no. 11, pp. 2915–2921, 2000.

[SCH 04] SCHREIER H.W., GARCIA D., SUTTON M.A., "Advances in light microscope stereo vision", *Experimental Mechaniques*, vol. 44, no. 3, pp. 278–288, 2004.

[SCR 07] SCRIVENS W.A., LUO Y., SUTTON M.A., COLLETTE S.A., MYRICK M.L., MINEY P., COLAVITA P.E., REYNOLDS A.P., LI X., "Development of patterns for digital image correlation measurements at reduced length scales", *Experimental Mechanics*, vol. 47, no. 1, pp. 63–77, 2007.

[SIE 07] SIEBERT T., BECKER T., SPILTTHOF K., NEUMANN I., KRUPKA R., "High-speed digital image correlation: error estimations and applications", *Optical Engineering*, vol. 46, no. 5, pp. 051004, 2007.

[SOP 01] SOPPA E., DOUMALIN P., BINKELE P., WIESENDANGER T., BORNERT M., SCHMAUDER S., "Experimental and numerical characterisation of in-plane deformation in two-phase materials", *Computational Materials Science*, vol. 21, no. 3, pp. 261–275, 2001.

[SUT 83] SUTTON M. A., WOLTERS W.J., PETERS W.H., RANSON W.F., MCNEILL S.R., "Determination of displacements using an improved digital correlation method", *Image and Vision Computing*, vol. 1, no. 3, pp. 133–139, 1983.

[SUT 86] SUTTON M.A., CHENG M., PETERS W.H., CHAO Y.J., MCNEILL S.R., "Application of an optimized digital correlation method to planar deformation analysis", *Image and Vision Computing*, vol. 4, no. 3, pp. 143–150, 1986.

[SUT 92] SUTTON M.A., TURNER J.L., CHAO Y.J., BRUCK H.A., CHAE T.L., "Experimental investigations of three-dimensional effects near a crack tip using computer vision", *International Journal of Fracture*, vol. 53, no. 3, pp. 201–228, 1992.

[SUT 06] SUTTON M.A., LI N., GARCIA D., CORNILLE N., ORTEU J.-J., MCNEILL S.R., SCHREIER H.W., LI X., "Metrology in a scanning electron microscope: theoretical developments and experimental validation", *Measurement Science and Technology*, vol. 17, no. 10, pp. 2613–2622, 2006.

[SUT 07a] SUTTON M.A., LI N., JOY D.C., REYNOLDS A.P., LI X., "Scanning electron microscopy for quantitative small and large deformation measurements. Part I: SEM imaging at magnifications from 200 to 10,000", *Experimental Mechanics*, vol. 47, no. 6, pp. 775–787, 2007.

[SUT 07b] SUTTON M.A., YAN J., DENG X., CHENG C.-S., ZAVATTIERI P., "Three-dimensional digital image correlation to quantify deformation and crack-opening displacement in ductile aluminum under mixed-mode I/III loading", *Optical Engineering*, vol. 46, no. 5, 051003, 2007.

[SUT 08a] SUTTON M.A., KE X., LESSNER S.M., GOLDBACH M., YOST M., ZHAO F., SCHREIER H.W., "Strain field measurements on mouse carotid arteries using microscopic three dimensional digital image correlation", *Journal of Biomedical Materials Research Part A*, vol. 84, no. 1, pp. 178–190, 2008.

[SUT 08b] SUTTON M.A., YAN J.H., TIWARI V., SCHREIER H.W., ORTEU J.-J., "The effect of out of plane motion on 2D and 3D digital image correlation measurements", *Optics and Lasers in Engineering*, vol. 46, no. 10, pp. 746–757, 2008.

[SUT 09] SUTTON M.A., ORTEU J.-J., SCHREIER H.W., *Image Correlation for Shape, Motion and Deformation Measurements – Basic Concepts, Theory and Applications*, Springer-Verlag, New York Inc., NY, 2009.

[TAT 03] TATSCHL A., KOLEDNIK O., "On the experimental characterization of crystal plasticity in polycrystals", *Materials Science and Engineering A – Structural Materials*, vol. 342, nos. 1–2, pp. 152–168, 2003.

[TER 09] TERZI S., SALVO L., SUERY M., LIMODIN N., ADRIEN J., MAIRE E., PANNIER Y., BORNERT M., BERNARD D., FELBERBAUM M., RAPPAZ M., BOLLER E., "In situ X-ray tomography observation of inhomogeneous deformation in semi-solid aluminium alloys", *Scripta Materialia*, vol. 61, no. 5, pp. 449–452, 2009.

[TIW 07] TIWARI V., SUTTON M.A., MCNEILL S.R., "Assessment of high speed imaging systems for 2D and 3D deformation measurements: methodology development and validation", *Experimental Mechaniques*, vol. 47, no. 4, pp. 561–579, 2007.

[TYS 02] TYSON J., SCHMIDT T., GALANULIS K., "Biomechanics deformation and strain measurements with 3D image correlation photogrammetry", *Experimental Techniques*, vol. 26, no. 5, pp. 39–42, 2002.

[VIA 05] VIALETTES P., SIGUIER J.-M., GUIGUE P., MISTOU S., DALVERNY O., KARAMA M., PETITJEAN F., "Etude par stéréo-corrélation de sous-ensembles de ballons stratosphériques pressurisés", *Instrumentation, Mesures, Métrologie*, vol. 4, nos. 3–4, pp. 125–145, 2005.

[WAN 09] WANG Y.Q., SUTTON M.A., BRUCK H.A., SCHREIER H.W., "Quantitative error assessment in pattern matching: effects of intensity pattern noise, interpolation, strain and image contrast on motion measurements", *Strain*, vol. 45, no. 2, pp. 160–178, 2009.

[WAT 01] WATTRISSE B., CHRYSOCHOOS A., MURACCIOLE J.-M., NÉMOZ-GAILLARD M.N., "Analysis of strain localization during tensile tests by digital image correlation", *Experimental Mechanics*, vol. 41, pp. 29–39, 2001.

[WIL 57] WILLIAMS M.L., "On the stress distribution at the base of a stationary crack", *Journal of Applied Mechanics*, vol. 24, pp. 109–114, 1957.

[YAN 10] YANG D.S., BORNERT M., GHARBI H., VALLI P., WANG L.L., "Optimized optical setup for DIC in rock mechanics", *Proceedings of the 14th International Conference on Experimental Mechanics*, EPJ Web of Conferences 6, 22019, Poitiers, France, 4–9 July 2010.

Chapter 7

From Displacement to Strain

7.1. Introduction

Most full-field measurement techniques yield the displacement field, whereas the strain field could be preferred for some applications. With the transformation from the displacement to the strain field comes the question of the differentiation of noisy experimental data, implying a particular treatment in order to reduce the effect of noise on the results. The standard methods are presented and discussed through simple examples, emphasizing the problems occurring with such differentiation operations. Then, the methods based on approximation techniques are detailed and the various errors are discussed so that some criteria for the choice of strain reconstruction parameters can be proposed.

7.2. From measurement to strain

7.2.1. *Three related steps*

Most methods for measuring strain based on digital images yield the displacement field for the region of interest. For some applications, such as the study of cracks corresponding to a displacement discontinuity (Chapter 14) and the identification of material properties through least-squares (Chapter 8), this kind of data is appropriate. Nonetheless, for other applications, such as the qualitative description of diffuse damage zones [FEI 09] and for identification based on the virtual fields method (Chapter 11), it is necessary to deduce the strain field from the displacement field. Such a task is not straightforward because it implies the evaluation of the gradient

---

Chapter written by Pierre FEISSEL.

of noisy data. This transformation amplifies the noise if it is not performed properly, which is discussed in section 7.3.2 through a simple example. Thus, it is necessary to filter the measurements in order to control the level of perturbation, the risk being the loss of mechanical information.

To deduce the strain field from the experimental images, we have to carry out, broadly, the three following steps:

1) measuring the displacements from the digital images;

2) filtering the displacements;

3) differentiating the displacements in order to obtain the strain field.

Depending on the chosen method, these three steps can correspond to three independent steps or can be merged into a smaller number of steps. For example, the use of an appropriate basis for the displacement field during the digital image correlation process [LEC 09] allows filtering from the displacement measurement step. For approaches based on diffuse approximation (DA), which will be described in detail in this chapter, the filtering and differentiation steps are grouped together in one single step, whereas the displacement measurement step stands alone and can be performed using any tool. However, each step implies some specific questions and, whether the steps are explicit or not, it is of interest to keep these in mind in order to control the errors.

The step corresponding to the measurement of the displacements provides a field that can be seen as the sum of the exact field, a systematic error and a random error. Dealing with the systematic error is not straightforward, implying an insight into the measurement means and taking into account information at the mechanical level, and will not be covered in this chapter. The reconstruction of the strain field from measured displacements aims at reconstructing the gradient of the exact field while limiting the effect of the errors.

In this chapter, the problems related to this strain reconstruction step are described. The aim is to emphasize the difficulties associated with this step and propose some answers and tools in order to better understand it. Finally, let us note that the approaches developed here are presented in the case of kinematic fields but can also be applied to thermal fields.

7.2.2. *Framework for the differentiation of displacement measurements*

The starting point is the measured displacement field, whatever the technique used to obtain it. These data are usually obtained on a discrete data grid that is generally regular (and bidimensional). The typical size of such a data grid is about 200×200, even if some techniques, like *ESPI* (Chapter 5), yield grids 10 times larger. We will consider regular data grids with the measured displacement, denoted by:

$$\tilde{\underline{u}}(\underline{x}_i), \quad i \in [1, N], \quad \text{where } \underline{x}_i \text{ is the position of the data point } i \qquad [7.1]$$

The displacement field is then described through its two components u and v. The gradient of $\underline{u}$ is usually calculated by evaluating the gradient of each component.

As mentioned above, the measured displacements can be seen as the sum of three terms:

– the mechanical displacement field whose gradient is sought;

– a systematic error, corresponding to the deterministic part of the error induced by the measurement device;

– a random error, corresponding to the non-deterministic error induced by the measurement device.

Errors of both types are specific to each displacement field measurement method. They have usually been characterized and an error estimator can be available for some methods. Nonetheless, the systematic error is seldom taken into account by the filtering tools. Presumably, the largest error on the strain is not due to the systematic error (even if, once the noise is properly filtered, we could address that problem). In this chapter, we will not take the systematic error into consideration, and the measurements will therefore be considered as the sum of an exact field whose symmetric gradient is to be estimated and a random field whose probabilistic characteristics are assumed to be known:

$$\tilde{\underline{u}}(\underline{x}_i) = \underline{u}_{ex}(\underline{x}_i) + \delta\underline{u}(\underline{x}_i), \quad i \in [1, N] \qquad [7.2]$$

In the following, the various fields at the data points, $\tilde{\underline{u}}(\underline{x}_i)$, $\underline{u}_{ex}(\underline{x}_i)$ and $\delta\underline{u}(\underline{x}_i)$, $\forall i \in [1, N]$, will be stacked in column vectors, respectively, denoted by: $\{\tilde{u}\}$, $\{u_{ex}\}$ and $\{\delta u\}$. A standard and often reasonable hypothesis is to consider the random field as Gaussian white noise; hence, its covariance matrix is written:

$$\text{cov}(\{\delta u\}) = \gamma^2 [I], \quad \text{where } [I] \text{ is the identity matrix} \qquad [7.3]$$

From these data, we wish to reconstruct the (infinitesimal) strain field, given as:

$$\underline{\underline{\varepsilon}} = \frac{1}{2}\left(\nabla\underline{u} + \nabla^t\underline{u}\right) \qquad [7.4]$$

The strain tensor can be represented as a vector collecting the components of $\underline{\underline{\varepsilon}}$ in the $(\vec{e}_X, \vec{e}_Y)$ basis associated with the data grid:

$$\underline{\varepsilon} = \begin{bmatrix} \varepsilon_{XX} \\ \varepsilon_{YY} \\ \varepsilon_{XY} \end{bmatrix} \qquad [7.5]$$

7.2.3. *The main families of methods for differentiating data*

The problem of the reconstruction of the gradient consists of estimating the derivatives of the displacement components, obtained on a regular grid of data points. A short description of the various usual methods is proposed in the following.

7.2.3.1. *Finite differences*

When dealing with discrete data, the most straightforward approach to estimate a gradient is the use of finite differences. This actually consists of the discretization of the derivating operators based on the use of Taylor expansions between the data points. These approaches were initiated to solve ordinary and partial differential equations, and are not dedicated to the treatment of measurement noise. It is possible to control the filtering by tuning the step of the numerical schemes, as discussed in section 7.3.2, but the means to control the noise remain limited.

7.2.3.2. *Image processing*

The displacement field is a data field, which depends on two variables. It can thus be considered as an image and all the image denoising tools can be used. Nonetheless, these denoising algorithms are aimed at goals such as edge and form detection, which differ from the goals associated with the study of kinematic fields.

Among the image- and signal-processing algorithms, the tool usually used for the smoothing of displacement fields and strain reconstruction is the convolution of the measurements with a Gaussian kernel. Such an approach is detailed and applied to a simple one-dimensional (1D) case in section 7.3.4, and can easily be extended to the two-dimensional (2D) case. Let us also note that it is not mandatory for the kernel to be Gaussian, as presented in section 7.4.4.

The parameter that enables filtering tuning is the span of the Gaussian, which can be characterized by its standard deviation. As detailed in section 7.3.4, one of the problems arising with such kernel convolutions is related to reconstruction around the edge of the measurement domain and, in the case of a Gaussian kernel, to the fact that its span is not bounded.

7.2.3.3. *Methods based on interpolation/approximation*

This last group of methods will be detailed in the following. It corresponds to the approximation methods that consist of projecting the measurements on a given basis and then only differentiating the basis functions (these methods are detailed in section 7.4). Through the choice of regular basis functions, we can perform a satisfactory filtering of the noise; nonetheless, the basis must be rich enough to describe well the mechanical field whose gradient is sought. The projection of the measurements on the basis is usually performed using a least-squares minimization.

To use such approaches, we have to answer the following two questions:

– How do we choose the basis to project the measurements on?

– Which least-squares formulation should we use? In particular, the formulation can be local or global, weighted or not.

7.2.4. *Quality of the reconstruction*

During filtering, a compromise is sought between reducing the effect of noise on the strain field and losing mechanical information due to an excessive smoothing of the physical field to be reconstructed. We must, therefore, characterize both effects and quantify them. To this end, we can reconcile two complementary points of view: one from the *measurements* and another from the *computation*.

7.2.4.1. *Resolution and spatial resolution*

Considering the whole process from the digital images to the strain as a measurement means that it is possible to apply the ideas of resolution and spatial resolution, which are standard for measurements. These quantifications of errors, introduced in Chapter 1, can be adapted to the case of strains [BER 09].

Resolution will thus characterize the level of strain that can be reliably detected at a given reconstruction point. Spatial resolution will then be associated with the spatial correlation of the reconstructed strain field and, in particular, will characterize the minimum distance between two strain zones that can be differentiated by observation. This distance depends partially on the strain field to be observed and, to avoid this drawback, we can also associate it with the span of the zone that is influenced by a local phenomenon. Some ways to characterize spatial resolution are given below:

– The study of the autocorrelation of the strain field reconstructed from white noise will give some information about the way the data will be correlated.

– For a strain field of a given shape, whose wavelength can be chosen in accordance with the chosen application, it is possible to determine the minimum length that allows two sources to be differentiated from a strain point of view.

– The reconstruction of the strain field from a displacement consisted of a Dirac function at a given measurement point which directly yields the span of the strain field associated with the data at this point. This can be related to the shape functions of the reconstruction operator, as presented in section 7.4.4.

7.2.4.2. *Random error and approximation error*

Another approach is to consider, at each reconstruction point, the true error, corresponding to the distance between the exact strain field and the reconstructed strain field. It can be split into two groups:

– An approximation error $\delta\varepsilon_k$, which corresponds to the systematic error of the reconstruction operator. This error is related to the loss of mechanical information about the reconstructed strain field.

– A random error $\delta\varepsilon_b$, which is related to the level of noise filtering.

Thus, the reconstructed strain can be expressed at each reconstruction point:

$$\underline{\varepsilon}_{ap}(\underline{x}) = \underline{\varepsilon}_{ex}(\underline{x}) + \delta\underline{\varepsilon}_k(\underline{x}) + \delta\underline{\varepsilon}_b(\underline{x}) \qquad [7.6]$$

In most cases, the reconstruction operator is a linear operator (e.g. a standard least-squares formulation leads to a linear system through the minimization of a quadratic criterion); therefore, the two types of error are not merged and can be studied separately:

– The approximation error $\delta\underline{\varepsilon}_k$ is associated with the reconstruction of the exact field only $\underline{u}_{ex}$.

– The random error $\delta\underline{\varepsilon}_b$ is related to the reconstruction of the measurement noise only $\delta\underline{u}$.

Considering a real case, such a split can not be performed due to the fact that the exact field is unknown, but it enables the characterization of the reconstruction of the strain, from a theoretical point of view or on the basis of numerical examples.

7.2.4.3. A link between the two approaches

Both proposed approaches focus on the quality of the reconstruction, and some links can be established between them.

Resolution is thus mainly related to the random error associated with the reconstruction operator (nonetheless, due to the approximation error, the reconstructed strain dynamics are smaller than those of the true strain field and should theoretically be taken into account). It will be characterized as the standard uncertainty of measurement, from a statistical point of view, as will be detailed in section 7.5.3.

The link between the spatial resolution and the approximation error is less straightforward because the latter is a local quantity. However, both are related to the loss of information regarding the mechanical field to be reconstructed. Spatial resolution should be compared to the *natural* spatial correlation of the strain field to be reconstructed, in order to discuss whether or not a local phenomenon can be observed. This correlation of the field to be reconstructed can be related, through a Taylor expansion, to the values of the local derivatives of the field, which are directly related to the approximation error (see section 7.5.2). This remark confirms the link between the two approaches even though it remains difficult to quantify, in particular, in a case where we aim at spatially distinguishing between two mechanical events; hence, there is the case of detecting phenomena not requiring, in particular, high precision in estimating their levels.

7.3. Differentiation: difficulties illustrated for a one-dimensional example

7.3.1. *A simple one-dimensional example*

To emphasize the problems related to the filtering of some measurements in order to obtain their derivatives, a first 1D example is discussed. Let us consider a set of measurements on an interval $[0, T]$ at given times t_i, regularly spaced with a time step Δt:

$$\tilde{f}_i = f_{\text{ex}}(t_i) + \delta f_i, \quad i \in [0, N] \quad \text{and} \quad t_i = i\frac{T}{N} \qquad [7.7]$$

where $f_{\text{ex}}(t)$ corresponds to the function whose derivative is to be estimated and $\delta f_i, i \in [0, N]$, corresponds to the sample of Gaussian white noise with standard deviation σ_f.

In this section, the chosen function $f_{\text{ex}}(t)$ is as follows:

$$f_{\text{ex}}(t) = \tan^{-1}\frac{t - \frac{T}{2}}{\tau}, \qquad \text{where} \quad \tau = \frac{1}{30} \quad \text{and} \quad T = 1 \qquad [7.8]$$

The interval $[0, T]$ is discretized with $N = 1,000$ and the standard deviation σ_f is chosen as 10% of the mean magnitude of f_{ex}. An example of measurements for $\tilde{f}$ is given in Figure 7.1; the choice of this function is driven by the fact that its derivative evolves rapidly around $t = T/2$, which implies that the smoothing is only moderate.

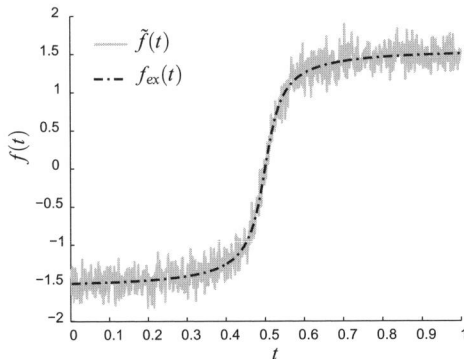

Figure 7.1. *Example of measurements for $\tilde{f}$*

From these measurements for $\tilde{f}$, we aim to reconstruct an approximation $f'_{ap}(t)$ of the derivative $f'_{ex}(t)$. As mentioned above, $f'_{ap}(t)$ can be split into three terms: the exact derivative, the approximation error and the random error:

$$f'_{ap} = f'_{ex} + \delta f'_k + \delta f'_b \qquad [7.9]$$

7.3.2. *Finite differences*

The first method applied to reconstruct f'_{ap} is based on a first-order finite differences scheme. f'_{ap} is reconstructed at each time step i as:

$$f'_{ap_i} = \frac{\tilde{f}_{i+1} - \tilde{f}_i}{\Delta t} \qquad [7.10]$$

From expression [7.7] of $\tilde{f}$ and the first-order Taylor expansion of f_{ex} around t_i, f'_{ap_i} is written as:

$$
\begin{aligned}
f'_{ap_i} &= \frac{f_{ex}(t_{i+1}) - f_{ex}(t_i)}{\Delta t} + \frac{\delta f_{i+1} - \delta f_i}{\Delta t} \\[2mm]
&= f'_{ex}(t_i) + \underbrace{\varepsilon(t)}_{\delta f_k} + \underbrace{\frac{\delta f_{i+1} - \delta f_i}{\Delta t}}_{\delta f_b} \quad \text{with} \quad \lim_{t \to t_i} \varepsilon(t) = 0
\end{aligned}
\qquad [7.11]
$$

where we can observe the two errors: the approximation error δf_k, related to the Taylor remainder of f_{ex}, and the random error δf_b. As Δt tends to 0, the approximation error tends to 0, whereas the random error tends to infinity (δf_i and δf_{i+1} being random values, they do not compensate in the expression of δf_b, and their magnitude is about that of the standard deviation of the measurement noise). This effect is outlined in Figure 7.2, where the uncertainty on the slope of the curve (represented by the angle between the two dotted lines) is represented for the same measurement uncertainty and two different time steps.

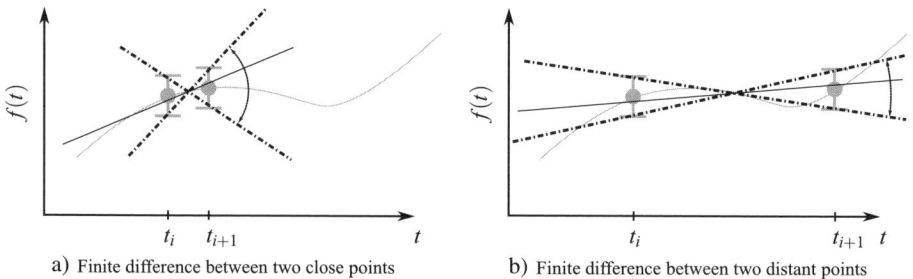

a) Finite difference between two close points   b) Finite difference between two distant points

Figure 7.2. *Estimation of the derivative by finite differences*

As a result, a small time step ensures a limited approximation error but implies a large random error. On the contrary, a larger time step (e.g. keeping the same finite differences scheme, but based on non-adjacent points) enables better noise filtering but increases the Taylor remainder and, thus, the approximation error.

7.3.3. *Global least squares – polynomial basis*

From the same measurements for $\tilde{f}$, the derivative of f_{ex} is sought through a polynomial approximation. $f_{\mathrm{ap}}(t)$ is thus sought as:

$$f_{\mathrm{ap}}(t) = \sum_{n=0}^{N_k} F_n t^n \qquad [7.12]$$

where f_{ap} is described through the coefficients $F_n, n \in [0, N_k]$, which can be determined by the minimization of a least-squares criterion (allowing the smallest distance between the measurements and the approximation):

$$\min_{F_1,\ldots,F_{N_k}} \sum_{i=0}^{N} \left(f_{\mathrm{ap}}(t_i) - \tilde{f}(t_i) \right)^2 \qquad [7.13]$$

Minimization [7.13] leads to a linear system from which the F_n are deduced. The reconstruction is performed on the measurements described in section 7.3.1 with and without measurement noise and for various degrees of the polynomial basis, from 1 to 120. We can then study both the random error and the approximation error.

Figure 7.3 shows the reconstructed derivative for the two types of measurements and a polynomial of degree 55. The derivative is represented on two separate plots, respectively, on the interval $[0, 1T; 0, 9T]$ and on the edge $[0, 9T; T]$, because the error on the edge is much larger than anywhere else. This is one of the main drawbacks of global least-squares with a polynomial basis: a (non-constant) truncated polynomial can hardly remain constant on a large interval. Furthermore, this edge effect becomes stronger with the choice of a high degree for the basis, which is necessary to describe well the richness of the function over the whole domain. As Figure 7.3 confirms, the use of a high-degree polynomial leads to local oscillations, even on the inner domain.

The study of the random error and the approximation error is therefore performed on the interval $[0, 1T; 0, 9T]$, omitting the edges where the reconstruction is not relevant. The errors presented in Figure 7.4 are relative errors, normalized with the mean value of f_{ex} on the studied domain. They are represented as a function of the degree of the basis. It is clear that the random error reaches reasonable values on f sooner than on f'. As the degree of the basis increases, the approximation error decreases, whereas the random error increases. In this example, the error level on the derivatives remains, in any case, high due to the fact there is no basis degree value for which both errors are small enough. The smallest error seems to be reached when both errors are equal, corresponding to a basis of $55°$. The corresponding functions are shown in Figure 7.3.

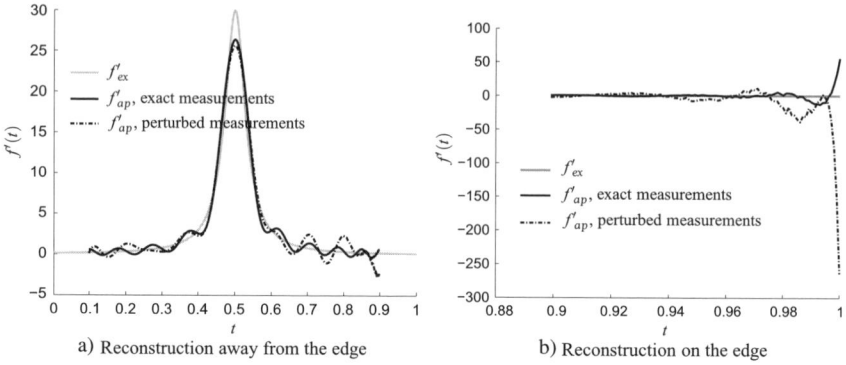

Figure 7.3. *Reconstructed derivatives – global 1D polynomial with 55°*

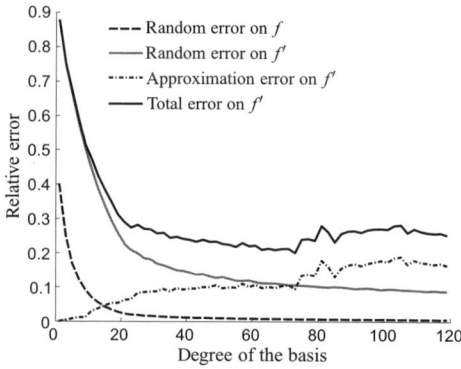

Figure 7.4. *Relative error on the reconstruction*

7.3.4. *Filtering through a convolution kernel*

Filtering through a convolution kernel is often used in signal processing. The idea is to perform a mean of the function to be approximated, with a weighting of the points depending on their distance from the reconstruction point. From a continuous point of view, we consider a time function $K(t)$ and perform the convolution between this function and the measured function $\tilde{f}(t)$ as follows:

$$f_{\mathrm{ap}}(t) = (K * \tilde{f})(t) = \int_{\mathbb{R}} K(t - \tau)\, \tilde{f}(\tau)d\tau \qquad [7.14]$$

Furthermore, $K(t)$ is to be normalized:

$$\int_{\mathbb{R}} K(\tau)d\tau = 1 \qquad [7.15]$$

The derivative of the approximated function stands as:

$$\frac{d}{dt}f_{\mathrm{ap}}(t) = \frac{d}{dt}(K * \tilde{f}) = K * \frac{d\tilde{f}}{dt} = \frac{dK}{dt} * \tilde{f} = \int_{\mathbb{R}} \frac{dK}{dt}(t - \tau)\,\tilde{f}(\tau)d\tau \quad [7.16]$$

Equation [7.16] means that the derivative of the approximated function corresponds to the convolution of $K(t)$ with the derivative of the measurements, which corresponds to the sought function. Equation [7.16] also means that it can be practically estimated through the convolution of the derivative of $K(t)$ with the measurements. In equation [7.14], the integral is performed on the whole of $\mathbb{R}$. In practice, $\tilde{f}$ is only defined on $[0, T]$. There are therefore two possibilities:

– Extending $\tilde{f}$ out of the domain $[0, T]$, for example through symmetry conditions at 0 and T, and through periodicity conditions. These conditions will affect the calculation of the derivatives and could lead to non-physical properties (e.g. a symmetry condition at $t = 0$ leads to $f'_{\mathrm{ap}}(0) = 0$);

– Choosing a function K with a bounded span (that will be assumed to be symmetric and centered on 0) with length $2T_K$, and limiting the reconstruction to the points t such that the span of $K(t - \tau)$ remains within the span of $\tilde{f}(\tau)$. It implies the evaluation of $f_{\mathrm{ap}}(t)$ only on $[T_K, T - T_K]$.

A standard choice for $K(t)$ is the use of a Gaussian, whose standard deviation will control the size of the zone contributing to the reconstruction at any point. The Gaussian, even if its values decrease rapidly, is not defined on a bounded span, giving rise to the problems mentioned above. A way to circumvent this drawback is to truncate $K(t)$, for example at three standard deviations, in order to recover a bounded span. It is this truncated function that needs to be normalized.

As the measurements are actually in a discrete grid of points, equations [7.14] and [7.16] must be rewritten in a discrete manner, which gives, for K with a bounded span:

$$\begin{cases} f_{\mathrm{ap}}(t_k) = \displaystyle\sum_{l=0}^{N_t} K(t_k - t_l)\tilde{f}(t_l) \\[3mm] f'_{\mathrm{ap}}(t_k) = \displaystyle\sum_{l=0}^{N_t} K'(t_k - t_l)\tilde{f}(t_l) \end{cases} \qquad \forall k \in \{N_K, N - N_K\}, \quad \text{where} \quad N_K = \frac{T_K}{dt}$$

$$[7.17]$$

In the following example, the convolution kernel is not a Gaussian, but a third degree polynomial with a bounded span, given as:

$$\begin{cases} K(t) = \left(1 - 3\left(\dfrac{t}{T_K}\right)^2 + 2\left(\dfrac{t}{T_K}\right)^3\right) & \text{if} \quad 0 \le \dfrac{t}{T_K} \le 1 \\[3mm] \quad\;\; = 0 \quad \text{if} \quad 1 \le \dfrac{t}{T_K} \\[3mm] K(-t) = K(t) \end{cases} \qquad [7.18]$$

This polynomial is close to a Gaussian, but the polynomial and its derivative are continuous at the boundaries of its span. It is shown in Figure 7.5, along with its first-order derivative.

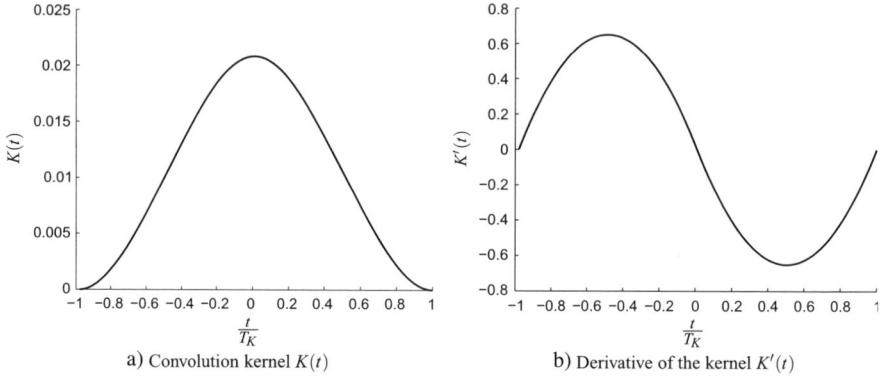

a) Convolution kernel $K(t)$ b) Derivative of the kernel $K'(t)$

Figure 7.5. *Cubic spline as a convolution kernel*

The measurements presented in section 7.3.1 are processed through a convolution based on the kernel proposed in equation [7.18], for various values of T_K (or $N_K = T_K/dt$). Figure 7.6(a) shows, for perturbed and unperturbed measurements, the reconstructed derivative for $N_K = 45$, which corresponds to the minimal total error in Figure 7.6(b). It can be noted that the reconstruction cannot be performed up to the boundary of the domain. When compared to the reconstruction based on global polynomial least-squares, the reconstruction from unperturbed data is much smoother and the drawback of local oscillations has disappeared. The approximation error is hence more limited and is concentrated close to $t = T/2$, where the derivative evolves rapidly compared to the span of K, T_K. A way to circumvent this problem could be to adapt the span of $K(t)$ with respect to the position of the reconstruction point.

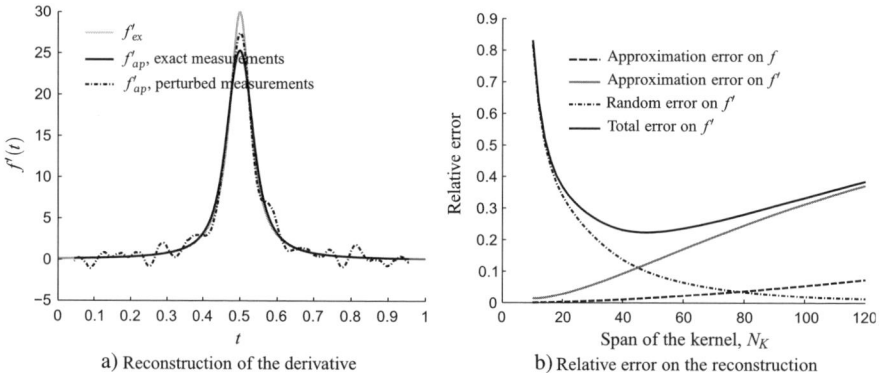

a) Reconstruction of the derivative   b) Relative error on the reconstruction

Figure 7.6. *Reconstructed derivative and relative error*

Figure 7.6(b) shows the relative errors on the whole reconstruction domain. When the parameter N_K increases, the random error decreases, whereas the approximation error increases, so that a compromise between the two types of error is to be found. On this plot, the minimum error is reached for values of N_K such that both errors have the same magnitude. Furthermore, the error on the function is smaller than the error on the derivative, which cannot be reduced below a given threshold.

7.4. Approximation methods

7.4.1. *General presentation*

The previous 1D example allowed the description of three approaches; the first approach was naive, the second approach was based on a global approximation and the third approach used a local mean. The latter gave the best results; nonetheless, the approximation methods, such as the second approach, remain interesting for at least two reasons:

– They are based on the minimization of a criterion and, hence, guarantee the reconstruction to be the best in a given manner.

– The basis that is used for the approximation can be chosen or enriched from mechanical considerations.

For these reasons, we will analyze the approximation methods in this section, keeping in mind that the approach based on a local mean was the one that yielded the best results. This leads us to propose two approximation methods and apply them to the numerical example of an open-hole tensile test on a linear isotropic elastic material. This example is interesting because the gradient of the strain field is large around the hole.

A brief presentation of these methods is already proposed in section 7.2.3.3 and, from a general point of view, the two main questions to address with these methods are the following:

– The choice of the approximation basis. The literature offers a large variety of bases for these methods: polynomials, splines, finite elements, etc. The choice of the basis will determine the quality of the reconstruction; if the basis is too poor, the reconstructed field will hardly reproduce the mechanical displacement, and if the basis is too rich, especially in high frequency, the filtering of the noise will be difficult. The relevancy of the reconstruction will largely depend on the choice of the basis.

– The choice of a global formulation or a local formulation for the least squares. For the same functional basis, the reconstructed displacement can be sought from all the data points or at each reconstruction point from the data around it (in a neighborhood to be defined).

The various approximation approaches will depend on the various choices concerning these two points [CLE 95, HIC 99, LIR 04].

The 1D example in section 7.3.3 showed that a global basis and a global formulation lead to large local oscillations when we want to reduce the approximation error. This means that it would be better to keep some of the treatment local. This can be performed by choosing basis functions whose spans are bounded or by choosing a local formulation. The two methods detailed hereafter are each based on one of these possibilities.

7.4.2. *Global least squares – Finite element basis*

This method is an approximation method based on global least squares over the whole measurement zone Ω. The approximation basis is chosen as a finite element basis [FEN 91] due to the fact that such functions have a bounded span (each span being limited to the elements including the node corresponding to a given shape function). The parameter to be tuned in order to control the filtering will be the mesh size, denoted by h (it would also be possible to change the degree of the shape functions). In the following examples, triangular elements with linear shape functions are used [AVR 10].

The reconstructed displacement field at any $\underline{x}$ is written as:

$$\underline{u}_{ap}(\underline{x}) = [\phi(\underline{x})] \{U\} \hspace{3cm} [7.19]$$

where $[\phi(\underline{x})]$ is the matrix collecting the shape functions evaluated at $\underline{x}$ and $\{U\}$ is the column vector of the nodal displacements. For a given mesh, the reconstructed displacement field is completely determined through $\{U\}$.

Let us denote by $[\Phi]$ the block matrix made up of the blocks of matrices $[\phi(\underline{x}_i)]$, $i \in [1, N]$, corresponding to the values of the shape functions at the measurement points. Hence, the reconstructed field can be expressed at the measurement points as:

$$\{u_{ap}\} = [\Phi] \{U\} \hspace{3cm} [7.20]$$

where $\{u_{ap}\}$ is the column vector collecting the values of $\underline{u}_{ap}$ at the data points: $\underline{u}_{ap}(\underline{x}_i)$, $i \in [1, N]$. $\{U\}$ can then be defined as the solution of the following minimization problem:

$$\min_{\{U\}} \sum_{i=1}^{N} \|[\phi(\underline{x}_i)]\{U\} - \tilde{\underline{u}}(\underline{x}_i)\|^2 \Leftrightarrow \min_{\{U\}} ([\Phi]^t\{U\} - \{\tilde{u}\})^t([\Phi]^t\{U\} - \{\tilde{u}\})$$

$$[7.21]$$

The minimization problem [7.21] leads to a linear system whose solution provides $\{U\}$:

$$\{U\} = [S]^{-1}[\Phi]^t\{\tilde{u}\}, \quad \text{where} \quad [S] = [\Phi]^t[\Phi] \tag{7.22}$$

The strain field can be directly deduced from $\underline{u}_{ap}(\underline{x})$ with the derivatives of the shape functions, leading to:

$$\underline{\varepsilon}_0(\underline{x}) = [b(\underline{x})]\{U\} \tag{7.23}$$

where $[b(\underline{x})]$ collects the components of the symmetrical gradient of $[\phi(\underline{x})]$ and $\underline{\varepsilon}_0(\underline{x})$ corresponds to the strain at point $\underline{x}$. The strain at the measurement points can be deduced by:

$$\{\varepsilon_0\} = [B]\{U\} \tag{7.24}$$

where $\{\varepsilon_0\}$ is the column vector formed from $\underline{\varepsilon}_0(\underline{x}_i)$, $i \in [1, N]$, and $[B]$ corresponds to the block matrix formed from blocks $[b(\underline{x}_i)]$, $i \in [1, N]$.

In the present case where the shape functions are linear, the field $\underline{\varepsilon}_0$ is discontinuous at the element boundaries. For some applications, a continuous strain field can be useful; such a field can be obtained by reprojecting in a least-squares manner the strain field on the shape functions. The approximated strain field $\underline{\varepsilon}_{ap}(\underline{x})$ is therefore introduced as:

$$\underline{\varepsilon}_{ap}(\underline{x}) = [\phi(\underline{x})]\{E\} \tag{7.25}$$

where $\{E\}$ corresponds to the vector of the nodal strains. These nodal strains $\{E\}$ are determined as the solution of the minimization problem:

$$\min_{\{E\}} \int_\Omega \|[b(\underline{x})]\{U\} - [\phi(\underline{x})]\{E\}\|^2 d\Omega \tag{7.26}$$

In equation [7.26], the integrals are estimated through numerical integration, here using one Gauss point per element, leading to a linear system involving matrices corresponding to standard mass matrices or stiffness matrices. $\{E\}$ is expressed as:

$$\{E\} = [S]^{-1}[S^\alpha][S]^{-1}[\Phi]^t\{\tilde{u}\} \quad \text{where} \quad [S^\alpha] = [B]^t[\Phi] \tag{7.27}$$

As a result, this approach reconstructs a continuous strain field, completely described by the nodal strain vector $\{E\}$. The parameters that can be tuned in order to control the reconstruction are the mesh (mainly the local mesh size h) and the degree of the shape functions (which will not be addressed in the following examples).

7.4.3. *Local least squares – polynomial basis*

A second group of methods relies on the use of local least squares [CLE 95] and is discussed here with the DA [NAY 91], which was first developed for solving partial differential equations, and then was applied to various domains such as metamodeling for optimization [BRE 05] or field transfer [BRA 08]. The main advantage of its use for approximation purposes is its flexibility and the large scope of possibilities, as well as its theoretical framework.

The basis functions are chosen as polynomials of one degree or more. The approach is based on weighted local least squares, which means that the reconstruction at any point is based on the data for a neighborhood of the point and that a larger weighting is given to the points closer to the reconstruction point. The size of the neighborhood is determined through the span of the weighting function and is denoted by R. This span is the main parameter to be tuned to find the compromise between approximation and filtering (or resolution and spatial resolution).

The idea of DA is to construct a continuous field, defined at any $\underline{x} = (x, y)$, from a discrete cloud of data points at positions $\underline{x}_i, i \in [1, N]$, with the following approach. Let us consider a function basis, here polynomial, stacked in a vector $p(\underline{x})$. For example, for a two degree polynomial basis:

$$p(\underline{x}) = <1 \quad x \quad y \quad \frac{x^2}{2} \quad xy \quad \frac{y^2}{2} > \qquad [7.28]$$

The diffuse field is then defined as:

$$u_{ap}(\underline{x}) = v(\underline{x}, \underline{x}), \quad \text{where} \quad v(\underline{x}, \underline{x}') = p(\underline{x}' - \underline{x}) \{a(\underline{x})\} \qquad [7.29]$$

where $\{a(\underline{x})\}$ is a coefficient vector depending on $\underline{x}$ and is determined through the local least squares associated with $\underline{x}$. The v function introduced here allows the presentation of DA centered on the reconstruction point (enabling us to establish a link with the Taylor expansion at the considered point [NAY 91]). For a one degree polynomial basis, $v(\underline{x}, \underline{x}')$ is written as:

$$v(\underline{x}, \underline{x}') = a_1(\underline{x}) + a_2(\underline{x})(x' - x) + a_3(\underline{x})(y' - y) \qquad [7.30]$$

At any reconstruction point $\underline{x}$, $a(\underline{x})$ is defined as the solution of the following local weighted least-squares problem, defined for the neighborhood $V(\underline{x})$ of $\underline{x}$:

$$\min_{a(\underline{x})} \frac{1}{2} \sum_{\underline{x}_i \in V(\underline{x})} w(\underline{x}, \underline{x}_i) \left(p(\underline{x} - \underline{x}_i) \{a(\underline{x})\} - \tilde{u}(\underline{x}_i) \right)^2 \qquad [7.31]$$

where $\underline{x}$ is a constant with respect to the minimization and $w(\underline{x}, \underline{x}_i)$ is a weighting function evaluated at each data point. In matrix form, equation [7.31] can be expressed as:

$$\min_{\{a\}} \frac{1}{2} [[P]\{a\} - \{\widetilde{U}_{\underline{x}}\}]^T [W][[P]\{a\} - \{\widetilde{U}_{\underline{x}}\}] \qquad [7.32]$$

where $[P]$ is a matrix collecting the lines $p(\underline{x} - \underline{x}_i)$, $\underline{x}_i \in V(\underline{x})$, $\{\widetilde{U}_{\underline{x}}\}$ is the vector collecting the measurements at the data points of the neighborhood $V(\underline{x})$ and $[W]$ is the diagonal weighting matrix, with $W_{ii} = w(\underline{x}, \underline{x}_i)$. $w(\underline{x}, \underline{x}_i)$ can be any positive function defined on a bounded domain, leading to a local reconstruction. Here, the weighting function is defined, taking advantage of the fact that the data grid is a regular square grid, as:

$$w(\underline{x}, \underline{x}_i) = w_{\text{ref}}\left(\frac{x - x_i}{R_x}\right) w_{\text{ref}}\left(\frac{y - y_i}{R_y}\right) \qquad [7.33]$$

where w_{ref} is a dimensionless function zeroing itself, as well as its first derivative, at 0 and 1 (this ensures the continuity of the DA and its first derivative); the cubic spline introduced in equation [7.18] is chosen as the w_{ref} function. Here, R_x and R_y are functions of $\underline{x}$ but not of $\underline{x}_i$, and are chosen as equal to the same value R, corresponding to the span of the weighting function. Such choices are not mandatory but simplify the numerical implementation of the method without too large a loss of generality.

$\{a(\underline{x})\}$ as the solution to [7.32] is then given by:

$$\{a\} = [M]\{\widetilde{U}_{\underline{x}}\}, \qquad \text{where} \quad [M] = [P^T W P]^{-1} P^T W \qquad [7.34]$$

If the basis is of one degree or more, the diffuse field and its first derivative are directly deduced from the first components of vector $\{a\}$:

$$u_{ap}(\underline{x}) = a_1(\underline{x}), \qquad \frac{\delta u_{ap}}{\delta x}(\underline{x}) = a_2(\underline{x}) \quad \text{and} \quad \frac{\delta u_{ap}}{\delta y}(\underline{x}) = a_3(\underline{x}) \qquad [7.35]$$

where $\delta u/\delta x$ corresponds to the diffuse derivative and is an approximation of the exact derivative. The reconstruction can be performed for each component of the measured displacement field $\tilde{u}(\underline{x}_i)$ and the strain field deduced from the first-order diffuse derivatives. The diffuse field is defined at any point by solving problem [7.32] at that point. Hence, some evaluation points of the diffuse field must be chosen, depending on the application. Here, the evaluation points are chosen as the data points of the measurement grid: $\underline{x}_i$, $i \in [1, N]$.

Finally, let us note that this approach enables the reconstruction to be performed up to the edge of the measurement domain, where problem [7.32] is still defined, even though the neighborhood is no longer symmetric with respect to the reconstruction point. To sum up, the parameters to be tuned to control the reconstruction are the degree of the basis and the span of the weighting function R.

7.4.4. *Three converging points of view*

When the reconstruction operator is linear, the reconstructed displacements and strains are linear functions of the measurements. Denoting by $\{u_{ap}\}$ the displacement vector and by $\{\varepsilon_{ap}\}$ the strain vector, both collecting the reconstructions at the reconstruction points, we can express the following matrix relationships:

$$\{u_{ap}\} = [R^u]\{\tilde{u}\} \quad \text{and} \quad \{\varepsilon_{ap}\} = [R^\varepsilon]\{\tilde{u}\} \tag{7.36}$$

Relationships [7.36] can be analyzed from three different points of view:

– The reconstruction approach, considering the reconstruction at a given point and expressing the contribution of the various data points. This is obtained by keeping one of the lines of the system [7.36]:

$$\varepsilon_{ap_i} = \sum_j R^\varepsilon_{ij}\tilde{u}_j = < R^\varepsilon_{i1}, \ldots, R^\varepsilon_{iN} > \{\tilde{u}\} \tag{7.37}$$

Such a relationship can be interpreted as a convolution product (see section 7.3.4).

– The data view, considering the contribution of one given datum point to the reconstruction at the various reconstruction points. This can be described by rewriting the system [7.36] column by column:

$$\{\varepsilon_{ap}\} = \sum_j \{\varepsilon^j_{ap}\}, \quad \text{where} \quad \{\varepsilon^j_{ap}\} = \begin{bmatrix} R^\varepsilon_{1j} \\ \ldots \\ R^\varepsilon_{Nj} \end{bmatrix} \tilde{u}_j \tag{7.38}$$

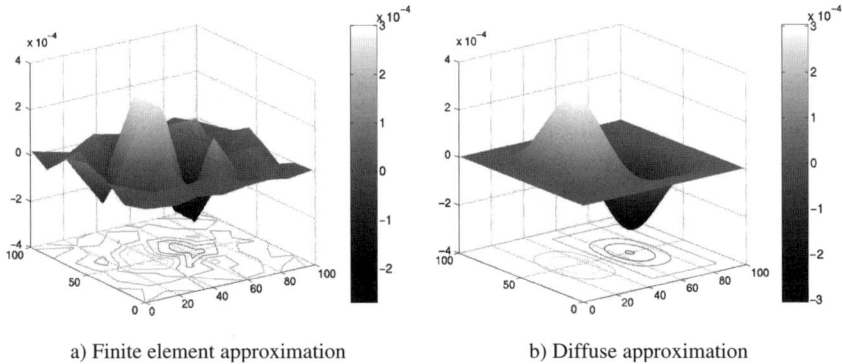

a) Finite element approximation   b) Diffuse approximation

Figure 7.7. *Shape functions associated with the reconstruction operator of the first-order derivative*

This leads to the shape functions of the reconstruction operator. These functions are illustrated for the two methods detailed above in Figure 7.7, for the first-order gradient. The spatial resolution is directly related to the span of the shape functions.

– The least-squares and approximation methods point of view. To each approximation method, there corresponds a convolution kernel and shape functions (even if they are not explicitly described). These methods can thus be considered as a way to construct shape functions or convolution kernels that are optimal in a given sense.

7.5. Behavior of the reconstruction methods

7.5.1. *Splitting the reconstruction error*

As mentioned in section 7.2.4.2, the reconstructed strain field can be split into three parts: the exact field, the reconstruction error and the random error:

$$\varepsilon_{ap}(\underline{x}) = \underline{\varepsilon}_{ex}(\underline{x}) + \delta\underline{\varepsilon}_{k}(\underline{x}) + \delta\underline{\varepsilon}_{b}(\underline{x}) \qquad [7.39]$$

A good reconstruction of the strain field is one such that the random error $\delta\underline{\varepsilon}_b$ is as low as possible while keeping a reasonable level for the approximation error $\delta\underline{\varepsilon}_k$. To characterize the quality of a reconstruction operator with respect to both errors, the operator can be applied to numerical examples, where knowledge of the exact field $\underline{\varepsilon}_{ex}$ enables the calculation of both types of error (something that cannot be performed on real data). For each error, a local norm of the error is defined, and its map will show the zones with large errors:

$$e_{\varepsilon}^{k}(\underline{x}) = \sqrt{\delta\varepsilon_{k_{XX}}^{2} + 2\,\delta\varepsilon_{k_{XY}}^{2} + \delta\varepsilon_{k_{YY}}^{2}} \quad \text{and} \quad e_{\varepsilon}^{b}(\underline{x}) = \sqrt{\delta\varepsilon_{b_{XX}}^{2} + 2\,\delta\varepsilon_{b_{XY}}^{2} + \delta\varepsilon_{b_{YY}}^{2}}$$

$$[7.40]$$

From these local errors, it is possible to define a mean error on a given area Ω' comprising N' reconstruction points:

$$e_{\varepsilon}^{\text{mean}}(\Omega') = \frac{1}{N'}\sum_{i=1}^{N'} e_{\varepsilon}(\underline{x}_i) \qquad [7.41]$$

Figure 7.8 shows the mean errors for a numerical example corresponding to a tensile test on an open-hole plate, with an elastic, homogeneous and isotropic constitutive law. The measurements were created from a finite element calculation, whose solution was projected onto a regular grid of points corresponding to the measurement grid. Furthermore, noise representing 5% of the mean displacement level was added to the finite element field (after the field transfer) to represent measurement perturbations. Figure 7.8(a) presents the mean errors over the whole measurement zone for the two proposed approximation approaches. Both methods yield similar levels of error, and the increase in the characteristic size R or h implies a decrease in the random error and an increase in the approximation error. Concerning the finite element approximation, the curve is less smooth due to the fact that the reconstruction is a function of the mesh itself (as is already observed in Figure 7.7). Figure 7.8(b) shows the mean error around the hole for polynomial diffuse approximations of degrees one–three. The increase in the degree of the basis implies a decrease in the approximation error but an increase in the random error. It seems that a good compromise can be found with the two degree basis, whose filtering ability is close to the degree 1 basis while reducing the approximation error.

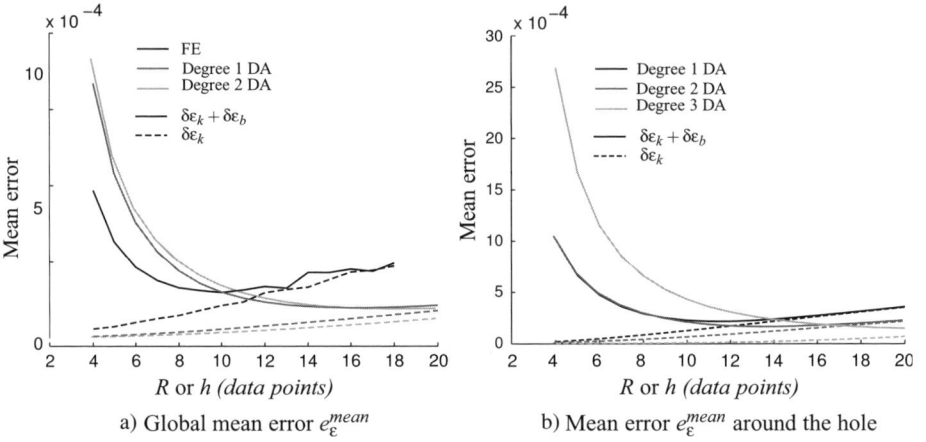

a) Global mean error e_ε^{mean} b) Mean error e_ε^{mean} around the hole

Figure 7.8. *Reconstruction error as a function of R and h*

This example underlines that we have to find filtering parameters that offer a good compromise between the two error types. In the case of real data, the two errors are not directly known and it is necessary to develop estimators for these errors without the knowledge of the exact field.

7.5.2. *Estimation of approximation error*

Such estimators for the approximation error are not straightforward and imply a specific development for each method. Considering the polynomial diffuse approximation, we can show that the approximation error is associated with the Taylor remainder of the expansion associated with the considered basis. In the particular case of a two degree basis and a one-variable function, the Taylor expansion can be written from the polynomial basis as:

$$u_{ex}(x_i) = p(x_i - x)^T \begin{bmatrix} u_{ex}(x) \\ \frac{du_{ex}}{dx}(x) \\ \frac{d^2 u_{ex}}{dx^2}(x) \end{bmatrix} + r(x_i) \qquad [7.42]$$

where r is the Taylor remainder.

Thus, from the decomposition [7.2] of the measurement field, the minimization [7.31] can be expressed, in the case of unperturbed measurements, as:

$$\min \frac{1}{2} \sum_{i \in V(x)} w(x, x_i) \left(p(x_i - x)^T \begin{bmatrix} a_1 - u_{ex}(x) \\ a_2 - \frac{du_{ex}}{dx}(x) \\ a_3 - \frac{d^2 u_{ex}}{dx^2}(x) \end{bmatrix} - r(x_i) \right)^2 \quad [7.43]$$

Vector $\delta \underline{a}_k$ such that $\delta \underline{a}_k = \begin{bmatrix} a_1 - u_{ex}(x) \\ a_2 - \frac{du_{ex}}{dx}(x) \\ a_3 - \frac{d^2 u_{ex}}{dx^2}(x) \end{bmatrix}$ collects the approximation errors

on the field and its derivatives and is the solution to problem [7.43]. This means that the approximation error corresponds to the reconstruction by diffuse approximation of the field of the remainder of the Taylor expansion at x.

The estimation of the approximation error implies, therefore, the estimation of higher-order derivatives in order to estimate the Taylor remainder. As the goal here is to estimate the first-order derivative (for the strain field), it seems inappropriate to develop special efforts for the estimation of higher-order derivatives. Such extra efforts could also lead to a better estimate of the first-order derivative, but the error estimator would no longer be valid [CLE 95]. Such estimators, still remain an open question, and we should rely for now on the pragmatic approach consisting of studying the *a priori* behavior of the filtering tools on representative numerical fields.

7.5.3. *Estimation of random error*

The random error of the reconstruction can be estimated more easily, at least, from a statistical point of view. This section presents, for the two detailed approaches, the way to deduce the covariance on the reconstructed strain from the knowledge of the covariance matrix of the measurement noise (in this chapter, it will be considered as a diagonal matrix, but the same method can be applied in other cases).

7.5.3.1. *Random error: finite elements least squares*

Random error can be characterized by the statistical relationship between the input noise and the output noise of the reconstruction operator. Assuming the input noise is white noise whose covariance is defined by [7.3], vector $\{U\}$ deduced from equation [7.22] is a random vector whose covariance is given by [AVR 10]:

$$\text{cov}(\{U\}) = \gamma^2 [S]^{-1} \quad [7.44]$$

Vector $\{E\}$ (equation [7.27]) is also a random vector, and its covariance is given by:

$$\text{cov}(\{E\}) = \gamma^2 [S]^{-1} [S^\alpha]^t [S]^{-1} [S^\alpha] [S]^{-1} \quad [7.45]$$

From numerical tests, it was verified that the off-diagonal terms of the covariances [7.44] and [7.45] are an order of magnitude lower than the diagonal terms and that the latter were all of the same order of magnitude. As a first-order approximation, it is, therefore, reasonable to consider:

$$\begin{cases} \mathrm{cov}(\{U\}) \cong \gamma^2 \alpha_U^2 [I] \\ \mathrm{cov}(\{E\}) \cong \gamma^2 \alpha_E^2 [I] \end{cases} \qquad [7.46]$$

where $[I]$ is the identity matrix and the coefficients α_U and α_E correspond to the sensitivity to noise. In practice, $\alpha_U \gamma$ corresponds to the mean standard deviation of $\{U\}$ and $\alpha_E \gamma$ corresponds to that of $\{E\}$. α_U and α_E are functions of the mesh size h. From the definition of $[S]$, equation [7.22], we can deduce, assuming the mesh is regular and is such that each node belongs to six elements:

$$\alpha_U \cong \frac{2.6}{h} \quad \text{and} \quad \alpha_E = \frac{0.12}{p_{\text{unit}} h^2} \qquad [7.47]$$

where p_{unit} is the spatial resolution of the measurements, that is the distance between two independent data points, see section 7.2.4.1 or Chapter 1.

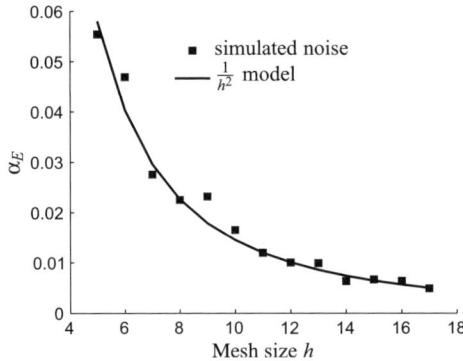

Figure 7.9. *Relative variance of the strain ε_{XX} as a function of the mesh size h*

The correctness of [7.47] has been validated numerically. Figure 7.9 shows the mean standard deviation from simulated data superimposed with the model of equation [7.47]; the two are in good agreement.

7.5.3.2. *Random error: diffuse approximation*

The same study is performed on the DA. Denoting by M_ε the strain reconstruction linear operator, the reconstructed strain field is given as:

$$\underline{\varepsilon}(\underline{x}) = M_\varepsilon \widetilde{U}_{\underline{x}} \qquad [7.48]$$

where $\widetilde{U}_{\underline{x}}$ collects the displacements at the data points contributing to the reconstruction at point $\underline{x}$.

Assuming the measurement noise is Gaussian white noise with standard deviation γ described by [7.3], the covariance of $\underline{\varepsilon}(\underline{x})$ is expressed as:

$$\text{cov}(\underline{\varepsilon}(\underline{x})) = \left\langle \underline{\varepsilon}(\underline{x})\underline{\varepsilon}(\underline{x})^T \right\rangle = \left\langle M_\varepsilon \widetilde{U}_{\underline{x}} \widetilde{U}_{\underline{x}}^T M_\varepsilon^T \right\rangle = \gamma^2 M_\varepsilon M_\varepsilon^T \qquad [7.49]$$

Therefore, it is possible to know the noise sensitivity of the reconstruction operator as well as the level of random error if the level of the measurement noise is known. Sensitivity will be a function of the span of the DA. In fact, the reconstruction operator M, equation [7.34] and the sensitivity matrix can be expressed introducing a dimensionless operator as:

$$M = D_R^{-1} M_{\text{adi}} \quad \Longrightarrow \quad \text{cov}(\{\delta a\}) = \sigma_b{}^2 D_R^{-1} \underbrace{M_{\text{adi}} M_{\text{adi}}^T}_{A} D_R^{-1} \qquad [7.50]$$

where M_{adi} and A take into account the shape of the reconstruction neighborhood, that is the number of points and their relative position. D accounts for the physical size of the neighborhood, corresponding to the span of the diffuse approximation R.

The variances on the displacement and its first-order derivative are established as:

– displacement – order 0 field: $\text{cov}(a_1) = A_{11}$

$\qquad\qquad\qquad\qquad\qquad\qquad\qquad\qquad\qquad\qquad\qquad\qquad$ [7.51]

– strain – order 1 field: $\text{cov}(a_2) = \dfrac{1}{R^2} A_{22}$

The terms of A tend to 0 almost as the inverse of the number of points contributing to the reconstruction (see the central limit theorem). When the span R increases, the physical size of the neighborhood increases, as does the number of contributing data points; both A and D are hence modified and tend to 0. As a result, the filtering level on the derivatives increases faster than that on the displacement (but its initial value is also much larger). This can be understood for the same reasons as for finite differences (Figure 7.2).

Furthermore, we can easily perform this study for reconstruction points close to the boundary of the domain, where the reconstruction neighborhood is no longer symmetric. Figure 7.10 shows the relative variance on ε_{XX} as a function of R and the offset of the reconstruction point with respect to the neighborhood. It confirms that the filtering is less effective near the edges of the domain. The evolution of the filtering as a function of R is linear on a *log-log* scale, confirming the power law deduced from [7.51].

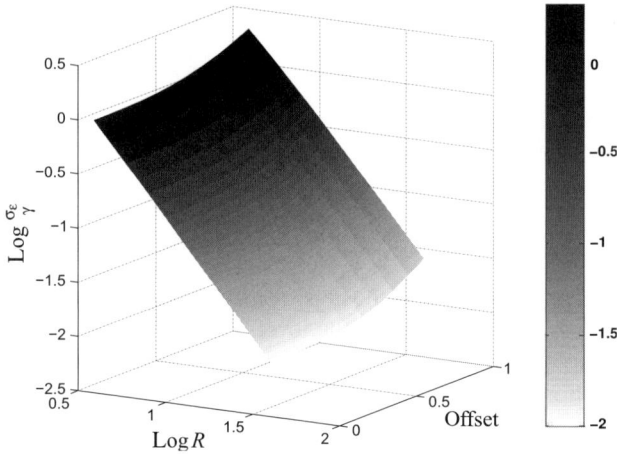

Figure 7.10. *Variance of ε_{XX} as a function of the span R and the offset, log-scale*

7.6. Selection criterion for the filtering parameters

In all the methods for the reconstruction of the strain, at least one parameter can be tuned to best adjust the filtering in order to obtain relevant mechanical information from the strain field. The choice of parameter(s) that can be spatially adapted remains a tricky question to address, and is discussed here through a few examples.

Some authors [BER 09] suggest guiding the choice of the reconstruction operator parameters on the basis of resolution and spatial resolution. Resolution is then characterized like random error is in the previous section, and spatial resolution is characterized due to quantitative tools based on the principles briefly presented in section 7.2.4.1. Such an approach enables us to estimate the link between, on the one hand, resolution and spatial resolution and, on the other hand, the reconstruction parameters; therefore, it can help in the choice, depending on the expected results and the *a priori* knowledge of the sought strain field. In this section, other criteria are presented that take advantage of splitting the error between random and approximation errors.

7.6.1. *Constant signal-to-noise ratio*

From the examples given in section 7.5.1, it seems that a suitable criterion for the choice of filtering parameters would be to find the parameters allowing an approximation error of the same magnitude as the random error, so that the total error would be close to its minimum. Nonetheless, this implies an error estimator for the approximation error, which is not trivial. A much simpler criterion is thus proposed

[AVR 10], which is based on the signal-to-noise ratio of the reconstructed strain. This criterion is adapted for tests with large local strains. In such a case, we would expect a lower filtering in the areas corresponding to these large local strains and the magnitude of the strain field is a good indicator of such zones.

A signal-to-noise ratio is introduced such that:

$$\frac{\varepsilon}{\sigma_\varepsilon} = \eta \qquad\qquad [7.52]$$

where

$- \varepsilon = \sqrt{\varepsilon_{XX}^2 + \varepsilon_{YY}^2 + 2\varepsilon_{XY}^2}$. ε is to be estimated. This can be performed through a first reconstruction with significant filtering.

$- \sigma_\varepsilon$ is defined as $\sigma_\varepsilon = \sqrt{\sigma_{XX}^2 + \sigma_{YY}^2 + 2\sigma_{XY}^2}$, where $\sigma_{o\bullet}$ corresponds to the standard deviation on $\varepsilon_{o\bullet}$ (therefore neglecting the coupling effect of the noise). σ_ε is directly related to h or R through equations [7.46] and [7.49] established in sections 7.5.3.1 and 7.5.3.2.

By choosing a target value of the signal-to-noise ratio η over the whole measurement zone, equation [7.52] yields a value of R or h at any point, and the strain reconstruction can be performed by adapting the filtering parameters spatially. The choice of η might not be straightforward and can be based on previous knowledge or a reasonable mean value coming from a global analysis of the whole measurement zone, for example by seeking the best ratio from a study such as the one performed to obtain Figure 7.8.

In the following example, corresponding to the same numerical tensile test on an open-hole plate as in section 7.5.1, the ratio η is chosen as equal to eight. The optimal mesh and the map of optimal spans, based on [7.52], are presented in Figure 7.11 (assuming the reconstruction behavior is the same near the edges, which is not true and could be modified by adapting R and h on the boundaries). The span map Figure 7.11(b) is not symmetric because it was established from noisy data.

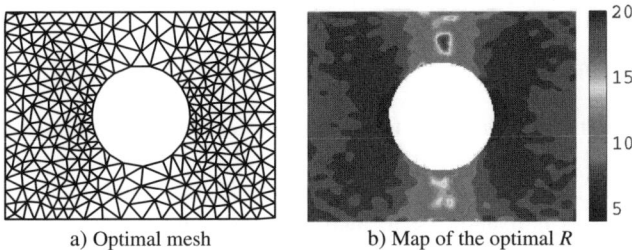

a) Optimal mesh   b) Map of the optimal R

Figure 7.11. *Optimal h and R for $\eta = 8$*

The associated reconstructions of the shear strain are shown in Figures 7.12(a) and (b) and can be compared to the exact field in Figure 7.12(c). Concerning the finite element approximation, we can see the effect of the mesh on the reconstructed field, which implies some repeatability problems, in particular, for identification purposes [AVR 08]. For DA, the reconstruction is smoother, even if some local oscillations associated with the noise remain, coming from a lower spatial resolution leading to faster spatial evolutions. In this example, a larger span all over the zone would actually lead to more accurate results, underlining some limitations of this parameter choice criterion (the choice is based on the first-order derivative, whereas section 7.5.2 showed that the approximation error for a two degree polynomial diffuse approximation was related to the third-order Taylor remainder). Nonetheless, this example illustrates the way the filtering parameters can be spatially tuned.

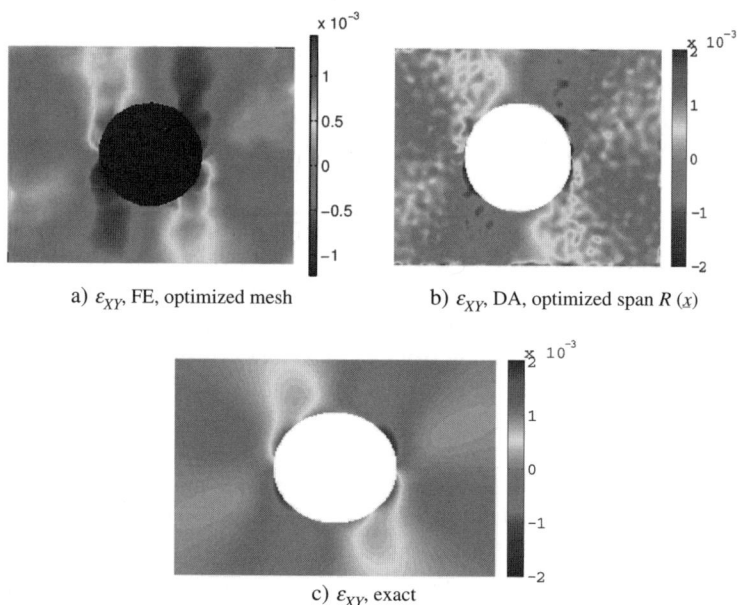

a) ε_{XY}, FE, optimized mesh   b) ε_{XY}, DA, optimized span $R(x)$

c) ε_{XY}, exact

Figure 7.12. *Reconstructed shear strain with the optimized parameters*

7.6.2. A pragmatic criterion

As a first interpretation of the full-field measurement, we can use virtual gauges corresponding to a subpart of the region of interest. The mean strains under the virtual gauge are then estimated and can give global information about the behavior of the specimen, as is usually performed with standard physical extensometry. The virtual gauge can also be used to extract the standard deviation of the measurements under it [FEI 09]. This standard deviation comes from two different sources: the first source is related to the random error associated with the strain reconstruction and the second source is related to the heterogeneity of the mechanical strain field. As illustrated hereafter, the study of this standard deviation can validate the statistical noise filtering model and enable us to propose a pragmatic criterion for the choice of filtering parameters.

The example chosen is a real tensile test performed on an interlock composite, where the full-field measurements are performed by digital image correlation with *Correli-Q4* [BES 06] at the *RVE* scale and then filtered by DA in order to reconstruct the strain field. Therefore, there are two parameters to be chosen in the strain reconstruction process: the element size ℓ in *Correli-Q4* and the span R of the DA. First, the reconstruction is performed on two snapshots taken at the beginning of the test, without any loading, allowing the study of the reconstruction random error. Figure 7.13(a) shows the standard deviation for ε_{XX} under a virtual gauge as a function of R (in pixels of the original image) and for various values of mesh size ℓ, as well as the theoretical filtering curve (section 7.5.3.2) whose first point was fitted to the experimental magnitude level. Both curves have the same behavior, thus validating the theoretical filtering, the experimental curve being lower than the theoretical curve because the latter does not take into account filtering due to the digital image correlation tool.

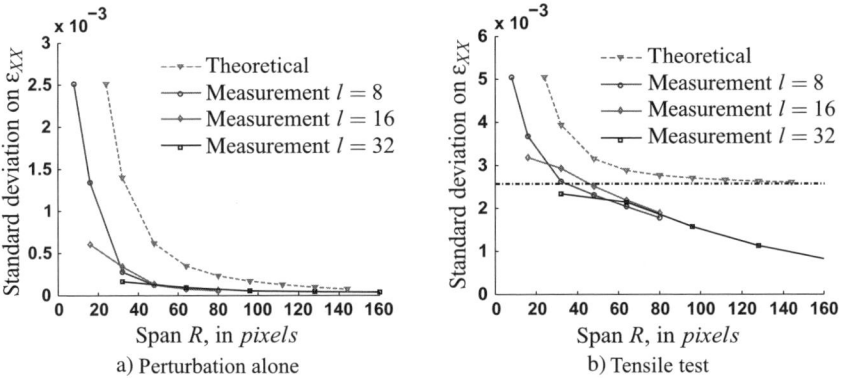

Figure 7.13. *Standard deviation on ε_{XX} as a function of span R for various digital image correlation mesh sizes (in pixels)*

The same study is performed between an unloaded state and a loaded state. Figure 7.13(b) shows again the evolution of the standard deviation under the virtual gauge as a function of R as well as the fitted theoretical filtering, corresponding to the contribution of the random error to the standard deviation. As the level of the standard deviation under the virtual gauge becomes smaller than the limit corresponding to the noise filtering (horizontal line), we can assume that the mechanical field is affected by the filtering. Around this line, there is a compromise to be found between the filtering and the loss of information regarding the deterministic field. Hence, it seems in the present example that a mesh size ℓ of 32 pixels is too large, and that a good compromise could be a mesh size ℓ of eight pixels and a diffuse approximation span R of 32 pixels.

7.7. Taking the time dimension into consideration

The effectiveness of the filtering depends, in particular, on the number of points actually contributing to the reconstruction at a given point (which depends on the span of the shape functions of the reconstruction operator, see section 7.4.4), but an increase in the number of data points in space will deteriorate the spatial resolution (or the approximation error). A simple way to improve the reconstruction, while maintaining the spatial resolution, is to perform a mean of several snapshots taken at the same load level, which will improve the signal-to-noise ratio. However, this implies stopping the loading momentarily, and this might be impossible or inappropriate. In tests where the time evolution is slow enough and the number of snapshots is large enough, we can choose to consider the time dimension in order to improve the filtering. This can be performed at the measurement stage by seeking the displacements as functions of both space and time [BES 12]. In thermography, Gaussian kernels are often used and can easily be extended to the time dimension, either by filtering successively in space and time [CHR 00], or by using a space–time Gaussian kernel. Berthel $et\ al.$ [BER 07] suggest extending the local least squares to both space and time.

Taking the time dimension into consideration is again illustrated with DA, which is extended to the three-dimensional (3D) case of space–time filtering. From a theoretical point of view, the approach remains the same, switching from a $2D$ to a $3D$ framework where the third variable is the time variable. The approximated field is sought as:

$$u_{ap}(\underline{X}) = p(\underline{X})^T a(\underline{X}), \quad \text{where} \quad \underline{X} = (\underline{x}, t) \qquad [7.53]$$

The weighting functions defining the reconstruction neighborhood are simply extended to the space–time case:

$$w(\underline{X}, \underline{X}_i) = w_{\text{ref}}\left(\frac{x - x_i}{R_x}\right) w_{\text{ref}}\left(\frac{y - y_i}{R_y}\right) w_{\text{ref}}\left(\frac{t - t_i}{R_t}\right) \qquad [7.54]$$

The remainder of the formulation is the same as in section 7.4.3, and the filtering behavior can be studied as in section 7.5.3.2, with two parameters to be tuned: the space span R_x and the time span R_t. The filtering level as a function of both spans is shown in Figure 7.14(a) with a log-scale. We can observe a power law and, for the same reason as the one leading to equations [7.51], the increase in the time span has a limited effect on the filtering compared to the effect of the space span.

From this surface, it is possible to draw isofiltering curves in the (R_x, R_t) plane, as presented in Figure 7.14(b). Such curves can guide the choice of the (R_x, R_t) couple, by guaranteeing a given filtering level while improving the spatial resolution. We can however note that, due to differences in filtering behavior with respect to the two parameters, a given filtering level can be achieved with reasonable values of time spans only for space spans that are not too small. The exploitation of these curves is shown in Figure 7.15, where the strain field ε_{XX} (normalized by the load level) for the tensile test on an interlock composite is shown for the space-only filtering approach, and where the space–time method has the same filtering level. It confirms the improvement of spatial resolution and encourages us, when possible, to take the time dimension into consideration.

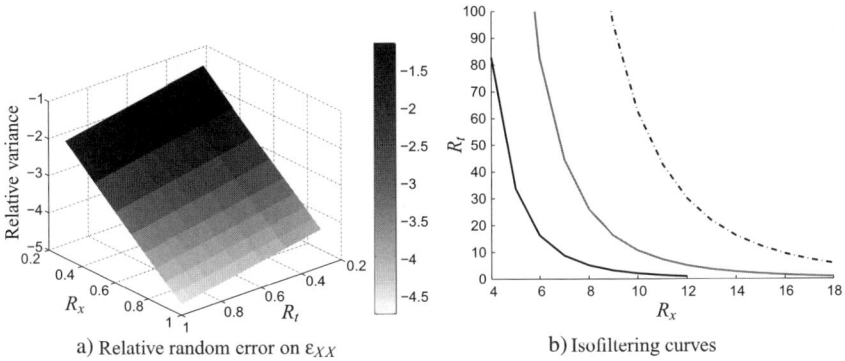

a) Relative random error on ε_{XX} b) Isofiltering curves

Figure 7.14. *Filtering of measurement noise for the space–time diffuse approximation*

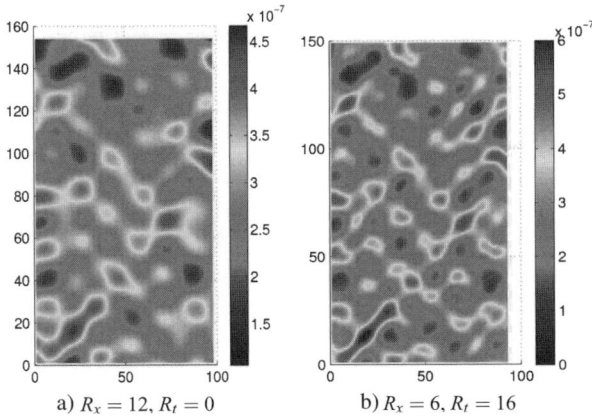

a) $R_x = 12, R_t = 0$ b) $R_x = 6, R_t = 16$

Figure 7.15. ε_{XX}: *isofiltering reconstructions a) space only and b) space–time*

7.8. Concluding remarks

This chapter has presented the main issues related to strain reconstruction from a displacement field. It has been shown that the control of noise is a key point in this transformation and that a compromise is to be found between noise filtering and limiting the loss of mechanical information. Furthermore, it can be observed that for the various methods it is always tricky, and sometimes impossible, to reconstruct proper strain fields at the boundary of the measurement domain.

A key issue concerning these reconstruction methods is the necessary choice of their parameters, and some criteria are needed to guide the decision. Some criteria have been suggested, but further development is still needed to propose error estimators enabling the use of really satisfying criteria.

From the standard approaches presented here, it is possible to suggest some prospects for improvement, for example, through the consideration of a larger amount of *a priori* information. This could be performed through the choice of an approximation basis relying on mechanical knowledge or by introducing, as is often done in inverse approaches, regularization strategies [TIK 77]. To conclude, the reconstruction of the strain field is just one step in the measurement process, and this should be borne in mind when addressing the subject.

7.9. Bibliography

[AVR 08] AVRIL S., FEISSEL P., PIERRON F., VILLON P., "Estimation of strain field from full-field displacement noisy data", *Revue Européenne de Mécanique Numérique*, vol. 17, nos. 5–7, pp. 857–868, 2008.

[AVR 10] AVRIL S., FEISSEL P., PIERRON F., VILLON P., "Comparison of two approaches for differentiating full-field data in solid mechanics", *Measurement Science and Technology*, vol. 21, no. 1, pp. 015703, 2010.

[BER 07] BERTHEL B., WATTRISSE B., CHRYSOCHOOS A., GALTIER A., "Thermographic analysis of fatigue dissipation properties of steel sheets", *Strain*, vol. 43, no. 3, pp. 273–279, 2007.

[BER 09] BERGE-GRAS R., MOLIMARD J., "Optimisation de la dérivation en fonction de la Résolution et de la Résolution Spatiale sur les déformations", 19e *Congrès Français de mécanique*, Marseille, France, AFM, August 2009.

[BES 06] BESNARD G., HILD F., ROUX S., "Finite-element displacement fields analysis from digital images: application to Portevin-Le Châtelier bands", *Experimental Mechanics*, vol. 46, pp. 789–803, 2006.

[BES 12] BESNARD G., LECLERC H., ROUX S., HILD F., "Analysis of image series through digital image correlation", *Journal of Strain Analysis*, vol. 47, no. 4, pp. 214–228, 2012.

[BRA 08] BRANCHERIE D., VILLON P., IBRAHIMBEGOVIC A., "On a consistent field transfer in nonlinear inelastic analysis and ultimate load computation", *Computational Mechanics*, vol. 42, no. 2, pp. 213–226, 2008.

[BRE 05] BREITKOPF P., NACEUR H., RASSINEUX A., VILLON P., "Moving least squares response surface approximation: formulation and metal forming applications", *Computers and Structures*, vol. 83, nos. 17–18, pp. 1411–1428, 2005.

[CHR 00] CHRYSOCHOOS A., LOUCHE H., "An infrared image processing to analyse the calorific effects accompanying strain localisation", *International Journal of Engineering Science*, vol. 38, no. 16, pp. 1759–1788, November 2000.

[CLE 95] CLEVELAND W., LOADER C., *Smoothing by Local Regression: Principles and Methods*, Springer, New York, NY, 1995.

[FEI 09] FEISSEL P., SCHNEIDER J., ABOURA Z., "Estimation of the strain field from full-field displacement noisy data: filtering through diffuse approximation and application to interlock graphite/epoxy composite", *17th International Conference on Composite Materials*, IOM, Edinburgh, UK, pp. 27–31, July 2009.

[FEN 91] FENG Z., ROWLANDS R., "Smoothing finite-element and experimental hybrid technique for stress analyzing composites", *Computers and Structures*, vol. 6, pp. 631–639, 1991.

[HIC 99] HICKERNELL F., HON Y., "Radial basis function approximations as smoothing splines", *Applied Mathematics and Computation*, vol. 102, no. 1, pp. 1–24, 1999.

[LEC 09] LECLERC H., PÉRIÉ J.-N., ROUX S., HILD F., "Integrated digital image correlation for the identification of mechanical properties", in GAGALOWICZ A., PHILIPS W. (eds), *MIRAGE 2009*, vol. LNCS 5496, Springer-Verlag, Berlin, Germany, pp. 161–171, 2009.

[LIR 04] LIRA I., CORDERO R., FRANÇOIS M., VIAL-EDWARDS C., "The uncertainty of experimental derivatives: application to strain measurement", *Measurement Science and Technology*, vol. 15, pp. 2381–2388, 2004.

[NAY 91] NAYROLES B., TOUZOT G., VILLON P., "La méthode des éléments diffus", *Comptes rendus de l'Académie des Sciences*, série 2, Mécanique, Physique, Chimie, Sciences de l'Univers, Sciences de la Terre, vol. 313, no. 2, pp. 133–138, 1991.

[TIK 77] TIKHONOV A., ARSENIN V., *Solutions of Ill-Posed Problems*, Winston and Sons, 1977.

Chapter 8

Introduction to Identification Methods

8.1. Introduction

The previous chapters of this book clearly show the richness, versatility and usefulness of kinematic or thermal full-field measurements. Such measurement methodologies yield a large amount of data, in practice often in the form of digitized image files. Experimental procedures based on the characterization of materials and structures have naturally evolved so as to accommodate this kind of data and exploit it to the full, in particular, when the measured kinematic or thermal fields are heterogeneous. The overall goal is to optimally exploit this kind of data to identify constitutive parameters and, in particular, to estimate as many parameters as possible using as few experiments as possible. Given both the obvious advantages and potentialities afforded by full-field measurements and the widespread needs they cover, their application to the characterization of the mechanical response of materials and structures has become a very active field of research. In this context, this chapter aims at presenting the problem of identification in general terms and surveying the main computational identification approaches applicable to heterogeneous field data.

8.2. Identification and inversion: a conceptual overview

8.2.1. *Inversion*

To know and understand better a physical system requires the gathering and exploitation of relevant experimental data. In many situations, the quantities that are actually being measured do not directly yield the sought information. Rather, the

---

Chapter written by Marc BONNET.

latter is hidden in the physical system under examination, the measurement being the consequence of a cause that is the real quantity of interest.

EXAMPLE 8.1.– A body whose material properties are characterized by a (thermal, electrostatic, etc.) heterogeneous conductivity coefficient $k(\boldsymbol{x})$ occupies the spatial region Ω. A flux q^D is prescribed on the boundary $\partial\Omega$. The functions u (temperature, electrostatic potential, etc.), q^D and k satisfy the equations:

$$\text{div}\,(k\boldsymbol{\nabla}u) = 0 \quad \text{in } \Omega \text{ (local equilibrium equation)}$$
$$k\boldsymbol{\nabla}u\cdot\boldsymbol{n} = q^D \quad \text{on } \partial\Omega \text{ (prescribed flux)}$$

[8.1]

The potential u, assumed to be measurable over $\partial\Omega$, is thus implicitly related to the sought conductivity k.

Exploiting the available experimental data (symbolically denoted by d) thus requires us to formulate a model describing the underlying physics so as to introduce a mathematical (and thus a quantitative) link to the hidden quantities (symbolically denoted by $\boldsymbol{\theta}$) of interest. The symbolic notation

$$\boldsymbol{g}(\boldsymbol{\theta}, \boldsymbol{d}) = \boldsymbol{0}$$

[8.2]

for the physical model then expresses the fact that $\boldsymbol{d}$ and $\boldsymbol{\theta}$ are related through the equations describing the relevant physics (such as those used in example 8.1, where $\boldsymbol{\theta} \equiv k$ and $\boldsymbol{d} \equiv u|_{\partial\Omega}$).

Models that accurately describe the mechanical response of solids or materials do not usually enable exact, analytical solutions due to the complexity of the considered configurations. The relevant equations are thus, as a rule, solved numerically. Computational mechanics and engineering has undergone a tremendous development over the last few decades, from the viewpoint of both computational power and algorithmic development.

Generally, the mathematical model is most frequently solved for the physical response of the system assuming the parameters that characterize its geometry, constitutive properties and kinematic constraints and the excitations (prescribed loads, displacements, temperatures, etc.) are known. However, in identification situations, the commonly available measured information pertains to the response of the system to given excitations. In other words, standard computational methods allow us to solve the *forward problem*, that is to find $\boldsymbol{d}$ from [8.2] with given $\boldsymbol{\theta}$. To evaluate unknown system parameters $\boldsymbol{\theta}$ from measurements of the response $\boldsymbol{d}$ entails solving the physical model [8.2] in a reverse fashion (relative to the standard situations), hence the term *inverse problem*.

In example 8.1, the forward problem consists of finding the potential u by solving equations [8.1] for given conductivity $k(\boldsymbol{x})$, geometry Ω and excitation q^D. A typical inverse problem consists of reconstructing the unknown conductivity field $k(\boldsymbol{x})$ from measured values of u on the boundary, the flux q^{D} being again considered as *a priori* known. Note that the measurement $u|_{\partial\Omega}$ depends on the conductivity k in a nonlinear way.

8.2.1.1. *Forward and inverse problems*

The concepts of forward and inverse problems may be conveniently formulated and explained by considering the action of solving the (mechanical, thermal, etc.) governing equations as the prediction of the response $\boldsymbol{d}$ (displacement, stress, temperature, etc.) of the system under consideration to applied excitations $\boldsymbol{X}$ (forces, sources, prescribed displacements, initial stresses, heat fluxes, etc.).

The system (e.g. the mechanical structure being tested, or the region where wave propagation or flow takes place, etc.) usually depends on parameters symbolically denoted by $\boldsymbol{\theta}$: geometry (the region in space occupied by the body), physical characteristics of the constitutive materials, kinematic constraints, etc. The forward problem consists of evaluating the response $\boldsymbol{d}$ given the excitation $\boldsymbol{X}$ and the system parameters $\boldsymbol{\theta}$. The mathematical model describing the relevant physics is usually such that the response $\boldsymbol{d}$ is an implicit function of $(\boldsymbol{X}, \boldsymbol{\theta})$:

$$\text{find } \boldsymbol{d} = \boldsymbol{d}(\boldsymbol{\theta}; \boldsymbol{X}) \text{ such that } \boldsymbol{g}(\boldsymbol{X}, \boldsymbol{\theta}, \boldsymbol{d}) = \boldsymbol{0} \quad (\text{with } \boldsymbol{X}, \boldsymbol{\theta} \text{ given}) \qquad [8.3]$$

In most cases, the forward problem is well-posed in the Hadamard sense, that is its solution (1) exists, (2) is unique and (3) depends continuously on the data. Condition 3 ensures that the response will be only moderately sensitive to small errors caused by, for example, discretization or imperfect data.

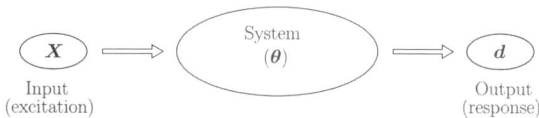

Figure 8.1. *Forward problem*

The inverse problem usually corresponds to situations where the system is at least partially unknown because of incomplete available information on features such as the geometry of the system, constitutive materials and initial conditions. To compensate for, and reconstruct, the missing information on the system parameters $\boldsymbol{\theta}$, supplementary (possibly partial) information about the response $\boldsymbol{d}$ must be sought in addition to the known excitations $\boldsymbol{X}$. The "inverse" qualifier serves as a reminder that the supplementary information is used in a reverse way relative to the usual solution methodologies applied to the physical model: from (partial) information about the response, we use the model equations backward to find hidden system characteristics that usually cannot be measured directly:

$$\text{find } \boldsymbol{\theta} \in \Theta \text{ such that } \quad g(\boldsymbol{X}, \boldsymbol{\theta}, \boldsymbol{d}^{\text{obs}}) = \boldsymbol{0} \qquad\qquad [8.4]$$

where Θ denotes the parameter space in which $\boldsymbol{\theta}$ is sought. The symbolic notation used in [8.3] and [8.4] fails to emphasize the fact that the forward and inverse problems have very different characteristics and mathematical properties. Inverse problems are often ill-posed in that at least one of the Hadamard well-posedness conditions is violated. In particular, any solution $\boldsymbol{\theta}$ is typically highly sensitive to experimental errors.

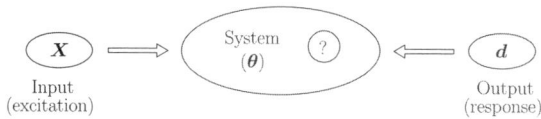

Figure 8.2. *Inverse problem*

8.2.1.2. *Reformulation as an optimization problem*

Practically, it is often neither convenient nor desirable to base the inversion of experimental data on solving exactly an equation of the type [8.4], which is usually multidimensional and nonlinear. The dimension of the data space $\mathcal{D}$ does not, in general, coincide with that of the parameter space Θ (such dimensions are, in practice, finite even though theoretical analyses often consider continuous models with observable data and unknown parameters modeled as functions). In fact, since reducing the adverse effect of uncertainties and making inversion methods more robust are main concerns, we often aim to have much more data than unknowns, making the observation equation [8.4] overdetermined. Therefore, the latter cannot, in general, be solved exactly (unless the data happen to verify certain solvability conditions), for at least two reasons: (1) the model equations $g(\boldsymbol{X}, \boldsymbol{\theta}, \boldsymbol{d})$ only approximately describe the actual physical behavior of the system and (2) the experimental data $\boldsymbol{d}^{\text{obs}}$ suffer from measurement uncertainties. To explain the general impossibility of solving exactly [8.4] in another way, we can note that the above considerations (1) and (2) imply that the observed data $\boldsymbol{d}^{\text{obs}}$ cannot, in general, be exactly reproduced by predictions $\boldsymbol{d}(\boldsymbol{\theta}; \boldsymbol{X})$ of the physical model [8.3].

These considerations often (although not always, as will be seen in section 8.4) suggest the reformulation of the data inversion problem as a minimization problem, whose typical form is:

$$\boldsymbol{\theta} = \arg\min_{\boldsymbol{\vartheta} \in \Theta} \mathcal{J}(\boldsymbol{\vartheta}), \quad \mathcal{J}(\boldsymbol{\vartheta}) = \|\boldsymbol{d}(\boldsymbol{\vartheta}; \boldsymbol{X}) - \boldsymbol{d}^{\text{obs}}\| \qquad\qquad [8.5]$$

where $\| \cdot \|$ denotes the norm to be specified. This norm is often chosen as the usual quadratic norm (also known as the L^2-norm), possibly weighted; this choice is often convenient because the L^2-norm, unlike other norms such as the L^1 and L^∞ norms, is differentiable. Problem [8.5] is in that case referred to as a least-squares minimization. It is important at this point to emphasize that the cost function $\mathcal{J}$ features an implicit dependence on $\boldsymbol{\theta}$ through the forward problem, which can be symbolized as:

$$\mathcal{J}(\boldsymbol{\theta}) = J(\boldsymbol{d}) \quad \text{with } \boldsymbol{g}(\boldsymbol{X}, \boldsymbol{\theta}, \boldsymbol{d}) = \boldsymbol{0}. \tag{8.6}$$

The cost function $\mathcal{J}$ does not have general properties that would *a priori* guarantee (i.e. for any inversion problem) that a local minimizer $\boldsymbol{\theta}$ is unique, or global.

Upon reformulation as a minimization task, seeking a solution $\boldsymbol{\theta}$ consists of minimizing the residual of the observation equation, rather than setting it to zero. This weakening of the concept of solution ensures its existence, and makes more practical sense than the (often unfeasible) task of attempting to exactly match imperfect observation by adjusting the parameters of an approximate model.

8.2.1.3. *Regularization*

As already mentioned, inverse problems are often mathematically ill-posed; in particular, a solution $\boldsymbol{\theta}$ is often highly sensitive to small changes to, or errors in, the experimental data. This property has strongly influenced the design of inversion methodologies since the pioneering works of [TIK 77] and [TWO 77], and is analyzed in many monographs, for example [ENG 96] and [HAN 98]. Such methodologies are based on the formulation and exploitation of *prior information* that is available in addition to experimental data:

– Quantitative prior information about the parameters to be identified (e.g. positiveness and variation range) may be specified via equality or inequality constraints that restrict the search space Θ.

– Qualitative prior information may be prescribed via a *stabilizing functional* $\mathcal{R}(\boldsymbol{\theta}) \geq 0$, expressing a requirement that a certain non-negative function of $\boldsymbol{\theta}$ be as small as possible. Classic examples include $\mathcal{R}(\boldsymbol{\theta}) = \|\boldsymbol{\theta} - \boldsymbol{\theta}_0\|$ (desired closeness of $\boldsymbol{\theta}$ to a reference value $\boldsymbol{\theta}_0$) and $\mathcal{R}(\boldsymbol{\theta}) = \|\boldsymbol{\nabla}\boldsymbol{\theta}\|$ (to avoid too-oscillatory solutions $\boldsymbol{\theta}$). This leads to the regularized form of the minimization [8.5]:

$$\boldsymbol{\theta} = \arg\min_{\boldsymbol{\vartheta} \in \Theta} \mathcal{J}(\boldsymbol{\vartheta}), \quad \mathcal{J}(\boldsymbol{\vartheta}) = J(\boldsymbol{d}(\boldsymbol{\vartheta}; \boldsymbol{X})) + \alpha \mathcal{R}(\boldsymbol{\vartheta}) \tag{8.7}$$

where $J(\cdot)$ again defines the distance between the measurements and their prediction by the forward problem, $\mathcal{R}$ is the stabilizing functional and $0 < \alpha \ll 1$ is the *regularization parameter*.

It is useful for the purpose of adjusting the inversion algorithm to have an available estimation δ of the measurement error. For example, some algorithms select the optimal value of the regularization parameter α on the basis of δ [TIK 95].

8.2.1.3.1. Regularization using total variation

When the dimension of the search space Θ is high, it is, in practice, necessary to use some form of regularization because the inversion is highly sensitive to experimental errors, even with linear forward models (which in that case are ill-conditioned). Such situations arise, for instance, in the reconstruction of heterogeneous material parameters (modulus, wave velocity, conductivity, damage, etc.), whose spatial discretization requires a large number of unknowns. A stabilizing functional of the form $\mathcal{R}(\boldsymbol{\theta}) = \|\boldsymbol{\theta} - \boldsymbol{\theta}_0\|$ may be useful, for example when attempting to reconstruct medium properties that deviate moderately from a given reference value $\boldsymbol{\theta}_0$. The choice $\mathcal{R}(\boldsymbol{\theta}) = \|\boldsymbol{\nabla}\boldsymbol{\theta}\|^2$ (with $\| \cdot \|$ denoting the L^2-norm of a square-integrable function) allows us to filter non-physical spatial oscillations that might affect the reconstruction of $\boldsymbol{\theta}$ due to the amplification of experimental uncertainties by the inversion algorithm. However, the squared norm, while very convenient from a computational standpoint because the corresponding stabilizing functional is differentiable, tends to yield oversmoothed solutions $\boldsymbol{\theta}$ that do not reproduce existing contrasts (e.g. when reconstructing piecewise homogeneous media). For this reason, it is often preferable for this kind of inversion problem to use stabilizing functionals of the form

$$\mathcal{R}(\boldsymbol{\theta}) = \left[\|\boldsymbol{\nabla}\boldsymbol{\theta}\|^2 + \eta\right]^{1/2} \quad 0 < \eta \ll 1$$

known as *total variation* functionals [ACA 94], which still filter random spatial oscillations while permitting a limited amount of contrast (the small parameter η serves to define a regularized form of the total variation functional that is differentiable at $\boldsymbol{\nabla}\boldsymbol{\theta} = 0$). This type of regularization is, for example, used in [EPA 08] for the reconstruction of three-dimensional (3D) heterogeneous moduli using seismic data.

8.2.1.4. *Bayesian formulations*

Other approaches adopt a probabilistic viewpoint to model prior information and various uncertainties. They proceed by constructing an *a posteriori* probability density function on $\boldsymbol{\theta}$ by considering the available prior information and the physical model as two independent sources of information [MEN 84, TAR 05, KAI 05]. The starting point for such formulations is the Bayes theorem, written as

$$f_{\Theta|D}(\boldsymbol{\theta}|\boldsymbol{d}^{\text{obs}})f_D(\boldsymbol{d}) = f_{D|\Theta}(\boldsymbol{d}^{\text{obs}}|\boldsymbol{\theta})f_\Theta(\boldsymbol{\theta}) \qquad [8.8]$$

where $f_{X|Y}(x|y)$ is the conditional probability density on x knowing y, and the density $f_X(x)$ is defined by the marginalization of $f(x, y)$ (and similarly by switching the roles of x and y). Here, the probability densities $f_\Theta(\boldsymbol{\theta})$, modeling the prior information on $\boldsymbol{\theta}$, and $f_{D|\Theta}(\boldsymbol{d}^{\text{obs}}|\boldsymbol{\theta})$, describing the forward physical model and enabling measurement or modeling uncertainties to be taken into account, are chosen *a priori*. The Bayes theorem [8.8] then yields

$$f_{\Theta|D}(\boldsymbol{\theta}|\boldsymbol{d}^{\text{obs}}) = \frac{f_{D|\Theta}(\boldsymbol{d}^{\text{obs}}|\boldsymbol{\theta})f_\Theta(\boldsymbol{\theta})}{\int f_{D|\Theta}(\boldsymbol{d}^{\text{obs}}|\boldsymbol{\theta})f_\Theta(\boldsymbol{\theta})\mathrm{d}\boldsymbol{\theta}} \qquad [8.9]$$

and therefore allows us to evaluate the probability of $\boldsymbol{\theta}$ knowing $\boldsymbol{d}^{\text{obs}}$, once all available (prior and model) information is taken into account. Practical information about $\boldsymbol{\theta}$ and the error sensitivity of its estimation are obtained by analyzing the posterior density $f_{\Theta|D}$. For instance, we may estimate $\boldsymbol{\theta}$ by seeking the value achieving the maximum likelihood, that is $\boldsymbol{\theta} = \arg\max_{\vartheta} f_{\Theta|D}(\vartheta|\boldsymbol{d}^{\text{obs}})$.

The simplest form of this approach corresponds to the case of a forward finite-dimensional linear model $\boldsymbol{g}(\boldsymbol{X}, \boldsymbol{\theta}, \boldsymbol{d}) = \boldsymbol{G}(\boldsymbol{X})\boldsymbol{\theta} - \boldsymbol{d}$ (where $\boldsymbol{G}$ is a matrix) and consists of setting $f_{\Theta}(\boldsymbol{\theta}) = \mathcal{N}(\boldsymbol{\theta}_0, \boldsymbol{C}_{\theta})$ and $f_{D|\Theta}(\boldsymbol{d}|\boldsymbol{\theta}) = \mathcal{N}(\boldsymbol{d}^{\text{obs}}, \boldsymbol{C}_D)$ (with $\mathcal{N}(\boldsymbol{x}, \boldsymbol{C})$ denoting the multidimensional Gaussian random variable with mean $\boldsymbol{x}$ and covariance matrix $\boldsymbol{C}$). The posterior conditional probability density function $f_{\Theta|D}(\boldsymbol{\theta}|\boldsymbol{d})$ then corresponds to the Gaussian variable $\mathcal{N}(\bar{\boldsymbol{\theta}}, \boldsymbol{C})$ with

$$\bar{\boldsymbol{\theta}} = \arg\min_{\theta} \left(\boldsymbol{G}\boldsymbol{\theta} - \boldsymbol{d}^{\text{obs}}\right)^{\text{T}} \boldsymbol{C}_D^{-1} \left(\boldsymbol{G}\boldsymbol{\theta} - \boldsymbol{d}^{\text{obs}}\right) + \left(\boldsymbol{\theta} - \boldsymbol{\theta}_0\right)^{\text{T}} \boldsymbol{C}_{\theta}^{-1} \left(\boldsymbol{\theta} - \boldsymbol{\theta}_0\right),$$
$$\boldsymbol{C} = \left[\boldsymbol{G}\boldsymbol{C}_D^{-1}\boldsymbol{G} + \boldsymbol{C}_{\theta}^{-1}\right]^{-1}$$

[8.10]

The Bayesian approach constitutes a form of regularization in that [8.9] simultaneously takes into account experimental data and prior information. For example, the quadratic cost function featured in [8.10] corresponds to a regularized cost function, where $\boldsymbol{C}_D \boldsymbol{C}_{\theta}^{-1}$ acts as a small regularization parameter α. Examples where such approaches are applied to the mechanics of materials and structures include [ARN 08], [DAG 07], [GOG 08] and [GUZ 02].

8.2.1.5. *Inversion versus identification*

A distinction is sometimes made between inverse problems and identification problems. In both cases, the aim is to determine certain quantities $\boldsymbol{\theta}$ that enter into the definition of the analyzed structure or material sample by exploiting experimental information concerning its response. The term *inverse problem* primarily refers to situations where the quantity to be reconstructed is mathematically formulated in terms of functions (such as heterogeneous moduli, time-dependent forces or sources, solution-dependent constitutive properties and domain shape or topology), whose discretization is expected to entail a large number of unknowns. Cases where we *a priori* consider a moderate number of unknowns, such as parameters entering into most usual constitutive models, are then referred to as *identification problems*. Both types of problems are solved using similar methods, often entailing the minimization of a cost function. Because of their smaller size, parameter identification problems are, on average, less sensitive to (experimental, modeling, etc.) uncertainties than inverse problems, and do not always require regularization (the cost functions used may thus differ at least in this respect).

8.2.2. *Constitutive parameter identification*

8.2.2.1. *Forward problem*

This chapter is primarily concerned with parameter identification problems for (linear or nonlinear) constitutive models. Consider, for instance, a solid body whose undeformed configuration occupies the domain Ω and whose mechanical state is governed in the small-deformation framework (chosen for expository convenience, but not mandatory) by the equilibrium equations

$$\begin{cases} \operatorname{div}\boldsymbol{\sigma} = \mathbf{0} & \text{in } \Omega & \text{[8.11a]} \\ \boldsymbol{\sigma}.\boldsymbol{n} = \bar{\boldsymbol{T}} & \text{on } S_f & \text{[8.11b]} \end{cases}$$

(assuming no body forces are present) and the kinematic compatibility equations

$$\begin{cases} \boldsymbol{\varepsilon} = \boldsymbol{\varepsilon}[\boldsymbol{u}] = \dfrac{1}{2}(\boldsymbol{\nabla}\boldsymbol{u} + \boldsymbol{\nabla}^t\boldsymbol{u}) & \text{in } \Omega & \text{[8.12a]} \\ \boldsymbol{u} = \bar{\boldsymbol{u}} & \text{on } S_u & \text{[8.12b]} \end{cases}$$

(where $\boldsymbol{u}$ is the displacement, $\boldsymbol{\varepsilon}$ is the linearized stress tensor, $\boldsymbol{\sigma}$ is the Cauchy stress tensor and $\boldsymbol{n}$ is the outward unit normal to $\partial\Omega$). The surfaces S_u (supporting prescribed displacements $\bar{\boldsymbol{u}}$) and S_f (supporting prescribed tractions $\bar{\boldsymbol{T}}$) are such that $S_u \cup S_f = \partial\Omega$ and $S_u \cap S_f = \emptyset$, so as to define well-posed boundary conditions.

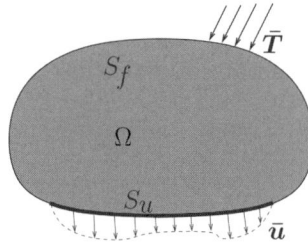

Figure 8.3. *Forward problem: notations*

For elastic linear material properties, the constitutive relation has the well-known form

$$\boldsymbol{\sigma} = \boldsymbol{\mathcal{A}}(\boldsymbol{\theta}):\boldsymbol{\varepsilon} \quad \text{in } \Omega \qquad \text{[8.13]}$$

where the elasticity tensor $\boldsymbol{\mathcal{A}}$ may be constant (homogeneous material) or spatially variable (heterogeneous material, e.g. due to damage or defects). In the context of this chapter, the elasticity tensor depends on a vector of parameters $\boldsymbol{\theta} = \{\theta_1, \ldots, \theta_M\}$, which may be emphasized in [8.13] with the notation $\boldsymbol{\mathcal{A}} = \boldsymbol{\mathcal{A}}(\boldsymbol{\theta})$ (e.g. $\boldsymbol{\theta} = \{E, \nu\}$ and $M = 2$ for homogeneous isotropic elasticity).

The set of equations [8.11b]–[8.13] may be replaced by the weak formulation

$$\int_\Omega \varepsilon[\tilde{u}] : \mathcal{A}(\boldsymbol{\theta}) : \varepsilon[u] \mathrm{d}V - \int_{S_f} \tilde{u}.T \mathrm{d}S = 0 \qquad \text{for all } \tilde{u} \text{ k.a. with } \mathbf{0} \qquad [8.14]$$

A Galerkin discretization using the finite element interpolation functions as an approximation and trial space leads to the standard finite element matrix equation

$$\boldsymbol{K}(\boldsymbol{\theta})\boldsymbol{U} = \boldsymbol{F} \qquad\qquad [8.15]$$

where $\boldsymbol{K}(\boldsymbol{\theta})$ is the stiffness matrix, here restricted to the unconstrained degrees of freedom (DOFs), $\boldsymbol{F}$ is the vector of generalized (nodal) forces resulting from the applied excitations [8.11b] and [8.12b] and the vector $\boldsymbol{U}$ gathers the unknown DOFs. The notation $\boldsymbol{K}(\boldsymbol{\theta})$ emphasizes the obvious but essential fact that the stiffness matrix depends on the constitutive parameters (in this case, elastic moduli).

For more complex constitutive properties (plasticity, damage, etc.) that are history-dependent, we often have to solve a (time-discrete) evolution problem using an incremental and iterative algorithm (typically involving an implicit treatment such as the radial return algorithm [SIM 98, BES 01]). Such treatment is based on a spatially continuous weak formulation of the form

$$\int_\Omega \varepsilon[\tilde{u}] : \boldsymbol{\sigma}[u_n; u_{n-1}, \mathcal{S}_{n-1}, \boldsymbol{\theta}] \mathrm{d}V - \int_{S_f} \tilde{u}.T_n \mathrm{d}S = 0$$

$$\text{for all } \tilde{u} \text{ k.a. with } \mathbf{0} \qquad\qquad [8.16]$$

where T_n denotes the applied loading at time t_n, $u_n = u(\cdot, t_n)$ is the unknown displacement at time t_n and $\boldsymbol{\sigma}[u_n; u_{n-1}, \mathcal{S}_{n-1}]$ denotes the stress at time t_n predicted by the (time-discretized) constitutive model for an assumed value of the displacement u_n and knowing the displacement u_{n-1} and all other mechanical quantities $\mathcal{S}_{n-1}$ (strains, stresses and internal variables) at time t_{n-1}. For finite element discretization in space, the incremental weak formulation [8.16] takes the form

$$\boldsymbol{R}_n(\boldsymbol{U}_n; \boldsymbol{U}_{n-1}, \mathcal{S}_{n-1}, \boldsymbol{\theta}) = \mathbf{0} \qquad\qquad [8.17]$$

It is solved by an iterative Newton–Raphson-type method, the iteration $i+1$ of the solution being obtained by solving the linear system

$$\boldsymbol{K}_n^{(i)}(\boldsymbol{\theta})\boldsymbol{U}_n^{(i)} = \boldsymbol{K}_n^{(i)}(\boldsymbol{\theta})\boldsymbol{U}_n^{(i-1)} - \boldsymbol{R}_n(\boldsymbol{U}_n^{(i-1)}; \boldsymbol{U}_{n-1}, \mathcal{S}_{n-1}, \boldsymbol{\theta}) \qquad [8.18]$$

where $\boldsymbol{K}_n^{(i)}(\boldsymbol{\theta}) \equiv \partial \boldsymbol{R}_n / \partial \boldsymbol{U}_n(\boldsymbol{U}_n^{(i-1)}; \boldsymbol{U}_{n-1}, \mathcal{S}_{n-1}, \boldsymbol{\theta})$ is the tangent stiffness matrix, also known as the "consistent tangent operator" [SIM 98].

For a given set of parameters $\boldsymbol{\theta}$, formulations such as [8.14] and [8.16] define the *forward problem*, allowing the computation of the response of a structure whose material has known properties. Solving the forward problem uses classic computational structural mechanics approaches, the main approach of which is the finite element method (FEM).

8.2.2.2. *Identification problem*

In cases where parameters $\boldsymbol{\theta}$, associated with a constitutive model or with other characteristics of the sample being analyzed such as internal defects, are unknown, the forward problem [8.14] or [8.16] cannot be solved as it is. Finding the parameters $\boldsymbol{\theta}$ requires supplementary information, obtained from experiments, in addition to the boundary data $\bar{\boldsymbol{u}}, \bar{\boldsymbol{T}}$ entering into the definition of [8.14] or [8.16]. Such supplementary data may, in particular, consist of kinematic field measurements.

It is worth emphasizing, moreover, that the available experimental information regarding applied loads is limited, for many practical identification situations, to resultant loads or couples. Depending on the chosen solution approach, we will then either treat distributed applied loads as supplementary unknowns, define kinematic boundary conditions from the measured kinematic fields or use virtual or adjoint fields so that the formulation of the identification would involve only the (known) resultant load.

8.3. Numerical methods based on optimization

For the sake of generality, consider the task of identifying parameters for a nonlinear, incremental constitutive model using a sequence of kinematic field measurements, obtained at various stages of the loading history, applied to the examined sample. Assume for definiteness that the kinematic response $\boldsymbol{d}_n^{\text{obs}}$ is measured at *all* discrete time instants $t_0 = 0, t_1, \ldots, t_N = T$ introduced for the time-marching algorithm. Identification then consists of determining $\boldsymbol{\theta}$ so as to achieve the best fit between the experimental data $\boldsymbol{d}_n^{\text{obs}}$ and its computed prediction $\boldsymbol{d}_n(\boldsymbol{\theta})$. This approach naturally leads to the problem of minimizing a cost function of the form

$$\mathcal{J}(\boldsymbol{\theta}) = \sum_{n=1}^{N} J(\boldsymbol{U}_n(\boldsymbol{\theta})) \qquad [8.19]$$

where $\boldsymbol{U}_n$ depends on $\boldsymbol{\theta}$ through the equilibrium problem [8.16]. For instance, for the least-squares method, the function J is defined by

$$J(\boldsymbol{V}) = \frac{1}{2}\|\boldsymbol{P}\boldsymbol{V} - \boldsymbol{d}_n^{\text{obs}}\|^2 \qquad [8.20]$$

8.3.1. *Gradient-based methods*

The evaluation of each cost function may entail substantial computational work because it requires a complete analysis of the structure, possibly under dynamic and/or nonlinear conditions. Therefore, we often prefer to rely on gradient-based optimization algorithms that allow us to reduce the total number of evaluations of

$\mathcal{J}(\boldsymbol{\theta})$. For a detailed presentation of the main optimization algorithms, the reader may refer to many monographs [FLE 00, NOC 06].

Gradient-based optimization algorithms assume that both the cost function $\mathcal{J}$ and its gradient $\nabla_\theta \mathcal{J}$ can be numerically evaluated for a given $\boldsymbol{\theta}$. They are usually designed so that each iteration ensures that the cost function is decreased. Classic algorithms of this kind include the steepest descent method (historically the first of its kind, but currently more or less abandoned due to its too-slow convergence), conjugate gradient and quasi-Newton. The latter consists, in fact, of solving the necessary first-order optimality condition $\nabla_\theta \mathcal{J} = \mathbf{0}$ using the Newton–Raphson method, where the true Hessian $\boldsymbol{H} = \nabla_\theta^2 \mathcal{J}$, *a priori* required by the Newton–Raphson method, is replaced by the (positive definite) approximation of the inverse Hessian that is updated (using the BFGS or DFP formula [BER 99, FLE 00, NOC 06]) after each iteration.

Other algorithms exploit the nonlinear least-squares cost function structure [8.19–8.20] often used for identification problems. They are also based on solving equation $\nabla_\theta \mathcal{J} = \mathbf{0}$ by the Newton–Raphson method, using an approximation of the Hessian $\boldsymbol{H}$ in order to avoid the often complex task of its numerical evaluation. The Gauss–Newton algorithm uses the approximation $\boldsymbol{H} \approx (\nabla_\theta \boldsymbol{U})^\mathrm{T} \nabla_\theta \boldsymbol{U}$, in which second-order derivatives of the residuals $\boldsymbol{PU} - \boldsymbol{d}^{\mathrm{obs}}$ are dropped (this approximation is correct at the local minimum reached upon convergence provided that the converged residuals themselves are small). The matrix $(\nabla_\theta \boldsymbol{U})^\mathrm{T} \nabla_\theta \boldsymbol{U}$ may, however, be non-invertible, or ill-conditioned. The Levenberg–Marquardt algorithm addresses this issue by using in the Gauss–Newton algorithm an approximate Hessian of the form $\boldsymbol{H} \approx (\nabla_\theta \boldsymbol{U})^\mathrm{T} \nabla_\theta \boldsymbol{U} + \alpha \boldsymbol{I}$, where the parameter $\alpha > 0$ is updated after each iteration (this approximation is thus positive definite by construction). The Gauss–Newton and Marquardt–Levenberg algorithms require repeated evaluations of the complete Jacobian matrix $\nabla_\theta \boldsymbol{U}$, that is of all the partial derivatives of the solution $\boldsymbol{U}$ to the forward problem rather than the partial derivatives of the cost function $\mathcal{J}(\boldsymbol{\theta})$ only.

It is sometimes useful, or even necessary, to reduce the search space Θ by prescribing constraints (e.g. moduli are positive and the Poisson ratio must belong to a certain interval), if only to avoid the occurrence of non-physical forward problems caused by forbidden values of $\boldsymbol{\theta}$ that may otherwise be reached by the iterative algorithm. Many constrained optimization algorithms are available; readers can refer to monographs such as [BER 99], [FLE 00] and [NOC 06]. Some of these are based on iteratively solving the Karush–Kuhn–Tucker necessary optimality conditions (generally using gradients of $\mathcal{J}$ and the functions defining the constraints, and approximate Hessians). Others consist of recasting the constrained optimization in terms of a sequence of unconstrained optimization problems *via* the introduction of penalty, barrier, etc., functions. For example, an augmented Lagrangian method was implemented for the combined shape and material identification of elastic inclusions in [BON 09b].

Various approaches, briefly reviewed next, are available to evaluate the gradient $\nabla_\theta \mathcal{J}$ or the Jacobian matrix $\nabla_\theta U$. A detailed presentation of parameter sensitivity analysis is, for example, available in [KLE 97].

8.3.1.1. *Numerical differentiation*

Numerical differentiation is based on the approximate formula

$$\nabla_\theta \mathcal{J} \cdot \Delta\boldsymbol{\theta} \approx \mathcal{J}(\boldsymbol{\theta} + \Delta\boldsymbol{\theta}) - \mathcal{J}(\boldsymbol{\theta}) \qquad [8.21]$$

for small but finite increments $\Delta\boldsymbol{\theta}$ of $\boldsymbol{\theta}$. If $\boldsymbol{\theta} \in \mathbb{R}^p$, one evaluation of the full gradient $\nabla_\theta \mathcal{J}$ then requires $1+p$ forward solutions, corresponding to the unperturbed configuration $\Delta\boldsymbol{\theta} = \mathbf{0}$ (for the evaluation of $\mathcal{J}(\boldsymbol{\theta})$) and perturbed configurations defined by $\Delta\boldsymbol{\theta} = (\Delta\theta_1, 0, \ldots, 0), \ldots, (0, \ldots, 0, \Delta\theta_p)$. Similarly, we may use a numerical differentiation of the forward solution for the purpose of setting up the Jacobian matrix:

$$\nabla_\theta U_n \cdot \Delta\boldsymbol{\theta} \approx U_n(\boldsymbol{\theta} + \Delta\boldsymbol{\theta}) - U_n(\boldsymbol{\theta})$$

This approach is simple to implement and does not require modifications to the forward solution code (it is thus said to be *non-intrusive*). It is, however, computationally expensive because a gradient evaluation entails p complete forward solutions, each new choice for $\boldsymbol{\theta}$ redefining the mechanical system anew and implying a complete analysis. Thus, it is sometimes desirable to replace numerical differentiation by other approaches based on a preliminary analytical differentiation.

8.3.1.2. *Direct differentiation of $\mathcal{J}$*

The direct differentiation approach consists of evaluating the gradient of $\mathcal{J}$ by means of the chain rule:

$$\nabla_\theta \mathcal{J} = \sum_{n=1}^{N} \nabla J(U_n) \cdot \nabla_\theta U_n \qquad [8.22]$$

This expression requires the derivatives $\nabla_\theta d_k$ of the kinematic field that solves the forward incremental problem [8.17]. Differentiating the latter with respect to $\boldsymbol{\theta}$ yields:

$$K_n(\boldsymbol{\theta})\nabla_\theta U_n = -\nabla_{\mathcal{S}_{n-1}} R_n \nabla_\theta \mathcal{S}_{n-1} - \nabla_\theta R_n \quad (0 = 1, 2, \ldots, N-1) \ [8.23]$$

The above derivative equations define a linear incremental problem, whose governing matrix is the tangent stiffness $K_n(\boldsymbol{\theta})$ reached on convergence of the Newton–Raphson algorithm [8.18]. A natural time-stepping procedure then consists, for each n $(0 = 1, 2, \ldots, N-1)$, of solving [8.17] (which requires an iterative algorithm) and then the linear problem [8.23]. The fields U_n and $\nabla_\theta U_n$ then enable the evaluation of the corresponding contribution to both $\mathcal{J}$ and $\nabla_\theta \mathcal{J}$. Moreover, this approach allows the computation of Jacobian matrices of the form $\nabla_\theta U_n$, required by the Gauss–Newton or Marquardt–Levenberg methods.

8.3.1.3. *Adjoint state*

Another strategy for evaluating the gradient of $\mathcal{J}$ without recourse to numerical differentiation consists of adopting the viewpoint of minimizing $\mathcal{J}$ subject to the constraint defined by the forward problem [8.17]. Accordingly, we introduce the Lagrangian:

$$\mathcal{L} = \sum_{n=1}^{N}\Big\{ J(\boldsymbol{U}_n) + \tilde{\boldsymbol{U}}_n^{\mathrm{T}}\boldsymbol{R}_n(\boldsymbol{U}_n; \boldsymbol{U}_{n-1}, \mathcal{S}_{n-1}, \boldsymbol{\theta})$$

$$+ \tilde{\mathcal{S}}_n^{\mathrm{T}}[\mathcal{S}_n - \mathcal{P}(\boldsymbol{U}_n, \boldsymbol{U}_{n-1}, \mathcal{S}_{n-1}, \boldsymbol{\theta})] \Big\}$$ [8.24]

where the relation $\mathcal{S}_n - \mathcal{P}(\boldsymbol{U}_n, \boldsymbol{U}_{n-1}, \mathcal{S}_{n-1}, \boldsymbol{\theta}) = 0$ expresses in symbolic notation the process of updating the mechanical quantities at $t = t_n$ once $\boldsymbol{U}_n$ is found, and $\tilde{\boldsymbol{U}}_n, \tilde{\mathcal{S}}_n$ are Lagrange multipliers. The first-order variation of $\mathcal{L}$ then reads

$$\delta\mathcal{L} = \boldsymbol{\nabla}_{\mathcal{S}_n}\mathcal{L}\cdot\delta\mathcal{S}_n + \boldsymbol{\nabla}_{U_n}\mathcal{L}\cdot\delta\boldsymbol{U}_n + \boldsymbol{\nabla}_{\tilde{\mathcal{S}}_n}\mathcal{L}\cdot\delta\tilde{\mathcal{S}}_n + \boldsymbol{\nabla}_{\tilde{U}_n}\mathcal{L}\cdot\delta\tilde{\boldsymbol{U}}_n + \boldsymbol{\nabla}_\theta\mathcal{L}\cdot\delta\boldsymbol{\theta}$$

The cofactors $\boldsymbol{\nabla}_{\tilde{\mathcal{S}}_n}\mathcal{L}$, $\boldsymbol{\nabla}_{\tilde{U}_n}\mathcal{L}$ vanish whenever $\boldsymbol{U}_n, \mathcal{S}_n$ satisfy the forward problem equations. The cofactors $\boldsymbol{\nabla}_{\mathcal{S}_n}\mathcal{L}$, $\boldsymbol{\nabla}_{U_n}\mathcal{L}$ are given by the formulas

$$\boldsymbol{\nabla}_{\mathcal{S}_n}\mathcal{L} = \tilde{\boldsymbol{U}}_{n+1}^{\mathrm{T}}\boldsymbol{\nabla}_{\mathcal{S}_n}\boldsymbol{R}_{n+1} - \tilde{\mathcal{S}}_{n+1}^{\mathrm{T}}\boldsymbol{\nabla}_{\mathcal{S}_n}\mathcal{P} + \tilde{\mathcal{S}}_n$$ [8.25a]

$$\boldsymbol{\nabla}_{U_n}\mathcal{L} = \boldsymbol{\nabla}J(\boldsymbol{U}_n) + \tilde{\boldsymbol{U}}_n^{\mathrm{T}}\boldsymbol{K}_n + \tilde{\boldsymbol{U}}_{n+1}^{\mathrm{T}}\boldsymbol{K}_{n+1,n} - \tilde{\mathcal{S}}_n^{\mathrm{T}}\mathcal{P}_n - \tilde{\mathcal{S}}_{n+1}^{\mathrm{T}}\mathcal{P}_{n+1,n}$$ [8.25b]

having set $\boldsymbol{K}_{n+1,n}:=\boldsymbol{\nabla}_{U_n}\boldsymbol{R}_{n+1}$, $\mathcal{P}_n:=\boldsymbol{\nabla}_{U_n}\mathcal{P}(\boldsymbol{U}_n, \boldsymbol{U}_{n-1}, \mathcal{S}_{n-1}, \boldsymbol{\theta})$, and $\mathcal{P}_{n,n-1}:=\boldsymbol{\nabla}_{U_{n-1}}\mathcal{P}(\boldsymbol{U}_n, \boldsymbol{U}_{n-1}, \mathcal{S}_{n-1}, \boldsymbol{\theta})$. They can be made to vanish by a judicious selection of the Lagrange multipliers $\tilde{\boldsymbol{U}}_n, \tilde{\mathcal{S}}_n$. To this end, we can note that for $n = N$, no quantity bearing the subscript $N+1$ should appear in expressions [8.25a] and [8.25b], which leads to the equalities

$$0 = \boldsymbol{\nabla}_{\mathcal{S}_N}\mathcal{L} = \tilde{\mathcal{S}}_N, \qquad 0 = \boldsymbol{\nabla}_{U_N}\mathcal{L} = \boldsymbol{\nabla}J(\boldsymbol{U}_N) + \tilde{\boldsymbol{U}}_N^{\mathrm{T}}\boldsymbol{K}_N - \tilde{\mathcal{S}}_N^{\mathrm{T}}\mathcal{P}_N$$

$$\Longrightarrow \quad \tilde{\mathcal{S}}_N = 0, \ \tilde{\boldsymbol{U}}_N = -\boldsymbol{K}_N^{-1}\boldsymbol{\nabla}J(\boldsymbol{U}_N)$$ [8.26]

Moreover, setting $\boldsymbol{\nabla}_{\mathcal{S}_n}\mathcal{L}$, $\boldsymbol{\nabla}_{U_n}\mathcal{L}$ to zero for $n < N$ yields:

$$\tilde{\mathcal{S}}_n^{\mathrm{T}} = \tilde{\mathcal{S}}_{n+1}^{\mathrm{T}}\boldsymbol{\nabla}_{\mathcal{S}_n}\mathcal{P} - \tilde{\boldsymbol{U}}_{n+1}^{\mathrm{T}}\boldsymbol{\nabla}_{\mathcal{S}_n}\boldsymbol{R}_{n+1}$$ [8.27a]

$$\boldsymbol{K}_n\tilde{\boldsymbol{U}}_n = \tilde{\mathcal{S}}_n^{\mathrm{T}}\mathcal{P}_n + \tilde{\mathcal{S}}_{n+1}^{\mathrm{T}}\mathcal{P}_{n+1,n} - \boldsymbol{\nabla}J(\boldsymbol{U}_n) - \tilde{\boldsymbol{U}}_{n+1}^{\mathrm{T}}\boldsymbol{K}_{n+1,n}$$ [8.27b]

A backward *adjoint state* is hence defined by (1) initialization [8.26] and (2) (backward) transition ([8.27a] and [8.27b]). The latter is linear irrespective of the possible nonlinearity of the forward problem. Equation [8.27b] uses the tangent stiffness matrix $\boldsymbol{K}_n$, reached on convergence of the Newton–Raphson step [8.18].

This allows for a very efficient computation of the adjoint state, provided that the converged tangent stiffness matrix has been stored, preferably in a factored form. In view of the backward nature of the adjoint problem, it is necessary to store *all* tangent stiffness matrices $\boldsymbol{K}_n$ and $\boldsymbol{\nabla}_{\mathcal{S}_{n-1}}\boldsymbol{R}_n$ arising in the course of solving the forward incremental problem, as well as the complete forward solution history $\boldsymbol{U}_n, \mathcal{S}_n$, *before* solving the adjoint problem.

This approach finally leads to the result

$$\boldsymbol{\nabla}_\theta \mathcal{J} = \boldsymbol{\nabla}_\theta \mathcal{L} = \sum_{n=1}^{N} \left\{ \tilde{\boldsymbol{U}}_n^{\mathrm{T}} \boldsymbol{\nabla}_\theta \boldsymbol{R}_n (\boldsymbol{U}_n; \boldsymbol{U}_{n-1}, \mathcal{S}_{n-1}, \boldsymbol{\theta}) \right.$$

$$\left. - \tilde{\mathcal{S}}_n^{\mathrm{T}} \boldsymbol{\nabla}_\theta \mathcal{P} (\boldsymbol{U}_n, \boldsymbol{U}_{n-1}, \mathcal{S}_{n-1}, \boldsymbol{\theta}) \right\} \qquad [8.28]$$

where $\boldsymbol{U}_n, \mathcal{S}_n$ is the forward solution and $\tilde{\boldsymbol{U}}_n, \tilde{\mathcal{S}}_n$ is the adjoint solution.

For linear constitutive behavior and equilibrium problems, the cost function depends only (assuming no regularization term) on the solution $\boldsymbol{U}(\boldsymbol{\theta})$ of the elastic equilibrium problem [8.15]: $\mathcal{J}(\boldsymbol{\theta}) = J(\boldsymbol{U})$. The gradient of $\mathcal{J}$ and the adjoint solution $\tilde{\boldsymbol{U}}$ are then defined simply by

$$\boldsymbol{\nabla}_\theta \mathcal{J} = \tilde{\boldsymbol{U}}^{\mathrm{T}} \boldsymbol{\nabla}_\theta \boldsymbol{K}(\boldsymbol{\theta}) \boldsymbol{U}, \qquad \tilde{\boldsymbol{U}} = -\boldsymbol{K}^{-1} \boldsymbol{\nabla} J \qquad [8.29]$$

It can thus be seen that adjoint state methods exploit a shortcut (the adjoint state) that enables us to completely avoid computing the Jacobian matrix $\boldsymbol{\nabla}_\theta \boldsymbol{U}$ associated with the solution. They are designed to maximize the evaluation efficiency for the gradient of cost functions (such as $\mathcal{J}(\boldsymbol{\theta})$) and are not suited to the evaluation of Jacobian matrices used in Gauss–Newton or Marquardt–Levenberg methods.

8.3.2. *Other methods*

8.3.2.1. *No-derivative methods*

In addition to previously discussed methods that exploit gradient information regarding the solution or the cost function, no-derivative minimization methods are also available. The theory and algorithms for no-derivative minimization methods are less developed than those for the more widely used gradient-based methods. No-derivative methods, which include the Nelder–Mead simplex algorithm [NEL 65, LAG 98] (not to be confused with the simplex method of linear programming), are inefficient for high-dimensional optimization problems.

8.3.2.2. *Evolutionary algorithms*

These algorithms aim at performing a global exploration of the search space Θ, and are designed by analogy with (and using the terminology of) Darwinian evolution;

see [MIC 96], [LER 07] or [BUR 08] for applications to identification. Among their advantages is the ability to identify multiple optimal solutions (and hence remove the dependency on an initial guess which traditional optimization algorithms suffer from), and to solve optimization problems that are ill-suited to classic algorithms due to the presence of combinatorial features. This, of course, comes with a price: such algorithms necessitate very large numbers of forward solutions. This has prompted investigations into the combined use of evolutionary algorithms and model reduction methods such as the *proper orthogonal decomposition* (POD) [BRI 07].

8.3.2.3. *Topological sensitivity*

Classic iterative inversion methods sometimes require significant computational work. In the context of flaw identification, where objects are sought whose geometry (location, size, shape) is unknown and whose support is small relative to the size of the sample being tested, alternative approaches aiming at the construction of a defect indicator function have recently been proposed. In particular, the concept of topological sensitivity, which aims to evaluate the asymptotic behavior of the featured cost function as the characteristic size of a trial defect becomes vanishingly small, leads to a global probing approach that is approximate but computationally fast (with a computational cost of the order of one forward solution) [AMS 05, BON 09a, BEL 09]. This approach is not restricted to specific types of cost functions or data, and is, in particular, easily amenable to the exploitation of full-field measurements.

8.4. Methods specifically designed for full-field measurements: an overview

8.4.1. *Finite element model updating*

Finite element model updating (FEMU) is primarily aimed at the identification of constitutive parameters by using equation [8.15], or similar equations, based on an FEM model of the sample. FEMU is usually based on the minimization of a discrepancy between a measured quantity and its prediction by the model for an assumed value of θ. Such a discrepancy is usually defined for measured displacements, strains or forces.

The cost functions introduced in this context are defined in direct relation to observable quantities. The latter may, advantageously, take the form of kinematic *fields*, but this is not necessary. Indeed, any information that is supplementary relative to boundary conditions ensuring the well-posedness of a forward problem with known material properties may, in this framework, be exploited. In cases where overdetermined data are available (e.g. the simultaneous knowledge of forces and displacements over some part of the sample under examination), part of the data may be considered as contributing to the boundary conditions, the remaining part being considered as overdetermined (the misfit with its simulation being used to define the cost function), with variations arising according to which part of the data is considered as overdetermined (e.g. cost functions focusing on either kinematic fields or forces). The choice of variant may depend on the precise nature of the available data and be determined by considerations such as algorithmic robustness, computational efficiency and ease of implementation.

A detailed presentation of FEMU is given in Chapter 9. Methodologies reviewed in section 8.3 are particularly relevant to FEMU. Applications of FEMU to constitutive parameter identification are the subject of many investigations, such as [FOR 04], [KAJ 04], [LEC 07], [COO 07], [MAH 02], [MAH 96], [PAG 07] and [SIL 09] (to cite just a few examples). FEMU in the context of structural dynamics and vibrations is also an active research topic; see, for instance, the survey article [FRI 95].

8.4.2. *Constitutive relation error*

Constitutive relation error (CRE) is an energy measure of the discrepancy between a stress field τ given *a priori* and another stress field evaluated from a given displacement field v using a constitutive model. For example, for a linear elastic constitutive model defined by the (possibly heterogeneous) elasticity tensor $\mathcal{A}$, the CRE between τ and v is defined by

$$\mathcal{E}(v, \tau, \mathcal{A}) = \frac{1}{2} \int_{\Omega} (\tau - \mathcal{A} : \varepsilon[v]) : \mathcal{A}^{-1} : (\tau - \mathcal{A} : \varepsilon[v]) \, \mathrm{d}V \qquad [8.30]$$

Note that using the compliance tensor $\mathcal{A}^{-1}$ for the purposes of weighting lends units of energy to $\mathcal{E}(v, \tau, \mathcal{A})$.

The CRE concept, which was initially introduced for linear elasticity by [LAD 83] in connection with error estimation for the FEM, quickly turned out to be very useful for model updating [REY 90, CHO 96, BAR 04]. More general formulations of CRE, applicable to incremental nonlinear constitutive models, have been proposed in [LAD 99], based on Drucker stability inequality, and in [MOË 99], based on free energy and dissipation potentials. The following remarks explain the usefulness of the CRE concept for identification:

1) The solution (u, σ) to a forward problem for a linear elastic solid is characterized by

$$(u, \sigma) = \underset{(v, \tau) \in \mathcal{C}(\bar{u}) \times \mathcal{S}}{\arg \min} \mathcal{E}(v, \tau, \mathcal{A}) \qquad \text{and} \quad \mathcal{E}(u, \sigma, \mathcal{A}) = 0 \qquad [8.31]$$

where $\mathcal{S}$ and $\mathcal{C}(\bar{u})$ denote the spaces of kinematically and statically admissible fields, respectively, corresponding to well-posed boundary conditions.

2) For a constitutive parameter identification problem for which overdetermined data are used, it is possible to modify the definitions of the admissible field spaces

$\mathcal{S}$ and $\mathcal{C}(\bar{u})$ so as to include *all* the available experimental information. The constitutive parameter identification problem then typically takes the form

$$\boldsymbol{\theta}^{\star} = \underset{\theta \in \mathbb{A}}{\arg \min}\, J(\boldsymbol{\theta}) \qquad \text{with} \quad J(\boldsymbol{\theta}) = \underset{(v,\tau) \in \mathcal{C} \times \mathcal{S}}{\min}\, \mathcal{E}(\boldsymbol{v}, \boldsymbol{\tau}, \boldsymbol{\mathcal{A}}(\boldsymbol{\theta})) \qquad [8.32]$$

where $\mathbb{A}$ is the set of physically admissible parameters $\boldsymbol{\theta}$. Equation [8.32] summarizes the CRE approach, which alternates minimizations with respect to (1) the admissible fields $(\boldsymbol{v}, \boldsymbol{\tau})$ (with fixed $\boldsymbol{\theta}$) and (2) the parameters $\boldsymbol{\theta}$ (with fixed admissible fields).

The concept of CRE is, in principle, applicable to any identification problem for which overdetermined data are available, that is which does not specifically require full-field measurements. This versatility has given rise to many applications in the context of model updating. However, the CRE concept is generally applicable to full-field measurements, as explained in Chapter 10.

8.4.3. *Methods based on equilibrium satisfaction*

This class of approach specifically relies on the experimental availability of a kinematic *field* (displacement $\bar{u}$ or strain $\bar{\varepsilon}$), or of a sequence of such fields when considering the identification of parameters for incremental constitutive models. Assuming, for simplicity, quasi-static conditions and no body forces, the local equilibrium equation

$$\operatorname{div} \boldsymbol{\sigma}[\bar{\varepsilon}; \boldsymbol{\theta}] = \mathbf{0}$$

yields, at any point of the sample where $\bar{\varepsilon}$ is known, an equality that must be satisfied by *any* constitutive model predicting $\boldsymbol{\sigma}$ for known $\bar{\varepsilon}$ (the necessity of knowing experimentally the *field* $\bar{\varepsilon}$ is, in particular, a consequence of the fact that the equilibrium equation involves spatial derivatives of $\bar{\varepsilon}$ via the divergence operator). We may thus conceivably identify, for example, parameters $\boldsymbol{\theta}$ associated with a model, or some spatial distribution of heterogeneous properties, by enforcing satisfaction of local equilibrium. Measurement or modeling errors generally implying the unfeasibility of exact equilibrium satisfaction, we may instead consider minimizing a global equilibrium residual $J(\boldsymbol{\theta})$, such as

$$J(\boldsymbol{\theta}) = \int_{\Omega} \|\operatorname{div} \boldsymbol{\sigma}[\bar{\varepsilon}; \boldsymbol{\theta}]\|^2 \, \mathrm{d}V$$

In principle, the formulation of an equilibrium residual requires a 3D kinematic field measurement, which may be obtained by means of some recently developed experimental methods. Measurements of kinematic fields on sample surfaces are otherwise commonly done, for instance, using digital image correlation. Such measurements must then be extrapolated to the whole 3D sample by exploiting kinematic modeling assumptions pertaining to, for example, plane-strain or plane-stress settings, thin or elongated structures, etc.

8.4.3.1. *Local form: the equilibrium gap method*

The enforcement of local equilibrium equations underlies the approach initiated in [CLA 04] for the identification of spatial distributions of a scalar damage variable $0 \leq D(\boldsymbol{x}) \leq 1$ [LEM 90] such that

$$\mathcal{A}(\boldsymbol{x}) = (1 - D(\boldsymbol{x}))\mathcal{A}_0 \qquad [8.33]$$

($\mathcal{A}_0$ denoting the elasticity tensor of the undamaged material) from displacement fields measured by means of digital image correlation. In this approach, whose detailed presentation is the subject of Chapter 12, the local equilibrium is exploited through equations generated by the FEM (i.e. the differential equilibrium equations written in weak form using locally supported trial functions) rather than pointwise local equations; the FE-generated equations may be considered local because they act over the element length scale h. The FE mesh is defined so that its nodes coincide with measurement points for the displacement field. The equilibrium equation associated with the mth DOF (assumed to be unloaded) thus has the form

$$\sum_{e|m\in E_e} (1 - D_e)\{\boldsymbol{e}_m\}^{\mathrm{T}}[\boldsymbol{K}_{e0}]\{\boldsymbol{u}_e\} = 0 \qquad [8.34]$$

where $\{\bar{\boldsymbol{u}}_e\}$ denotes the restriction to element E_e of the measured displacement, $[\boldsymbol{K}_{e0}]$ is the element stiffness for the undamaged element and D_e is the (unknown) value of the damage variable in E_e. The set of all possible equations [8.34] is usually overdetermined. It is thus solved for $\{D\}$ in the least-squares sense and subject to the constraints $0 \leq \{D\} \leq 1$. This work has since then been extended to the identification of parameters associated with damage laws [PÉR 09].

Other procedures based on the satisfaction of local field equations by an experimentally known field variable have been proposed, in connection with, for example, the identification of heterogeneous thermal diffusivities [BAM 09] or medical applications of elastography [SIN 05] based on 3D displacement field measurements using magnetic resonance imaging (MRI). They directly exploit finite-difference approximations of the local field equations and do not resort to weak finite element-type formulations. The main difficulty raised by this kind of approach lies in the fact that local field equations require (usually second-order) partial derivatives of the measured field quantity. Numerical differentiation of sampled data unfortunately often causes significant amplification of the original measurement errors. Following a similar approach, the identification of heat sources from infrared thermography data, treated as the right-hand side of the local heat diffusion equation (written in 2D form after integration along the thickness so as to exploit data available on the sample surface), has been investigated in [MOR 07]; there, the measured temperature field is replaced in the field equation by its projection onto a predefined finite-dimensional approximation space.

8.4.3.2. *Weak form: the virtual fields method*

This approach relies on the experimental availability of the strain field $\bar{\varepsilon}$ (possibly via the approximate differentiation of a measured displacement field) and assumes that loading conditions are known (we may, however, manage by using information on global resultant loads by using well-chosen virtual fields). The identification of constitutive parameters θ then exploits the equilibrium equation in weak (virtual work) form, for example

$$- \int_\Omega \boldsymbol{\sigma}[\bar{\varepsilon}; \boldsymbol{\theta}] : \boldsymbol{\varepsilon}[\boldsymbol{u}^\star] \, \mathrm{d}V + \int_{S_f} \boldsymbol{T}.\boldsymbol{u}^\star \, \mathrm{d}S = 0 \qquad [8.35]$$

where $\boldsymbol{u}^\star$ denotes the virtual field admissible with zero kinematic data and which is otherwise arbitrary, under quasi-static conditions and assuming zero body forces. Each possible choice of virtual field $\boldsymbol{u}^\star$ thus yields a scalar equation that must be verified by the constitutive model $\boldsymbol{\sigma}[\bar{\varepsilon}; \boldsymbol{\theta}]$ that predicts the stress value for given strain and constitutive parameters.

The virtual fields method, initiated in [GRÉ 89], consists of exploiting an identity of type [8.35] with judiciously chosen virtual fields according to the specificities of the identification problem at hand and presumes an *a priori* chosen constitutive model whose parameters are to be identified. This approach is detailed more in Chapter 11. Many applications and extensions of the virtual fields method have been investigated, for example by [CHA 06] (anisotropic elasticity with damage) or by [GRÉ 06, AVR 08] (elastoplasticity).

It is worth noting that the FE-based equilibrium gap method coincides with the virtual fields method if each virtual field is chosen as the interpolation function associated with a finite element DOF. Moreover, it is shown in [AVR 07] that the stationarity conditions for cost functionals associated with FEMU, CRE or equilibrium gap can be interpreted in terms of the virtual fields method for specific suitably chosen virtual fields.

8.4.4. *Reciprocity gap*

The reciprocity gap method mainly concerns situations where the field measurements are available at the boundary. Denoting by $(\hat{\boldsymbol{u}}, \hat{\boldsymbol{T}})$ the displacement and density fields on the boundary, we can define a reciprocity gap functional based on the virtual power principle. For example, we can consider measurements performed on a solid Ω whose behavior is defined by the elasticity tensor field $\boldsymbol{A}(\boldsymbol{x})$, which we are seeking to identify. An auxiliary displacement $\boldsymbol{u}^\star$, often called "adjoint", is defined. This displacement is generated, in a solid of identical geometry Ω but for a fictitious reference material characterized by the elastic tensor $\boldsymbol{A}^\star(\boldsymbol{x})$, by a force density $\boldsymbol{T}^\star$ applied on $\partial\Omega$. By combining the equations given by the virtual work principle for

$(\hat{u}, u^\star)$ and $(u^\star, \hat{u})$ (or else, equivalently, by writing the Maxwell–Betti reciprocity theorem for states $\hat{u}$ and $u^\star$), we obtain

$$\int_\Omega \varepsilon[u] : [\mathcal{A} - \mathcal{A}^\star] : \varepsilon[u^\star] \, \mathrm{d}V = \int_{\partial\Omega} (\hat{T}.u^\star - T^\star.\hat{u}) \, \mathrm{d}S \equiv R(\mathcal{A}^\star, u; u^\star) \quad [8.36]$$

Identity [8.36] provides an independent scalar equation that must be verified by the unknown elasticity tensor field $\mathcal{A}(x)$, or by parameters $\boldsymbol{\theta}$ defining it, for each choice of adjoint loading $T^\star$; this defines the essence of the reciprocity gap method, which is presented in detail in Chapter 13. We can consider the reciprocity gap method as a variant of the virtual fields method for which the kinematic fields are only available at the boundary. In the absence of arguments enabling the kinematic extrapolation of the data to Ω in its entirety, the real displacement field u is *a priori* unknown in Ω and must be reconstructed together with $\mathcal{A}$, at least on the geometrical support of the contrast $\mathcal{A} - \mathcal{A}^\star$. A linearized version of [8.36] for weak contrasts ($\|\mathcal{A} - \mathcal{A}^\star\| \ll \|\mathcal{A}^\star\|$) enables the theoretical analysis of the identifiability of heterogeneous elastic moduli [IKE 90] or of heterogeneous stiffnesses in plate bending [IKE 93], the identifiability condition generally being the knowledge of the Dirichlet–Neumann map, that is, of all the possible $(\hat{u}, \hat{T})$ pairs. Reciprocity gap functionals are also useful for identifying cracks [AND 97, BEN 99] or discrete point sources [ELB 00].

8.5. Conclusion

In this introductory chapter on identification, we proposed an overview of the concepts of inversion and identification, and the numerical solution methods commonly used in this context. Finally, the main methodologies devoted to constitutive parameter identification using full-field kinematic measurements are briefly described; they will be analyzed in more detail in Chapters 9 to 13.

8.6. Bibliography

[ACA 94] ACAR R., VOGEL C.R., "Analysis of bounded variation penalty methods for ill-posed problems", *Inverse Problems*, vol. 10, pp. 1217–1229, 1994.

[AMS 05] AMSTUTZ S., HORCHANI I., MASMOUDI M., "Crack detection by the topological gradient method", *Control and Cybernetics*, vol. 34, pp. 81–101, 2005.

[AND 97] ANDRIEUX S., BEN ABDA A., BUI H.D., "Sur l'identification de fissures planes *via* le concept d'écart à la réciprocité en élasticité", *C.R. Acad. Sci. Paris, série II*, vol. 324, pp. 1431–1438, 1997.

[ARN 08] ARNST M., CLOUTEAU D., BONNET M., "Identification of non-parametric probabilistic models from measured transfer functions", *Computer Methods in Applied Mechanics and Engineering*, vol. 197, pp. 589–608, 2008.

[AVR 07] AVRIL S., PIERRON F., "General framework for the identification of constitutive parameters from full-field measurements in linear elasticity", *International Journal of Solids and Structures*, vol. 44, pp. 4978–5002, 2007.

[AVR 08] AVRIL S., PIERRON F., PANNIER Y., ROTINAT R., "Stress reconstruction and constitutive parameter identification in plane-stress elastoplastic problems using surface measurements of deformation fields", *Experimental Mechanics*, vol. 48, pp. 403–419, 2008.

[BAM 09] BAMFORD M., BATSALE J.C., FUDYM O., "Nodal and modal strategies for longitudinal thermal diffusivity profile estimation: application to the non-destructive evaluation of SiC/SiC composites under uniaxial tensile tests", *Infrared Physics and Technology*, vol. 52, pp. 1–13, 2009.

[BAR 04] BARTHE D., DERAEMAEKER A., LADEVÈZE P., LE LOCH S., "Validation and updating of industrial models based on the constitutive relation error", *AIAA Journal*, vol. 42, pp. 1427–1434, 2004.

[BEL 09] BELLIS C., BONNET M., "Crack identification by 3D time-domain elastic or acoustic topological sensitivity", *Comptes Rendus Mécanique*, vol. 337, pp. 124–130, 2009.

[BEN 99] BEN ABDA A., BEN AMEUR H., JAOUA M., "Identification of 2D cracks by boundary elastic measurements", *Inverse Problems*, vol. 15, pp. 67–77, 1999.

[BER 99] BERTSEKAS D., *Nonlinear Programming*, Athena Scientific, Nashua, NH, 1999.

[BES 01] BESSON J., CAILLETAUD G., CHABOCHE J.L., FOREST S., *Mécanique non linéaire des matériaux*, Hermès, Paris, 2001.

[BON 09a] BONNET M., "Higher-order topological sensitivity for 2-D potential problems. Application to fast identification of inclusions", *International Journal of Solids and Structures*, vol. 46, pp. 2275–2292, 2009.

[BON 09b] BONNET M., GUZINA B.B., "Elastic-wave identification of penetrable obstacles using shape-material sensitivity framework", *Journal of Computational Physics*, vol. 228, pp. 294–311, 2009.

[BRI 07] BRIGHAM J.C., AQUINO W., "Surrogate-model accelerated random search algorithm for global optimization with applications to inverse material identification", *Computer Methods in Applied Mechanics and Engineering*, vol. 196, pp. 4561–4576, 2007.

[BUR 08] BURCZYŃSKI T., KUŚ W., "Identification of material properties in multi-scale modelling", *Journal of Physics: Conference Series*, vol. 135, pp. 012025, 2008.

[CHA 06] CHALAL H., AVRIL S., PIERRON F., MERAGHNI F., "Experimental identification of a damage model for composites using the grid technique coupled to the virtual fields method", *Composites: Part A*, vol. 37, pp. 315–325, 2006.

[CHO 96] CHOUAKI A., LADEVÈZE P., PROSLIER L., "An updating of structural dynamic model with damping", in DELAUNAY D., RAYNAUD M., WOODBURY K. (eds), *Inverse Problems in Engineering: Theory and Practice*, American Society of Mechanical Engineers, pp. 335–342, 1996.

[CLA 04] CLAIRE D., HILD F., ROUX S., "A finite element formulation to identify damage fields", *International Journal for Numerical Methods in Engineering*, vol. 61, pp. 189–208, 2004.

[COO 07] COOREMAN S., LECOMPTE D., SOL H., VANTOMME J., DEBRUYNE D., "Elasto-plastic material parameter identification by inverse methods: calculation of the sensitivity matrix", *International Journal of Solids and Structures*, vol. 44, pp. 4329–4341, 2007.

[DAG 07] DAGHIA F., DE MIRANDA S., UBERTINI F., VIOLA E., "Estimation of elastic constants of thick laminated plates within a Bayesian framework", *Composite Structures*, vol. 80, pp. 461–473, 2007.

[ELB 00] EL BADIA A., HA DUONG T., "An inverse source problem in potential analysis", *Inverse Problems*, vol. 16, pp. 651–663, 2000.

[ENG 96] ENGL H.W., HANKE M., NEUBAUER A., *Regularization of Inverse Problems*, Kluwer, Dordrecht, 1996.

[EPA 08] EPANOMERITAKIS I., AKCELIK V., GHATTAS O., BIELAK J., "A Newton-CG method for large-scale three-dimensional elastic full-waveform seismic inversion", *Inverse Problems*, vol. 24, pp. 034015, 2008.

[FLE 00] FLETCHER R., *Practical Methods of Optimization*, 2nd ed., John Wiley & Sons, New York, 2000.

[FOR 04] FORESTIER R., Développement d'une méthode d'identification de paramètres par analyse inverse couplée avec un modèle éléments finis 3D, PhD Thesis, Ecole Nationale Supérieure des Mines de Paris, France, 2004.

[FRI 95] FRISWELL M., MOTTERSHEAD J.E., "Finite element model updating in structural dynamics", *Solid Mechanics and its Applications*, Springer, 1995.

[GOG 08] GOGU C., HAFTKA R., LE RICHE R., MOLIMARD J., VAUTRIN A., SANKAR B., "Comparison between the basic least squares and the Bayesian approach for elastic constants identification", *Journal of Physics: Conference Series*, vol. 135, pp. 012045, 2008.

[GRÉ 89] GRÉDIAC M., "Principe des travaux virtuels et identification", *Comptes Rendus de l'Académie des Sciences*, vol. II/309, pp. 1–5, 1989.

[GRÉ 06] GRÉDIAC M., PIERRON F., "Applying the virtual fields method to the identification of elastoplastic constitutive parameters", *International Journal of Plasticity*, vol. 22, pp. 602–627, 2006.

[GUZ 02] GUZINA B.B., LU A., "Coupled waveform analysis in dynamic characterization of lossy solids", *Journal of Engineering Mechanics, ASCE*, vol. 128, pp. 392–402, 2002.

[HAN 98] HANSEN P.C., *Rank-Deficient and Discrete Ill-Posed Problems*, SIAM, Philadelphia, PA, 1998.

[IKE 90] IKEHATA M., "Inversion formulas for the linearized problem for an inverse boundary value problem in elastic prospection", *SIAM Journal on Applied Mathematics*, vol. 50, pp. 1635–1644, 1990.

[IKE 93] IKEHATA M., "An inverse problem for the plate in the Love-Kirchhoff theory", *SIAM Journal on Applied Mathematics*, vol. 53, pp. 942–970, 1993.

[KAI 05] KAIPIO J., SOMERSALO E., *Statistical and Computational Inverse Problems*, Springer-Verlag, Berlin, 2005.

[KAJ 04] KAJBERG J., LINDKVIST G., "Characterization of materials subjected to large strains by inverse modelling based on in-plane displacement fields", *International Journal of Solids and Structures*, vol. 41, pp. 3439–3459, 2004.

[KLE 97] KLEIBER M., HIEN T.D., ANTUNEZ H., KOWALCZYK P., *Parameter Sensitivity in Nonlinear Mechanics: Theory and Finite Element Computations*, John Wiley & Sons, New York, 1997.

[LAD 83] LADEVÈZE P., LEGUILLON D., "Error estimate procedure in the finite element method and applications", *SIAM Journal on Numerical Analysis*, vol. 20, pp. 485–509, 1983.

[LAD 99] LADEVÈZE P., *Nonlinear Computational Structural Mechanics: New Approaches and Non-Incremental Methods of Calculations*, Mechanical Engineering series, Springer-Verlag, Berlin, 1999.

[LAG 98] LAGARIAS J.C., REEDS J.A., WRIGHT M.H., WRIGHT P.E., "Convergence properties of the Nelder-Mead simplex method in low dimensions", *SIAM Journal on Optimization*, vol. 9, pp. 112–147, 1998.

[LEC 07] LECOMPTE D., SMITS A., SOL H., VANTOMME J., VAN HEMELRIJCK D., "Mixed numerical-experimental technique for orthotropic parameter identification using biaxial tensile tests on cruciform specimens", *International Journal of Solids and Structures*, vol. 44, pp. 1643–1656, 2007.

[LEM 90] LEMAÎTRE J., CHABOCHE J.L., *Mechanics of Solid Materials*, Cambridge University Press, Cambridge, UK, 1990.

[LER 07] LE RICHE R., SCHOENAUER M., SEBAG M., "Un état des lieux de l'optimisation évolutionnaire et de ses implications en sciences pour l'ingénieur", in BREITKOPF P., KNOPF-LENOIR C. (eds), *Modélisation Numérique: défis et perspectives, vol. 2*, Traité Mécanique et Ingénierie des Matériaux, Hermès, Paris, pp. 187–259, 2007.

[MAH 96] MAHNKEN R., STEIN E., "A unified approach for parameter identification of inelastic material models in the frame of the finite element method", *Computer Methods in Applied Mechanics and Engineering*, vol. 136, pp. 225–258, 1996.

[MAH 02] MAHNKEN R., "Theoretical, numerical and identification aspects of a new model class for ductile damage", *International Journal of Plasticity*, vol. 18, pp. 801–831, 2002.

[MEN 84] MENKE W., *Geophysical Data Analysis: Discrete Inverse Theory*, Academic Press, New York, 1984.

[MIC 96] MICHALEWICZ Z., *Genetic Algorithms + Data Structures = Evolution Programs*, Springer, Berlin, 1996.

[MOË 99] MOËS N., LADEVÈZE P., DOUCHIN B., "Constitutive relation error estimators for (visco)plastic finite element analysis with softening", *Computer Methods in Applied Mechanics and Engineering*, vol. 176, pp. 247–264, 1999.

[MOR 07] MORABITO A.E., CHRYSOCHOOS A., DATTOMA V., GALIETTI U., "Analysis of heat sources accompanying the fatigue of 2024 T3 aluminium alloys", *International Journal of Fatigue*, vol. 29, pp. 977–984, 2007.

[NEL 65] NELDER J.A., MEAD R., "A simplex method for function minimization", *The Computer Journal*, vol. 7, pp. 308–313, 1965.

[NOC 06] NOCEDAL J., WRIGHT S., *Numerical Optimization*, Springer, Berlin, 2006.

[PAG 07] PAGNACCO E., MOREAU A., LEMOSSE D., "Inverse strategies for the identification of elastic and viscoelastic material parameters using full-field measurements", *Materials Science and Engineering: A*, vol. 452–453, pp. 737–745, 2007.

[PÉR 09] PÉRIÉ J.-N., LECLERC H., ROUX S., HILD F., "Digital image correlation and biaxial test on composite material for anisotropic damage law identification", *International Journal of Solids and Structures*, vol. 46, pp. 2388–2396, 2009.

[REY 90] REYNIER M., Sur le contrôle de modélisations éléments finis: recalage à partir d'essais dynamiques, PhD Thesis, UPMC, Paris, 1990.

[SIL 09] SILVA G.H.C., Identification of material properties using finite elements and full-field measurements with focus on the characterization of deterministic experimental errors, PhD Thesis, Ecole Nationale Supérieure des Mines de Saint-Etienne, 2009.

[SIM 98] SIMO J.C., HUGHES T.J.R., *Computational Inelasticity*, Springer, New York, 1998.

[SIN 05] SINKUS R., TANTER M., XYDEAS T., CATHELINE S., BERCOFF J., FINK M., "Viscoelastic shear properties of in vivo breast lesions measured by MR elastography", *Magnetic Resonance Imaging*, vol. 23, pp. 159–165, 2005.

[TAR 05] TARANTOLA A., *Inverse Problem Theory and Methods for Model Parameter Estimation*, SIAM, Philadelphia, PA, 2005.

[TIK 77] TIKHONOV A.N., ARSENIN V.Y., *Solutions to Ill-Posed Problems*, Winston-Wiley, New York, 1977.

[TIK 95] TIKHONOV A.N., GONCHARSKI A.V., STEPANOV V.V., YAGODA A.G., *Numerical Methods for the Solution of Ill-Posed Problems*, Kluwer Academic Publishers, Dordrecht, 1995.

[TWO 77] TWOMEY S., *Introduction to the Mathematics of Inversion in Remote Sensing and Indirect Measurements*, Dover Publications, New York, 1977.

Chapter 9

Parameter Identification from Mechanical Field Measurements using Finite Element Model Updating Strategies

9.1. Introduction

In recent decades, experimental devices and methodologies for mechanical testing have evolved toward the use of full-field measurements. The data they generate are therefore introduced in identification methods and enable the evaluation of reliable constitutive material parameters. These methods, known as "constitutive equation gap method" (Chapter 10), "virtual fields method" (Chapter 11), "equilibrium gap method" (Chapter 12), "reciprocity gap method" (Chapter 13) and "finite element model updating (FEMU) method" are all derived from the well-known principles of continuum mechanics (Chapter 8). In this chapter, we will focus on the FEMU strategy, which consists of updating the parameters of a finite element model in order to minimize the difference between measured and simulated fields. This method can be applied at the structural scale and can handle complex (three-dimensional, 3D) loads and geometrical shapes using various measurements, such as displacements, strains and loads, even if they are acquired over a limited area of the mechanical part. Moreover, it is independent of the type of mechanical behavior. At first, this mixed method consisted of coupling finite element (FE) analyses with mechanical tests and experimental measurements (usually displacements from a linear variable differential

Chapter written by Emmanuel PAGNACCO, Anne-Sophie CARO-BRETELLE and Patrick IENNY.

transformer). Nowadays, with full-field measurements, this combined numerical–experimental method opens up a very broad range of experimental investigation possibilities, such as:

– new experimental designs that enable (possibly multiaxial) loading conditions much closer to those of practical industrial applications, or specific loading that induces response fields in the numerical model that are more sensitive to parameter changes;

– the study of responses at high strain levels, which could appear in a localized area from a non-homogeneous test (Chapter 14);

– the investigation of the mechanical properties of multibody systems or structural materials such as composites. The extreme anisotropy and/or non-homogeneous distribution of material properties over the specimen produces non-homogeneous stress and strain fields in the investigated volume;

– the analysis of multi-scale modeling, using optical and/or electronic microscopy (Chapter 15).

The measurement of deformation or strain fields enables analyses of more complex tests. Moreover, contactless measurements of the surface field by optical methods enable some varieties of tests which are difficult to handle using more traditional sensors (testing of small-scale structures, protected surrounding conditions, etc.). However, testing has to be modeled with reasonable accuracy in order to produce a realistic numerical simulation. For example, the numerical model of the specimen being tested requires the proper modeling of both its initial shape and the boundary conditions. Nevertheless, field measurements enable testing freedom: spurious bending effects or boundary conditions, such as gaps or sliding, can be allowed.

The first part of this chapter introduces the basic concepts of the FE method and its application to solids and structures (finite element analysis, FEA). This allows us to introduce the theory of inverse analysis by updating the constructed FE model, when modeling assumptions are made. This inverse problem is an optimization problem in which several cost functions can be considered for its formulation. A summary of the literature gives some insights into common inverse formulations for the FEMU method. When applicable, gradient-based methodologies (possibly combined with an evolutionary strategy) are preferred because of their robustness and convergence speed. In the final part of this chapter, we will summarize several kinds of identification problem presented in the literature and addressed by the FEMU method. Investigated material behavior laws (e.g. anisotropy, nonlinearity, temperature or velocity dependence, damage) as well as the number of parameters and conducted tests (traction, compression and bending) are summarized. This leads us to discuss some thoughts on identification accuracy before closing the chapter with a summary of the major identification keys.

9.2. Finite element method

9.2.1. *Principles of the method*

The finite element method [BAT 96, BEL 01] transforms the spatial partial differential equations over a smooth manifold, having a compact support and boundary conditions, into algebraic equations. In mechanics, it is based, firstly, on the principle of virtual work/power or, more generally, on the principles of variational formulations, and, secondly, it is based on appropriate approximations defined over subdomains. It enables continuous fields to be approached by discrete numerical fields having existence conditions that are less restrictive than those of initial fields.

Let us consider a solid occupying, in a reference configuration given by its initial position $\mathbf{x}_0$, a domain $\Omega_0 \in \mathbb{R}^3$ with its (sufficiently regular) boundary denoted by $\partial\Omega_0$. When it is subjected to loads, it undergoes a deformation motion specified by the displacement field $\mathbf{u}$ in Ω_t, which is the deformed counterpart of Ω_0. Thus, the position in the current configuration is given by $\mathbf{x}(t) = \mathbf{x}_0 + \mathbf{u}(\mathbf{x}, t)$. When body forces are not considered by the model, the principle of virtual power states that:

$$\int_{\Omega_t} \mathbf{D}^* : \boldsymbol{\sigma} \mathrm{d}V + \int_{\Omega_t} \dot{\mathbf{u}}^* . \rho . \ddot{\mathbf{u}} \mathrm{d}V - \oint_{\partial\Omega_t} \dot{\mathbf{u}}^* . \boldsymbol{\sigma} . \mathbf{n} \mathrm{d}S = 0 \qquad [9.1]$$

$\forall t, \dot{\mathbf{u}}^*$ sufficiently regular

where $\boldsymbol{\sigma}$ is the Cauchy stress tensor, ρ is the material density, $\dot{\mathbf{u}}^*$, $\mathbf{D}^* = \frac{1}{2}\left(\nabla\dot{\mathbf{u}}^* + \nabla\dot{\mathbf{u}}^{*\mathrm{T}}\right)$ refer to a virtual velocity and strain rate, respectively, and $\mathbf{n}$ is the outward normal unit vector to $\partial\Omega_t$. The current stress state can be related to a strain state or a strain history state (defined from $\mathbf{u}$ and possibly its temporal derivatives) by choosing an appropriate strain measurement and a constitutive law, representative of the material behavior. When a linear material behavior is assumed without prestress, the constitutive relationship takes the form:

$$\boldsymbol{\sigma} = \mathcal{A} : \varepsilon \qquad [9.2]$$

where the tensor $\mathcal{A}$ involves material parameters that we can collect in a vector $\boldsymbol{\theta}$. In practice, tensors quantifying stress (strain) and material behavior for 3D solids are conveniently represented as a 6D vector and a 6×6 matrix, respectively, in an orthonormal coordinate system in the current configuration $\mathbf{x}(t)$. Hence, equation [9.2] is rewritten in the matrix–vector form:

$$\begin{bmatrix} \sigma_{xx} \\ \sigma_{yy} \\ \sigma_{zz} \\ \sigma_{xy} \\ \sigma_{xz} \\ \sigma_{yz} \end{bmatrix} = \mathbf{A} \begin{bmatrix} \varepsilon_{xx} \\ \varepsilon_{yy} \\ \varepsilon_{zz} \\ \gamma_{xy} = 2\varepsilon_{xy} \\ \gamma_{xz} = 2\varepsilon_{xz} \\ \gamma_{yz} = 2\varepsilon_{yz} \end{bmatrix}, \mathbf{A} = \begin{bmatrix} a_{xxxx} & a_{xxyy} & a_{xxzz} & a_{xxxy} & a_{xxxz} & a_{xxyz} \\ a_{yyxx} & a_{yyyy} & a_{yyzz} & a_{yyxy} & a_{yyxz} & a_{yyyz} \\ a_{zzxx} & a_{zzyy} & a_{zzzz} & a_{zzxy} & a_{zzxz} & a_{zzyz} \\ a_{xyxx} & a_{xyyy} & a_{xyzz} & a_{xyxy} & a_{xyxz} & a_{xyyz} \\ a_{xzxx} & a_{xzyy} & a_{xzzz} & a_{xzxy} & a_{xzxz} & a_{xzyz} \\ a_{yzxx} & a_{yzyy} & a_{yzzz} & a_{yzxy} & a_{yzyz} & a_{yzyz} \end{bmatrix}$$

$$[9.3]$$

Depending on the choice of material parameters, there are several possibilities to express the matrix $\mathbf{A}$. For parameter identification, a linear relationship between these parameters enables a better convergence of the numerical procedure. An adequate choice is $\boldsymbol{\theta} = \{K, G\}$, with K being the bulk modulus and G the shear modulus, when considering a 3D isotropic linear elastic behavior, leading to:

$$\mathbf{A}(K,G) = \begin{bmatrix} K + \frac{4}{3}G & K - \frac{2}{3}G & K - \frac{2}{3}G & 0 & 0 & 0 \\ K - \frac{2}{3}G & K + \frac{4}{3}G & K - \frac{2}{3}G & 0 & 0 & 0 \\ K - \frac{2}{3}G & K - \frac{2}{3}G & K + \frac{4}{3}G & 0 & 0 & 0 \\ 0 & 0 & 0 & G & 0 & 0 \\ 0 & 0 & 0 & 0 & G & 0 \\ 0 & 0 & 0 & 0 & 0 & G \end{bmatrix}$$

If an orthotropic linear elastic behavior is assumed for a 2D mechanical test involving a plane stress state (in the O_{xy} plane), an adequate choice of parameters is $\boldsymbol{\theta} = \{Q_{xx}, Q_{yy}, Q_{yy}, G_{xy}\}$ such that:

$$\begin{bmatrix} \sigma_{xx} \\ \sigma_{yy} \\ \sigma_{xy} \end{bmatrix} = \mathbf{A} \begin{bmatrix} \varepsilon_{xx} \\ \varepsilon_{yy} \\ \gamma_{xy} \end{bmatrix}, \quad \mathbf{A} = \begin{bmatrix} Q_{xx} & Q_{xy} & 0 \\ Q_{xy} & Q_{yy} & 0 \\ 0 & 0 & G_{xy} \end{bmatrix} \tag{9.4}$$

with the two Poisson's ratios $\nu_{xy} = Q_{xy}/Q_{yy}$ and $\nu_{yx} = Q_{xy}/Q_{xx}$ and the two elastic moduli $E_x = (1 - \nu_{xy}\nu_{yx})Q_{xx}$ and $E_y = (1 - \nu_{xy}\nu_{yx})Q_{yy}$. Assuming an isotropic behavior, this relationship can be reduced to only two parameters, Q_{xx} and G_{xy}, such that:

$$\begin{bmatrix} \sigma_{xx} \\ \sigma_{yy} \\ \sigma_{xy} \end{bmatrix} = \mathbf{A} \begin{bmatrix} \varepsilon_{xx} \\ \varepsilon_{yy} \\ \gamma_{xy} \end{bmatrix}, \quad \mathbf{A} = \begin{bmatrix} Q_{xx} & Q_{xx} - 2G_{xy} & 0 \\ Q_{xx} - 2G_{xy} & Q_{xx} & 0 \\ 0 & 0 & G_{xy} \end{bmatrix}$$

An approximate form of principle [9.1] can be expressed by introducing a representation of real and virtual kinematic fields in an adequate finite-dimensional space (spanned by the finite element shape functions $\mathbf{N}$):

$$\dot{\mathbf{U}}^{*T}\mathbf{F}_{int} + \dot{\mathbf{U}}^{*T}\mathbf{M}\ddot{\mathbf{U}} - \dot{\mathbf{U}}^{*T}\mathbf{F}_{ext} = 0 \qquad \forall t, \dot{\mathbf{U}}^* \tag{9.5}$$

In this equation, $\mathbf{F}_{int}$ is a vector of internal forces related to the current stress state and possibly to its history, and $\mathbf{M}$ is the symmetric global mass matrix of the mechanical system. The vector $\mathbf{F}_{ext}$ is associated with generalized nodal forces having action and reaction forces, while the vector $\mathbf{U}$ gathers together the degrees of freedom (dofs) introduced by FE discretization (usually nodal displacements) through the relationships:

$$\mathbf{u}(\mathbf{x}, t) = \mathbf{N}\mathbf{u}(\mathbf{x}).\mathbf{U}(t), \quad \dot{\mathbf{u}} = \mathbf{N}.\dot{\mathbf{U}}, \quad \ddot{\mathbf{u}} = \mathbf{N}.\ddot{\mathbf{U}}$$

assuming that the same FE shape functions are chosen for real and virtual quantities. Thus, introducing $\mathbf{B} = \nabla\mathbf{N}$, virtual velocities are expressed in the form:

$$\mathbf{D}^* = \mathbf{B}.\dot{\mathbf{U}}^*$$

which leads to:

$$\mathbf{F}_{\text{int}} = \int_{\Omega_t} \mathbf{B}^{\mathrm{T}}\boldsymbol{\sigma}\mathrm{d}V$$

while:

$$\mathbf{F}_{\text{ext}} = \oint_{\partial\Omega_t} \mathbf{N}^{\mathrm{T}}\boldsymbol{\sigma}.\mathbf{n}\mathrm{d}S \quad \text{and} \quad \mathbf{M} = \int_{\Omega_t} \mathbf{N}^{\mathrm{T}}.\rho.\mathbf{N}\mathrm{d}V$$

Finally, equation [9.5] leads to the global discrete dynamic equilibrium of the solid, such that:

$$\mathbf{M}\ddot{\mathbf{U}} = \mathbf{F}_{\text{ext}} - \mathbf{F}_{\text{int}} \tag{9.6}$$

Moreover, when inertial effects can be neglected, equilibrium can be reduced to a static problem:

$$\mathbf{F}_{\text{int}} = \mathbf{F}_{\text{ext}} \tag{9.7}$$

which, in the case of small perturbations and linear material behavior is written in a more practical form:

$$\mathbf{K}.\mathbf{U} = \mathbf{F}_{\text{ext}} \tag{9.8}$$

where $\mathbf{F}_{\text{int}} = \mathbf{K}.\mathbf{U}$ and $\mathbf{K} = \int_{\Omega_0} \mathbf{B}^{\mathrm{T}}\mathbf{A}.\mathbf{B}\mathrm{d}V$ is a constant stiffness matrix, independent of $\mathbf{U}$. Dynamic equilibrium for small perturbations where damping is neglected can then be expressed in the form:

$$\mathbf{M}\ddot{\mathbf{U}} + \mathbf{K}.\mathbf{U} = \mathbf{F}_{\text{ext}} \tag{9.9}$$

for which the associated homogeneous equation defines the modal analysis problem:

$$\left(\mathbf{K} - \omega_l^2\mathbf{M}\right)\phi_l = \mathbf{0} \tag{9.10}$$

assuming that a normalization condition is provided for a well-posed problem, leading to $\phi_l^{\mathrm{T}}\mathbf{M}\phi_l = \mu_l$. In these equations, ω_l is the angular frequency, ϕ_l is the lth normal mode corresponding to ω_l and μ_l is its modal mass.

Thus, only loading conditions, structural models (mesh and mechanical behavior assumptions) and material properties compose the finite element model, and equation [9.6] expresses the overall balance of inertial, external and internal forces for the discrete solid at every time in algebraic form. Note that the choice of the principle of virtual power leads to finite elements in their simplest form, the so-called "displacement form". Note also that in structural mechanics, $\mathbf{U}$, $\mathbf{F}_{\text{int}}$ and $\mathbf{F}_{\text{ext}}$ include rotational dofs, internal and external moments, respectively.

9.2.2. *The "direct mechanical problem" and finite element analysis*

The standard problem of continuum mechanics consists of finding the displacement field, strain and stress over time t, from the initial shape of the solid, the material behavior with its constitutive parameters, and a set of boundary and initial conditions. The forward or direct problem, described in the literature [BAT 96, BEL 01], is well-posed for a suitable choice of boundary and initial conditions. For this, the surface boundary contour is decomposed into two disjoint open parts S_{tu} and S_{tr}, such that $\partial\Omega_t = S_{tu} \cup S_{tr}$, with respect to a kinematic condition type prescribed for S_{tu} and mechanical condition $\bar{\mathbf{f}}_{s_r}$ applied to S_{tr} (this function can be null over a large portion of the surface). In solid mechanics, the kinematic conditions generally focus on displacements while $\bar{\mathbf{f}}_{s_r} = \sigma.\mathbf{n}$ corresponds to the traction imposed over time. This spatial partition leads to the introduction of an unknown surface traction $\bar{\mathbf{f}}_{s_u}$, such that $\bar{\mathbf{f}}_{s_u} = \sigma.\mathbf{n}$ over S_{tu}, in order to impose the kinematic condition on this area. For structural mechanics problems, these conditions also concern rotations and moments.

The variational form [9.1] leads to the definition of a set of kinematically admissible displacement fields over which $\mathbf{u}$ must satisfy:

$$\int_{\Omega_t} \mathbf{D}^* : \sigma dV + \int_{\Omega_t} \dot{\mathbf{u}}^*.\rho.\ddot{\mathbf{u}} dV - \oint_{S_{tr}} \dot{\mathbf{u}}^*.\bar{\mathbf{f}}_{s_r} dS - \oint_{S_{tu}} \dot{\mathbf{u}}^*.\bar{\mathbf{f}}_{s_u} dS = 0 \tag{9.11}$$

$\forall t, \dot{\mathbf{u}}^*$ sufficiently regular

Since closed-form solutions to these direct problems are only available for a few particular cases, the principle of the finite element method provides a set of algebraic equations that can be efficiently solved numerically. To obtain this set of equations, the vector of external forces is first expressed as the sum of the known discrete action forces $\bar{\mathbf{F}} = \oint_{S_{tr}} \mathbf{N}^T.\bar{\mathbf{f}}_{s_r} dS$ and unknown reactions $\mathbf{R} = \oint_{S_{tu}} \mathbf{N}^T.\bar{\mathbf{f}}_{s_u} dS$, such that:

$$\mathbf{F}_{\text{ext}} = \bar{\mathbf{F}} + \mathbf{R}$$

while the vector $\mathbf{U}$ is beforehand split into two complementary sub-vectors such that $\mathbf{U} = [\mathbf{U}_i \quad \mathbf{U}_c]^T$, where $\mathbf{U}_c$ embeds the dofs corresponding to S_{tu}. This enables us to express the imposed kinematic conditions by a discretized and linearized constraint as $\mathbf{C}_c\mathbf{U}_c = \bar{\mathbf{U}}$ or:

$$\mathbf{C}.\mathbf{U} = \bar{\mathbf{U}} \tag{9.12}$$

thereby completing the system of equation [9.6] or [9.7], for $\mathbf{C} = [\mathbf{C}_i \quad \mathbf{C}_c]$ with $\mathbf{C}_i = \mathbf{0}$. In this expression, the matrix $\mathbf{C}_c$ can have a Boolean form for a simple condition on each dof, or a more complicated form when relationships between dofs have to be imposed.

Hence, FEA consists first of determining unknown quantities $\mathbf{U}_i$ and $\mathbf{R}$ to approach the mechanical problem solution from the finite element model when taking into account initial conditions, known forces $\bar{\mathbf{F}}$ and kinematic conditions at each time step. For the linear static problem [9.8], a simple strategy consists of introducing the vector of Lagrange multipliers $\boldsymbol{\lambda}$ that leads to the augmented system:

$$\begin{bmatrix} \mathbf{K} & \mathbf{C}^{\mathrm{T}} \\ \mathbf{C} & \mathbf{0} \end{bmatrix} \begin{bmatrix} \mathbf{U} \\ \boldsymbol{\lambda} \end{bmatrix} = \begin{bmatrix} \bar{\mathbf{F}} \\ \bar{\mathbf{U}} \end{bmatrix} \qquad [9.13]$$

where the reaction forces are expressed by the quantity $\mathbf{R} = -\mathbf{C}^{\mathrm{T}}\boldsymbol{\lambda}$. By splitting the stiffness matrix $\mathbf{K}$, $\bar{\mathbf{F}}$ and $\mathbf{R}$ into:

$$\mathbf{K} = \begin{bmatrix} \mathbf{K}_{ii} & \mathbf{K}_{ic} \\ \mathbf{K}_{ci} & \mathbf{K}_{ii} \end{bmatrix}$$

$\bar{\mathbf{F}} = [\bar{\mathbf{F}}_i \quad 0]^{\mathrm{T}}$ and $\mathbf{R} = [0 \quad \mathbf{R}_c]^{\mathrm{T}}$, the sequential set of equations is reduced to:

$$\begin{cases} \mathbf{C}_c \mathbf{U}_c = \bar{\mathbf{U}} \\ \mathbf{K}_{ii} \mathbf{U}_i = \bar{\mathbf{F}}_i - \mathbf{K}_{ic} \mathbf{U}_c \\ \mathbf{R}_c = \mathbf{K}_{ci} \mathbf{U}_i + \mathbf{K}_{cc} \mathbf{U}_c \end{cases} \qquad [9.14]$$

For the nonlinear mechanical problem [9.7], it is common[1] to linearize the equation via a Newton–Raphson strategy, since the vector of internal forces is defined in the current stress state and depends upon the current displacement field. This leads to the introduction of the tangent stiffness matrix $\mathbf{K}_T = (\partial \mathbf{F}_{\mathrm{int}}/\partial \mathbf{U}) - (\partial \mathbf{F}_{\mathrm{ext}}/\partial \mathbf{U})$ in the place of the stiffness matrix $\mathbf{K}$ of the linear problem, which relates incremental solutions $\Delta \mathbf{U}$ and $\Delta \mathbf{R}$ to incremental load conditions $\Delta \bar{\mathbf{F}}$ and $\Delta \bar{\mathbf{U}}$, instead of $\mathbf{U}$ and $\mathbf{R}$ involved in previous equations. In this case, several iterations are generally required between each new load increment, since $\mathbf{K}_T$ is nonlinearly dependent on $\mathbf{U}$.

Once the numerical displacement field $\mathbf{U}$ and reaction forces $\mathbf{R}$ are obtained for each load step or time increment, strain and stress states can be stored to complete the analysis of the forward mechanical problem. The finite element method is therefore a combination of (1) an approximation of a variational form and (2) a numerical resolution of this discrete form, such as [9.13] or [9.14]. In this work, it is important to distinguish these two key points, since the second point is not systematically involved when solving an inverse problem using the FEMU method.

---

1 Such a linearization strategy concerns both static and implicit dynamic resolution schemes (such as Newmark or HHT schemes), but it is not required for an explicit dynamic scheme (central finite difference scheme) that is used to deal with high-speed problems or highly nonlinear problems.

9.3. Updating a finite element model for parameter identification

9.3.1. *Theory*

The FEMU method consists of determining the value(s) of parameters of an FE numerical model in order to reproduce a given known state. It is usually used for identifying the constitutive parameters of a material constitutive law, qualifying the loading conditions of a structural test, or optimizing a structural shape or topology. Its principle is shown in Figure 9.1, compared with classic analytical identification. The mechanical test of the studied structure is simulated using FEA with known data (shape and boundary conditions, denoted by the set ϖ). Experimental data are usually discrete data (defined in terms of a time step or load) and FEA generates results that must be adapted to be compared with experimental data. Observable data extracted from experiments (e.g. measurements of deformations, strains, forces) and numerical data from the FE analysis are denoted by $\mathbf{m}(t)$ and $\mathbf{h}_\varpi(\boldsymbol{\theta}, t)$, respectively, at each time step $t \in [0, T]$. $\boldsymbol{\theta}$ is the set of parameters to extract (e.g. material parameters, stiffness, local load) and $\boldsymbol{\theta}^{(0)}$ is the guess of starting point (variable often required to initiate the optimization process). The parameters $\boldsymbol{\theta}$, data $\mathbf{m}(t)$ and output variables $\mathbf{h}_\varpi(\boldsymbol{\theta}, t)$ are stored in column vectors:

$$\mathbf{m}(t) = [m_1(t), m_2(t), \dots, m_n(t)]^{\mathrm{T}}$$

$$\mathbf{h}_\varpi(\boldsymbol{\theta}, t) = [h_{\varpi_1}(\boldsymbol{\theta}, t), h_{\varpi_2}(\boldsymbol{\theta}, t), \dots, h_{\varpi_n}(\boldsymbol{\theta}, t)]^{\mathrm{T}} = \mathbf{P}.\mathbf{U} \qquad [9.15]$$

$$\boldsymbol{\theta} = [\theta_1, \theta_2, \dots, \theta_p]^{\mathrm{T}}$$

where n and p denote, respectively, the number of measurements and the number of parameters to be identified, and $\mathbf{P}$ is a post-processed FE output operator that enables a direct comparison with experimental data.

Considering the wide range of testing possibilities and measurable quantities, a variety of identification situations may be encountered. It is thus possible, but by no means necessary, to use field measurements in finite element models. The FEMU method, thanks to its flexibility, can be applied with any kind of data (supernumerary or not).

The user is free to choose both input and output variables and to associate them with an optimization criterion. The latter can be either algebraic (least-squares criterion, weighted or not, for example) or probabilistic (as the likelihood criterion). For the first type of criterion[2], we seek parameters $\boldsymbol{\theta}$ that minimize the distance between $\mathbf{m}(t)$ and $\mathbf{h}_\varpi(\boldsymbol{\theta}, t), \forall t \in [0, T]$. This minimization can be written formally through the functional J, such that $J : P \to \mathbb{R}$:

---

2 The maximum value or the absolute value norm is also a possible choice [PAN 98, TAR 87].

$$\min_{\boldsymbol{\theta} \in P} J(\boldsymbol{\theta}) = \min_{\boldsymbol{\theta} \in U} \sum_{t=0}^{T} \frac{1}{2} \mathbf{j}_{\varpi}(\boldsymbol{\theta}, t)^T \mathbf{V} \mathbf{j}_{\varpi}(\boldsymbol{\theta}, t) \qquad [9.16]$$

where $P \subset \mathbb{R}^p$ and

$$\mathbf{j}_{\varpi}(\boldsymbol{\theta}, t) = \mathbf{m}(t) - \mathbf{h}_{\varpi}(\boldsymbol{\theta}, t) = [j_{\varpi_1}, j_{\varpi_2}, \ldots, j_{\varpi_n}] \qquad [9.17]$$

with $j_{\varpi_i} : P \times [0, T] \to \mathbb{R}^N \; \forall i \in [0, \ldots, n]$ and $\mathbf{V}$ is a weighting $n \times n$ positive definite matrix.

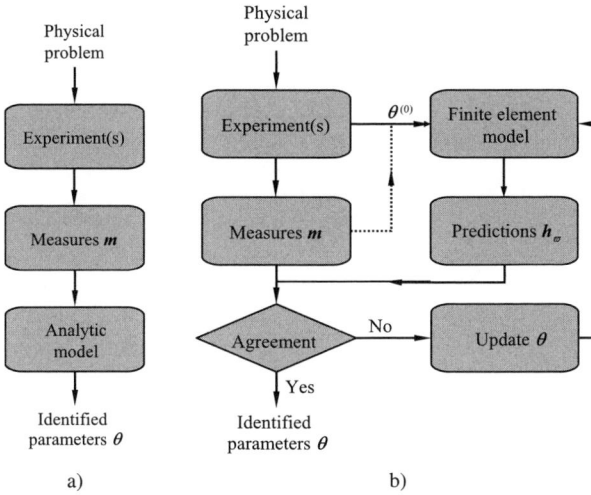

Figure 9.1. *Schematic view of (a) analytical and (b) FEMU identification*

The second criterion, based on an estimate of the minimum variance, involves an assumed probability law associated with a measurement error $\mathbf{e}(t)$:

$$\mathbf{m}(t) = \mathbf{h}_{\varpi}(\boldsymbol{\theta}, t) + \mathbf{e}(t)$$

The optimal parameters minimize the quadratic expression:

$$\min_{\boldsymbol{\theta} \in P} J(\boldsymbol{\theta}) = \min_{\boldsymbol{\theta} \in P} \sum_{t=0}^{T} \frac{1}{2} \mathbf{j}_{\varpi}(\boldsymbol{\theta}, t)^T \mathbf{V} \mathbf{j}_{\varpi}(\boldsymbol{\theta}, t) + (\boldsymbol{\theta}^{(0)} - \boldsymbol{\theta})^T \mathbf{W}(\boldsymbol{\theta}^{(0)} - \boldsymbol{\theta}) \quad [9.18]$$

where $P \subset \mathbb{R}^p$ and $\mathbf{j}_{\varpi}(\boldsymbol{\theta}, t)$ are described in [9.17], $\boldsymbol{\theta}^{(0)}$ is an initial guess for the parameter $\boldsymbol{\theta}$, and $\mathbf{V}$ and $\mathbf{W}$ are $n \times n$ weighting matrices. This criterion is widely used in medical applications [HEN 91, OOM 93, VAN 94] and in the FEMU method applied to structural dynamic material [COL 74, FRI 95].

Both these minimization problems are solved by an iterative procedure: find $\boldsymbol{\theta}$ solution of [9.16] (or [9.18]), such that:

$$\boldsymbol{\theta}^{(k+1)} = f(\boldsymbol{\theta}^{(k)}) \quad \text{and} \quad \boldsymbol{\theta} = \lim_{k \to +\infty} \boldsymbol{\theta}^{(k)} \qquad [9.19]$$

The expression of f depends on the chosen resolution algorithm (see section 9.3.2).

In most cases, inverse problems using field measurements are used in the identification of the material properties of flat samples. In fact, the FEMU method is particularly suitable for the identification of properties from inhomogeneous response fields. This may be the case of specimens composed of heterogeneous materials (i.e with several phases, or in the case of a composite) or specimens whose shape generates heterogeneous stresses, even when they are composed of homogeneous materials. Figure 9.2 shows examples of samples with non-standard shapes, typically used to generate heterogeneous mechanical fields.

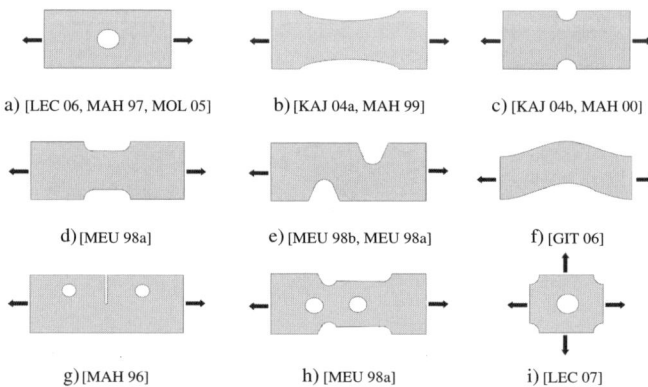

Figure 9.2. Shapes of specimens used for parameter identification by FEMU identification in various references

However, since the method proposed in this chapter is not limited by the shapes under investigation, the identification of material parameters can also be performed directly at the structural scale. This is the case, for example, in [KAV 72], where an axisymmetric shell (a frustum) composite material is selected, or in [BUR 03] where a 3D model is constructed to identify the elastic parameters of a human pelvis.

9.3.2. Objective functions and minimization procedure

The inverse problem usually consists of identifying $\boldsymbol{\theta}$ from measured data. In the literature, various forms of the cost function $\mathbf{j}_{\varpi}$ in [9.16] or [9.18] (and also for the weighting matrices $\mathbf{V}$ and $\mathbf{W}$) can be found, depending on the application.

For a least-squares criterion, if measurements concern quantities with several physical dimensions, it is possible to tune their relative impact through the use of the weighting matrices.

In practice, for each mechanical test, kinematic measurements provide data on either Ω_t or $\partial\Omega_t$, or only on a part of them, and the formulation of the problem must be adapted to take into account the available experimental information. Denoting by $\hat{\bullet}$ the measured quantities, it is possible, depending on the available and selected experimental data, to impose conditions, such as $\mathbf{U}(t) = \hat{\mathbf{U}}(t)$ in Ω_t or $\mathbf{U}(t) = \hat{\mathbf{U}}(t)$ on $\partial\Omega_t$ or S_{t_u} with $\mathbf{F}_{\text{ext}}(t) = \hat{\mathbf{F}}(t)$ in Ω_t or $\mathbf{F}_{\text{ext}}(t) = \hat{\mathbf{F}}(t)$ on $\partial\Omega_t$ or S_{t_r} in the inverse problem.

9.3.2.1. *Objective functions*

Historically, the first objective function found in the literature is based on both the data of the full-field displacements $\hat{\mathbf{U}}$ and the full-field forces $\hat{\mathbf{F}}$ in Ω_t:

$$\mathbf{j}_{\varpi}(\boldsymbol{\theta}, t) = \hat{\mathbf{F}}(t) - \mathbf{F}_{\text{int}}(\boldsymbol{\theta}, t) \ \forall t \in [0, T] \quad \text{and} \quad \mathbf{V} = \mathbf{I} \tag{9.20}$$

such that $\mathbf{F}_{\text{ext}}(t) = \hat{\mathbf{F}}(t)$ (see equation [9.7]). This leads to a non-iterative (thus direct) inverse problem, known as the "force balance method" (also known as the "input residual method" or "equilibrium method") [KAV 71, FRI 95, PAG 07]. In [9.20], $\mathbf{I}$ denotes the identity matrix. Since $\mathbf{F}_{\text{int}}$ is evaluated from the proposition $\mathbf{U} = \hat{\mathbf{U}}$ in Ω_t, the "direct problem" of the FE analysis is never solved. Evaluating $\mathbf{j}_{\varpi}$ in [9.20] is thus very efficient from a computational point of view, necessitating only an FE model. Moreover, if the parameterization of the problem is linear in $\boldsymbol{\theta}$ (e.g. the bulk and shear moduli, i.e. $\boldsymbol{\theta} = [K, G]$ for problems devoted to the identification of isotropic material properties), the minimization problem associated with this objective function is an overdetermined system of equations of the form [KAV 71]:

$$\frac{\partial[\mathbf{K}(\boldsymbol{\theta}).\hat{\mathbf{U}}]}{\partial\boldsymbol{\theta}}.\boldsymbol{\theta} = \hat{\mathbf{F}} \tag{9.21}$$

which can be solved using a least-squares method in only one step, without reference to any starting guess $\boldsymbol{\theta}^{(0)}$. Consequently, neither assumptions about $\boldsymbol{\theta}^{(0)}$, nor FE analysis, nor iterations are required in this method.

In another common but more recent approach [COL 74], the objective function, is based only on the displacement data, while the force data are used within the FE analysis as a boundary condition:

$$\mathbf{j}_{\varpi}(\boldsymbol{\theta}, t) = \mathbf{j}_{\varpi_u}(\boldsymbol{\theta}, t) = \hat{\mathbf{U}}(t) - \mathbf{U}_{\varpi}(\boldsymbol{\theta}, t) \ \text{and} \ \mathbf{V} = \mathbf{V}_u \tag{9.22}$$

This leads to the so-called "displacement method" or "output residual method". In this relationship, $\mathbf{U}_{\varpi}$ is evaluated from the FE analysis and is commonly arranged as $\mathbf{U}_{\varpi} = [(U_{1_x}, U_{1_y}), (U_{2_x}, U_{2_y}), \ldots, (U_{n_x}, U_{n_y})]^{\mathrm{T}}$ and $\hat{\mathbf{U}} = [(\hat{U}_{1_x}, \hat{U}_{1_y}),$

$(\hat{U}_{2_x}, \hat{U}_{2_y}), \ldots, (\hat{U}_{n_x}, \hat{U}_{n_y})]$ for n measured nodal data[3]. This can either be the total number of dofs describing the finite element mesh of the specimen, or a selection of well-chosen dofs, limited to a specific area, for example. This is an advantage of this method over the force method, since full-field kinematic measurements of all dofs are not required everywhere in Ω_t.

Minimization is always performed iteratively, and any possible nonlinearity dependency on $\boldsymbol{\theta}$ does not matter. Moreover, another advantage of this method over the force method is that it is less sensitive to noise measurements, according to the numerical experimentation proposed in [COT 84]. In addition, considering the examples shown in Figure 9.2, the plane stress assumptions adopted in the FE model lead to a limited number of dofs, which enables efficient FE analyses.

Since this method requires an FE analysis, displacement boundary conditions (derived either from theoretical assumptions or measured data) should usually be prescribed over a portion of the boundary to obtain a well-posed problem. However, this choice may well have important consequences on the solution of the inverse problem. Thus, some authors [PAG 06b] propose an alternative strategy for solving the FE analysis, without displacement boundary conditions, leading to a self-balancing problem. In this strategy, movements of rigid bodies are filtered and force loads are added to the set of unknown parameters.

On the other hand, the function considered in [9.22] can be normalized (or not) by the choice of a suitable weighting matrix. For example, $\mathbf{V}_u = \mathbf{I}$ in [MAH 96], $\mathbf{V}_u = \mathbf{V}_u(t) = (\hat{\mathbf{U}}(t) - \mathbf{U}^{(0)}(\boldsymbol{\theta}))^{-1}\mathbf{I}$ in [MAH 97] (where the upper index $(\bullet)^{(j)}$ refers to the number of iterations of the identification process) and $\mathbf{V}_u = \mathbf{V}_u(t) = \hat{\mathbf{U}}(t)^{-1}\mathbf{I}$ in [GEN 06a]. A wide spectrum of problems can be studied (with various physical meanings) by choosing these matrices $\mathbf{V}_u$. Thus, $\mathbf{j}_{\varpi}$, introduced in [9.22], is homogeneous to an energy, where $\mathbf{V}_u = \mathbf{K}$, and becomes close to $\mathbf{j}_{\varpi}$ [9.20] if $\mathbf{V}_u = \mathbf{K}^2$.

We also find in the literature cost functions based on strains or stresses. In [KAV 72], the strain field is obtained by a field of strain gauges, while in [LEC 07] displacements of the optical measurements are derived. The objective function is then:

$$\mathbf{j}_{\varpi}(\boldsymbol{\theta}, t) = \mathbf{j}_{\varpi_\epsilon}(\boldsymbol{\theta}, t) = \hat{\boldsymbol{\varepsilon}}(t) - \boldsymbol{\varepsilon}_{\varpi}(\boldsymbol{\theta}, t) \text{ and } \mathbf{V} = \mathbf{V}_\epsilon \qquad [9.23]$$

where $\boldsymbol{\varepsilon}_{\varpi}(t) = [(\varepsilon_{1_x}, \varepsilon_{1_y}, \varepsilon_{1_{xy}}), (\varepsilon_{2_x}, \varepsilon_{2_y}, \varepsilon_{2_{xy}}), \ldots, (\varepsilon_{n_x}, \varepsilon_{n_y}, \varepsilon_{n_{xy}})]^{\mathrm{T}}$, $\hat{\boldsymbol{\varepsilon}}(t) = [(\hat{\varepsilon}_{1_x}, \hat{\varepsilon}_{1_y}, \hat{\varepsilon}_{1_{xy}}), (\hat{\varepsilon}_{2_x}, \hat{\varepsilon}_{2_y}, \hat{\varepsilon}_{2_{xy}}), \ldots, (\hat{\varepsilon}_{n_x}, \hat{\varepsilon}_{n_y}, \hat{\varepsilon}_{n_{xy}})]^{\mathrm{T}}$ for 2D tests, with n

---

3 In 3D cases $\mathbf{U}_{\varpi} = [(U_{1_x}, U_{1_y}, U_{1_z}), \ldots, (U_{n_x}, U_{n_y}, U_{n_z})]$ and $\hat{\mathbf{U}} = [(\hat{U}_{1_x}, \hat{U}_{1_y}, \hat{U}_{1_z}), \ldots, (\hat{U}_{n_x}, \hat{U}_{n_y}, \hat{U}_{n_z})]$, with subscripts $(\bullet)_x$, $(\bullet)_y$ and $(\bullet)_z$ referring to the coordinate system.

corresponding to the number of elements of the specimen mesh, and $\mathbf{V}_\epsilon = \mathbf{V}_\epsilon(t) = \mathrm{diag}(\hat{\dot{\epsilon}}(t))^{-1}$. A stress formulation is adopted in [MOL 05]:

$$\mathbf{j}_\varpi(\boldsymbol{\theta}, t) = \mathbf{j}_{\varpi_\sigma}(\boldsymbol{\theta}, t) = \hat{\boldsymbol{\sigma}}(t) - \boldsymbol{\sigma}_\varpi(\boldsymbol{\theta}, t) \ \text{ and } \ \mathbf{V} = \mathbf{V}_\sigma \qquad [9.24]$$

and, in [CAR 06], maximum in-plane shear stresses are considered.

Other studies [KAJ 04a, CUG 06, GIT 06] use an objective function based on the force response (as an implicit parameter function) at each loading step, while the measured displacements are used as realistic boundary conditions only on $\partial\Omega_t$:

$$\mathbf{j}_\varpi(\boldsymbol{\theta}, t) = \mathbf{j}_{\varpi_F}(\boldsymbol{\theta}, t) = \hat{\mathbf{F}}(t) - \mathbf{R}_\varpi(\boldsymbol{\theta}, t) \ \forall t \in [0, T] \text{ and } \mathbf{V} = \mathbf{V}_F \qquad [9.25]$$

where $\mathbf{j}_{\varpi_F}(\boldsymbol{\theta}, t)$ has two or three components, depending on whether it is a 2D or 3D simulation. The matrix $\mathbf{V}_F$ is most often used to normalize the functional (using $\mathbf{V}_F = \mathrm{diag} \ (\hat{\mathbf{F}}_e)^{-1}$, where $\hat{\mathbf{F}}_e(T)$ is the yield force reached at the end of the numerical simulation [KAJ 04a]), or to tune the weights of the misfits between measured and calculated quantities. For example, Figure 9.3 shows a comparison between experimental and simulated displacement fields (with the cost function in equation [9.25]) for a biaxial test on a composite material [CLA 04, PAG 06a].

Recently, some authors have combined the objective functions described above: they associate $\mathbf{j}_{\varpi_u}$, $\mathbf{j}_{\varpi_\epsilon}$ and $\mathbf{j}_{\varpi_F}$ by scaling them with the difference between the maximum and the mean value of the current quantities [KAJ 04a]. In [PAD 07], the authors have combined equations [9.24] and [9.22]. In [MAH 97, MAH 99, MAH 00, GIT 06], the objective function includes $\mathbf{j}_{\varpi_u}$ and $\mathbf{j}_{\varpi_F}$, such that $\mathbf{V}_u = w_u\mathbf{I}$ and $\mathbf{V}_F = w_F\mathbf{I}$, where the weighting factors w_u and w_F are chosen such that $\mathbf{j}_{\varpi_u}^T(\boldsymbol{\theta}, t)\mathbf{V}_u\mathbf{j}_{\varpi_u}(\boldsymbol{\theta}, t)$ and $\mathbf{j}_{\varpi_F}^T(\boldsymbol{\theta}, t)\mathbf{V}_F\mathbf{j}_{\varpi_F}(\boldsymbol{\theta}, t)$ are of the same order of magnitude. In [HOC 03], $\mathbf{j}_{\varpi_F}$ and $\mathbf{j}_{\varpi_\epsilon}$ are used together with $\mathbf{V}_F = w_F\mathbf{I}$ and $\mathbf{V}_\epsilon = w_\epsilon\mathbf{I}$, such that the force contribution (which provides the best overall response) is predominant.

In [KAV 72] and [LAT 06], the authors have used – weighted or not weighted – least-squares estimates to assess both the stress and elasticity tensors from the kinematic measurements and applied loads. Also, in [GEY 03], a variational approach (called the "constitutive equation gap") minimizes a function that is separately convex with respect to the stress and elasticity tensors:

$$\mathbf{j}_\varpi(\boldsymbol{\sigma}, \mathbf{A}) = \boldsymbol{\sigma}_\varpi - \mathbf{A}\varepsilon(\hat{U}) \ \text{ and } \ \mathbf{V} = \mathbf{A}^{-1} \qquad [9.26]$$

where $\mathbf{A}$ is the elasticity tensor. This approach is detailed in Chapter 10.

In the medical field, these techniques are also widely used to characterize the mechanical properties of biological materials. Some authors [OOM 93, VAN 94, MEU 98a] use the criterion of the minimum variance and $\mathbf{j}_\varpi(\boldsymbol{\theta}, t) = \mathbf{j}_{\varpi_u}(\boldsymbol{\theta}, t)$ in equation [9.18]. The weighting matrices $\mathbf{V}_u$ and $\mathbf{W}$ are diagonal and related to the statistical data with respect to errors. Their components are approximately the inverse square of the expected errors concerning the movement and starting guess for parameters. Large values for these matrices indicate a greater level of confidence.

Finally, it can be noted that field measurements are used in a large proportion for the identification of elastic parameters, but the method is not limited to this choice. In [LEM 03, PAD 07, PAG 05], the authors have used the objective functions [9.22] and/or [9.23] to identify loads applied to the structure and its associated stiffness. In [CAR 06], θ refers to the applied stress field around a hole, in order to evaluate the residual stresses in the material using the objective function [9.22].

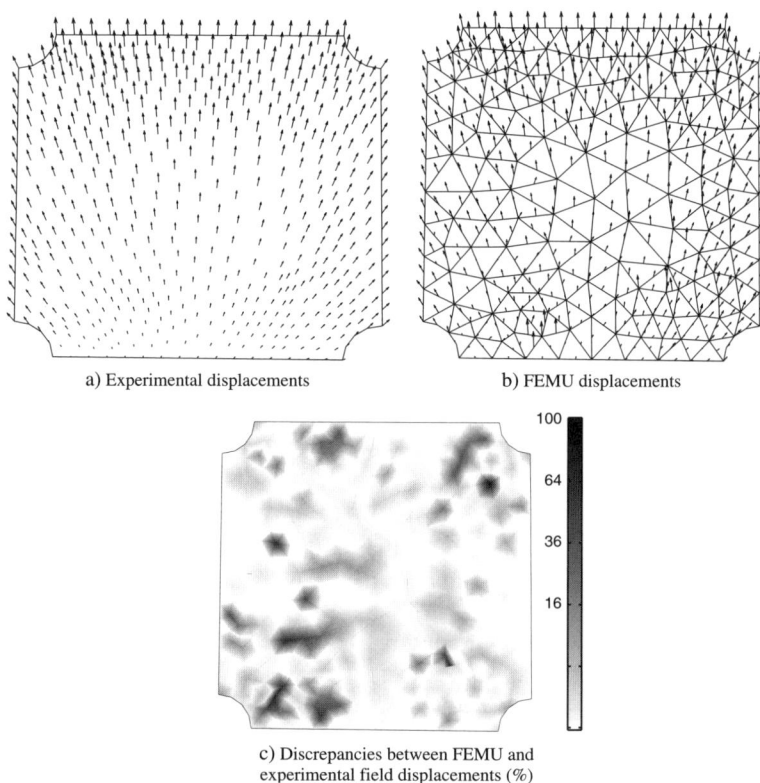

a) Experimental displacements   b) FEMU displacements

c) Discrepancies between FEMU and
experimental field displacements (%)

Figure 9.3. *Displacement fields and errors after updating
the FE model for a biaxial test*

9.3.2.2. *Minimization strategies*

In most cases, methods appropriate to least-squares problems are used to minimize functional J in [9.16], for example: the Gauss–Newton method [OOM 93, VAN 94, MEU 98a, FOR 02, LEC 07] or its alternative, the Levenberg–Marquardt method

[MOL 05, CUG 06, GIT 06], which reduces (compared with the Gauss–Newton method) potential instabilities due to the ill-conditioning of matrices (by filtering the lowest singular values of the iteration matrix). Both are suitable for cost function calculations using a low residual and the specific form of the gradient of J in the iterative solution. All these methods therefore adopt an iterative procedure for their resolution:

$$\boldsymbol{\theta}^{(k+1)} = \boldsymbol{\theta}^{(k)} + \rho(\boldsymbol{\theta}^{(k)}, \mathbf{d}^{(k)}) \cdot \mathbf{d}^{(k)} \qquad\qquad [9.27]$$

where $\rho(\boldsymbol{\theta}^{(k)}, \mathbf{d}^{(k)})$ and $\mathbf{d}^{(k)}$ depend on the chosen method. The starting guess $\boldsymbol{\theta}^{(0)} \in \mathbb{R}^p$ has to be specified.

When the objective function is sufficiently regular, these methods are very effective in terms of computation time. A modified version of the Gauss–Newton method, called the Gauss–Newton corrected method, allows convergence when the largest residuals are encountered. For these problems, the Quasi-Newton approximation (the best known being the Broyden–Fletcher–Goldfarb–Shanno (BFGS) approximation) does not exploit matrix sparsity and a modified version of this approximation, proposed by Bertsekas, can be defined [GIL 06, MAH 96]. In most situations, the uniqueness of a minimum of the objective function is not proven and the results of these methods depend strongly on the starting guesses.

In general, problems formulated as least squares are mathematically ill-posed, nonlinear and highly ill-conditioned. The objective function can be highly irregular, and it is possible that the methods mentioned above will not work well: the iteration matrix $(\partial \mathbf{h}_{\varpi}/\partial \boldsymbol{\theta})$ evaluated at every step of the analysis appears most often unstable, whether it is calculated analytically [MAH 96, MAH 97] or numerically (by finite differences, for example) [MEU 98a, GIT 06]. The use of direct search methods (without the gradient function) can be a remedy, because the value of the function is evaluated and compared sequentially to find the optimum. In this context, the simplex method is deterministic and is well-suited for determining some parameters [KAJ 04a, KAJ 04b]. It consists of evaluating the cost function of a simplex of $\mathbb{R}^p$ and translating this simplex based on the results obtained. The starting point of these methods is the original simplex.

Stochastic algorithms are also direct search methods. The most widely used methods in the field of identification are genetic algorithms (using, for example, an artificial neuronal network) and simulated annealing methods. Both are based on an analogy with biological systems. A thermodynamic process is the basis of the simulated annealing method [GEN 06a, GEN 06b], while genetic algorithms use the concept of natural selection [BUR 03, BRU 08]. Such algorithms do not require starting guesses, unlike those based on gradient methods, but simply a set of permissible parameters. However, although these algorithms are particularly robust, they remain very expensive in computation time since they require a large number of analyses. Determining the eligibility area is crucial in terms of computation time and the accuracy of the resulting solution, and the use of parallel computing, may be necessary to obtain a solution in a reasonable time. A common application of genetic algorithms is determining the initial set of parameters used by other (gradient-based) optimization methods, particularly when the functional to minimize is not convex or if the numerical problem is ill-conditioned.

Finally, to conclude this section, the convexity of the functional used in the variational approach [9.26] enables the use of relaxation algorithms, since this property implies the uniqueness of the solution [LAT 08]. When the algorithm is the minimum variance, the minimization of the functional J in [9.18] can be carried out iteratively using the following scheme:

$$\boldsymbol{\theta}^{(k+1)} = \boldsymbol{\theta}^{(k)} + \mathbf{H}^{(k+1)}(\mathbf{m}(t) - \mathbf{h}_{\varpi}(\boldsymbol{\theta}, t)) \qquad [9.28]$$

where $\mathbf{H}^{(k+1)}$ is a weighting matrix depending on error covariance and iterative matrices.

9.3.3. Structural sensitivities

In the area of a static and transient response, current interest is focused on sensitivity analysis (to shape variation, material response, boundary conditions, etc.). Two general approaches can be adopted to calculate these sensitivities: differentiation of the continuum equations or differentiation of the discrete equations. With regard to FEA, it is more common to derive discrete equations, whether analytically, semi-analytically or numerically (using, for example, finite difference methods). In the case of linear static analysis, analytical differentiation of [9.14], considering θ_j (the jth component of $\boldsymbol{\theta}$), leads to the following first-order sensitivity equations:

$$\begin{cases} \dfrac{\partial \mathbf{U}_c}{\partial \theta_j} = 0 \\[2mm] \mathbf{K}_{ii}\dfrac{\partial \mathbf{U}_i}{\partial \theta_j} = \dfrac{\partial \mathbf{F}_i}{\partial \theta_j} - \dfrac{\partial \mathbf{K}_{ic}}{\partial \theta_j}\mathbf{U}_c - \dfrac{\partial \mathbf{K}_{ii}}{\partial \theta_j}\mathbf{U}_i \\[2mm] \dfrac{\partial \mathbf{R}_c}{\partial \theta_j} = \dfrac{\partial \mathbf{K}_{ci}}{\partial \theta_j}\mathbf{U}_i + \mathbf{K}_{ci}\dfrac{\partial \mathbf{U}_i}{\partial \theta_j} + \dfrac{\partial \mathbf{K}_{cc}}{\partial \theta_j}\mathbf{U}_c \end{cases}$$

where, in the first equation, $\mathbf{U}_c$ (which come from $\bar{\mathbf{U}}$) are assumed to be independent of θ_j. The second equation of the previous system can easily be successively differentiated (using Leibnitz's formula for a product derivative) to give:

$$\mathbf{K}_{ii}\frac{\partial^q \mathbf{U}_i}{\partial \theta_j^q} = \frac{\partial^q \mathbf{F}_{\text{ext}}}{\partial \theta_j^q} - \frac{\partial^q \mathbf{K}_{ic}}{\partial \theta_j^q}\mathbf{U}_c$$

$$- \left(q\frac{\partial \mathbf{K}_{ii}}{\partial \theta_j}\frac{\partial^{q-1} \mathbf{U}_i}{\partial \theta_j^{q-1}} + \frac{q(q-1)}{1.2}\frac{\partial^2 \mathbf{K}_{ii}}{\partial \theta_j^2}\frac{\partial^{q-2} \mathbf{U}_i}{\partial \theta_j^{q-2}} + \ldots + \frac{\partial^q \mathbf{K}_{ii}}{\partial \theta_j^q}\mathbf{U}_i \right)$$

$\forall q > 1$, which leads to the expression of $\partial^q \mathbf{U}_i / \partial \theta_j^q$ that can be calculated from known sensitivities of order less than q. Moreover, it is also interesting to note that for a linear stiffness matrix (considering parameters θ_j), the previous expression can be simplified as:

$$\mathbf{K}_{ii} \frac{\partial^q \mathbf{U}_i}{\partial \theta_j^q} = \frac{\partial^q \mathbf{F}_{ext}}{\partial \theta_j^q} - \frac{\partial^q \mathbf{K}_{ic}}{\partial \theta_j^q} \mathbf{U}_c - q \frac{\partial \mathbf{K}_{ii}}{\partial \theta_j} \frac{\partial^{q-1} \mathbf{U}_i}{\partial \theta_j^{q-1}}$$

showing that $\partial^2 \mathbf{U}_i / \partial \theta_j^2 \neq \mathbf{0}$ while $\partial^2 \mathbf{K}_{ii} / \partial \theta_j^2 = \mathbf{0}$.

Furthermore, it is necessary to consider a specific case. For example, under small strain hypothesis and for an orthotropic elastic material behavior under a plane stress assumption, stiffness sensitivity can be estimated by:

$$\frac{\partial \mathbf{K}}{\partial \theta_j} = \int_{\Omega_0} \mathbf{B}^T \frac{\partial \mathbf{A}}{\partial \theta_j} \mathbf{B} dV$$

where $\mathbf{A}$ is extracted from [9.4] and $\boldsymbol{\theta} = \{Q_{xx}, Q_{yy}, Q_{yy}, G_{xy}\}$, with:

$$\frac{\partial \mathbf{A}}{\partial Q_{xx}} = \begin{bmatrix} 1 & 0 & 0 \\ 0 & 0 & 0 \\ 0 & 0 & 0 \end{bmatrix} \quad \frac{\partial \mathbf{A}}{\partial Q_{xy}} = \begin{bmatrix} 0 & 1 & 0 \\ 1 & 0 & 0 \\ 0 & 0 & 0 \end{bmatrix}$$

$$\frac{\partial \mathbf{A}}{\partial Q_{yy}} = \begin{bmatrix} 0 & 0 & 0 \\ 0 & 1 & 0 \\ 0 & 0 & 0 \end{bmatrix} \quad \frac{\partial \mathbf{A}}{\partial G_{xy}} = \begin{bmatrix} 0 & 0 & 0 \\ 0 & 0 & 0 \\ 0 & 0 & 1 \end{bmatrix}$$

For modal analysis, equation [9.10] and the normalization condition are reconsidered. By differentiating both these relationships once, the angular frequency and the normal-mode first-order sensitivities are readily expressed and organized in the following form:

$$\begin{bmatrix} \mathbf{K} - \omega_l^2 \mathbf{M} & -\mathbf{M}\phi_l \\ -\phi_l^T \mathbf{M} & 0 \end{bmatrix} \begin{bmatrix} \frac{\partial \phi_l}{\partial \theta_j} \\ \frac{\partial \omega_l^2}{\partial \theta_j} \end{bmatrix} = \begin{bmatrix} -\left(\frac{\partial \mathbf{K}}{\partial \theta_j} - \omega_l^2 \frac{\partial \mathbf{M}}{\partial \theta_j} \right) \phi_l \\ \frac{1}{2} \phi_l^T \mathbf{M} \phi_l \end{bmatrix} \qquad [9.29]$$

This is the simplest strategy to obtain normal mode sensitivities, since the matrix expressed in equation [9.29] is invertible,[4] while the matrix $(\mathbf{K} - \omega_l^2 \mathbf{M})$, on which it depends, is not invertible. However, if only angular frequency sensitivity is expected, it is possible to reduce this expression to a simpler expression:

$$\left(\phi_l^T \mathbf{M} \phi_l \right) \frac{\partial \omega_l^2}{\partial \theta_j} = \phi_l^T \left(\frac{\partial \mathbf{K}}{\partial \theta_j} - \omega_l^2 \frac{\partial \mathbf{M}}{\partial \theta_j} \right) \phi_l$$

4 Note that when there are repeated angular frequencies, the expanded system has to be formed by considering the set of corresponding normal modes instead of only the normal mode involved here, which expresses the sensitivity for all these modes.

9.4. Applications, results and accuracy

9.4.1. *Full-field measurements for the FEMU method*

The existing optical methods and their performance are reported in the first six chapters of this book, so here we will limit ourselves to applications of these techniques when they are coupled to the FEMU method. We can distinguish two classes of application of this method, depending on the scale of observation: studying local phenomena at a high strain gradient or considering global observations to a greater or lesser extent.

In most cases, thin specimens are tested in biaxial planes (flat measurements) or swelling tests (out-of-plane measurements); the assumption of plane stress is generally accepted. Moreover, isotropic strain states are assumed in a direction normal to the loads, which eliminates the need for a measurement through the solid volume in the direction normal to the plane observed. In these cases, it is essential to align both the camera and the loading axis of the sample and to monitor or measure the specimen shape and boundary conditions associated with the corresponding measured fields [GIT 06].

The estimation using a FEMU method of boundary conditions, in terms of amplitude and distribution of contact forces between the various constituents of a 3D composite structure, was carried out in [PAD 07]. There has also been some interesting work on shape optimization and/or boundary conditions to generate a displacement field that is more sensitive to specific material parameters [BRU 08]. For example, for a biaxial flexure test disk, circular loading is replaced by three-point loading to make this test more sensitive to the Poisson ratio of the material. It is important to note that the interpretation of such tests becomes more complex and thus necessary to simulate them numerically. The analysis of the experimental setup is performed numerically using FE software with a strategy similar to that adopted for shape optimization.

Again at the macroscopic scale, some authors have studied heterogeneous tests in which spatial resolution still allows the measurement of strain gradients. Field measurements are typically used when tests are performed on thin specimens containing holes (or of similar type) [MAH 96, MOL 05, LEC 06]. This kind of analysis is found in the evaluation of internal stress states using the hole technique associated with a moiré measurement: the inverse or backward problem requires the observed displacement fields and knowledge of the elastic material properties to evaluate the boundary conditions in the form of applied principal stresses [CAR 06]. Other studies have focused on local phenomena such as necking, which appears during uniaxial tensile tests, cracking tests (CT specimen), which produce large strain gradients, or even tear tests or tensile tests of bimaterials [MAH 96, GIT 06, FOR 02, LAT 08].

In all cases, the resolution of the measured displacements (or signal/noise ratio) increases with the level of deformation. Identification would be more accurate for plastic behavior (strain levels greater than 1) even if the presence of a strain gradient is a source of significant errors (especially in the vicinity of an interface between two materials or near a singularity). Thus, in all cases, the displacement measurements are filtered through calibration functions (such as splines or weighted sums of a set of orthogonal polynomials). Displacements are usually extracted from experimental positions of the integration points, and the FE model, where the strain fields are introduced to minimize the functional, numerical and experimental strains, is calculated with the same post-processor as that generally introduced in FE software. For these two classes of studies, optical methods are improved to ensure better resolution of the measured fields: for example, to improve the sensitivity of interferometric methods, the phase variation due to displacement is measured by applying a temporal phase-shifting algorithm.

At the microscopic scale, the mechanical tests are performed either with a mechanical cell introduced into a scanning electron microscope (SEM) or in front of a camera with a microscopic lens [HOC 03, KAJ 04b, CUG 06]. In all cases, the optical method is combined with a tracking method based on changes in contrast or digital image correlation. These techniques, analyzing sequences of images acquired during the test, allow the instrumentation of dynamic testing techniques such as high-speed microscopic photography [KAJ 04b]. Working with an SEM also enables microstructural characterization. Examples of applications are the study of a solder joint used to connect electronic components, or the study of polycrystal mechanics at the grain scale to improve polycrystalline models. A microscopic extensometer can be associated with the measurement of the local orientation of grains in SEM backscatter analysis techniques [HOC 03].

9.4.2. *Application to the material behavior*

As the quantity $\mathbf{h}_{\varpi}$ in equation [9.17] is evaluated numerically and iteratively, a large class of behavior can be identified. Moreover, there is no restriction on the choice of model for the material behavior, which can be linear elastic isotropic as well as viscohyperelastic or anisotropic viscoplastic at large strains.

Within the framework of linear elasticity, some authors have been interested in isotropic materials [BUR 03, PAG 05, CAR 06, GEN 06a, CUG 06, LAT 08] while others have considered anisotropic behavior [KAV 72, BRU 02, LEC 05, MOL 05, SIL 07]. The parameters describing viscoelasticity can be identified both for long times (see [LEM 02] for applications on wood panels) and short times, typically in studies involving vibroacoustic tests (see [PAG 07] for a polymer application). Also in the context of linear elasticity, the team of Professor Oomens (Eindhoven University) has applied these approaches in the biomedical field on human skin *in vivo* [MEI 97] or on dog skin *in vitro* [OOM 93, VAN 94]; these two materials are assumed to be anisotropic. Burczynski [BUR 03] undertook the identification of an elastic 3D human pelvis. In all cases, the number of parameters to be identified never exceeds 6. In [KAV 71] and [GEN 06a], the identification of parameters from a hyperelastic law (Mooney–Rivlin-type) resulting from biaxial tests has been reported. In [GIT 06], the addition of a viscoelastic response to the above behavior is used for an elastomer with an inhomogeneous field test. A set of seven parameters is identified. Assuming elastic parameters known from homogeneous tests, anisotropic fatigue damage parameters have been identified for short glass fiber-reinforced polyamide; in this case, the cost function combines [9.22] and [9.23] (see [MER 11]).

Many studies in the literature have also been dedicated to the identification of plasticity phenomena (or viscoplasticity) for materials (usually metal) whose elastic behavior is known. These studies include [MAH 96] and [MAH 97], where the plasticity of the material is modeled through a flow described by a criterion depending on the second invariant of stresses J_2 and coupled to nonlinear isotropic hardening. This work was extended to the modeling of damage according to Gurson [MAH 99, MAH 00]. In [MEU 98a, MEU 98b, CUG 06] and [POT 11], the parameters of plasticity models (von Mises or Hill associated with isotropic or kinematic hardening) are identified from tensile tests performed on specimens of various shapes (notched or drilled). Other authors have used flat tensile tests to identify the plasticity piecewise linear and parabolic hardening law [KAJ 04a]. Finally, an extension of this work, involving the parameters of a viscoplastic law, has been carried out for a high strain rate and presented in [KAJ 04b]. A comprehensive study of isotropic behavior described by 11 parameters has been conducted in [FOR 02]. In the framework of civil engineering, many studies concerning fracture phenomena have been performed (evaluation of the mean elastic properties of a polymer concrete around the crack tip through a standard single-edge-cracked three-point bend specimen [NUN 11] or the parameter identification of a cohesive zone model [BOL 02, SHE 11], for example). In [FED 09], the material identification of an adhesive layer within a joined assembly through a nonlinear law (mixed-mode cohesive) has been studied. Finally, at a microscopic level, several hardening laws have been identified from tensile tests on a polycrystal [HOC 03].

Considering structural dynamics, an inventory of FEMU methods [DAS 07] shows the interest of their coupling with the full-field measurements now available in dynamic testing (holographic or interferometry techniques (electronic/digital speckle-pattern interferometry ESPI/DSPI) and scanning laser Doppler vibrometry (SLDV)). These non-contact measurements can improve the speed of the experimental protocol with a high spatial and frequency resolution without adding weight to the tested structures, which is crucial for dynamic testing. To date, many studies have been devoted to the measurement of reliable operating deflection shapes, and most authors have sought to transform the optical measurements into digital data by using FE models [LIN 01, MIC 01, PIR 03, SIM 03]. For instance, in [MIC 01, PIR 03], modal analysis coupled with correlation tools can process large amounts of data from ESPI (typically one hundred thousand). To update the FE model of a car window, a photogrammetry technique is used [LIN 01] both to evaluate the shape and to validate the numerical model by comparing roughly the operating deflection shapes obtained from out-of-plane measurements (measured by pulsed laser ESPI) and the normal vibration modes, obtained by FE analysis.

More recently, measurements of dynamic full fields have enabled the identification of material parameters. The elastic properties of multilayer composites were evaluated [CUG 07] after identifying the modes and frequencies from the full-frequency response fields of thin and thick plates. In this study, the experimental technique combines a speaker and a scanning laser vibrometer, while the FE model is built from advanced shell elements, obtained from a higher order kinematic theory through the thickness. The properties are estimated by a least-squares algorithm using a residual formed from the combination of several criteria (concerning the frequencies, modes and orthogonality obtained by modal analysis with the location of antiresonances). The complete procedure leads to the identification of six parameters for a single test. The ability to determine the viscoelastic parameters of isotropic medium-thick plates by a single test has also been evaluated by [MOR 06, PAG 07]. Response field measurements are carried out using a scanning laser vibrometer and an automatic impactor. In this work, an extension of the force and displacement methods, i.e. a force method weighted by displacements, is proposed to address directly the frequency response fields in the identification procedure, without using a modal analysis. In the frequency domain, the viscoelastic parameters are the real and imaginary parts of the Young's modulus and Poisson's ratio, with their dependency on frequency without modeling it (the parameters are considered constant by frequency ranges).

9.4.3. *Identification accuracy*

While the previous methods can be used to validate a constitutive model, the fact remains that there is an indirect relationship between the errors in the choice of the constitutive law and those of the expected response of the material. A preliminary numerical modeling of experimental tests should be performed adequately to validate the model and the identification protocol.

The stability of the numerical identification from the mesh refinement and from the choice of starting guess for parameters must then be verified. However, it appears that the choice of initial parameters generally does not affect the values of the parameters identified. This is certainly due to *a priori* knowledge of parameter boundaries, an area where there is only one single local minimum. Genetic algorithms for various optimization tests have been carried out by varying the range of parameters identified. These studies have shown that these methods are insensitive to the choice of these ranges [GEN 06a], which is certainly the consequence of the large quantity of experimental data from full-field measurements. The Kalman filter technique can be a useful alternative; the sequence of experimental data is filtered and parameter estimations are gradually improved by their mean values [BOL 02].

When the stability of the numerical scheme is checked, the accuracy of the response must be confirmed. Unfortunately, the parameters describing the nonlinear behavior are not usually compared with those obtained from more conventional, macro-homogeneous tests (or reference). This makes it difficult to assess the quality of results. Thus, the authors have often linked this accuracy to a good agreement between experimental and simulated data. Other authors have analyzed the pattern of the residual formed by these data. In [MEU 98a], the authors have validated their results by comparing the measurement noise (defined by the signal/noise ratio) with the standard deviation of this distribution. Unfortunately, this information leads to a global criterion. In [MOL 05], the conditioning of the Gauss–Newton iteration matrix $\nabla^T \mathbf{j}_{\varpi} \nabla \mathbf{j}_{\varpi}$ defines an accuracy coefficient that allows the authors to choose among different objective functions. Other approaches have been used to study the accuracy of the results by adding noisy displacement fields [HEN 91, OOM 93, VAN 94] or a controlled bias [KAJ 04a, KAJ 04b, GIT 06].

To complete these analyses, the correlation between parameters can also be studied using the correlation matrix:

$$C_{ij} = \cos\left(\frac{\partial \mathbf{h}_{\varpi}}{\partial \boldsymbol{\theta}_i}, \frac{\partial \mathbf{h}_{\varpi}}{\partial \boldsymbol{\theta}_j}\right)$$

If all off-diagonal terms are close to 1, this indicates that the parameters are strongly correlated and the identification may be incorrect [FOR 02].

9.5. Conclusion

The common problem in continuum mechanics, called the direct or forward problem, refers to the determination of outputs such as displacement, strain and stress fields. For a well-posed problem, the solution is obtained from the knowledge of inputs such as the solid shape, the constitutive parameters and a set of boundary conditions. The identification of constitutive parameters, boundary conditions or stress distributions is a different problem, called the inverse or backward problem. Among the numerical methods proposed for solving these inverse problems, the FEMU method is used extensively due to its versatility. This method is very flexible and can handle a large number of situations and data. This chapter has examined the studies based on this method where field measurements are used to identify the parameters characterizing the mechanical behavior of materials or a set of boundary conditions.

The tremendous expansion in the development of full-field kinematic measurements now provides a wealth of information for the characterization of the mechanical responses of structures and materials. These 2D or 3D kinematic measurements allow tests to be planned and performed close to service life conditions by allowing multiaxial loading, which results in inhomogeneous stress states. The FEMU method is then well-suited for parameter identification in these situations involving complex tests associated most often with partial experimental data. This presentation has provided an overview of the applications of this method identification using data from full-field measurements. The method has proven effective in the characterization of highly anisotropic and inhomogeneous materials as well as structural components. However, some important tasks remain to be addressed in the future to make it accurate and usable by every potential user.

The accuracy of the method obviously depends on many factors, including (but not only) optical components, image resolution and image processing. However, from an experimental standpoint, it is also essential to have reliable data on the shape of the test sample and the test conditions. Numerical investigations must be conducted to determine precisely the influence of these test conditions and to assess the maximum sensitivity of the testing to determine the parameters. Combined with experimental data, the resolution of the measured data must satisfy a compromise with the spatial resolution, especially for heterogeneous strain fields. A major source of error may be that resulting from an incorrect model choice for the tested material. Another source of error may also result from the numerical simulation, which may be inaccurate, false or differing, especially in the case of nonlinear problems. In all cases, it is difficult to distinguish errors due to those measurements from the numerical simulations, due to a strongly nonlinear relationship between these two entities. The objective function and optimization algorithm can also play an important role in this method: robustness, convergence, the stability and the dependence of the solution using starting guesses for parameters are questions that must be addressed for the satisfactory use of the FEMU method.

9.6. Bibliography

[BAT 96] BATHE K., *Finite Element Procedures*, Prentice Hall, 1996.

[BEL 01] BELYTSCHKO T., LIU W., MORAN B., *Nonlinear Finite Elements for Continua and Structures*, John Wiley & Sons, 2001.

[BOL 02] BOLZON G., FEDELE R., MAIER G., "Parameters identification of a cohesive crack model by Kalman filter", *Computer Methods in Applied Mechanics and Engineering*, vol. 191, pp. 2847–2871, 2002.

[BRU 02] BRUNO L., FURGIUELE F., PAGNOTTA L., POGGIALINI A., "A full-field approach for the elastic characterization of anisotropic materials", *Optics and Lasers in Engineering*, vol. 37, pp. 417–431, 2002.

[BRU 08] BRUNO L., FURGIUELE F. M., PAGNOTTA L., POGGIALINI A., STIGLIANO G., "Elastic characterization of orthotropic plates of any shape via static testing", *International Journal of Solids and Structures*, vol. 45, nos. 3–4, pp. 908–920, February 2008.

[BUR 03] BURCZYNSKI T., "Evolutionary computation in mechanics", *IACM Expressions, Bulletin for the International Association for Computational Mechanics*, pp. 4–9, 2003.

[CAR 06] CARDENAS-GARCIA J., PREIDIKMAN S., "Solution of the moiré hole drilling method using a finite-element-method-based approach", *International Journal of Solids and Structures*, vol. 43, nos. 22–23, pp. 6751–6766, November 2006.

[CLA 04] CLAIRE D., HILD F., ROUX S., "A finite element formulation to identify damage fields: the equilibrium gap method", *International Journal for Numerical Methods in Engineering*, vol. 2, pp. 189–208, 2004.

[COL 74] COLLINS J., HART G., KENNEDY B., "Statistical identification of structures", *AIAA Journal*, vol. 12, no. 2, pp. 185–190, 1974.

[COT 84] COTTIN N., FELGENHAUER H., NATKE H., "On the parameter identification of elastomechanical systems using input and output residuals", *Ingenieur-Archiv*, vol. 54, pp. 378–387, 1984.

[CUG 06] CUGNONI J., BOTSIS J., SIVASUBRAMANIAM V., JANCZAK-RUSCH J., "Experimental and numerical studies on size and constraining effects in lead-free solder joints", *Fatigue and Fracture of Engineering Materials and Structures*, vol. 30, no. 2, pp. 387–399, 2006.

[CUG 07] CUGNONI J., GMUR T., SCHORDERET A., "Inverse method based on modal analysis for characterizing the constitutive properties of thick composite plates", *Computers and Structures*, vol. 85, nos. 17–18, pp. 1310–1320, September 2007.

[DAS 07] DASCOTTE E., "Model updating for structural dynamics: past, present and future outlook", *International Conference on Engineering Dynamics (ICED)*, cdrom, Carvoeiro, Algarve, Portugal, 12 pages, 16–18 April 2007.

[FED 09] FEDELE R., RAKA B., HILD F., ROUX S., "Identification of adhesive properties in GLARE assemblies using digital image correlation", *Journal of the Mechanics and Physics of Solids*, vol. 57, pp. 1003–1016, 2009.

[FOR 02] FORESTIER R., MASSONI E., CHASTEL Y., "Estimation of constitutive parameters using an inverse method coupled to a 3D finite element software", *Journal of Materials Processing Technology*, vols. 125–126, pp. 594–601, 2002.

[FRI 95] FRISWELL M., MOTTERSHEAD J., *Finite Element Model Updating in Structural Dynamics*, Kluwer Academic Publishers, 1995.

[GEN 06a] GENOVESE K., LAMBERTI L., PAPPALETTERE C., "Mechanical characterization of hyperelastic materials with fringe projection and optimization techniques", *Optics and Lasers in Engineering*, vol. 44, no. 5, pp. 423–442, May 2006.

[GEN 06b] GENOVESE K., LAMBERTI L., PAPPALETTERE C., "Identification of mechanical properties of bovine bones by combining PS-ESPI and global optimization", *Speckle06, Nîmes, Proceedings of SPIE*, vol. 6341, pp. 634108.1–634108.7, 2006.

[GEY 03] GEYMONAT G., PAGANO S., "Identification of mechanical properties by displacement field measurement: a variational approach", *Meccanica*, vol. 38, pp. 535–545, 2003.

[GIL 06] GILL P.E., MURRAY W., WRIGHT M.H., *Practical Optimization*, Elsevier Academic Press, London, 2006.

[GIT 06] GITON M., CARO-BRETELLE A.S., IENNY P., "Hyperelastic behaviour identification by a forward problem resolution: application to a tear test of a silicone-rubber", *Strain*, vol. 42, no. 4, pp. 291–297, November 2006.

[HEN 91] HENDRICKS M., Identification of the mechanical properties of solid materials, PhD Thesis, Eindhoven University of Technology, 1991.

[HOC 03] HOC T., GÉLÉBART L., CRÉPIN J., ZAOUI A., "A procedure for identifying the plastic behavior of single crystals from the local response of polycrystals", *Acta Materialia*, vol. 51, pp. 5477–5488, 2003.

[KAJ 04a] KAJBERG J., LINDKVIST G., "Characterization of materials subjected to large strains by inverse modelling based on in-plane displacement fields", *International Journal of Solids and Structures*, vol. 41, pp. 3439–3459, 2004.

[KAJ 04b] KAJBERG J., SUNDIN K., MELIN L., STÅHLE P., "High strain rate tensile and viscoplastic parameter identification using micoscopic high-speed photography", *International Journal of Plasticity*, vol. 20, pp. 561–575, 2004.

[KAV 71] KAVANAGH K., CLOUGH R., "Finite element applications in the characterization of elastic solids", *International Journal of Solids and Structures*, vol. 7, pp. 11–23, 1971.

[KAV 72] KAVANAGH K., "Extension of classical experimental techniques for characterizing composite-material behavior", *Experimental Mechanics*, vol. 12, no. 1, pp. 50–56, January 1972.

[LAT 06] LATOURTE F., CHRYSOCHOOS A., PAGANO S., WATRISSE B., "Identification of elastoplastic parameter distribution using digital image correlation", *Photomechanics 2006, International Conference on Full-Field Measurement Techniques and their Applications in Experimental Solid Mechanics*, Clermont-Ferrand, France, 10–12 July 2006.

[LAT 08] LATOURTE F., CHRYSOCHOOS A., PAGANO S., WATRISSE B., "Elastoplastic behavior identification for heterogeneous loadings and materials", *Experimental Mechanics*, vol. 48, no. 4, pp. 435–449, 2008.

[LEC 05] LECOMPTE D., SOL H., VANTOMME J., HABRAKEN A., "Identification of elastic orthotropic material parameters based on ESPI measurements", *SEM Annual Conference & Exposition on Experimental and Applied Mechanics*, CDROM, 7 pages, 7–9 June 2005.

[LEC 06] LECOMPTE D., SMITS A., BOSSUYT S., SOL H., VANTOMME J., VAN HEMELRIJCK D., HABRAKEN A., "Quality assessment of speckle patterns for digital image correlation", *Optics and Lasers in Engineering*, vol. 44, pp. 1132–1145, November 2006.

[LEC 07] LECOMPTE D., SMITS A., SOL H., VANTOMME J., VAN HEMELRIJCK D., "Mixed numerical–experimental technique for orthotropic parameter identification using biaxial tensile tests on cruciform specimens", *International Journal of Solids and Structures*, vol. 44, no. 5, pp. 1643–1656, March 2007.

[LEM 02] LE MAGOROU L., BOS F., ROUGER F., "Identification of constitutive laws for wood-based panels by means of an inverse method", *Composite Science and Technology*, vol. 62, no. 4, pp. 591–596, 2002.

[LEM 03] LEMOSSE D., PAGNACCO E., "Identification de propriétés matériaux à partir de mesures de champs de déplacements statiques et en présence d'une distribution d'efforts inconnue", *Colloque National en Calcul des Structures de Giens*, Giens, CDROM, 15 pages, 2003.

[LIN 01] LINET V., LEPAGE A., SOL A., BOHINEUST X., "Validation and improvement of body panels FE models from 3d-shape and vibration measurements by optical methods", *SAE Noise & Vibration Conference & Exposition*, no. 2001-01-1536, CDROM, Grand Traverse, MI, 8 pages, April 2001.

[MAH 96] MAHNKEN R., STEIN E., "A unified approach for parameter identification of inelastic material models in the frame of the finite element method", *Computer Methods in Applied Mechanics and Engineering*, vol. 136, nos. 3–4, pp. 225–258, September 1996.

[MAH 97] MAHNKEN R., STEIN E., "Parameter identification for finite deformation elasto-plasticity in principal directions", *Computer Methods in Applied Mechanics and Engineering*, vol. 147, nos. 3–4, pp. 17–39, 1997.

[MAH 99] MAHNKEN R., "Aspects on the finite-element implementation of the Gurson model including parameter identification", *International Journal of Plasticity*, vol. 15, nos. 3–4, pp. 1111–1137, September 1999.

[MAH 00] MAHNKEN R., "A comprehensive study of a multiplicative elastoplasticity model coupled to damage including parameter identification", *Computers and Structures*, vol. 74, nos. 3–4, pp. 179–200, September 2000.

[MEI 97] MEIJER R., DOUVEN L., OOMENS C., "Characterisation of anisotropic and non-linear behaviour of human skin in-vivo", *Computer Methods in Biomechanics and Biomedical Engineering*, vol. 1, pp. 13–27, 1997.

[MER 11] MERAGHNI F., NOUNI H., BOURGEOIS N., CZARNOTA C., LORY P., "Parameters identification of fatigue damage model for short glass fiber reinforced polyamide (PA6-GF30) using digital image correlation", *Procedia Engineering*, vol. 10, pp. 2110–2116, 2011.

[MEU 98a] MEUWISSEN M.H.H., OOMENS C.W.J., BAAIJENS F.P.T., PETTERSON R., JANSSEN J.D., "Determination of the elasto-plastic properties of aluminium using a mixed numerical–experimental method", *Journal of Materials Processing Technology*, vol. 75, nos. 1–3, pp. 204–211, March 1998.

[MEU 98b] MEUWISSEN M., An inverse method for the mechanical characterisation of metals, PhD Thesis, TU Eindhoven, 1998.

[MIC 01] MICHOT S., COGAN S., FOLTÊTE E., RAYNAUD J., PIRANDA J., "Apports des techniques de mesures holographiques dans le domaine de la localisation de défauts en dynamique des structures", *2ème Colloque d'Analyse Vibratoire Expérimentale*, CDROM, Blois, 9 pages, 13–14 November 2001.

[MOL 05] MOLIMARD J., LE RICHE R., VAUTRIN A., LEE J., "Identification of the four orthotropic plate stiffnesses using a single open-hole tensile test", *Journal of SEM*, vol. 45, no. 5, pp. 404–411, 2005.

[MOR 06] MOREAU A., PAGNACCO E., BORZA D., LEMOSSE D., "An evaluation of different mixed experimental/numerical procedures using FRF for the identification of viscoelastic materials", *International Conference on Noise and Vibration Engineering, ISMA 2006*, CDROM, Leuven, 15 pages, 2006.

[NUN 11] NUNES L., REIS J., DA COSTA MATTOS H., "Parameters identification of polymer concrete using a fracture mechanics test method and full-field measurements", *Engineering Fracture Mechanics*, vol. 78, pp. 2957–2965, 2011.

[OOM 93] OOMENS C., VAN RATINGEN M., JANSSEN J., KOK J., HENDRIKS M., "A numerical–experimental for a mechanical characterization of biological materials", *Journal of Biomechanics*, vol. 26, nos. 4–5, pp. 617–621, 1993.

[PAD 07] PADMANABHAN S., HUBNER J., KUMAR A., IFJU P., "Load and boundary conditions calibration using full-field strain measurement", *Experimental Mechanics*, vol. 46, no. 5, pp. 569–578, 2007.

[PAG 05] PAGNACCO E., LEMOSSE D., HILD F., AMIOT F., "Inverse strategy from displacement field measurement and distributed forces using FEA", *SEM Annual Conference and Exposition on Experimental and Applied Mechanics*, CDROM, 7 pages, 7–9 June 2005.

[PAG 06a] PAGANO S., "A review of identification methods based on full fields measurements", *Photomechanics 2006, International Conference on Full-Field Measurements Techniques and their Applications in Experimental Solid Mechanics*, Clermont-Ferrand, France, 10–12 July 2006.

[PAG 06b] PAGNACCO E., LEMOSSE D., BORZA D., "A coupled FE based inverse strategy from displacement field measurement subject to an unknown distribution of forces", *Photomechanics 2006, International Conference on Full-Field Measurements Techniques and Their Applications in Experimental Solid Mechanics*, Clermont-Ferrand, France, 10–12 July 2006.

[PAG 07] PAGNACCO E., MOREAU A., LEMOSSE D., "Inverse strategies for the identification of elastic and viscoelastic material parameters using full-field measurements", *Materials Science and Engineering: A*, vols. 452–453, pp. 737–745, April 2007.

[PAN 98] PANG T., TIN-LOI F., "A penalty interior point algorithm for a parameter identification problem in elastoplasticity", *Mechanics of Structures and Machines*, vol. 29, pp. 85–99, 1998.

[PIR 03] PIRANDA J., FOLTÊTE E., RAYNAUD J.L., MICHOT S., LEPAGE A., LINET V., "New trends in modal analysis and model updating using electronic speckle pattern interferometry", in KURKA P., FLEURY A. (eds), *X Diname*, Ubatuba - SP - Brazil, pp. 107–117, 10–14 March 2003.

[POT 11] POTTIER T., TOUSSAINT F., VACHER P., "Contribution of heterogeneous strain field measurements and boundary conditions modelling in inverse identification of material parameters", *European Journal of Mechanics A/Solids*, vol. 30, pp. 373–382, 2011.

[SHE 11] SHEN B., PAULINO G., "Identification of cohesive zone model and elastic parameters of fiber-reinforced cementitious composites using digital image correlation and a hybrid inverse technique", *Cement and Concrete Composites*, vol. 33, pp. 572–585, 2011.

[SIL 07] SILVA G., LE RICHE R., MOLIMARD J., VAUTRIN A., GALERNE C., "Identification of material properties using FEMU: application to the open hole tensile test", *Journal of Applied Mechanics and Materials*, vols. 7–8, pp. 73–78, September 2007.

[SIM 03] SIMON D., GOLINVAL J.C., "Use of whole-field displacement measurements for model updating of blades", *IMAC XXI*, no. 131, CDROM, 10 pages, 3–6 February 2003.

[TAR 87] TARANTOLA A., *Inverse Problem Theory: Methods for Data Fitting and Model Parameter Estimation*, Elsevier, 1987.

[VAN 94] VAN RATINGEN M., Mechanical identification of inhomogeneous solids: a mixed numerical experimental approach, PhD Thesis, Eindhoven University of Technology, 1994.

Chapter 10

Constitutive Equation Gap

10.1. Introduction

In this chapter, we examine the concept of constitutive equation gap (CEG) as a tool for the identification of parameters associated with constitutive models for solid materials. The concept of CEG is based, in its simplest form (small-strain hypothesis, equilibrium, linear elastic behavior), on cost functions of the form:

$$E(v, \tau, \mathcal{B}) = \frac{1}{2} \int_{\Omega} (\tau - \mathcal{B} : \varepsilon[v]) : \mathcal{B}^{-1} : (\tau - \mathcal{B} : \varepsilon[v]) \, \mathrm{d}V$$

(where v is a displacement field, τ a stress field and $\mathcal{B}$ a possibly heterogeneous elasticity tensor), expressing a quadratic gap in the verification of the constitutive law. Weighting by the tensor $\mathcal{B}$ confers units of energy to the CEG.

The concept of CEG was initially proposed for error estimation in the finite element method (FEM) [LAD 83]. It then turned out to also be a powerful tool for identification purposes, especially with many applications in model updating. Essentially, equivalent concepts have been proposed in other contexts for solving inversion problems, such as the electrostatic energy functionals of Kohn and Vogelius [KOH 84, KOH 90]. Two important characteristics of CEG functionals are (1) their strong and clear physical meaning, and (2) their additive character with respect to the structure, allowing the definition of local error indicators over substructures.

As will be discussed in this chapter for different situations, the minimization of a CEG is in principle applicable to any identification problem for which overdetermined

---

Chapter written by Stéphane PAGANO and Marc BONNET.

data are available. In particular, the latter need not necessarily consist of full-field measurements. In fact, many applications of the CEG concern structural model updating from vibrational data (see section 10.5). The CEG method is nonetheless well suited to the exploitation of full-field measurements.

Initially, we will focus on the identification of heterogeneous linear elastic properties under static (i.e. equilibrium) conditions and within the small-strain assumption (section 10.2). An extension of this formulation to the identification of heterogeneous elastoplastic constitutive parameters is then described (section 10.3). The algorithms presented in these two sections exploit kinematic field measurements in a plane, and assume either plane-strain or plane-stress conditions. The remainder of the chapter will be devoted to a more succinct discussion of the possibilities offered by the construction of CEG functionals based on Legendre–Fenchel error density (section 10.4), and a synthetic presentation of CEG formulations suitable for structural dynamics and vibrations (section 10.5).

10.2. CEG in the linear elastic case: heterogeneous behavior and full-field measurement

The equilibrium of a generic elastic solid Ω of boundary $\partial\Omega$ is governed by three sets of equations, namely the equilibrium equations:

$$\begin{cases} \operatorname{div} \boldsymbol{\sigma} = 0 & \text{in } \Omega, \\ \boldsymbol{\sigma}.\boldsymbol{n} = \bar{\boldsymbol{T}} & \text{on } S_f, \end{cases}$$
[10.1]

the kinematic compatibility conditions:

$$\begin{cases} \boldsymbol{\varepsilon}[\boldsymbol{u}] = \dfrac{1}{2}\left(\nabla\boldsymbol{u} + \nabla^t\boldsymbol{u}\right) & \text{in } \Omega, \\ \boldsymbol{u} = \bar{\boldsymbol{u}} & \text{on } S_u \end{cases}$$
[10.2]

and the constitutive equation:

$$\boldsymbol{\sigma} = \boldsymbol{\mathcal{A}}(\boldsymbol{x}):\boldsymbol{\varepsilon}[\boldsymbol{u}]$$
[10.3]

where $\boldsymbol{u}$ is the in-plane displacement, $\boldsymbol{\varepsilon}[\boldsymbol{u}]$ the linearized strain tensor associated with $\boldsymbol{u}$, $\boldsymbol{\sigma}$ the Cauchy stress tensor and $\boldsymbol{n}$ the outward unit normal on $\partial\Omega$. Quantities $\bar{\boldsymbol{u}}$ and $\bar{\boldsymbol{T}}$ appearing in the boundary conditions [10.1] and [10.2] denote prescribed displacements and tractions, respectively. Surfaces S_u and S_f are such that $S_u \cup S_f = \partial\Omega$ and $S_u \cap S_f = \emptyset$, so as to define well-posed boundary conditions. The elasticity tensor $\boldsymbol{\mathcal{A}}$ may be constant (homogeneous material) or space-dependent (heterogeneous material). If the elastic properties are isotropic, they are described in terms of two independent moduli, such as Lamé constants λ, μ, or Young's modulus E and Poisson's ratio ν.

The usual problem, often referred to as direct, consists of computing fields u, σ knowing the geometry Ω of the solid, the elasticity tensor $\mathcal{A}$ and well-posed boundary data $\bar{u}, \bar{T}$. Exact solutions are limited to very specific classes of geometries and boundary conditions; the forward problem is usually solved numerically, most often using the FEM.

The identification of constitutive parameters is another type of problem, referred to as inverse. Incomplete knowledge of the tensor-valued function $\mathcal{A}(x)$, or of moduli used in the definition of $\mathcal{A}$, must be offset by overdetermined data. In other words, the boundary data $\bar{u}, \bar{T}$ appearing in [10.1] and [10.2] must be supplemented by experimental information. Such additional data can take various forms.

In this chapter, the focus is on situations where full-field measurements are available either in Ω or on $\partial\Omega$. The three most frequently encountered situations are then:

1) Measurements $\hat{u}$ of the displacement field u and complete well-posed boundary data $\bar{u}, \bar{T}$ are available. The strain tensor components are then obtained by differentiating the displacement field and are therefore regarded as known.

2) Measurements $\hat{u}$ of the displacement field u and boundary data $\bar{u}$ are available, together with only incomplete static boundary data; for example, only resultant forces associated with $\bar{T}$ are known.

3) Overdetermined boundary data are available (e.g. $\bar{T}$ known on S_f and $\bar{u}$ known on $\partial\Omega$), with displacement and strain fields inside the solid remaining to be determined.

This chapter focuses on situations of type 1) or 2) above. Case 3) is mentioned in section 10.5, and also corresponds to a data structure suitable for the reciprocity gap method [AND 92, AND 96] (see Chapter 13).

The CEG measures the distance between a stress field τ and another stress field resulting from the application of a constitutive model to a displacement field v. The CEG between τ and v, under the assumption of linear elastic behavior characterized by the (possibly heterogeneous) elasticity tensor $\mathcal{B}$, is defined by:

$$E(v, \tau, \mathcal{B}) = \frac{1}{2} \int_{\Omega} (\tau - \mathcal{B}:\varepsilon[v]):\mathcal{B}^{-1}:(\tau - \mathcal{B}:\varepsilon[v]) \, \mathrm{d}V. \qquad [10.4]$$

The presence of the compliance tensor $\mathcal{B}^{-1}$ in the integral confers units of energy to $E(v, \tau, \mathcal{B})$. It can actually be shown that:

$$E(v, \tau, \mathcal{B}) = \mathcal{P}(v, \mathcal{B}) + \mathcal{P}^{\star}(\tau, \mathcal{B}), \qquad [10.5]$$

with $\mathcal{P}$ and $\mathcal{P}^{\star}$, respectively, denoting the potential energy and complementary energy.

The usefulness of the concept of CEG is emphasized by the following remarks:

1) For a well-posed boundary value problem such as that defined by equations [10.1], [10.2] and [10.3] ($\bar{u}, \bar{T}, \mathcal{A}$ known, u, ε, σ unknown), spaces $\mathcal{C}$ of *kinematically admissible* displacement fields and $\mathcal{S}$ of *statically admissible* stress fields are defined by:

$$\mathcal{C}(\bar{u}, S_u) = \left\{ v | v_i \in \mathrm{H}^1(\Omega), v = \bar{u} \text{ on } S_u \right\}, \tag{10.6a}$$

$$\mathcal{S}(\bar{T}, S_f) = \left\{ \tau \in \mathrm{H}_{\mathrm{div}}(\Omega), \text{ div } \tau = 0 \text{ in } \Omega, \text{ and } \tau.n = \bar{T} \text{ on } S_f \right\}, \tag{10.6b}$$

where the space $\mathrm{H}_{\mathrm{div}}(\Omega)$ of tensor-valued fields is defined by:

$$\mathrm{H}_{\mathrm{div}}(\Omega) = \left\{ \tau | \tau_{ij} = \tau_{ji}, \tau_{ij} \in \mathrm{L}^2(\Omega) , \tau_{ij,j} \in \mathrm{L}^2(\Omega) \right\}.$$

The solution (u, σ) of the boundary value problem is then characterized by:

$$(u, \sigma) = \underset{(v,\tau) \in \mathcal{C} \times \mathcal{S}}{\arg \min} E(v, \tau, \mathcal{A}) \quad \text{and} \quad E(u, \sigma, \mathcal{A}) = 0. \tag{10.7}$$

2) For a constitutive parameter identification problem for which (for instance) the elasticity tensor $\mathcal{A}$ is unknown, the definitions [10.6a] and [10.6b] of spaces of admissible fields can be modified to include all available experimental information on the displacements and stresses. The elasticity tensor $\mathcal{A}$ can then be identified by minimizing the CEG:

$$\mathcal{A} = \arg \min_{\mathcal{B}} \mathcal{E}(\mathcal{B}) \quad \text{with} \quad \mathcal{E}(\mathcal{B}) = \min_{(v,\tau) \in \mathcal{C} \times \mathcal{S}} E(v, \tau, \mathcal{B}) \tag{10.8}$$

This minimization problem is the essence of the constitutive equation gap method (CEGM). It consists of an alternating directions method for which a partial minimization with respect to (v, τ) is followed by a partial minimization with respect to $\mathcal{B}$.

Two variants of the CEGM will now be described. Both require the availability of a measurement $\bar{u} \in \left(\mathrm{H}^1(\Omega) \right)^2, \Omega \subset \mathbb{R}^2$ of the displacement field u.

10.2.1. *First variant: exact enforcement of kinematic measurements*

In this variant, the measured displacement field $\bar{u}$ is introduced directly into the CEG functional. In addition, the values g_i of linear observation functionals L_i of the traction vector $\tau.n$ (where n is the outward normal vector) are assumed to be known:

$$L_1(\tau.n) = g_1 \text{ on } \Gamma_1, \ldots, L_N(\tau.n) = g_N \text{ on } \Gamma_N, \tag{10.9}$$

where $\Gamma_i \subset \partial\Omega$. To exploit experimental data, two instances of such functionals occur commonly:

1) $L(\tau.n) \equiv \tau.n = g$ (in particular, $g = 0$ for the stress-free part of the boundary);

2) $L(\tau.n) \equiv \int_\Gamma \tau.n \, dl = R$ (R is the known resultant load associated with a distributed loading over a part Γ of the boundary).

All this leads us to seek $\mathcal{A}$ as a solution to the minimization problem:

$$\mathcal{A} = \arg\min_{\mathcal{B}\in\mathcal{A}} \mathcal{E}(\mathcal{B}) \qquad \text{with} \quad \mathcal{E}(\mathcal{B}) = \min_{\tau\in\mathcal{S}} E(\bar{u}, \tau, \mathcal{B}). \qquad [10.10]$$

In this case, the space of admissible stress fields is defined as:

$$\mathcal{S}(\bar{T}, S_f) = \left\{ \tau \in \mathrm{H_{div}}(\Omega), \mathrm{div}\, \tau = 0 \text{ in } \Omega, \right.$$
$$\left. L_1(\tau.n) = g_1, \dots, L_N(\tau.n) = g_N \right\} \quad [10.11]$$

instead of [10.6b]. Each functional $L_i : \left(\mathrm{H}^{-1/2}(\Gamma_i)\right)^2 \to E_i$ is linear and continuous (where E_i is a Banach space). $\mathcal{S}$ is assumed to be non-empty (from an abstract viewpoint, an open question is to find suitable conditions on the data $g_1, \dots, g_N$ in such a way that this assumption be satisfied).

Two possible simple choices may be considered for the space of admissible elasticity (stiffness or compliance) tensors $\mathcal{A}$:

$$\mathcal{A}_1 = \left\{ B \in \mathbb{R}^{3\times3}; B_{ij} = B_{ji}, Bx\!:\!x \geq \alpha|x|^2, \alpha > 0, \forall x, \right.$$
$$\left. Bx\!:\!y \leq M|x||y|, M > 0, \forall x, y \right\},$$

$$\mathcal{A}_2 = \left\{ B_{ij} \text{ piecewise linear}; B_{ij} = B_{ji}, Bx\!:\!x \geq \alpha|x|^2, \alpha > 0, \forall x, \right.$$
$$\left. Bx\!:\!y \leq M|x||y|, M > 0, \forall x, y \right\}.$$

The most general choice would be:

$$\mathcal{A}_3 = \left\{ B \in (\mathrm{L}^\infty(\Omega))^{3\times3}; B_{ij} = B_{ji}, Bx\!:\!x \geq \alpha|x|^2, \alpha > 0, \forall x, \right.$$
$$\left. Bx\!:\!y \leq M|x||y|, M > 0, \forall x, y \right\}.$$

With these notations, the following relations can be established for the CEG functional $E(\bar{u}, ., .) : \mathcal{S} \times \mathcal{A} \to \mathbb{R}$.

PROPOSITION 10.1.– The following properties for the CEG functional hold true:

1) $E(\bar{u}, \tau, \mathcal{B}) \geq 0, \forall(\tau, \mathcal{B}) \in \mathcal{S} \times \mathcal{A}$,

2) $E(\bar{u}, \tau^\star, \mathcal{B}^\star) = 0$ if and only if [10.3] is true with $\sigma = \tau^\star$ and $\mathcal{A} = \mathcal{B}^\star$,

3) functional $E(\bar{u}, ., .)$ is convex on $\mathcal{S} \times \mathcal{A}$.

PROOF.– Properties 1) and 2) are obvious, since the elasticity tensor $\mathcal{B}$ belongs to space $\mathcal{A}$. The convexity of $E(\bar{u}, ., .)$ remains to be established. To this end, the CEG functional is first rewritten as:

$$E(\bar{u}, \tau, \mathcal{B}) = \int_\Omega \left\{ \frac{1}{2}\varepsilon[\bar{u}] : \mathcal{B} : \varepsilon[\bar{u}] + \frac{1}{2}\tau : \mathcal{B}^{-1} : \tau - \tau : \varepsilon[\bar{u}] \right\} dV \qquad [10.12]$$

and the convexity of function $\psi(\tau, \mathcal{B}) := 1/2\tau : \mathcal{B}^{-1} : \tau$ on $\mathcal{S} \times \mathcal{A}$ is investigated. A first-order Taylor expansion yields:

$$\psi(\tau, \mathcal{B}) = \psi(\tau_0, \mathcal{B}_0) + \psi'(\tau_0, \mathcal{B}_0; \tau - \tau_0, \mathcal{B} - \mathcal{B}_0) + \psi(\tau - \mathcal{B} : \mathcal{B}_0^{-1} : \tau_0, \mathcal{B}),$$

where $\psi'(\tau_0, \mathcal{B}_0; \tau, \mathcal{B}) = -1/2\tau_0 : (\mathcal{B}_0^{-1} : \mathcal{B} : \mathcal{B}_0^{-1}) : \tau_0 + \tau : \mathcal{B}_0^{-1} : \tau_0$. The desired convexity property follows from the non-negativity of ψ by virtue of the definition of the set $\mathcal{A}$.   □

ALGORITHM 10.1. ALTERNATE DIRECTION METHOD.– The functional $E(\bar{u}, ., .)$ is minimized on $\mathcal{S} \times \mathcal{A}$ using an alternate direction method:

- Initialization of the algorithm with $(\sigma^0, \mathcal{A}^0)$
- For $(\sigma^n, \mathcal{A}^n)$ known, find σ^{n+1} and $\mathcal{A}^{n+1}$ by successively solving

    Step 1: $E(\bar{u}, \sigma^{n+1}, \mathcal{A}^n) \leq E(\bar{u}, \tau, \mathcal{A}^n), \quad \forall \tau \in \mathcal{S}$

    Step 2: $E(\bar{u}, \sigma^{n+1}, \mathcal{A}^{n+1}) \leq E(\bar{u}, \sigma^{n+1}, \mathcal{B}), \quad \forall \mathcal{B} \in \mathcal{A}$

- Convergence test.

1) Partial minimization of $E(\bar{u}, ., \mathcal{A}^n)$ with respect to the admissible stress

In Step 1, the stress solution σ^{n+1} is constrained by the equilibrium equations [10.1] and the linear observations [10.9]. Several methods can be used for this purpose, and three possible representations of σ^{n+1} are presented in the following:

First possibility: stress-based method using Q1 element. Minimization is performed by using a Q1 finite element interpolation of the admissible stress fields. For a plane-stress problem, the unknowns at step $n + 1$ are $(\sigma_{xx}^{n+1}, \sigma_{yy}^{n+1}, \sigma_{xy}^{n+1})$ at each node. Enforcing the equilibrium equations and resultant load measurements contributing to the definition of $\mathcal{S}$ via Lagrange multipliers leads to the solution of the variational problem:

$$\inf_{\tau \in H_{\mathrm{div}}(\Omega)} \sup_{\gamma, \lambda_i} E(\bar{u}, \tau, \mathcal{B}) + \int_\Omega \gamma . \operatorname{div} \tau \, dV + \sum_{i=1}^N \lambda_i [L_i(\tau.n) - g_i]. \quad [10.13]$$

The fact that Lagrange multipliers are involved leads to a substantial increase in the size of the governing linear system to be solved for the computation of the stresses, and a deterioration of its condition number. Instead, the computation of the stresses can be made more direct by resorting to an Airy stress function.

Second possibility (Airy function): this formulation is naturally equilibrated [GER 86]. On each element of the mesh, the Airy function is assumed to be a third-degree polynomial with respect to each coordinate:

$$\varphi(x,y) = \sum_{i=0}^{3}\sum_{j=0}^{3} a_{ij} x^i y^j. \qquad [10.14]$$

Consistent with this choice, the three stress components can be directly derived from this potential:

$$\tau_{xx}(x,y) = \varphi_{,yy} = \sum_{i=0}^{3}\sum_{j=2}^{3} j(j-1) a_{ij} x^i y^{j-2},$$

$$\tau_{yy}(x,y) = \varphi_{,xx} = \sum_{i=2}^{3}\sum_{j=0}^{3} i(i-1) a_{ij} x^{i-2} y^j, \qquad [10.15]$$

$$\tau_{xy}(x,y) = -\varphi_{,xy} = -\sum_{i=1}^{3}\sum_{j=1}^{3} ij a_{ij} x^{i-1} y^{j-1}.$$

Whatever the choice of coefficients a_{ij}, div $\tau = 0$. The variational problem is similar to [10.13] without the term enforcing local equilibrium. At each node, the unknowns, formulated in terms of the stress function, are $(\varphi, \varphi_{,x}, \varphi_{,y}, \varphi_{,xy})$.

The present choice [10.14] of a bicubic potential enables a sufficiently rich representation of the stresses, since they are at least piecewise linear functions of each Cartesian coordinate. Note that this special interpolation method does not achieve inter-element stress continuity:

– the stress component σ_{xx}^{n+1} is continuous only along the axis Ox,

– the stress component σ_{yy}^{n+1} is continuous only along the axis Oy,

– the stress component σ_{xy}^{n+1} is continuous.

Stress continuity can also be imposed, for example, using Lagrange multipliers.

Third possibility (displacement-based representation): stress fields are represented in the form $\sigma^{n+1} = A^n : \varepsilon[w]$ in terms of displacement fields w (built using a finite element interpolation), and constraints on σ^{n+1} involved in the definition of S are enforced via a weak formulation associated with the finite element basis.

2) Partial minimization of $E(\bar{u}, \sigma^{n+1}, .)$ over the admissible elasticity tensors

The minimization of $\mathcal{E}(\mathcal{B})$ is explicit. In the examples in this chapter, material properties are assumed to be constant in each element of the mesh. The capability of grouping elements into subsets ω_i with constant material properties has also been implemented. This possibility is particularly interesting when large sub-regions made up of some specific material exist in Ω (as with, for example, coarse-grained polycrystalline steels and composite materials). In such situations, the components of the elasticity tensor are given by very simple, explicit, formulae.

Two cases will be studied hereafter, depending on whether we are considering (in the framework of plane elasticity) elastic behavior with cubic symmetry (three independent parameters) or isotropic symmetry (two independent parameters). The identification of parameters for cubic elasticity will be presented first, isotropic elasticity then being a special case. Similar explicit expressions are also available for three-dimensional elasticity [CON 95].

Elasticity with cubic symmetry: for elastic behavior with cubic symmetry, in plane stress, the elasticity tensor $\mathcal{A}$ can be written in terms of three parameters as:

$$\mathcal{A} = \begin{bmatrix} a_1 & a_2 & 0 \\ a_2 & a_1 & 0 \\ 0 & 0 & a_3 \end{bmatrix}, \qquad [10.16]$$

with $a_1 = E/1 - \nu^2$, $a_2 = -\nu a_1$, $a_3 = 2G$ and E, ν, G, respectively, denoting the Young's modulus, Poisson's ratio and shear modulus. The determination of E, ν, G is made simple by expressing tensors $\varepsilon[\bar{u}]$ and σ in the principal directions of the elasticity tensor. The CEG functional then takes a simple form, and for each subdomain ω_i, the entries of $\mathcal{A}$ (directly linked to E^i, ν^i, G^i) are found to be given by:

$$2a_1^i = \sqrt{\frac{\int_{\omega_i} (\tau_{xx} + \tau_{yy})^2 \, \mathrm{d}V}{\int_{\omega_i} (\varepsilon_{xx} + \varepsilon_{yy})^2 \, \mathrm{d}V}} + \sqrt{\frac{\int_{\omega_i} (\tau_{xx} - \tau_{yy})^2 \, \mathrm{d}V}{\int_{\omega_i} (\varepsilon_{xx} - \varepsilon_{yy})^2 \, \mathrm{d}V}}, \qquad [10.17a]$$

$$2a_2^i = \sqrt{\frac{\int_{\omega_i} (\tau_{xx} + \tau_{yy})^2 \, \mathrm{d}V}{\int_{\omega_i} (\varepsilon_{xx} + \varepsilon_{yy})^2 \, \mathrm{d}V}} - \sqrt{\frac{\int_{\omega_i} (\tau_{xx} - \tau_{yy})^2 \, \mathrm{d}V}{\int_{\omega_i} (\varepsilon_{xx} - \varepsilon_{yy})^2 \, \mathrm{d}V}}, \qquad [10.17b]$$

$$2a_3^i = \sqrt{\frac{\int_{\omega_i} \tau_{xy}^2 \, \mathrm{d}V}{\int_{\omega_i} \varepsilon_{xy}^2 \, \mathrm{d}V}}. \qquad [10.17c]$$

Isotropic elasticity: for isotropic elastic behavior, parameters E and ν can be determined using a technique identical to that previously developed, noting that now $a_3 = a_1 - a_2$. The two independent entries of $\mathcal{A}$ are obtained as:

$$2a_1^i = \sqrt{\frac{\int_{\omega_i} (\tau_{xx} + \tau_{yy})^2 \, \mathrm{d}V}{\int_{\omega_i} (\varepsilon_{xx} + \varepsilon_{yy})^2 \, \mathrm{d}V}} + \sqrt{\frac{\int_{\omega_i} (\tau_{xx} - \tau_{yy})^2 + \tau_{xy}^2 \, \mathrm{d}V}{\int_{\omega_i} (\varepsilon_{xx} - \varepsilon_{yy})^2 + \varepsilon_{xy}^2 \, \mathrm{d}V}}, \qquad [10.18a]$$

$$2a_2^i = \sqrt{\frac{\int_{\omega_i}(\tau_{xx}+\tau_{yy})^2\,dV}{\int_{\omega_i}(\varepsilon_{xx}+\varepsilon_{yy})^2\,dV}} - \sqrt{\frac{\int_{\omega_i}(\tau_{xx}-\tau_{yy})^2 + \tau_{xy}^2\,dV}{\int_{\omega_i}(\varepsilon_{xx}-\varepsilon_{yy})^2 + \varepsilon_{xy}^2\,dV}}.$$ [10.18b]

For identification purposes, one advantage of isotropic elasticity over cubic elasticity appears in the previous system of equations: the identification of all elastic coefficients on each subdomain ω_i does not require the shear strain to be locally non-zero.

10.2.2. *Second variant: enforcement of measurements by kinematic penalization*

It is not necessarily desirable to impose exactly the kinematic measurements, particularly if they are corrupted by measurement noise. For this reason, it is interesting to consider the variant based on the functional:

$$F(\boldsymbol{v},\boldsymbol{\tau},\boldsymbol{\mathcal{B}}) = \alpha E(\boldsymbol{v},\boldsymbol{\tau},\boldsymbol{\mathcal{B}}) + \frac{\beta}{2}\|\boldsymbol{v}-\bar{\boldsymbol{u}}\|^2,$$ [10.19]

where α and β are positive weighting coefficients [CAL 02]. In [10.19], the distance $\|\boldsymbol{v}-\bar{\boldsymbol{u}}\|$ must be defined such that the two terms of F have comparable magnitudes. For example, $\|\boldsymbol{w}\|^2 = E\int_\Omega \varepsilon[\boldsymbol{w}]:\varepsilon[\boldsymbol{w}]\,dV$ if we focus on cases where $\boldsymbol{\mathcal{A}}$ is isotropic (i.e. defined in terms of Young's modulus E and Poisson's ratio ν), or $\|\boldsymbol{w}\|^2 = \gamma\boldsymbol{w}.\boldsymbol{w}$ with a coefficient γ to be chosen appropriately.

In line with previous work on CEG-based approaches for model updating, such as [REY 90] and [DER 02], the two partial minimizations are carried out successively. The first minimization (location step) gives $(\boldsymbol{u},\boldsymbol{\sigma})$ with $\boldsymbol{\sigma}\in\mathcal{S}$, with $\mathcal{S}$ again defined by [10.11]. Numerically, the stress fields are obtained using one of the previously discussed representations. As with the first variant, the partial minimization of $E(\boldsymbol{v},\boldsymbol{\tau},\boldsymbol{\mathcal{B}})$ with respect to $\boldsymbol{\mathcal{B}}$ (correction step) is then explicit (provided that the coefficient β does not depend on $\boldsymbol{\mathcal{B}}$).

Functionals F of the form [10.19] are convex over $\mathcal{C}\times\mathcal{S}\times\mathcal{A}$. As explained in [GEY 02] and [GEY 03], another family of functionals $\tilde{F}(\boldsymbol{v},\boldsymbol{\tau},\boldsymbol{\mathcal{S}})$, separately convex over $\mathcal{C}\times\mathcal{S}$ and $\mathcal{A}$, can be defined from [10.19] in terms of the compliance tensor $\boldsymbol{\mathcal{S}}=\boldsymbol{\mathcal{B}}^{-1}$ rather than the stiffness tensor $\boldsymbol{\mathcal{B}}$:

$$\tilde{F}(\boldsymbol{v},\boldsymbol{\tau},\boldsymbol{\mathcal{S}}) = F(\boldsymbol{v},\boldsymbol{\tau},\boldsymbol{\mathcal{S}}^{-1}), \quad (\boldsymbol{v},\boldsymbol{\tau},\boldsymbol{\mathcal{S}})\in\mathcal{C}\times\mathcal{S}\times\mathcal{A}.$$ [10.20]

10.2.3. *Comments*

REMARK 10.1.– The works [KOH 84, KOH 90] on conductivity imaging are based on a CEG-type functional adapted to electrostatic constitutive behavior. They also consider specific situations of overdetermined complete data on the boundary (potential and flux both known over the whole boundary $\partial\Omega$), and proceed by minimizing the energy gap between Dirichlet and Neumann solutions associated with a given conductivity field. A similar approach is followed for the identification of elastic moduli fields in [CON 95].

REMARK 10.2.– Functionals of the form [10.19] correspond to the *modified CEG* previously introduced for model updating (see references [REY 90, BEN 95, LAD 99, BUI 00, DER 01, DER 02] in section 10.5). They have the advantage of being applicable to arbitrary, possibly quite scarce, data (introduced via the second term of [10.19]).

REMARK 10.3.– Reference [AVR 07] proposes an interpretation of the minimization of the CEG as a particular form of the virtual fields method.

REMARK 10.4.– The authors of [CHA 99], considering the problem of identification of heterogeneous moduli in terms of the classical minimization approach for a cost function $J(v)$ expressing the L^2-norm of the misfit between the experimental displacement data and their simulation v (i.e. defined as the second term of the functional [10.19]), proposed to include the constraint defined by the model equations in the form of a penalty via the sum $\mathcal{P} + \mathcal{P}^\star$ of the potential and complementary energies, whose optimal value (zero) is known beforehand. Given relation [10.5], this led them to propose the penalized formulation:

$$\mathcal{A} = \lim_{\eta \to 0} \; \operatorname*{arg\,min}_{v \in \mathcal{C}, \tau \in \mathcal{S}, \mathcal{B} \in A} \left(J(v) + \frac{1}{\eta} E(v, \tau, \mathcal{B}) \right),$$

wherein the functional used is equal (up to a multiplication factor and for a fixed value of the penalty parameter η) to $F(v, \tau, \mathcal{B})$ as defined by [10.19]. This observation thus gives a useful interpretation of the latter, and in particular of the compromise parameter β.

REMARK 10.5.– An original application of the CEG, proposed by [AND 06] and [AND 08], concerns data completion, that is the reconstruction of data on the boundary, given ill-posed boundary conditions of Cauchy type (simultaneous knowledge of displacements and tractions on part of the boundary, no boundary data on another part). An optimal control type of approach, based on the minimization of a CEG cost function (energy norm of the difference between Dirichlet and Neumann solutions), leads to a completion algorithm whose convergence is much faster than alternating methods such as that of Kozlov *et al.* [KOZ 91].

10.2.4. *Some numerical examples*

The algorithm is illustrated by three examples; the first two are scalar problems whereas the last example is a vectorial case. The first two problems are solved with Algorithm 10.1 with the functional $\tilde{F}$ defined by [10.20].

EXAMPLE 10.1. IDENTIFICATION OF A CONDUCTIVITY FIELD.– To test the performance of Algorithm 10.1, we consider a scalar problem with "perfect measurements" on the unit square $\Omega = (0.1) \times (0.1)$ discretized with square elements. For instance, the solution to this conductivity problem is important in hydrology and has been solved in Kohn and Lowe [KOH 88]. At the nodal points, the values of the measured field $u^\star = x + y + (1/3)(x^3 + y^3)$, $f = -\text{div}(\nabla u^\star / a_{ex})$ are assumed to be known, where $a_{ex} = 1/(1 + y^2)$ is the inverse of the conductivity. On the boundary, the given surface force is $g = (\partial_n u^\star)/a_{ex}$. We assume that $\mathcal{E} = \mathcal{E}_1$ and $L_1(\boldsymbol{\sigma}.\mathbf{n}) \equiv \boldsymbol{\sigma}.\mathbf{n} = g$ on all sides of the square Ω. The observed convergence for the functional $\tilde{F}$ is $\|a^n - a_{ex}\|_{L^2} \sim O(h^{0.57})$ and $\tilde{F}(\boldsymbol{\sigma}^n, a_n) \sim O(h^{1.16})$, where h is the characteristic diameter of a finite element. By applying the algorithm to F, the same type of convergence is obtained. Figure 10.1 shows plots of $\tilde{a} - a_{ex}$, $a - a_{ex}$ and $\tilde{a} - a$, where $\tilde{a}$ and a, respectively, result from the minimization of $\tilde{F}$ and F.

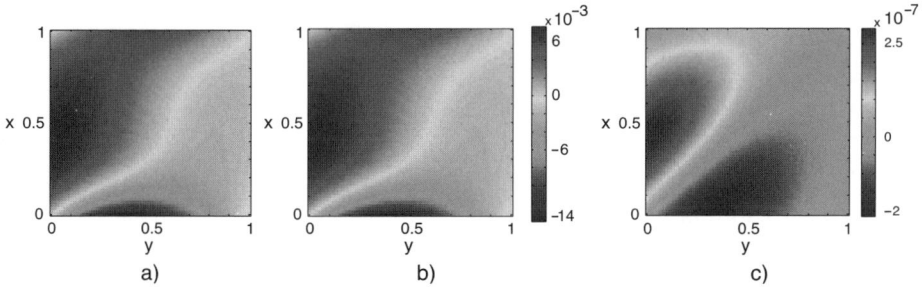

Figure 10.1. *a) $a - a_{ex}$, b) $\tilde{a} - a_{ex}$ and c) $a - \tilde{a}$, computed using 400 quadrangular finite element, convergence threshold $\varepsilon_c = 10^{-6}$. For a color version of this figure, see www.iste.co.uk/gh/solidmech.zip*

EXAMPLE 10.2. PROBLEM WITH A SINGULARITY.– This second example tests the ability of Algorithm 10.1 to capture local singularities or diffuse damage. Situations where the true conductivity inverse a_{ex} is either $a_{ex} = 1/(\varepsilon + (y - 1/2)^2)$, with ε a positive constant (local singularity case) or $a_{ex} = 1/(1 - D)(1 + y^2)$, with $D \in (0, 1)$ (diffuse damage case) are thus considered. As in Example 10.1, Ω is the unit square discretized with square elements. Algorithm 10.1 was found to require more iterations as ε decreases, and in fact failed to converge for $\varepsilon \leq 0.005$ (see Figure 10.2); improved results for $\varepsilon = 0.005$ were obtained using a refined mesh. Figure 10.2 shows plots of $\tilde{a} - a_{ex}$, $a - a_{ex}$ and $\tilde{a} - a$. Finally, concerning the diffuse damage case, the number of iterations was found to increase as D gets closer to 1 (10 iterations for $D = 0$, 12 for $D = 0.5$ and 86 for $D = 0.999$).

EXAMPLE 10.3. TENSILE TEST.– A comparison of different identification methods has been undertaken as a joint effort by several research groups [AVR 08]. This comparison concerns, in particular, the results obtained in identifying elastic parameters in various experimental situations. The simplest of these situations is a tensile test on a 2024 aluminum alloy. The specimen is a bar of section

4.8 mm × 4.8 mm. Two elastic parameters are identified: Young's modulus E and Poisson's ratio ν. For this test, the displacement fields were provided by F. Hild of Laboratoire de Mécanique et Technologie (LMT Cachan). The images were taken by means of an 8-bit camera ($1,008 \times 1,016$ pixels) equipped with a long-distance microscope. A 2×2 mm region of interest was observed. A series of 21 images was taken while an increasing load was applied to the specimen. In addition to the displacement fields and force measurements, reference elastic properties were provided. They were obtained by a linear regression on all the tensile tests using strain gage measurements, as $E_{\text{ref}} = 76 \pm 0.5$ GPa, $\nu_{\text{ref}} = 0.33 \pm 0.01$.

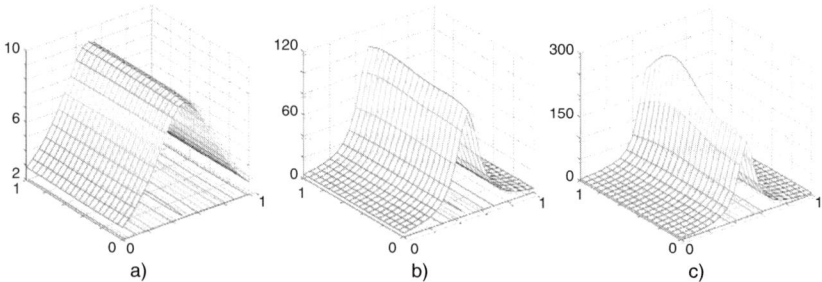

Figure 10.2. *Identified values of a_{ex} for different values of ε: 0.1, 0.01 and 0.005 (400 quadrangular elements, convergence threshold $\varepsilon_c = 10^{-6}$). For a color version of this figure, see www.iste.co.uk/gh/solidmech.zip*

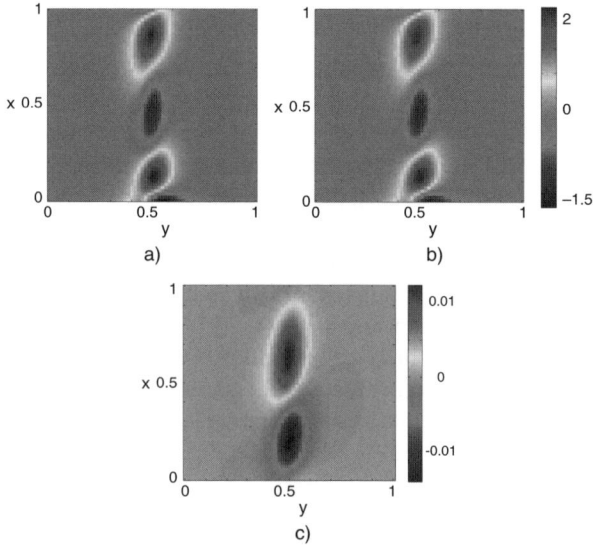

Figure 10.3. *a) $a - a_{ex}$, b) $\tilde{a} - \tilde{a}_{ex}$ and c) $a - \tilde{a}$, computed using 400 quadrangular elements, convergence threshold $\varepsilon_c = 10^{-6}$. For a color version of this figure, see www.iste.co.uk/gh/solidmech.zip*

For the last image, corresponding to the maximum value $(2,003$ N) of the applied load, the displacement field is shown in Figure 10.4. The components u_x and u_y, respectively, show the stretching of the bar and the contraction of its section. The fields are given at 15×15 points; they are consistent with the identification method.

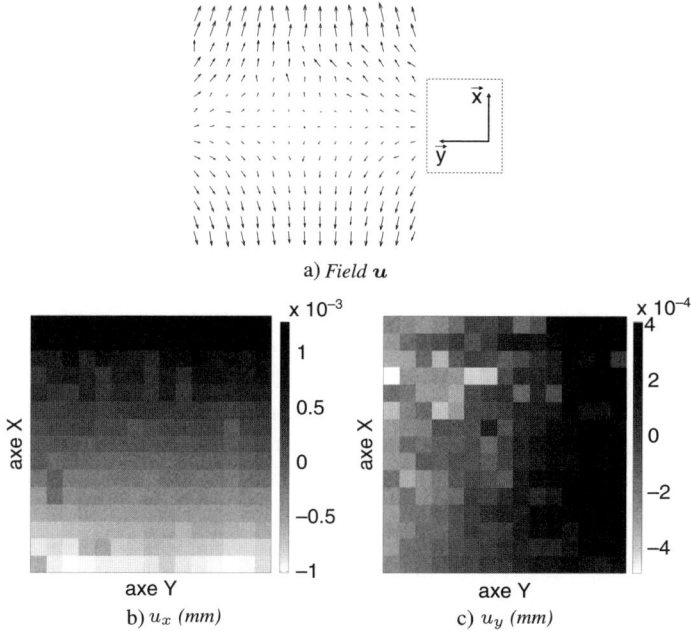

a) *Field* $\boldsymbol{u}$

b) u_x *(mm)*   c) u_y *(mm)*

Figure 10.4. *In-plane displacement components for $F = 2,003$ N*

The results for the identified elastic parameters are given in Figure 10.5. For each image i, a set of parameters (E_i, ν_i) is identified, whose values are then compared to the reference values. For low load levels, the identified parameters are not very close to the reference parameters, but the identification improves as the applied force increases. More generally, for low strain amplitudes (in the linear elastic range), the signal-to-noise ratio improves with the load level.

By averaging the last five results $(i = 17, \ldots, 21)$, the following identified values are obtained: $E_{id} = 78.3$ GPa, $\nu_{id} = 0.343$. Comparing these values with the reference values, the relative discrepancies found are 3% for the Young's modulus and 4% for the Poisson's ratio. The obtained values are satisfactory and we see that it is advisable for this type of test to use a large number of image-force couples. Indeed, from one image to another, the identified elastic coefficients may vary significantly. This variability is associated here with experimental noise that corrupts the data and hinders identification.

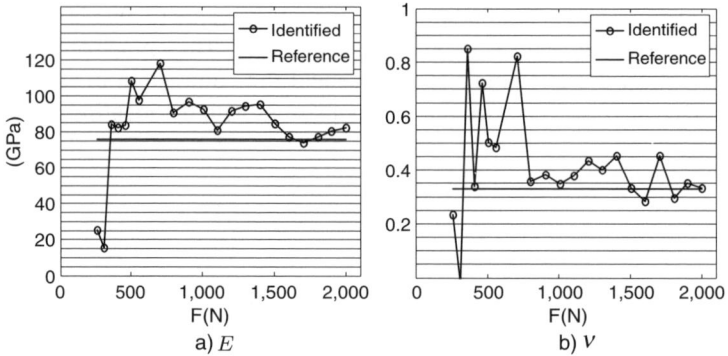

Figure 10.5. *Values of identified elastic coefficients*

To conclude this test, homogeneous properties were identified from a simple tensile test. A comparison of different identification methods (in particular the virtual fields method and FEM updating) was conducted as part of the benchmark organized by members of the GdR 2519 consortium. Identified values using other methods were in the 70–75 GPa range for the Young's modulus, and in the 0.3–0.36 range for the Poisson's ratio. Overall, a good agreement between the results obtained by all methods was thus reached. This experiment did not enable the demonstration of the main advantages of the identification method developed in this chapter, for instance its ability to yield both a stress field and a distribution of material properties. Situations that better exploit this potential are considered in the following section.

10.3. Extension to elastoplasticity

10.3.1. *Formulation*

In this section, identification assumes elastoplastic constitutive properties following the linear kinematic hardening model of Prager. Naturally, this simple model cannot reproduce all the complexity of the development of plasticity. It is expressed in terms of two plastic parameters: a yield stress σ_0 and a constant hardening modulus k.

The objective here is to develop a method to identify local parameters of this model and characterize its performance for identification. The use of models that are more representative of the actual behavior of metallic materials is possible, but will not be discussed in this section. The chosen model may be defined synthetically by the four equations:

$$\boldsymbol{\sigma} = \boldsymbol{\mathcal{A}}(\boldsymbol{\varepsilon} - \boldsymbol{\varepsilon}^{\mathrm{p}}), \qquad\qquad\qquad [10.21a]$$

$$f(\boldsymbol{\sigma}, \boldsymbol{X}) = (\boldsymbol{\sigma} - \boldsymbol{X})_{\mathrm{eq}} - \sigma_0 \leq 0, \qquad\qquad\qquad [10.21b]$$

$$\dot{\varepsilon}^{\text{P}} = \dot{\gamma}\frac{\partial f}{\partial \boldsymbol{\sigma}}, \qquad\qquad\qquad\qquad\qquad\qquad [10.22a]$$

$$\dot{\boldsymbol{X}} = \frac{2}{3}k\dot{\varepsilon}^{\text{P}}, \qquad\qquad\qquad\qquad\qquad\qquad [10.22b]$$

where $\boldsymbol{\mathcal{A}}$ is the elasticity tensor, ε^{P} the plastic part of the strain tensor, γ the plastic multiplier, f the yield function and $\boldsymbol{X}$ the backstress tensor.

The Prager model described above can be recast in incremental form using an implicit time discretization [SIM 98]. In what follows, subscripts n and $n+1$ indicate values at the initial and final instants of the current, $(n+1)$th, time step. For a plastic step for which the initial hardening is zero (i.e. $\boldsymbol{X}_n = 0$ and $\varepsilon_n^{\text{p}} = 0$), we can establish an explicit formula for the elastoplastic secant tensor $\boldsymbol{\mathcal{A}}_p^s$ [SIM 98]:

$$\boldsymbol{\sigma}_{n+1} = \left[\boldsymbol{\mathcal{A}}_p^s\right]_{n+1} : \varepsilon_{n+1}, \quad \left[\boldsymbol{\mathcal{A}}_p^s\right]_{n+1} = \left[\boldsymbol{\mathcal{A}}^{-1} - \frac{3\Delta\gamma(\sigma_0)}{3 + 2k\Delta\gamma(\sigma_0)}\boldsymbol{P}\right]^{-1}, \qquad [10.23]$$

where $\Delta\gamma$ is the plastic multiplier increment and $\boldsymbol{P}$ a mapping matrix. In other cases, the relationship between $\boldsymbol{\sigma}_{n+1}$ and ε_{n+1} is not available in explicit form, but an elastoplastic tangent tensor $\boldsymbol{\mathcal{A}}_p^t$ can be defined such that $\left[\boldsymbol{\mathcal{A}}_p^t\right]_{n+1} = \mathrm{d}\boldsymbol{\sigma}/\mathrm{d}\varepsilon|_{n+1}$.

We will consider two kinds of formulation for the mechanical problem: a standard formulation and an incremental formulation. The former is associated with a global load step for which the strain is obtained between an undeformed reference state and a deformed state. The latter is associated with a load increment for which the strain is computed between two successive deformed states.

The elastic identification problem, thus, consists of finding the elasticity tensor $\boldsymbol{\mathcal{A}}^s$ and the stress $\boldsymbol{\sigma}$ that satisfy the equilibrium equation [10.24a], the constitutive equation [10.24b] and the global equilibrium [10.24c] for a load step during which no hardening occurs. Plastic identification consists of finding, during a plastic evolution, the elastoplastic tangent stiffness tensor $\boldsymbol{\mathcal{A}}_p^t$ and the stress increment $\Delta\boldsymbol{\sigma}$ so as to verify the governing equations [10.25a], [10.25b] and [10.25c] of the incremental problem, applied to a plastic increment:

Standard formulation:

$$\text{div } \boldsymbol{\sigma} = 0 \qquad\qquad \text{in } \Omega, \qquad\qquad\qquad [10.24a]$$

$$\boldsymbol{\sigma} = \boldsymbol{\mathcal{A}}^s : \varepsilon[\bar{\boldsymbol{u}}] \qquad\qquad \text{in } \Omega, \qquad\qquad\qquad [10.24b]$$

$$L_i(\boldsymbol{\sigma}.\boldsymbol{n}) = g_i \qquad\qquad \text{on } \Gamma_i. \qquad\qquad\qquad [10.24c]$$

Incremental formulation:

$$\text{div } \Delta\boldsymbol{\sigma} = 0 \qquad\qquad \text{in } \Omega, \qquad\qquad [10.25\text{a}]$$

$$\Delta\boldsymbol{\sigma} = \boldsymbol{\mathcal{A}}^t : \varepsilon[\Delta\bar{\boldsymbol{u}}] \qquad\qquad \text{in } \Omega, \qquad\qquad [10.25\text{b}]$$

$$L_i(\Delta\boldsymbol{\sigma}.\boldsymbol{n}) = \Delta g_i \qquad\qquad \text{on } \Gamma_i. \qquad\qquad [10.25\text{c}]$$

Within the elastic domain, the behavior is described by an elastic stiffness tensor $\boldsymbol{\mathcal{A}}$, that can be obtained from either the "standard formulation" or the "incremental formulation". In theory, $\boldsymbol{\mathcal{A}} = \boldsymbol{\mathcal{A}}^s = \boldsymbol{\mathcal{A}}^t$. In practice, in order to use data for identification with an optimal signal-to-noise ratio, a secant formulation for the identification of elastic parameters is chosen.

Similarly to what was done for the elastic case, it is possible to write the CEG $E(\bar{\boldsymbol{u}}, \boldsymbol{\tau}, \boldsymbol{\mathcal{B}}^s)$ for the standard formulation, and the CEG $E(\bar{\boldsymbol{u}}, \Delta\boldsymbol{\tau}, \boldsymbol{\mathcal{B}}^t)$ for the incremental formulation. Stress and stress increment fields, respectively, belong to spaces $\mathcal{S}$ and $\Delta\mathcal{S}$, defined similarly to [10.6b]. The results given in Proposition 10.1 are still valid in both cases.

10.3.2. *Numerical method*

The minimization of the CEG functional with respect to stress fields is carried out analogously to the elastic case [LAT 07, LAT 08]. The computation of material property distributions remains to be addressed.

10.3.2.1. *Plastic detection*

Given the impossibility of identifying locally more than three parameters simultaneously, the identification of elastic parameters and plastic parameters is performed separately, with plastic parameters identified only after the elastic parameters are determined. The first loading step must, therefore, be assumed to be purely elastic. The elastic parameters identified are then considered as "reference" parameters, denoted by $\boldsymbol{\mathcal{A}}^r$. Among the following loading steps, a distinction must be made between purely elastic and plastic steps. To detect that a step is plastic, elastic identification is performed in order to compare the elastic tensor $\boldsymbol{\mathcal{A}}$ thus obtained with the reference elastic tensor $\boldsymbol{\mathcal{A}}^r$. The step is then deemed plastic whenever this difference is too large. It is possible to compare tensors obtained after either the complete loading or just one load increment.

10.3.2.2. *Computation of plastic material properties*

Concerning the determination of plastic material properties, two cases may arise depending on whether a secant or tangent problem is addressed. Both situations also involve the elastic parameters. Both types of plastic problem are described in the following, aiming at the determination of two plastic unknowns: the hardening modulus k and the yield stress σ_0.

Secant problem: for the secant problem, we seek to determine the elastoplastic secant tensor $\boldsymbol{B}_p^s$, which is possible when the backstress tensor $\boldsymbol{X}_n$ is zero:

$$\left[\boldsymbol{B}_p^s\right]_{n+1} = \left[\boldsymbol{A}^{-1} - \frac{3\Delta\gamma(\sigma_0)}{3 + 2k\Delta\gamma(\sigma_0)}\boldsymbol{P}\right]^{-1}. \qquad [10.26]$$

Just as with the elastic tensor, the plastic secant tensor can be written (assuming plane-stress conditions) in a form similar to the elastic tensor:

$$\boldsymbol{B}_p^s = \begin{bmatrix} b_{ps1} & b_{ps2} & 0 \\ b_{ps2} & b_{ps1} & 0 \\ 0 & 0 & b_{ps3} \end{bmatrix}. \qquad [10.27]$$

Equation [10.26] allows each of the three independent entries of $\boldsymbol{B}_p^s$ to be written in terms of k and σ_0. Upon diagonalizing $\boldsymbol{B}_p^s$ and enforcing the stationarity of $E(\bar{\boldsymbol{u}}, \boldsymbol{\tau})$ with respect to the eigenvalues of $\boldsymbol{B}_p^s$, the three equations obtained involve a plastic variable $K_p(k, \sigma_0, \boldsymbol{\tau})$. The induced coupling between k and σ_0 thus does not allow their simultaneous determination by considering only the secant problem associated with the first plastic step.

Tangent problem: the tangent problem can be written for a plastic loading increment, knowing the backstress $\boldsymbol{X}_n$ computed at the previous step. For incremental plasticity, the elastoplastic tangent tensor is given by [10.23]. The expression of this tensor is not simple enough to provide the proof that the three plasticity equations are dependent. However, in simple situations (e.g. uniaxial tension), a numerical calculation enables us to plot the functional associated with the tangent problem, whose minimization yields the sought plastic parameters (see Figure 10.6 for the case of a tensile test on a steel sample whose properties are $E = 200$ GPa, $\nu = 0.3$, $k = 10$ GPa and $\sigma_0 = 300$ MPa). For this situation, the functional has an elongated valley that prevents the identification of the parameter σ_0, whereas the hardening modulus k can be identified.

To identify both coefficients k and σ_0, it is thus necessary to consider two successive plastic loading steps, denoted n and $n + 1$. Both steps are schematically depicted in Figure 10.7, with the tangent and secant moduli used for each step. The first plastic step n is best exploited by means of a secant problem, which allows larger strains to be considered than for the tangent problem, and thus to enjoy a better signal-to-noise ratio for identification purposes. Since the yield threshold is reached during this first step, the yield stress σ_0 is to be identified from a secant formulation associated with the CEG functional E_s defined by:

$$E_s(\bar{\boldsymbol{u}}, \boldsymbol{\tau}, \sigma_0, k) = E(\bar{\boldsymbol{u}}, \boldsymbol{\tau}, \boldsymbol{B}_p^s(\boldsymbol{\tau}, \sigma_0, k)). \qquad [10.28]$$

Then, the second plastic step $n + 1$ is used to identify the hardening modulus k, a tangent problem associated with the CEG functional E_t defined by:

$$E_t(\bar{u}, \Delta\tau, \sigma_0, k) = E(\bar{u}, \Delta\tau, \mathcal{B}_p^t(\tau, \sigma_0, k)). \qquad [10.29]$$

The fact that the tangent and secant elastoplastic tensors depend on the stress tensor (or stress increment) somewhat complicates the resulting algorithm (Algorithm 10.2).

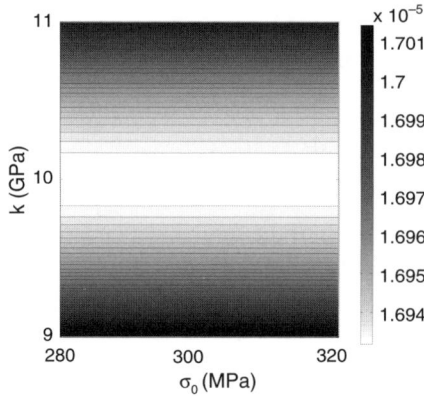

Figure 10.6. *Values of the functional $E(\bar{u}, \Delta\tau, \mathcal{B}^t)$ for a tensile test*

ALGORITHM 10.2.– Two-step plastic algorithm, with enforced kinematic measurements $\bar{u}$:

    – Plastic initialization of the algorithm
    $\sigma_0^0 = (\sigma_{n-1})_{II}$
    k^0 estimated from an elastic tangent computation
    $E_t(\bar{u}, \Delta\sigma_n, \sigma_0^0, k^0) \le E_t(\bar{u}, \Delta\tau, \sigma_0^0, k^0), \quad \forall \Delta\tau \in \Delta\mathcal{S}$
    – (For known $\sigma_n^i, \Delta\sigma_{n+1}^i, \sigma_0^i, k^i$), successively determine
    $(\sigma_n^{i+1}, \Delta\sigma_{n+1}^{i+1}, \sigma_0^{i+1}, k^{i+1})$ by solving:

        Step 1: Plastic computation step $n + 1$, incremental formulation
        $E_t(\bar{u}, \Delta\sigma_{n+1}^{i+1}, \sigma_0^i, k^i) \le E_t(\bar{u}, \Delta\tau, \sigma_0^i, k^i), \quad \forall \Delta\tau \in \Delta\mathcal{S}$
        $E_t(\bar{u}, \Delta\sigma_{n+1}^{i+1}, \sigma_0^i, k^{i+1}) \le E_t(\bar{u}, \Delta\sigma_{n+1}^{i+1}, \sigma_0^i, \kappa), \quad \forall\kappa$

        Step 2: Plastic computation step n, standard formulation
        $E_s(\bar{u}, \sigma_n^{i+1}, \sigma_0^i, k^{i+1}) \le E_s(\bar{u}, \tau, \sigma_0^i, k^{i+1}), \quad \forall\tau \in \mathcal{S}$
        $E_s(\bar{u}, \sigma_n^{i+1}, \sigma_0^{i+1}, k^{i+1}) \le E_s(\bar{u}, \sigma_n^{i+1}, s_0, k^{i+1}), \quad \forall s_0$
    – Convergence test.

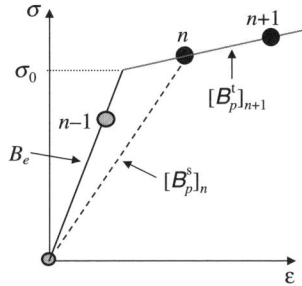

Figure 10.7. *Curve $\sigma - \varepsilon$*

10.4. Formulations based on the Legendre–Fenchel transform

The formulation developed in section 10.3 for elastoplastic identification is based on an incremental version of the linear elastic CEG. A generalization to nonlinear behavior of the linear elastic CEG, which is more consistent from a theoretical point of view but so far seldom exploited for identification purposes, exploits the Legendre–Fenchel transform. For example, for nonlinear elastic behaviors defined by the convex free energy density $\psi(\varepsilon)$, we set:

$$E(v, \tau, \psi) = \int_\Omega \big(\psi(\varepsilon[v]) + \psi^\star(\tau) - \tau : \varepsilon[v]\big) \, \mathrm{d}V, \qquad [10.30]$$

where the potential $\psi^\star(\tau)$, given by:

$$\psi^\star(\tau) = \sup_\varepsilon \big(\tau : \varepsilon - \psi(\varepsilon)\big), \qquad [10.31]$$

is the complementary energy density, that is the Legendre–Fenchel transform of $\psi(\varepsilon)$. Defining the Legendre–Fenchel gap $e(\varepsilon, \tau)$ by:

$$e(\varepsilon, \tau) = \psi(\varepsilon) + \psi^\star(\tau) - \tau : \varepsilon, \qquad [10.32]$$

classical convex analysis results yield:

$$e(\varepsilon, \tau) \geq 0 \ \text{(for all } \varepsilon, \tau\text{)}; e(\varepsilon, \tau) = 0 \Longleftrightarrow \tau = \frac{\partial \psi}{\partial \varepsilon}.$$

The choice made in [10.30] to use $e(\varepsilon[v], \tau)$ as CEG density is thus fully justified, the latter quantity being non-negative and vanishing when τ and ε are linked by the (nonlinear elastic) constitutive model. Of course, choosing the potential $\psi(\varepsilon)$ as quadratic and convex gives the linear elastic CEG [10.4].

The definition of the CEG functional using the Legendre–Fenchel density can be generalized to standard generalized constitutive models [HAL 75, GER 83], defined by means of a free-energy density ψ and a dissipation potential φ. This approach is used in [MOË 99] for error estimations in finite element calculations in nonlinear conditions. The CEG functional associated with a standard generalized elastoplastic model is defined by:

$$E(\boldsymbol{v}, \boldsymbol{\varepsilon}^{\mathrm{p}}, p, \boldsymbol{\tau}, R; \boldsymbol{\theta}) = \int_{\Omega} \Big\{ \big[\psi(\boldsymbol{\varepsilon}[\boldsymbol{v}] - \boldsymbol{\varepsilon}^{\mathrm{p}}, p) + \psi^{\star}(\boldsymbol{\tau}, R) - \boldsymbol{\tau} : (\boldsymbol{\varepsilon}[\boldsymbol{v}] - \boldsymbol{\varepsilon}^{\mathrm{p}})\big]_{t=T}$$

$$+ \int_{0}^{T} \big[\varphi(\boldsymbol{\varepsilon}^{\mathrm{p}}, \dot{p}) + \varphi^{\star}(\boldsymbol{\sigma}, R) - \boldsymbol{\tau} : \dot{\boldsymbol{\varepsilon}}^{P} + R\dot{p}\big] \, \mathrm{d}t \Big\} \, \mathrm{d}V, \quad [10.33]$$

where $\boldsymbol{\varepsilon}^{\mathrm{p}}$ is the plastic part of the strain, p the cumulative plastic strain, R the current yield stress and $\boldsymbol{\theta}$ a vector of (possibly heterogeneous) constitutive parameters.

The identification of $\boldsymbol{\theta}$ can then be formulated as:

$$\boldsymbol{\vartheta} = \arg \min_{\theta} \mathcal{E}(\boldsymbol{\theta}),$$

where $\mathcal{E}(\boldsymbol{\theta})$ is defined by a partial minimization of the CEG functional [10.33]:

$$\mathcal{E}(\boldsymbol{\theta}) = \min_{(v, \varepsilon^{P}, \dot{p}) \in \mathcal{C}^{\mathrm{EP}}, \, (\tau, R) \in \mathcal{S}^{\mathrm{EP}}} E(\boldsymbol{v}, \boldsymbol{\varepsilon}^{\mathrm{p}}, \dot{p}, \boldsymbol{\tau}, R; \boldsymbol{\theta})$$

and the admissible spaces $\mathcal{C}^{\mathrm{EP}}, \mathcal{S}^{\mathrm{EP}}$ are defined in terms of the spaces $\mathcal{C}$ and $\mathcal{S}$ of kinematically and statically admissible fields by:

$$\mathcal{C}^{\mathrm{EP}} = \big\{ \boldsymbol{v}, \boldsymbol{\varepsilon}^{\mathrm{p}}, p \mid \boldsymbol{v} \in \mathcal{C}, \ \boldsymbol{\varepsilon}^{\mathrm{p}} : \mathbf{1} = 0, \ \sqrt{(2/3)\boldsymbol{\varepsilon}^{\mathrm{p}} : \boldsymbol{\varepsilon}^{\mathrm{p}}} = \dot{p} \big\},$$

$$\mathcal{S}^{\mathrm{EP}} = \big\{ \boldsymbol{\tau}, R \mid \boldsymbol{\tau} \in \mathcal{S}, \ \|\boldsymbol{\tau}\|_{\mathrm{eq}} - R - R_0 \le 0 \big\}.$$

For isotropic elastoplastic behavior defined by an initial yield stress R_0, the von Mises criterion, an associated flow rule and linear isotropic hardening h, we have $\boldsymbol{\theta} = (\mu, \nu, R_0, h)$ and the CEG functional is given (with $\boldsymbol{A} = \boldsymbol{A}(\mu, \nu)$ and $\boldsymbol{\varepsilon}^{\mathrm{E}}[\boldsymbol{v}] = \boldsymbol{\varepsilon}[\boldsymbol{v}] - \boldsymbol{\varepsilon}^{\mathrm{p}}$) by:

$$E(\boldsymbol{v}, \boldsymbol{\varepsilon}^{\mathrm{p}}, p, \boldsymbol{\tau}, R; \boldsymbol{\theta}) = \int_{\Omega} \Big\{ \Big[\frac{1}{2}(\boldsymbol{\tau} - \boldsymbol{A} : \boldsymbol{\varepsilon}^{\mathrm{E}}[\boldsymbol{v}]) : \boldsymbol{A}^{-1} : (\boldsymbol{\tau} - \boldsymbol{A} : \boldsymbol{\varepsilon}^{\mathrm{E}}[\boldsymbol{v}])$$

$$+ \frac{1}{2h}(R - hp)^2 \Big]_{t=T} + \int_{0}^{T} \big[R_0\dot{p} - \boldsymbol{\sigma} : \dot{\boldsymbol{\varepsilon}}^{\mathrm{p}} + R\dot{p} \big] \, \mathrm{d}t \Big\} \, \mathrm{d}V.$$

A different approach, also based on a gap built from the Fenchel–Legendre transform, is proposed in [HAD 07a, HAD 07b] to address situations where overdetermined measurements on the boundary are available. The proposed functional measures the difference between two solutions, defined for the same constitutive model and respectively associated with kinematic or static boundary data. The authors have primarily focused on the identification of viscoelastic constitutive parameters. The numerical implementation of this approach demonstrates, in particular, the improvement of local convexity (in the constitutive parameter space) of the proposed functional compared to a least squares functional measuring the boundary displacement misfit.

10.5. Suitable formulations for dynamics or vibration

Applications of the concept of CEG to identification problems initially focused on model updating using vibrational data (experimental information on natural frequencies and modal displacements, possibly after processing the measured time response). This is partly motivated by the fact that many structures in operational conditions are subjected to dynamic loading or vibrations. Early work in this direction, carried out at LMT Cachan, includes the thesis [REY 90].

Formulation in the frequency domain: model updating using the CEG in the framework of conservative dynamic vibrations is typically based on a functional of the form:

$$\mathcal{E}(\boldsymbol{A}, \rho) = \min_{\boldsymbol{v}, \boldsymbol{\tau}, \boldsymbol{\gamma}} \sum_{\hat{\omega} \text{ measured}} \left\{ E_{\hat{\omega}}(\boldsymbol{v}, \boldsymbol{\tau}, \boldsymbol{\gamma}, \boldsymbol{A}, \rho) + \frac{\beta}{2} \int_D a(\boldsymbol{v} - \bar{\boldsymbol{u}}, \boldsymbol{v} - \bar{\boldsymbol{u}}) \, \mathrm{d}V \right\}, \quad [10.34]$$

defined by the partial minimization of a *modified CEG* functional:

$$E_{\hat{\omega}}(\boldsymbol{v}, \boldsymbol{\tau}, \boldsymbol{\gamma}, \boldsymbol{A}, \rho) = \frac{\alpha}{2} \int_\Omega (\boldsymbol{\tau} - \boldsymbol{A} : \varepsilon[\boldsymbol{v}]) : \boldsymbol{A}^{-1} : (\boldsymbol{\tau} - \boldsymbol{A} : \varepsilon[\boldsymbol{v}]) \, \mathrm{d}V$$

$$+ \frac{1 - \alpha}{2} \int_\Omega \frac{1}{\rho \hat{\omega}^2} \| \boldsymbol{\gamma} + \rho \hat{\omega}^2 \boldsymbol{v} \|^2 \, \mathrm{d}V.$$

This functional, in fact, considers two constitutive equations: relation $\sigma = \boldsymbol{A} : \varepsilon$ associated with linear elasticity, and $\boldsymbol{\gamma} = -\rho \omega^2 \boldsymbol{u}$, linking the acceleration quantity density $\boldsymbol{\gamma}$ to the displacement $\boldsymbol{u}$. The dynamic admissibility constraint reads, in weak form:

$$\mathcal{R}(\boldsymbol{\tau}, \boldsymbol{\gamma}; \boldsymbol{w}) := \int_\Omega (\boldsymbol{\tau} : \varepsilon[\boldsymbol{w}] + \boldsymbol{\gamma} . \boldsymbol{w}) \, \mathrm{d}V = 0 \quad (\forall \boldsymbol{w} \in \mathcal{V}),$$

and can be combined with $E_{\hat{\omega}}(\boldsymbol{v}, \boldsymbol{\tau}, \boldsymbol{\gamma}, \boldsymbol{A}, \rho)$ using a Lagrangian $\mathcal{L} = E_{\hat{\omega}} - \mathcal{R}$. The stationarity equations for $\mathcal{L}$ are:

$$0 = \alpha \int_\Omega \varepsilon[\boldsymbol{w}] : \boldsymbol{A} : \varepsilon[\tilde{\boldsymbol{u}}] \, \mathrm{d}V - \int_\Omega \rho \hat{\omega}^2 \boldsymbol{w} \tilde{\boldsymbol{u}} \, \mathrm{d}V + \beta \int_D a(\boldsymbol{u} - \bar{\boldsymbol{u}}, \tilde{\boldsymbol{u}}) \, \mathrm{d}V \quad (\forall \tilde{\boldsymbol{u}} \in \mathcal{V}),$$

$$[10.35a]$$

$$0 = \int_\Omega \left[\varepsilon[\boldsymbol{u} - \frac{1}{\alpha} \boldsymbol{w}] : \boldsymbol{A} : \varepsilon[\tilde{\boldsymbol{w}}] - \rho \hat{\omega}^2 \left(\boldsymbol{u} + \frac{1}{(1 - \alpha)} \boldsymbol{w} \right) \tilde{\boldsymbol{w}} \right] \mathrm{d}V \quad (\forall \tilde{\boldsymbol{w}} \in \mathcal{V}),$$

$$[10.35b]$$

$$\boldsymbol{\sigma} = \boldsymbol{\mathcal{A}}\varepsilon[\boldsymbol{u} - \frac{1}{\alpha}\boldsymbol{w}], \qquad\qquad\qquad\qquad\qquad\qquad [10.35c]$$

$$\boldsymbol{\gamma} = -\rho\hat{\omega}^2[\boldsymbol{u} + \frac{1}{(1-\alpha)}\boldsymbol{w}]. \qquad\qquad\qquad\qquad\qquad [10.35d]$$

Equations [10.35a] and [10.35b] lead to a coupled system for the unknowns $\boldsymbol{u}$, $\boldsymbol{w}$, with $\boldsymbol{\sigma}$, $\boldsymbol{\gamma}$ given explicitly by [10.35a] and [10.35b]. The CEG densities associated with the elastic and kinetic constitutive relations are given by:

$$\frac{1}{\alpha}\varepsilon[\boldsymbol{w}]:\boldsymbol{\mathcal{A}}:\varepsilon[\boldsymbol{w}], \qquad \rho\|\boldsymbol{w}\|^2.$$

Many numerical experiments [REY 90, BEN 95, LAD 99, BUI 00, DER 01, DER 02] have emphasized an important property of CEG densities, namely their tendency to reach their largest values in "incorrectly modeled areas" whose material properties present a discrepancy $\Delta\rho$, $\Delta\boldsymbol{\mathcal{A}}$ with respect to the assumed values $\boldsymbol{\mathcal{A}}$, ρ, especially if their spatial size is small compared to the analyzed structure (defects).

Concerning the identification of constitutive parameters under dynamic conditions, a "relaxed" version of the CEG has been proposed more recently for transient dynamic conditions [FEI 06, NGU 08]. The idea is to avoid the exact enforcement (via admissible spaces $\mathcal{C}$, $\mathcal{S}$) of kinematic and dynamic data, which may be very noisy in dynamic tests. With reference to, for example, split Hopkinson pressure bar testing, simultaneous experimental knowledge of all forces and velocities on the sample boundary is assumed. The principle of this relaxed approach (explained here in a static framework for simplicity) is then based on the definition of $\mathcal{C}$, $\mathcal{S}$ in terms of fictitious data (displacements and forces) $\boldsymbol{\xi}$, $\boldsymbol{\varphi}$:

$$\mathcal{C}(\boldsymbol{\xi}, \partial\Omega) = \{\boldsymbol{v} \mid v_i \in \mathrm{H}^1(\Omega), \; \boldsymbol{v} = \boldsymbol{\xi} \text{ sur } \partial\Omega\},$$

$$\mathcal{S}(\boldsymbol{\varphi}, \partial\Omega) = \{\boldsymbol{\tau} \in \mathrm{H}_{\mathrm{div}}(\Omega), \; \boldsymbol{\tau}.\boldsymbol{n} = \boldsymbol{\varphi} \text{ sur } \partial\Omega\}.$$

Considering for clarity the identification of elastic moduli, a modified CEG functional $\mathcal{H}(\boldsymbol{\mathcal{A}})$ is defined as:

$$\mathcal{H}(\boldsymbol{\mathcal{A}}) = \min_{(\boldsymbol{\xi}, \boldsymbol{\varphi})}\left(\min_{(\boldsymbol{v}\in\mathcal{C}(\boldsymbol{\xi}, \partial\Omega), \boldsymbol{\tau}\in\mathcal{S}(\boldsymbol{\varphi}, \partial\Omega))} H(\boldsymbol{v}, \boldsymbol{\tau}, \boldsymbol{\xi}, \boldsymbol{\varphi}, \boldsymbol{\mathcal{A}})\right), \qquad [10.36]$$

with

$$H(\boldsymbol{v}, \boldsymbol{\tau}, \boldsymbol{\xi}, \boldsymbol{\varphi}, \boldsymbol{\mathcal{A}}) = \frac{1}{2}\int_{\Omega}(\boldsymbol{\tau} - \boldsymbol{\mathcal{A}}:\varepsilon[\boldsymbol{v}]):\boldsymbol{\mathcal{A}}^{-1}:(\boldsymbol{\tau} - \boldsymbol{\mathcal{A}}:\varepsilon[\boldsymbol{v}]) \, \mathrm{d}V$$

$$+ \int_{S}\left(\frac{A}{2}\|\bar{\boldsymbol{u}} - \boldsymbol{\xi}\|_{U}^2 + \frac{B}{2}\|\bar{\boldsymbol{T}} - \boldsymbol{\varphi}\|_{T}^2\right)\mathrm{d}S.$$

Coefficients A and B ensure the dimensional consistency of the functional $\mathcal{H}$ and enable the two types of data to be weighted according to their expected quality. This approach implicitly performs a smoothing of noisy data $\bar{u}, \bar{T}$. Numerical experiments in dynamic identification [FEI 06, NGU 08] have shown that this approach can operate satisfactorily for high levels of data noise, for which the performance of other functionals (including least squares misfit functionals acting directly on the boundary data) markedly deteriorates.

10.6. Conclusions

The identification of parameters featured in constitutive laws is an interesting and non-trivial problem. The experimentalist measures certain "outputs" corresponding to certain "inputs". Before the advent of kinematic field measurements, experiments were generally designed in such a way that the mechanical state could be assumed homogeneous in the region of interest where gauges are applied. For homogeneous isotropic materials, such experiments allow the identification of the Young's modulus and the Poisson's ratio. For general homogeneous anisotropic materials, the identification of all elastic constants needs a more sophisticated analysis.

Over the past two decades, completely new experimental techniques enabling the measurement of displacement fields have appeared. Such techniques are particularly interesting when multiaxial loading along two or three perpendicular directions is applied and strain field heterogeneity occurs. In such situations, a displacement field measurement is needed to check and quantify the heterogeneity of the strain field. The identification methods presented in this chapter use the vast amount of information contained in these fields. We have particularly emphasized the identification problem in heterogeneous linear elastic conditions, illustrating this approach with both academic examples and a case study with available experimental data. The identification algorithm used has been presented and discussed in detail to facilitate its subsequent use by the reader. An interesting aspect of local identification is the study of the sensitivity to contrasts of material properties. As expected, errors in the identification increase with the contrast, because of strong gradients induced in the stress and strain fields, which must then be approximated with the highest possible accuracy by the identification method.

The CEG elastic identification procedure also enables us to establish energy balances in polycyclic fatigue. Stress fields are identified over a mechanical cycle so as to estimate the mechanical energy supplied locally during a fatigue cycle. The originality of this method is to consider small variations in a secant elastoplastic stiffness tensor, mainly due to microplasticity mechanisms, as small variations in an elastic stiffness tensor. The latter is determined for each loading step during the mechanical cycle, and locally since no assumption is made about constitutive homogeneity. The spatial and temporal variability of material behavior is described as accurately as possible. The method allows the identification of stress fields associated with a heterogeneous material. A preliminary study involving this type of energy balance in the context of fatigue is presented in Chapter 16.

Kinematic measurements using, say, digital image correlation enable data to be obtained on in-plane displacement fields. CEG-based identification methods have therefore so far been mostly developed for configurations for which the assumption of plane stress can be applied, the material properties also being identified in a two-dimensional domain. The identification of three-dimensional heterogeneous properties is an interesting extension of the methods presented in this chapter. Kinematic field measurements in the entire volume are in fact made possible by tomography techniques, which are beginning to be used for mechanical tests (see Chapter 6) but remain difficult to implement. Meanwhile, it is also important to note that the CEG (especially in its "modified" form) can be used with data obtained using other measurement methods.

Finally, the case of elastoplasticity has been considered using different CEG-based approaches. The first approach, an incremental version of the linear elastic CEG, raises some issues when the plastic zones evolve rapidly during loading. In an interesting extension of the identification method, the identification of hardening laws with more parameters, such as those involving an exponential term accounting for the observed saturation of hardening steels, could be considered. The second approach, based on the Legendre–Fenchel transformation, should help to overcome these difficulties. It is more consistent from a theoretical standpoint, but its practical implementation remains to be achieved. The formalism and the underlying theoretical framework should give rise to yet more interesting generalizations of the concept of CEG.

10.7. Bibliography

[AND 92] ANDRIEUX S., BEN ABDA A., "Identification de fissures planes par une donnée au bord unique; un procédé direct de localisation et d'identification", *Comptes Rendus Mécanique*, vol. 315, pp. 1323–1328, 1992.

[AND 96] ANDRIEUX S., BEN ABDA A., "Identification of planar cracks by complete overdetermined data: inversion formulae", *Inverse Problems*, vol. 12, pp. 553–563, 1996.

[AND 06] ANDRIEUX S., BARANGER T., BEN ABDA A., "Solving Cauchy problems by minimizing an energy-like functional", *Inverse Problems*, vol. 22, pp. 115–133, 2006.

[AND 08] ANDRIEUX S., BARANGER T., "An energy error-based method for the resolution of the Cauchy problem in 3D linear elasticity", *Computer Methods in Applied Mechanics and Engineering* , vol. 197, pp. 902–920, 2008.

[AVR 07] AVRIL S., PIERRON F., "General framework for the identification of constitutive parameters from full-field measurements in linear elasticity", *International Journal of Solids and Structures*, vol. 44, pp. 4978–5002, 2007.

[AVR 08] AVRIL S., BONNET M., BRETELLE A., GRÉDIAC M., HILD F., IENNY P., LATOURTE F., LEMOSSE D., PAGNACCO E., PIERRON F., "Overview of identification methods of mechanical parameters based on full-field measurements", *Experimental Mechanics*, vol. 48, pp. 381–402, 2008.

[BEN 95] BEN ABDALLAH J., Inversion gaussienne appliquée à la correction paramétrique de modèles structuraux, PhD Thesis, Ecole Polytechnique, Paris, France, 1995.

[BUI 00] BUI H.D., CONSTANTINESCU A., "Spatial localization of the error of constitutive law for the identification of defects in elastic bodies", *Archives of Mechanics*, vol. 52, pp. 511–522, 2000.

[CAL 02] CALLOCH S., DUREISSEIX D., HILD F., "Identification de modèles de comportement de matériaux solides: utilisation d'essais et de calculs", *Technologies et Formations*, vol. 100, pp. 36–41, 2002.

[CHA 99] CHAVENT G., KUNISCH K., ROBERTS J.E., "Primal-dual formulations for parameter estimation problems", in RAUPP M., JR J.D., KOILLER J., MENZALA G. (eds), *Computational and Applied Mathematics*, vol. 18, pp. 173–229, 1999.

[CON 95] CONSTANTINESCU A., "On the identification of elastic moduli from displacement-force boundary measurements", *Inverse Problems in Engineering*, vol. 1, pp. 293–315, 1995.

[DER 01] DERAEMAEKER A., Sur la maîtrise des modèles en dynamique des structures à partir de résultats d'essais, PhD Thesis, ENS Cachan, 2001.

[DER 02] DERAEMAEKER A., LADEVÈZE P., LECONTE P., "Reduced bases for model updating in structural dynamics based on constitutive relation error", *Computer Methods in Applied Mechanics and Engineering*, vol. 191, pp. 2427–2444, 2002.

[FEI 06] FEISSEL P., ALLIX O., "Modified constitutive relation error identification strategy for transient dynamics with corrupted data: the elastic case", *Computer Methods in Applied Mechanics and Engineering*, vol. 196, pp. 1968–1983, 2006.

[GER 83] GERMAIN P., NGUYEN Q.S., SUQUET P., "Continuum thermodynamics", *ASME Journal of Applied Mechanics*, vol. 50, pp. 1010–1020, 1983.

[GER 86] GERMAIN P., *Mécanique*, Cours de l'Ecole Polytechnique, 1986.

[GEY 02] GEYMONAT G., HILD F., PAGANO S., "Identification of elastic parameters by displacement field measurement", *Comptes Rendus de l'Académie des Sciences Paris, Series II*, vol. 330, pp. 403–408, 2002.

[GEY 03] GEYMONAT G., PAGANO S., "Identification of mechanical properties by displacement field measurement: a variational approach", *Meccanica*, vol. 38, pp. 535–545, 2003.

[HAD 07a] HADJ-SASSI K., Une stratégie d'estimation conjointe des paramètres et de l'état de structures à comportement non linéaire. Assimilation de données et erreur en loi de comportement, PhD Thesis, Ecole Polytechnique, Paris, France, 2007.

[HAD 07b] HADJ-SASSI K., ANDRIEUX S., "Une nouvelle fonctionnelle d'énergie incrémentale totale pour le contrôle des parties réversibles et dissipatives des matériaux standards", *Huitième Colloque National en Calcul des Structures*, Hermes, pp. 255–261, 2007.

[HAL 75] HALPHEN B., NGUYEN Q.S., "Sur les matériaux standards généralisés", *Journal de Mécanique*, vol. 14, pp. 39–63, 1975.

[KOH 84] KOHN R., VOGELIUS M., "Determining conductivity by boundary measurements", *Communications on Pure and Applied Mathematics*, vol. 37, pp. 289–298, 1984.

[KOH 88] KOHN R.V., LOWE B.D., "A variational method for parameter identification", *Mathematical Modelling and Numerical Analysis*, vol. 1, pp. 119–158, 1988.

[KOH 90] KOHN R., MCKENNEY A., "Numerical implementation of a variational method for electrical impedance tomography", *Inverse Problems*, vol. 6, pp. 389–414, 1990.

[KOZ 91] KOZLOV V.A., MAZ'YA V.G., FOMIN A.F., "An iterative method for solving the Cauchy problem for elliptic equations", *Computational Mathematics and Mathematical Physics*, vol. 31, pp. 45–52, 1991.

[LAD 83] LADEVÈZE P., LEGUILLON D., "Error estimate procedure in the finite element method and applications", *SIAM Journal on Numerical Analysis*, vol. 20, pp. 485–509, 1983.

[LAD 99] LADEVÈZE P., CHOUAKI A., "Application of *a posteriori* error estimation for structural model updating", *Inverse Problems*, vol. 15, pp. 49–58, 1999.

[LAT 07] LATOURTE F., Identification des paramètres d'une loi de Prager et calcul de champs de contraintes dans des matériaux hétérogènes, PhD Thesis, University of Montpellier 2, France, 2007.

[LAT 08] LATOURTE F., CHRYSOCHOOS A., PAGANO S., WATTRISSE B., "Elastoplastic behavior identification for heterogeneous loadings and materials", *Experimental Mechanics*, vol. 48, pp. 435–449, 2008.

[MOË 99] MOËS N., LADEVÈZE P., DOUCHIN B., "Constitutive relation error estimators for (visco)plastic finite element analysis with softening", *Computer Methods in Applied Mechanics and Engineering*, vol. 176, pp. 247–264, 1999.

[NGU 08] NGUYEN H.-M., ALLIX O., FEISSEL P., "A robust identification strategy for rate-dependent models in dynamics", *Inverse Problems*, vol. 24, pp. 065006, 2008.

[REY 90] REYNIER M., Sur le contrôle de modélisations éléments finis: recalage à partir d'essais dynamiques, PhD Thesis, UPMC, Paris, France, 1990.

[SIM 98] SIMO J.C., HUGHES T.J.R., *Computational Inelasticity*, Springer-Verlag, 1998.

Chapter 11

The Virtual Fields Method

11.1. Introduction

The virtual fields method (VFM) is one of the techniques developed to identify the parameters governing constitutive equations, the experimental data processed for this purpose being displacement or strain fields. This chapter shows one of its main advantages which is the fact that, in several cases, the sought parameters can be found directly from the measurements, without calculating the stress using a numerical tool such as a finite element program.

The VFM relies on the principle of virtual work (PVW), which is written with particular virtual fields. This method was first proposed in statics, within the framework of linear anisotropic elasticity. This simple case serves here as an example to present the main features of the method in the case of constitutive equations, which linearly depend on the constitutive parameters. Some other more complex cases are also examined, such as dynamic loads, viscoelasticity, hyperelasticity and elastoplasticity. These cases are presented more briefly at the end of the chapter, but references where more complete information is available are also given. Some experimental examples also illustrate the feasibility of this approach.

11.2. General principle

The PVW represents, in fact, the weak form of the local equations of equilibrium, which is classically introduced in the mechanics of deformable media [DYM 73].

Chapter written by Michel GRÉDIAC, Fabrice PIERRON, Stéphane AVRIL, Evelyne TOUSSAINT and Marco ROSSI.

Assuming small strains, the PVW can be written as follows for any domain defined by its volume V and its external boundary S:

$$-\int_V \sigma : \varepsilon^\star dV + \int_{S_f} \bar{\mathbf{T}}.\mathbf{u}^\star dS + \int_V \mathbf{f}.\mathbf{u}^\star dV = \int_V \rho\gamma.\mathbf{u}^\star dV \quad \forall \mathbf{u}^\star \mathrm{KA} \quad [11.1]$$

where σ is the Cauchy stress tensor, $\varepsilon^\star$ is the virtual strain tensor derived from the virtual displacement field $\mathbf{u}^\star$, $\bar{\mathbf{T}}$ is the traction acting over a certain subdomain (denoted by S_f) of the external boundary S, $\mathbf{f}$ is the volume force acting at each point of the volume (denoted by V), ρ is the mass per unit volume and γ is the acceleration. A very important property is, in fact, that the above equation is satisfied for any kinematically admissible (KA) virtual field $\mathbf{u}^\star$. By definition, a KA virtual field must satisfy the displacement boundary conditions of the actual displacement field in order to cancel the contribution of the resulting forces along the portion of the boundary along which actual displacements are prescribed. It must be pointed out that this requirement is not really necessary in all cases, but this point is not discussed here for the sake of simplicity. KA virtual fields are also assumed to be C_0 functions.

The first step of the VFM consists of introducing the constitutive equations. Hence, the stress tensor σ disappears to the benefit of the actual strain tensor ε, which is deduced from the measurements. In general, the constitutive equations can be written as:

$$\sigma = g(\varepsilon) \qquad\qquad\qquad\qquad [11.2]$$

where g is a certain function of the actual strain components. g also depends on the constitutive parameters that characterize the material. These parameters are, in fact, the unknowns of the problem addressed here. A general assumption is to consider that the volume force $\mathbf{f}$ is null. In this case, equation [11.1] becomes:

$$-\int_V g(\varepsilon) : \varepsilon^\star dV + \int_{S_f} \bar{\mathbf{T}}.\mathbf{u}^\star dS = \int_V \rho\gamma.\mathbf{u}^\star dV \quad \forall \mathbf{u}^\star \mathrm{KA} \qquad [11.3]$$

This equation being satisfied for any KA virtual field, any new KA virtual field provides a new equation. The VFM relies on this property by writing equation [11.3] above with a set of KA virtual fields chosen *a priori* [GRÉ 89]. The number of virtual fields and their type depend on the nature of the constitutive equation used to define g in equation [11.2]. Two different cases can be distinguished.

Case 1: the constitutive equations linearly depend on the sought parameters. Linear elasticity, some types of hyperelasticity, viscoelasticity and some cases of laws modeling damage belong to this case. Writing equation [11.3] with as many virtual fields as unknowns leads to a system of linear equations that provides the sought parameters after inversion. It has been observed that the equations of the system are generally independent if the actual strain fields that are processed are heterogeneous and if the virtual fields that are used in the different equations are independent. In the case of linear elasticity, we can best choose the virtual fields following the procedure described in section 11.3.

Case 2: the constitutive equations are nonlinear with respect to their governing parameters. In this case, identification is carried out by minimizing a cost function derived from equation [11.3] (see section 11.4.2).

Case 1 is addressed preferentially in this chapter because it is the simplest case and because it has led to the greatest number of applications up to now. Case 2 is more complex. It is discussed in a shorter format at the end of the chapter. A more complete description of the method in this case can be found in [PIE 12].

11.3. Constitutive equations depending linearly on the parameters: determination of the virtual fields

11.3.1. *Introduction*

The case of constitutive equations depending linearly on the parameters is interesting because retrieving the parameters is a direct calculation, as briefly explained above. A key issue is, however, the choice of the virtual fields. This point is addressed in the following sections with increasing complexity: the determination is first carried out intuitively, then using a general approach based on the so-called "special" virtual fields and, finally, by minimizing the noise effect on the identified parameters. The method is developed here in the simple case of linear orthotropic elasticity, for a state of plane stress, in order to underline the different steps of the method. The particular case of orthotropy is also interesting for the following reasons: four independent stiffnesses are to be determined, while a classic tensile test provides a maximum of three linear equations because of the homogeneity of the strain fields expected in the specimen. This is the reason why various tests must be carried out for anisotropic materials. The advantage of the VFM is that these four unknowns are directly extracted from heterogeneous strain fields, as explained below. Other cases, such as plate bending, large deformations and constitutive equations that do not depend linearly on the constitutive parameters, are also briefly addressed in this chapter.

11.3.2. *Developing the PVW*

Using the usual contracting rules for indexes (Voigt notation), the constitutive equations can be written as follows in the (1,2) plane:

$$\sigma_i = Q_{ij}\varepsilon_j, i, j = 1, 2, 6 \tag{11.4}$$

where the Q_{ij} parameters are the unknown stiffnesses. The state of stress being defined in the (1,2) plane only, the actual strain components are assumed to be constant through the thickness of the specimen, thus leading the volume integrals involved in equation [11.3] to become surface integrals. This assumption is necessary in the usual case for which measurements are available on the external surface of the specimen only. Some recent measurement methods provide three-dimensional (3D) measurements, so this assumption is no longer necessary in this case. The case of statics is considered here ($\gamma = 0$ for the sake of simplicity). Introducing equation [11.4] into equation [11.3] and simplifying by the thickness of the specimen lead to:

$$\int_S Q_{ij}\varepsilon_j\varepsilon_i^* dS = \int_{L_f} \bar{\mathbf{T}}.\mathbf{u}^* dl \quad \forall \mathbf{u}^\star \text{KA} \tag{11.5}$$

where S is the area of the zone over which integration is performed and L_f is the line along which the load is applied. If the material under consideration is heterogeneous in the plane of the specimen, the sought parameters (namely the Q_{ij}s in this case) can be defined as functions of the coordinates. In this case, the sought parameters are those governing these functions. Another possibility consists of defining the material properties by subdomains, as explained in section 11.3.3. In the simple case of a homogeneous material, the sought parameters are constant; so they can be factorized in equation [11.5]. Thus:

$$Q_{ij}\int_S \varepsilon_j\varepsilon_i^* dS = \int_{L_f} \bar{\mathbf{T}}.\mathbf{u}^* dl \quad \forall \mathbf{u}^\star \text{KA} \tag{11.6}$$

Any new virtual field produces a new equation of the same type as equation [11.6]. Writing this equation with as many virtual fields as unknowns leads to a linear system where the Q_{ij}s are the unknowns. In practice, the integrals involved in the system are approximated by discrete sums; thus:

$$\int_S \varepsilon_j\varepsilon_i^* dS \simeq S_a \sum_{k=1}^n \varepsilon_j(x_k,y_k)\varepsilon_i(x_k,y_k)^* \tag{11.7}$$

where n is the number of measurement points and S_a is the area of the small surfaces over which measurements are performed, as in the usual case of CCD cameras. This system can finally be written as follows:

$$\mathbf{PQ} = \mathbf{R} \tag{11.8}$$

where $\mathbf{P}$ is a square matrix, $\mathbf{Q}$ is a vector gathering the unknown Q_{ij}s and $\mathbf{R}$ is a vector whose components correspond to the external virtual work per unit thickness calculated for each virtual field. The equations of the system are generally independent if the actual strain field is heterogeneous and if the virtual strain fields are independent [GRÉ 02a]. Interestingly, it can be noted that heterogeneity is here an advantage, whereas it is a drawback with the usual approach because the latter relies on homogeneous strain fields. Virtual fields must be chosen from among the infinite set of KA virtual fields. Note, however, that non-KA virtual fields can also be used theoretically, but it requires the load acting where the prescribed displacements are applied to be known. At this stage, an important question is to know how best the virtual fields should be chosen and therefore how a criterion should be defined to find them.

In the first attempt, the virtual fields were chosen empirically [GRÉ 90] in such a way that the equations of system [11.6] were at least partially uncoupled. The first improvement was to define an objective criterion to determine them. A first strategy consisted of choosing the virtual fields in such a way that the conditioning of the system was the best possible, thus leading to the so-called *special* virtual fields. Such virtual fields cause matrix $\mathbf{P}$ involved in equation [11.8] to be equal to the identity matrix $\mathbf{I}$. Because there exist an infinite number of such virtual fields, another criterion was introduced to choose the best virtual field: the sensitivity to noise of the sought parameters. These issues are addressed below.

11.3.3. *Special virtual fields*

The equations developed here assume that the material is linear elastic, orthotropic and homogeneous (the VFM was first proposed for composite materials). Therefore, there are four unknowns in the problem, Q_{11}, Q_{22}, Q_{12} and Q_{66}, provided the orthotropy directions are known *a priori*. Equation [11.6] can be rewritten as:

$$Q_{11} \underbrace{\int_S \varepsilon_1 \varepsilon_1^* dS}_{=1} + Q_{22} \underbrace{\int_S \varepsilon_2 \varepsilon_2^* dS}_{=0} + Q_{12} \underbrace{\int_S (\varepsilon_1 \varepsilon_2^* + \varepsilon_2 \varepsilon_1^*) dS}_{=0}$$

$$+ Q_{66} \underbrace{\int_S \varepsilon_6 \varepsilon_6^* dS}_{=0} = \int_{L_f} \bar{\mathbf{T}}(M, \mathbf{n}).\mathbf{u}^*(M) dl \qquad [11.9]$$

If matrix $\mathbf{P}$ of the above linear system is equal to matrix $\mathbf{I}$, the integrals in equation [11.9] are successively equal to one, the other integrals being equal to zero. For instance, in order to retrieve Q_{11}, the virtual field is defined in such a way that the integral that is the coefficient of Q_{11} in the equation above is equal to one, the other integrals all being equal to zero. Changing by turns the location of numeral one in the equation provides successively Q_{22}, Q_{12} and Q_{66}. With such virtual fields, system [11.8] becomes:

$$\mathbf{Q} = \mathbf{R} \qquad [11.10]$$

because $\mathbf{P} = \mathbf{I}$. The sought parameters are directly determined if the special virtual fields are known. There are two procedures to construct such virtual fields:

– The first procedure consists of defining the unknown virtual fields over the whole domain, for instance by using polynomials. For the current in-plane problem, the two components of the virtual displacement $\mathbf{u}^*$ can be expanded as:

$$\begin{cases} u_1^* = \displaystyle\sum_{i=0}^{m}\sum_{j=0}^{n} A_{ij}x_1^i x_2^j \\ u_2^* = \displaystyle\sum_{i=0}^{p}\sum_{j=0}^{q} B_{ij}x_1^i x_2^j \end{cases} \qquad\qquad [11.11]$$

The A_{ij} and B_{ij} coefficients, therefore, define the special virtual fields.

– The second procedure is to define the virtual fields by subdomains, in a way similar to the finite element method (FEM). This leads to "virtual elements" whose nodal virtual displacements define the virtual displacements at any point, using suitable shape functions defined by subdomains (see section 11.3.5 for more details). Thus:

$$\mathbf{u}^* = \mathbf{\Phi}\mathbf{U}^* \qquad\qquad [11.12]$$

where $\mathbf{u}^*$ is a vector gathering the virtual displacements along directions 1 and 2, $\mathbf{\Phi}$ is a $2 \times N$ matrix that contains the shape functions of the virtual elements and $\mathbf{U}^*$ is a vector containing the virtual nodal displacements.

The coefficients that define these special virtual fields are directly the coefficients of either the set of functions defining the basis on which the virtual displacement fields are developed or the nodal virtual displacements, depending on how the virtual displacement fields are expressed: either with a basis of functions defined over the whole domain or by subdomains. The following procedure is used to find these coefficients:

– One of the coefficients in equation [11.9] is set to one, the others to zero. In the example above, this leads to four linear equations where the unknowns are the parameters that define the special virtual fields.

– The virtual fields being KA, they must be null on the part of the boundary where the displacement is known and the applied force is unknown. This property leads to a set of additional linear equations, whose number depends on the problem to be solved.

Some other conditions can be added. For instance, if measurements are not available for a part of the tested specimen, we can impose that the virtual displacement field be rigid-body-like there in order to cancel out the contribution of this part in the equations [GRÉ 02b].

An important point must be underlined at this stage. On the one hand, the number of equations used to define the special virtual fields is limited by the two above constraints. On the other hand, the number of parameters that define such special virtual fields remains theoretically unlimited because this number merely depends on m, n, p and q in equation [11.11] in the case of virtual fields defined over the whole surface, or on the number of virtual nodes for virtual fields defined by subdomains. If the number of parameters defining these fields is too large, the number of equations available to determine these fields may potentially become too small, thus leading to an infinite number of special virtual fields. This new degree of freedom in the choice of the virtual fields can be used to minimize an important criterion for identification purposes: the influence of measurement noise on the identified parameters. The method used to minimize this effect is described in the following section.

11.3.4. *Virtual fields optimized with respect to measurement noise*

Because of measurement noise, ε_1, ε_2 and ε_6 used in the identification procedure are different from the actual values. The noise is, in fact, added to the actual (and "true") strain components. Hence, the PVW is rigorously satisfied only if this noise is subtracted from the measured strain components. Thus, the expression given by equation [11.10] can, in fact, be written as:

$$(\mathbf{I} - \mathbf{E})\mathbf{Q} = \mathbf{R} \qquad\qquad [11.13]$$

where $\mathbf{E}$ is a matrix reflecting the difference with the equation that must be rigorously satisfied, this difference being due to measurement noise. Thus, the identification parameters become:

$$\mathbf{Q} = \mathbf{R} + \mathbf{E}\mathbf{Q} \qquad\qquad [11.14]$$

In the following, it is assumed that the errors are random processes [SOO 93, AVR 04]. Thus, the error $\mathbf{E}\mathbf{Q}$ on the identified parameters is a random vector that can be analyzed using standard statistical tools. Assuming the mean of each component is null, the variance can be written as:

$$\mathcal{V}(\mathbf{Q}) = \mathcal{E}(\mathbf{E}\mathbf{Q}\mathbf{Q}^t\mathbf{E}^t) \qquad\qquad [11.15]$$

where $\mathcal{E}(\mathbf{X})$ is the expectation of the random vector $\mathbf{X}$ and $\mathbf{X}^t$ is the transpose of vector $\mathbf{X}$. $\mathcal{V}(\mathbf{Q})$ is a matrix, whose diagonal contains the variance of each identified parameter Q_{ij}, in other words indicators on the uncertainty of the identification. The off-diagonal components are covariances that are not used here. For the sake of simplicity, the constitutive material is still assumed to exhibit an orthotropic, linear elastic response. Minimizing the diagonal terms of $\mathcal{V}(\mathbf{Q})$ leads to four special virtual fields $\mathbf{u}^{*[1]}$, $\mathbf{u}^{*[2]}$, $\mathbf{u}^{*[3]}$ and $\mathbf{u}^{*[4]}$, which are used to identify the four unknown parameters Q_{11}, Q_{22}, Q_{12} and Q_{66}. The noise is assumed to be white noise, so an independent copy of this noise is added to each of the three in-plane strain components used as input data for the identification procedure. In this case, it has been shown that the virtual fields $\mathbf{u}^{*[k]}$, $k = 1, \ldots, 4$, must minimize the following expression to be optimum [AVR 04]:

$$\mathcal{F}(\mathbf{u}^{*[k]}) = \frac{1}{2} \left([Q_{11}^2 + Q_{22}^2] \int_S [\varepsilon_1(\mathbf{u}^{*[k]})]^2 dS \right.$$

$$+ [Q_{22}^2 + Q_{12}^2] \int_S [\varepsilon_2(\mathbf{u}^{*[k]})]^2 dS + Q_{66}^2 \int_S [\varepsilon_6(\mathbf{u}^{*[k]})]^2 dS$$

$$\left. + 2[Q_{12}(Q_{11} + Q_{22})] \int_S [\varepsilon_1(\mathbf{u}^{*[k]})\varepsilon_2(\mathbf{u}^{*[k]})] dS \right) \qquad [11.16]$$

A virtual field that is *special*, KA, and that minimizes the cost function $\mathcal{F}$ is unique for a given basis of function. This virtual field is the solution to a problem of minimization under constraints where the virtual fields are *special* and KA are the constraints and $\mathcal{F}$ is the cost function to be minimized.

As mentioned above, the virtual fields can be expressed using different strategies: either on the basis of functions defined over the whole domain, such as polynomials, or piecewise, by subdomains (see section 11.3.5). The vectors whose components are either the coefficients of the basis functions in the first case or the nodal virtual displacements in the second case, which are determined by solving the minimization problem described above. This finally leads to the following equation [AVR 04]:

$$\left[\begin{array}{c|c} \mathbf{H} & \mathbf{A}^t \\ \hline \mathbf{A} & 0 \end{array} \right] \left\{ \begin{array}{cccc} \mathbf{U}^{*[1]} & \mathbf{U}^{*[2]} & \mathbf{U}^{*[3]} & \mathbf{U}^{*[4]} \\ \mathbf{\Lambda}^{[1]} & \mathbf{\Lambda}^{[2]} & \mathbf{\Lambda}^{[3]} & \mathbf{\Lambda}^{[4]} \end{array} \right\} = \left\{ \begin{array}{cccc} 0 & 0 & 0 & 0 \\ \mathbf{Z}^{[1]} & \mathbf{Z}^{[2]} & \mathbf{Z}^{[3]} & \mathbf{Z}^{[4]} \end{array} \right\} \qquad [11.17]$$

where:

– $\mathbf{U}^{*[1]}$, $\mathbf{U}^{*[2]}$, $\mathbf{U}^{*[3]}$ and $\mathbf{U}^{*[4]}$ are vectors containing the coefficients of the functions (of monomials if polynomials are used) or nodal virtual displacements. These virtual fields lead to the identification of Q_{11}, Q_{22}, Q_{12} and Q_{66}, respectively.

– $\mathbf{A}$ is a matrix containing the constraints.

– $\mathbf{H}$ is a matrix such that:

$$\mathcal{F}(\mathbf{u}^{*[k]}) = \frac{1}{2} \mathbf{U}^{*[k]t} \mathbf{H} \mathbf{U}^{*[k]} \qquad [11.18]$$

– $\mathbf{\Lambda}^{[1]}$, $\mathbf{\Lambda}^{[2]}$, $\mathbf{\Lambda}^{[3]}$ and $\mathbf{\Lambda}^{[4]}$ are the Lagrange multipliers corresponding to these constraints.

– $\mathbf{Z}^{[1]}$, $\mathbf{Z}^{[2]}$, $\mathbf{Z}^{[3]}$ and $\mathbf{Z}^{[4]}$ are the vectors that are associated with these constraints. Their components are null apart from one of them being equal to one to ensure that the virtual field to be found is *special*.

Solving equation [11.17] provides four vectors, $\mathbf{U}^{*[1]}, \mathbf{U}^{*[2]}, \mathbf{U}^{*[3]}$ and $\mathbf{U}^{*[4]}$, from which four special virtual fields are deduced: $\mathbf{u}^{*[1]}, \mathbf{u}^{*[2]}, \mathbf{u}^{*[3]}$ and $\mathbf{u}^{*[4]}$. These virtual fields lead to the identification of Q_{11}, Q_{22}, Q_{12} and Q_{66}, respectively. However, this problem is implicit: Q_{11}, Q_{22}, Q_{12} and Q_{66} are unknowns at the beginning of the procedure but they are involved in the expression of $\mathcal{F}(\mathbf{u}^{*[k]})$. This problem can be solved iteratively by replacing the unknowns by the values that are identified during the previous step. For the first iteration, any value can be used as long as it is realistic from a physical point of view. It has been shown in [AVR 04] that this procedure converges very quickly because only two or three iterations are required in practice.

The case for which virtual fields are defined with shape functions defined by subdomains is the most usual in practice (see section 11.3.5). We can even choose the same subdomains and the same functions to construct the actual fields from the experimental data [AVR 07]. This particular case is interesting because it has been shown that the linear systems provided by the VFM (equation [11.8] or equation [11.10]) are, for a set of particular virtual fields that must be defined for each situation, the same as those obtained using the finite element model updating technique [HEN 91], the equilibrium gap method [CLA 02] or the constitutive equation gap method [GEY 02], thus enabling easy comparison between these techniques in this particular case. Solving equation [11.17] provides four special virtual fields that directly lead to the sought stiffnesses. Because these four virtual fields minimize sensitivity to noise, they lead to the so-called *maximum likelihood* solution [AVR 07]. In this particular case, and for a homogeneous elastic linear material, it is shown in [AVR 07] that the only possibility to obtain such a solution is to apply the VFM with a set of virtual fields that are the solutions of equation [11.17]. The solutions obtained using the finite element model updating technique, the equilibrium gap method and the constitutive equation gap method are less robust in this case because their corresponding virtual fields obtained with this formalism are different from those that minimize equation [11.17]. In practice, however, it has been shown in [AVR 07] that the solution provided by the finite element model updating technique is close to the maximum likelihood solution [AVR 07].

11.3.5. *Virtual fields defined by subdomains*

As already mentioned above, it is possible to construct special virtual fields by subdomains. Such virtual fields can be used advantageously compared to virtual fields expanded with functions defined over the whole domain. This is due to several reasons [TOU 06]:

1) The degree of the shape functions used in each subdomain is generally lower than that of the polynomials used if the virtual fields are defined with the same polynomials over the whole domain. This causes, in practice, a lower influence of

noise on the identified values: the lower the degree of the polynomials, the better the robustness, as shown in [GRÉ 03].

2) The virtual fields must only be C_0 because discontinuities of the virtual field derivatives between subdomains do not cause any error in the expression of the PVW. This leads to more flexibility in the construction of the virtual fields.

3) Studying multimaterials is also much easier with such virtual fields if mechanical properties are defined piecewise, because virtual fields are easier to program in this case.

It is important to mention that virtual and actual displacement fields being potentially expanded in a different basis of functions (contrary to the FEM, for which both fields are generally expanded in the same basis of functions: the shape functions), nodes used to define the virtual fields may be outside the specimen.

By splitting S into p subdomains such that $S = S_1 \cup S_2 \cup \cdots S_p$, the left-hand side of equation [11.6] becomes:

$$\int_S Q_{ij}\varepsilon_j\varepsilon_i^* dS = \sum_{k=1}^{p} \int_{S_k} Q_{ij}^{(k)}\varepsilon_j\varepsilon_i^* dS = \sum_{k=1}^{p} Q_{ij}^{(k)} \int_{S_k} \varepsilon_j\varepsilon_i^* dS \qquad [11.19]$$

In each subdomain, the virtual field can be expressed as the product of the shape functions and the virtual nodal displacement, as in equation [11.12]. At this stage, we can note that the Q_{ij} parameters can be considered as unknown for each of the subdomains, thus facilitating the investigation of heterogeneous materials. The following expression is obtained after assembling the matrices defined in each of the subdomains:

$$\mathbf{AY}^* = \mathbf{B} \qquad [11.20]$$

where the components of vector $\mathbf{Y}^*$ are the degrees of freedom of all the nodes used to define the subdomains. Matrices $\mathbf{A}$ and $\mathbf{B}$ are obtained by assembling the subdomains used to mesh S. Vector $\mathbf{Y}^*$ has kN components, where N is the number of nodes and k is the number of degrees of freedom at each node. Furthermore, the number of equations available to solve the problem is equal to $s + b$, where s is the number of special virtual fields, which is equal to the number of unknown stiffnesses, and b is the number of equations reflecting the fact that the virtual field is KA. If $s + b$ is smaller than the number of degrees of freedom $k \times N$, matrix $\mathbf{A}$ becomes rectangular and additional conditions must be added to determine the nodal virtual fields. These additional conditions may be arbitrary or may come from other requirements, for instance the fact that the effect of measurement noise on the identified parameters should be minimized, as explained in section 11.3.4 [SYE 08].

11.3.6. *Examples*

11.3.6.1. *Introduction*

The objective here is to illustrate the method described above with some examples of experimental implementations.

11.3.6.2. *Plane orthotropic linear elastic response*

Many applications of the VFMs have dealt with the identification of parameters governing orthotropic elastic linear laws, especially for long-fiber composites. The first example given here concerns the Iosipescu test, suitable for measuring the shear modulus of composites (see ASTM D5379 standard). However, the notches of the initial specimen measurement, whose role was to concentrate the shear strain in the gauge section of the specimen, have been removed here because they are not necessary any more. Hence, this test becomes a shear/bending test on a rectangular specimen (see Figure 11.1). It has been shown that the strain field was sufficiently heterogeneous in the gauge section (that supporting the grid in Figure 11.1) to identify with one single test the four stiffnesses characterizing the orthotropic response of a composite plate [PIE 00]. Experimental displacement fields obtained here with the grid method have led to satisfactory results, as explained in detail in [CHA 06]. This test was also successfully used to identify the stiffnesses of maritime pine in the (L,R) plane [XAV 07]. The main advantages of this test are as follows:

– It requires small-sized specimens.

– As a result, it is also well suited for identifying the through-thickness properties of thick plates, for which reliable tests are scarce.

Figure 11.1. *Unnotched specimen in an Iosipescu fixture*

This example shows that it is possible to measure with one test the four stiffnesses characterizing an orthotropic material (instead of three tests with the standard approach).

The spirit of the following example remains much the same, but the idea is to choose the best surface location from which the parameters can be extracted. This example deals with the four through-thickness stiffnesses of a thick composite tube. This tube was manufactured using filament winding, so it has a cylindrical material symmetry through its thickness. This feature potentially induces some difficulties in the characterization in the (R, θ) plane. The idea was therefore to apply a compressive loading to a ring cut off from the tube (see Figure 11.2) and to measure, here again with the grid method, the displacement–strain only on a portion of the surface of the ring rather than on the whole surface. The corresponding zone can be seen in Figure 11.2. Using the whole section would, indeed, have led to not using a significant part of the pixels (the internal part of the ring, for instance). On the other hand, using only a portion enables us to use many more pixels in practice and, therefore, to optimize the spatial resolution of the displacement and strain measurements. The choice of the zone over which the measurements were performed was made by maximizing the identifiability of each of the four unknowns. It is also worth mentioning that two cameras were used at the same time (one for each side). This was shown to suppress the effect of some heterogeneities of the strain distribution through the thickness. This phenomenon unavoidably occurs in the specimen because of its high aspect ratio that makes it difficult to load it exactly in-plane. The strain fields measured on both sides are linearly interpolated through the thickness and integrated over the volume, as required to calculate the coefficients of the sought parameters in the PVW, as explained above. This precaution was necessary to obtain satisfactory results in practice, as shown in [MOU 06]. This approach has also recently been applied to orthotropic supraconductive pancakes that are elements of a large MRI magnet [KIM 10].

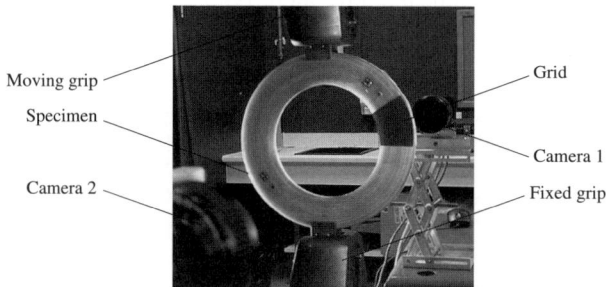

Figure 11.2. *Compressive test on a thick composite ring*

Some other geometries have also been tested. For instance, a T-shaped specimen under bending–tensile conditions was proposed in [GRÉ 98b]. Experimental validation was only moderately successful because of the quality of the measurements that were obtained at that time, but this configuration should certainly be reconsidered with more recent measurement tools, because it seemed to be promising. A three-point short beam bending test has also been investigated. It was observed, however, that it was not sufficiently heterogeneous to correctly provide the whole set of sought parameters [GRÉ 04]. Finally, a more long-term objective would be to design a novel test by using an optimization procedure, as suggested in section 11.3.7.3.

11.3.6.3. *Case of nonlinear response*

As mentioned above, it is possible to extend the method to any constitutive equations that linearly depend on its governing parameters. In the case of some long-fiber composites, it has been shown that the in-plane shear modulus can be expressed as a third-degree polynomial of the shear strain [BRO 00]:

$$\sigma_6 = Q_{66}\epsilon_6 - K\epsilon_6^3 \tag{11.21}$$

where K is the coefficient representing the reduction of the shear stiffness with the shear strain. Equation [11.9] becomes:

$$Q_{11}\int_S \varepsilon_1\varepsilon_1^* dS + Q_{22}\int_S \varepsilon_2\varepsilon_2^* dS + Q_{12}\int_S (\varepsilon_1\varepsilon_2^* + \varepsilon_2\varepsilon_1^*) dS$$
$$+ Q_{66}\int_S \varepsilon_6\varepsilon_6^* dS - K\int_S \varepsilon_6^3\varepsilon_6^* dS = \int_{L_f} \bar{\mathbf{T}}(M,\mathbf{n}).\mathbf{u}^*(M)dl \tag{11.22}$$

Therefore, it is possible to identify the four parameters driving the purely elastic response plus coefficient K with the approach described above in the purely elastic case. The same configuration as in Figure 11.1 has been used conclusively with simulated data in [GRÉ 01] and [CHA 04], with virtual fields being defined manually and with optimized virtual fields, respectively. The experimental validation performed in [CHA 06] has shown that it was, in fact, more suitable to introduce a threshold value corresponding to the onset of nonlinearity [LAD 92]:

$$\sigma_6 = Q_{66}\epsilon_6 - K\epsilon_6 < \epsilon_6 - \epsilon_6^0 >^+ \tag{11.23}$$

where $<>^+$ represents the positive part of $<>$, this quantity being null when $<>$ is negative. ε_6^0 is the threshold value. In this example, the initial stiffness values were obtained with special virtual fields constructed by subdomains and optimized with respect to sensitivity to noisy data. K and ε_6^0 were identified with virtual fields determined manually because the law is not linear with respect to ε_6^0. Results presented in [CHA 06] are very convincing, even if the method described in section 11.3.4 could have been used in this case, too, to find the virtual fields.

11.3.7. *Plate bending*

11.3.7.1. *Introduction*

From a historical point of view, the VFM was first applied to the case of plate bending [GRÉ 90, GRÉ 96a]. The calculations are very similar to the in-plane case.

A difference comes from the assumption made to transform the volume integrals involved in the PVW into surface integrals that involve only the measurements over one of the external surfaces of the plate specimen. Contrary to the previous cases where the through-thickness strain distribution is assumed to be constant, the classic Love–Kirkhhoff assumption [TIM 59] is used here. The parameters to be determined are then the bending stiffnesses.

11.3.7.2. *Principle*

The constitutive equations can be written as:

$$\begin{bmatrix} M_1 \\ M_2 \\ M_6 \end{bmatrix} = \begin{bmatrix} D_{11} & D_{12} & D_{16} \\ D_{12} & D_{22} & D_{26} \\ D_{16} & D_{26} & D_{66} \end{bmatrix} \begin{bmatrix} k_1 \\ k_2 \\ k_6 \end{bmatrix}$$ [11.24]

where D_{ij}s are the sought parameters, M_is are the generalized bending moments and k_is are the curvatures defined by:

$$k_1 = -\frac{\partial^2 u_3}{\partial x_1^2}; k_2 = -\frac{\partial^2 u_3}{\partial x_2^2}; k_6 = -2\frac{\partial^2 u_3}{\partial x_1 \partial x_2}$$ [11.25]

where u_3 represents the deflection of the plate. The PVW can be written as follows in this case:

$$\int_S D_{11} k_1 k_1^* dS + \int_S D_{22} k_2 k_2^* dS + \int_S D_{12} \left(k_1 k_2^* + k_2 k_1^* \right) dS + \int_S D_{66} k_6 k_6^* dS$$

$$+ \int_S D_{16} \left(k_1 k_6^* + k_6 k_1^* \right) dS + \int_S D_{26} \left(k_2 k_6^* + k_6 k_2^* \right) dS = W_e^*$$

[11.26]

Assuming the material is homogeneous, the D_{ij}s can be factorized in the integrals above and equations similar to those obtained in the in-plane case are obtained. Hence, the same formalism as above in terms of special virtual fields, virtual fields defined by subdomains or virtual fields that minimize the effect of noisy data can be applied in the case of plate bending. In [GRÉ 90] and [GRÉ 93], for instance, the virtual deflection field u_3^* is defined manually, whereas special virtual fields are used in [GRÉ 03]. In practice, the fields of actual curvatures are deduced from the fields of slopes obtained from measurements made by deflectometry [SUR 99] or using a double differentiation of the deflection field [BRU 05]. In [GRÉ 96a] and [GRÉ 99], the virtual fields are chosen manually and the invariants that drive the anisotropic linear elastic law are extracted directly. In [GRÉ 96b] and [GRÉ 98a], the load is no longer static, but dynamic. The initial terms (right-hand terms in equation [11.3]) must therefore be taken into account.

The examples shown above deal with homogeneous plates for which the D_{ij} parameters are the same over the whole surface of the specimen. More recently, the VFM has been applied to the case of impacted plates [KIM 09] for which the stiffnesses cannot be considered as constant over the whole surface of the plate. The idea was to describe their spatial evolution with a polynomial that affects the nominal stiffnesses using a multiplicative coefficient $k(x_1, x_2)$ lying between 0 and 1. The coefficients of this polynomial become the unknowns of the problem. It can easily be shown that, again, the formalism described above can also be applied to this type of problem. In [KIM 09], tests carried out on various types of plates exhibiting localized and controlled defects (missing plies introduced during manufacturing) showed that it is possible to localize defects from measurements made by deflectometry and to quantify the local stiffness reduction by comparing the identified values with the values deduced from classic lamination theory. As an example, Figure 11.3 shows the distribution of $k(x_1, x_2)$ in the case of an impacted plate subjected to a bending test (supported on three corners and loaded at its center). To conclude, it must be pointed out that in addition to the ability to localize the impact precisely, the quantification of its effect in terms of local stiffness reduction is also possible with this approach. We can therefore show that it is possible to go beyond the classic use of full-field measurements in non-destructive testing, which only provides the location of the damage.

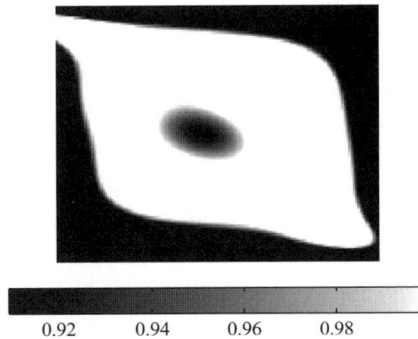

Figure 11.3. *Characterization of an impact on a composite plate: localization and distribution of coefficient $k(x, y)$ [KIM 09]*

11.3.7.3. *Optimization of test configuration*

The VFM can be applied to specimens of any shape. In addition, sensitivity to noise can be estimated *a priori* for any testing configuration, thus enabling the comparison of different specimen shapes or loading conditions, for instance. Because this sensitivity directly reflects the quality of identification, it constitutes a valuable indicator to compare various testing configurations and clearly paves the way for specimen measurement or loading condition optimization. This issue has been addressed in [SYE 07] concerning plate bending. Before optimizing the test, it is necessary to normalize this sensitivity because it depends on some of the test parameters: the stiffness itself (the higher the stiffness, the lower the sensitivity), the number of measurement points (the higher the number of measurement points, the better the sensitivity), the amplitude of the load or any quantity that is proportional to it, such as the deflection for a bent plate or the equivalent Von Mises strain (the greater the signal to be processed, the lower the noise for a given measurement resolution). Applying this normalization leads to a quantification that provides an objective comparison of various test configurations for a given material, and thus to an optimization of the test configuration using a finite element program, for instance. In practice, due to such calculations, we can make the best possible adjustment of parameters, such as the specimen measurement, the location of the load, the nature of the displacement boundary conditions or the orientation of the specimen in the testing device if the material is anisotropic. Concerning plate bending, the most influential parameters are the location of the load and the supports, the shape of the specimen and its orientation, albeit to a lesser extent. In [SYE 08] and [KIM 08], this type of approach is used to optimize the location of the load, using normalization with respect to the maximum deflection and to the average equivalent Von Mises strain, respectively. It seems that the results obtained in the second case are better.

Finally, the quality of the identification depends on various parameters related to the measurement techniques used. To better investigate this aspect, the whole measurement and identification chain was simulated using FEM and generating synthetic images that reproduce a real acquisition with a CCD camera. Details are given in [ROS 12b]. A full-field measurement technique was applied to the synthetic images to extract the strain fields and the VFM was used to identify the material properties. Looking at the identification error, it was shown that the VFM is very sensitive to the spatial resolution of the measurements. Indeed, it is advisable to use specimen shapes that approach the aspect ratio of the CCD chip to increase this resolution. Moreover, for the analyzed Iosipescu test, missing data close to the edges of the specimen highly deteriorate the quality of the identification. Noise and smoothing also play an important role. The VFM itself introduces a smoothing effect because the measured strain components are multiplied by the virtual strain components and integrated over the surface. Therefore, it can be demonstrated that when the level of noise is low, performing smoothing on the measured displacement data before applying the VFM increases the identification error. In fact, the perturbation error due to the noise decreases, but the approximation error increases because the smoothing acts as a low-pass filter.

11.3.7.4. *Measuring damping using dynamic tests*

The approach presented above can be extended to viscoelasticity. The objective is to measure the damping parameters gathered here in a matrix $\mathbf{C}$ in addition to the stiffnesses gathered in $\mathbf{Q}$. $\mathbf{C}$ is defined by the following equation:

$$\sigma = \mathbf{Q}\varepsilon + \mathbf{C}\dot{\varepsilon} \qquad [11.27]$$

where $\dot{\varepsilon}$ denotes the time derivative of ε. These damping parameters were measured in the case of isotropic plate bending in [GIR 05] and [GIR 10], subjected to harmonic loading by extending an approach developed in [GRÉ 96b] and [GRÉ 98a] in which bending stiffnesses of vibrating composite plates were measured. In [GIR 05], complex virtual fields were used but it has been shown in [GIR 10] that such complex virtual fields were not, in fact, necessary. An important remark is that these damping parameters were measured with tests performed out of resonance, which is unusual in dynamic testing. In [GIR 10], it is shown that in the case of isotropic plate bending, writing the PVW with two different virtual deflection fields denoted by $u_3^{*(1)}$ and $u_3^{*(2)}$ leads to:

$$
\begin{bmatrix}
A^{t(1)} & B^{t(1)} & \omega A^{t+\frac{T}{4}(1)} & \omega B^{t+\frac{T}{4}(1)} \\
A^{t+\frac{T}{4}(1)} & B^{t+\frac{T}{4}(1)} & \omega A^{t(1)} & \omega B^{t(1)} \\
A^{t(2)} & B^{t(2)} & \omega A^{t+\frac{T}{4}(2)} & \omega B^{t+\frac{T}{4}(2)} \\
A^{t+\frac{T}{4}(2)} & B^{t+\frac{T}{4}(2)} & \omega A^{t(2)} & \omega B^{t(2)}
\end{bmatrix}
\begin{Bmatrix}
Q_{11} \\
Q_{12} \\
C_{11} \\
C_{12}
\end{Bmatrix}
$$

$$
= \frac{12\rho\omega^2}{h^2}
\begin{Bmatrix}
\displaystyle\int_S u_3^{*(1)} u_3^t \, dS \\[2mm]
\displaystyle\int_S u_3^{*(1)} u_3^{t+\frac{T}{4}} \, dS \\[2mm]
\displaystyle\int_S u_3^{*(2)} u_3^t \, dS \\[2mm]
\displaystyle\int_S u_3^{*(2)} u_3^{t+\frac{T}{4}} \, dS
\end{Bmatrix} \qquad [11.28]
$$

where

$$
\begin{cases}
A^{t(n)} = \displaystyle\int_S \left(k_1^t k_1^{*(n)} + k_2^t k_2^{*(n)} + \frac{1}{2} k_6^t k_6^{*(n)} \right) \, dS \\[3mm]
A^{t+\frac{T}{4}(n)} = \displaystyle\int_S \left(k_1^{t+\frac{T}{4}} k_1^{*(n)} + k_2^{t+\frac{T}{4}} k_2^{*(n)} + \frac{1}{2} k_6^t k_6^{*(n)} \right) \, dS \\[3mm]
B^{t(n)} = \displaystyle\int_S \left(k_1^t k_1^{*(n)} + k_2^t k_2^{*(n)} - \frac{1}{2} k_6^t k_6^{*(n)} \right) \, dS \\[3mm]
B^{t+\frac{T}{4}(n)} = \displaystyle\int_S \left(k_1^{t+\frac{T}{4}} k_1^{*(n)} + k_2^{t+\frac{T}{4}} k_2^{*(n)} - \frac{1}{2} k_6^{t+\frac{T}{4}} k_6^{*(n)} \right) \, dS
\end{cases} \qquad [11.29]
$$

where n is equal to 1 for the first virtual field and 2 for the second virtual field. The k_is and the k_i^*s, $i = 1, 2, 6$, are the actual and virtual curvatures, respectively.

ρ is the mass per unit volume, h is the thickness of the plate and ω is defined by $2\pi f$, f being the loading frequency. Symbols X^t, where X is either one of the curvatures or the deflection, means that X are measured at a certain time t and $t+T/4$, which is quarter of a period later. Q_{11} and Q_{12} are the independent unknowns in this system, the shear stiffness Q_{66} being directly deduced from them. This linear system is obtained assuming the virtual deflection fields $u_3^{*(1)}$ and $u_3^{*(2)}$ are such that they are null where the load is applied. The main reason for this is that with the internal virtual work now being balanced by the virtual work of acceleration quantities, the contribution of the external virtual work is no longer necessary, so the load becomes only an inertial one. This is a significant advantage as force distributions in clamps or other connections applied to the plate are usually unknown. In vibration, such contact zones usually dissipate significant amounts of energy, which spoils material damping measurements with the usual procedures. The present technique is totally insensitive to such effects, as demonstrated in [GIR 10]. The canceling out of the external forces in the equations is obtained merely by choosing a virtual deflection field that is null over certain surfaces that include the physical connections between the device and the plate specimen. In this context, using piecewise virtual fields makes things easier by imposing a null virtual displacement on the virtual elements that include the connections. This freedom in the choice of the virtual field, which behaves here as a type of filter for the measurements, is one of the strengths of the VFM. Solving the linear system is then straightforward.

It can be observed that if the identification is performed within the framework of elasticity only [SUR 99, GRÉ 99], the C_i parameters are assumed to be null. Measurements are performed for a given value of time: $t+T/4$, with $t=0$ to measure the amplitude of the movement, and for a load frequency equal to one of the natural frequencies of the specimen. The error introduced, if damping is not really negligible, can easily be assessed by considering the first and third equations of system [11.28] [GIR 10], in which the last two columns have now disappeared, the corresponding terms being subtracted from the right-hand side of the initial system. Thus:

$$\begin{bmatrix} A^{t+\frac{T}{4}(1)} & B^{t+\frac{T}{4}(1)} \\ A^{t+\frac{T}{4}(2)} & B^{t+\frac{T}{4}(2)} \end{bmatrix} \begin{Bmatrix} Q_{11} \\ Q_{12} \end{Bmatrix}$$

$$= \frac{12\rho\omega^2}{h^2} \begin{Bmatrix} \int\int_S w^{*(1)} w^{t+\frac{T}{4}} \, dS - \omega A^{t(1)} C_{11} - \omega B^{t(1)} C_{22} \\ \int_S w^{*(2)} w^{t+\frac{T}{4}} \, dS - \omega A^{t(2)} C_{11} - \omega B^{t(2)} C_{22} \end{Bmatrix} \qquad [11.30]$$

The quantities $\omega A^{t(i)} C_{11} + \omega B^{t(i)} C_{22}$, $i = 1, 2$, are hidden for the user of the VFM who does not take damping into account. The parasitic effect of damping can easily be assessed with the system above [GIR 10].

Finally, performing tests in which only an inertial load is applied has recently shown to be promising in the case of impact, for which a high-speed camera must be used to obtain the actual displacement/strain fields [PIE 12].

11.3.8. *Large deformations: example of hyperelasticity*

11.3.8.1. *Introduction*

All the previous examples deal with small strains. This assumption is, however, not obligatory to apply the basic principles of the VFM: writing the PVW, introducing the constitutive equations and choosing particular virtual fields can be done for any strain amplitude as long as the theoretical framework in terms of strain tensor, stress tensor and nature of the constitutive equations is relevant. If the constitutive equation linearly depends on the constitutive parameters, the PVW provides a linear equation for these parameters, even though the strains are large and modeled with a nonlinear geometrical model. This case is presented in this section. If the constitutive equations do not linearly depend on the constitutive parameters, the VFM can still be used, but by applying the procedure described in section 11.4.

It is important to mention that in the case of large strains, the PVW can be written using three different forms: purely Eulerian quantities, purely Lagrangian quantities and mixed quantities [HOL 00a, PIE 12]. These three forms are equivalent, because any quantity can be expressed with respect to the others, but one of them is generally easier to use depending on the type of experimental data available (expressed with respect to the reference or the current configuration) and on the nature of the coordinates with which the constitutive equations are given. An example is briefly described below. Concerning the notations, uppercase letters refer to quantities defined with respect to the initial configuration whereas lowercase letters refer to quantities defined with respect to the current configuration.

11.3.8.2. *Identification of the Mooney law for elastomeric materials*

The first case deals with the characterization of elastomeric materials. The PVW expressed with mixed quantities is chosen here. This choice is justified by the fact that in the first study to be concerned with these theoretical developments [PRO 09], the displacement fields from which the displacement gradient components were deduced were available in the reference configuration, whereas the force was measured in the current configuration. Moreover, the finite element program that was used to simulate the procedure before applying it to real experimental data provided the Cauchy stresses. These latter quantities had therefore to be transformed into first Piola–Kirchhoff stresses. Assuming a particular case of in-plane stresses and that no volume force is applied, the PVW can be written as follows in statics:

$$\int_{S_0} \mathbf{\Pi} : \frac{\partial \mathbf{U}^*}{\partial X} \, dS = \int_{L_f} (\mathbf{\Pi}.\mathbf{n}).\mathbf{U}^* \, dL \quad \forall \, \mathbf{U}^* \text{KA} \qquad [11.31]$$

where S_0 is the surface of the specimen in the initial configuration, $\mathbf{\Pi}$ is the first Piola–Kirchhoff stress tensor, $\mathbf{U}^*$ is a KA virtual displacement field written with Lagrangian coordinates and $\mathbf{n}$ is the outward normal unit vector along which the load is applied. The Lagrangian coordinates are denoted by $X_i, i = 1, 2$.

As in the case of small strains, the VFM consists of writing the stress components as a function of the strain or displacement gradient components by introducing the constitutive equations. A set of particular KA virtual displacement fields is then chosen. The resulting equations lead to a linear system if the constitutive equations are linear with respect to the constitutive parameters. This case is examined here for the sake of simplicity.

Assuming the material is incompressible, the Cauchy stress tensor σ can be written as follows:

$$\sigma = -p\mathbf{I} + 2\mathbf{B}\frac{\partial W}{\partial \mathbf{B}} \qquad [11.32]$$

where W is the strain energy density, $\mathbf{B}$ is the left Cauchy–Green tensor defined by $\mathbf{B} = \mathbf{FF}^t$, where $\mathbf{F}$ is the deformation gradient tensor, and p is an arbitrary hydrostatic pressure that ensures the incompressibility of the material [TRU 65]. The Mooney law is considered here [MOO 40] with two constitutive parameters only to simplify the equations, but considering more parameters would lead to similar equations. The strain energy density can be written as:

$$W = C_1(I_1 - 3) + C_2(I_2 - 3) \qquad [11.33]$$

where I_1 and I_2 are, respectively, the first and second invariants of the left and right Cauchy–Green tensors $\mathbf{B}$ and $\mathbf{C}$. The first Piola–Kirchhoff stress tensor $\mathbf{\Pi}$ can be deduced from the Cauchy stress tensor σ using the following equation:

$$\mathbf{\Pi} = J\sigma\mathbf{F}^{-t} \qquad [11.34]$$

where $\det J = 1$ if the material is assumed to be incompressible, J being the Jacobian of the transformation. Combining the equations above leads to the following linear equation for the sought parameters C_1 and C_2:

$$C_1 \int_{S_0} \Theta : \frac{\partial \mathbf{U}^*}{\partial \mathbf{X}} dS + C_2 \int_{S_0} \Lambda : \frac{\partial \mathbf{U}^*}{\partial \mathbf{X}} dS = \int_{L_f} (\mathbf{\Pi}.\mathbf{n}).\mathbf{U}^* dL \qquad [11.35]$$

The detailed expressions of the Θ and Λ coefficients are given in [PRO 09]. It is observed that these coefficients are much more complex in the case of large deformations than in the case of small deformations. Hence, the best determination of the virtual displacement fields cannot be carried out using the strategy described in section 11.3.4. A simpler heuristic approach was used in [PRO 09]. Virtual fields defined by subdomains were first randomly created. These virtual fields were then paired to build up a linear system and the condition number of the matrix of each linear system was determined. The virtual fields leading to the highest condition number were chosen to perform the identification. Various simulations and results obtained with a cross-shaped specimen are shown in [PRO 09]. Another test giving rise to heterogeneous strain distributions but performed with a classic tensile machine is described in [GUÉ 09]. The C_1 and C_2 coefficients that are deduced in both cases are similar to those obtained with classic tests for the same material, thus validating the approach.

11.4. Case of constitutive equations that do not linearly depend on the constitutive parameters

11.4.1. *Introduction*

In the case of constitutive equations, which do not linearly depend on the parameters, it is no longer possible to obtain a linear system where these parameters are unknowns. Identification is carried out by minimizing the squared difference between the integrals corresponding to the internal and external virtual works in the PVW. This minimization is performed with respect to the sought parameters, in general for different load levels and with different virtual fields to enrich the cost function. For the sake of simplicity, only the main features of the method and some references are given hereafter in the cases of elastoplasticity and hyperelastic behavior.

11.4.2. *Elastoplasticity*

11.4.2.1. *Minimization of a cost function*

In elastoplasticity, equation [11.2] can be written as:

$$\sigma(\mathbf{X}, t) = Q[\varepsilon(t) - \varepsilon^p(\mathbf{X}, t)] \qquad [11.36]$$

where $\mathbf{X}$ is a vector containing the sought parameters, ε^p is the plastic strain tensor that depends on the parameters, $\varepsilon(t)$ is the total strain tensor measured at time t and Q is the stiffness tensor. An iterative procedure is necessary to identify $\mathbf{X}$ with the VFM because the stress tensor σ implicitly depends on the sought parameters through the flow rule. Hence, the test is divided into m loading stages with corresponding times denoted t_k, $k = 1, \ldots, m$, for which the total strain field is measured. A cost function to be minimized is defined using the following procedure:

− For each load step k, the plastic strain tensor $\varepsilon^p(X, t_k)$ is deduced from the strain calculated at the previous stage, from the total strain increment at time t_k and from the flow rule [GRÉ 06, SUT 96].

– The difference between the quantities involved in the PVW, denoted by $c_k(\mathbf{X})$, is written at stage k using a virtual field denoted by $\mathbf{u}^{*k}$, the measured total strain components and the sought parameters gathered in $\mathbf{X}$. For a quasi-static test, this quantity is equal to:

$$c_k(\mathbf{X}) = -Q_{ij} \int_S [\, \varepsilon_i(t_k) - \varepsilon_i^p(\mathbf{X}, t_k) \,] \varepsilon_j^{*k} dS + \int_{L_f} \bar{\mathbf{T}}(t_k).\mathbf{u}^{*k} dl \qquad [11.37]$$

– The $c_k(\mathbf{X})$ defined above are squared for all values of k stages, thus leading to the following cost function:

$$\mathcal{C}(\mathbf{X}) = \sum_{k=1}^{m} [c_k(\mathbf{X})]^2$$

$$= \sum_{k=1}^{m} \left[-Q_{ij} \int_S [\, \varepsilon_i(t_k) - \varepsilon_i^p(X, t_k) \,] : \varepsilon_j^{*k} dS + \int_{L_f} \bar{\mathbf{T}}(t_k).\mathbf{u}^{*k} dl \right]^2$$

$$[11.38]$$

It is worth mentioning that only one virtual field is used for each load step in the preceding expression. It is, however, possible to use more than one virtual field in order to enrich the cost function. The sought parameters gathered in $\mathbf{X}$ are identified by minimizing $\mathcal{C}(\mathbf{X})$. The calculation time is longer here compared to the linear case because of the iterative calculations performed to obtain the minimum for the cost function $\mathcal{C}(\mathbf{X})$. However, it is much lower than for the finite element model updating technique, as no direct problem has to be solved to evaluate the cost function.

11.4.2.2. Choice of virtual fields

During the first attempt at using the VFM in the case of constitutive equations that do not linearly depend on their constitutive parameters, the virtual fields were chosen manually, the main rule being to involve in the PVW the largest stress component present in the specimen and the loading force. A more systematic route, similar to that described in section 11.3.4 in the linear case, has been proposed in [PIE 10]. It has been shown that the effect of noise could be written as follows:

$$\mathcal{V}(c_k) = \mathbf{U}^{*[k]t} \left[\sum_{l=1}^{k} \mathbf{H}(t_l, \mathbf{X}) \mathbf{H}(t_l, \mathbf{X}) \right] \mathbf{U}^{*[k]} \qquad [11.39]$$

where $\mathbf{H}(t_l, \mathbf{X})$ is the tangent stiffness matrix at time t_l and $\mathbf{U}^{*[k]}$ is the vector containing the virtual nodal degrees of freedom used to define the sought optimized virtual fields. Concerning the minimization of the quantity defined in equation [11.39], it was confirmed in [AVR 08a] that the identification procedure is more robust using virtual fields obtained with this approach.

11.4.2.3. *Large strains*

In the case of large strains, the elastic part of the strain tensor can be neglected and the total strain tensor can be used directly to calculate the deviatoric stress tensor without any iterations. Because of this simplification, the VFM can be used to identify the parameters of complex plasticity models in a very fast way compared with inverse methodologies that involve the FEM. The same considerations reported in section 11.3.8 with respect to the stress and strain tensors to be used in the virtual work computation remain valid.

11.4.2.4. *Applications*

This version of the VFM has been applied to experimental data in various situations. These studies are described in detail in the following papers:

– Identification of the parameters governing the linear isotropic hardening of mild steel [PAN 06, AVR 08a]. In these studies, tensile tests were performed on specimens with two symmetrical notches. The displacement fields were measured using the grid method and strain fields were deduced by numerical differentiation. Six specimens were characterized. The parameters measured were those driving a linear isotropic model. These values are consistent with those available in the literature for such materials. Calculations carried out with the identified constitutive equations enable the reconstruction of the stress field in the specimens and the analysis of the plastic flow [AVR 08a].

– Identification of the parameters governing the isotropic and kinematic hardening of 316L steel using tensile/compressive tests [PIE 10]. Similar specimens as above were tested. The displacement fields were measured using the grid method. First, a law with six parameters (two elastic parameters plus four parameters governing the exponential isotropic hardening of the Voce model) describing the response of the specimens submitted to monotonic tensile tests was identified without using the unloading and compressive phases of the tests. The latter two phases were then considered and the five parameters governing a kinematic hardening law were identified. For these two identifications, the parameters found were in good agreement with those obtained with the usual tests [PIE 10].

– Identification of the elastoviscoplastic response of mild steel with high-speed tests [AVR 08b]. A tensile specimen with gradual cross-sectional variation was considered here. The minimal strain rate of the tests was 1 s^{-1}. The displacement fields were measured by digital image correlation (DIC). The number of frames per second was 4,796. The strain fields deduced from the measured displacement fields clearly showed that local strain rate heterogeneities were occurring, with variations lying between 0.1 and 10 s^{-1} in the same specimen. These heterogeneities are sufficient to identify simultaneously all the parameters governing the response of the material. The Perzyna model was used first to model the elastoviscoplastic response [PER 63]. However, this model did not predict the sudden and systematic drop in the loading force measured by the load cell. This phenomenon is due to a Lüders-type phenomenon. The Perzyna model was, therefore, modified to account for this and it led to a correct identification of the parameters [AVR 08b].

– Identification of the heterogeneous isotropic hardening of a welded joint. Displacement fields were measured by DIC on the heat-affected zone (HAZ) of a C100 steel welded joint [SUT 08]. Because of the strain fields, localizations were observed in this zone, thus rendering the identification problem quasi-unsolvable with classic tests and their associated identification procedures. In the current case, seven zones were defined and seven sets of parameters (yield stress and hardening parameter) were therefore identified (one set per zone) [SUT 08]. Consequently, this method enabled the simultaneous characterization of the base material, the HAZ, the joint itself and the transition zones located between them. In this case too, the results obtained are consistent with those obtained with indentation tests [SUT 08].

– The case of large deformations. This case is detailed in [ROS 12a], where the method was also extended to 3D displacement fields. It is also shown that when an anisotropic plasticity model is used (i.e. Hill48), the deviatoric stress tensor must be calculated in the local material coordinate system. Indeed, at large deformations the rotation is not negligible, and locally the orientation of the material texture may not coincide with the global coordinate system.

11.4.3. *Hyperelastic behavior*

The second case deals with anisotropic hyperelastic behavior. An example dealing with the characterization of the tissue of human artery walls is proposed to briefly illustrate the approach.

Different strain energy functions for the anisotropic hyperelastic behavior of arteries may be found in the literature. Some of them are phenomenological and others are based on the microstructure. The Holzapfel model [HOL 00b], which is of the latter type, is chosen here. This model assigns separate strain energy functions to the media and adventitia layers. The simplest form of the strain energy function may be written as:

$$\Phi = \frac{c}{2}(I_1-1)+\frac{k_1}{2k_2}[\exp(k_2(I_4-1)^2)-1]+\frac{k_1}{2k_2}[\exp(k_2(I_6-1)^2)-1] \quad [11.40]$$

where c and k_1 are stress-like material parameter, and k_2 is a dimensionless material parameter, I_4 and I_6 are the squares of the stretch component in the two directions of symmetrical tissue fibers. Hence, this model prescribes a fiber orientation angle ϕ for each layer, usually based on micrographic analysis. The fiber angles for the specimen used in this study were not known, and so the fiber angles are another material parameter to be determined from the experimental data.

The principle for identifying these parameters with the VFM may be summarized in two steps:

1) computing the Cauchy stress fields from equation [11.40] with the measured values of I_1, I_4 and I_6 and an estimation for the material parameters;

2) finding the material parameters that cause the Cauchy stress fields to be at equilibrium in the deformed configuration.

For the second point, as the internal virtual work cannot be expressed linearly with regard to the unknown material parameters, a cost function is defined as the quadratic gap between the internal virtual work and the external virtual work. This cost function is minimized by the simplex method. The chosen virtual fields and other details about the experiments can be found in [AVR 10] and [KIM 12]. This approach, based on the minimization of a certain cost function, is also used in the case of elastoplasticity that has been described in more detail in the previous section.

11.5. Conclusion

The main features of the VFM are described in this chapter. Concerning the theoretical aspects, particular emphasis was placed, for the sake of simplicity, on the case for which the constitutive equations linearly depend on the sought parameters. Because this case is the simplest case, a suitable strategy has been progressively established during the last decade to determine optimized virtual fields, these fields being defined as those leading to the most robust identification. This type of approach has been more recently extended to the case of constitutive equations that do not linearly depend on their governing parameters. Examples of cases in which this method has been applied illustrate its feasibility.

In general, it must be pointed out that the VFM is simple because its core principle is merely to apply the PVW with particular virtual fields. This simplicity makes the procedure very general and therefore applicable to many cases. It is also important to mention that relevant tools related to this method, such as the prediction of the sensitivity to noise of the sought parameters, are available in the linear case, thus enabling a straightforward comparison between different test configurations. In the same spirit, the versatility of the VFM has also led to distinguishing and quantifying the main sources of error in the identification procedure from full-field measurements. These aspects contribute to create a favorable framework from which various tests, optimized for identification purposes, will certainly be defined in the near future. Some further developments go beyond the identification technique itself, because they are directly related to the cases for which the VFM is used (or will be used in the near future), such as high-speed dynamic tests for which purely inertial loads are used or the identification of parameters from 3D measurements in the bulk with suitable measurement techniques.

11.6. Bibliography

[AVR 04] AVRIL S., GRÉDIAC M., PIERRON F., "Sensitivity of the virtual fields method to noisy data", *Computational Mechanics*, vol. 34, pp. 439–452, 2004.

[AVR 07] AVRIL S., PIERRON F., "General framework for the identification of constitutive parameters from full-field measurements in linear elasticity", *International Journal of Solids and Structures*, vol. 44, nos. 14–15, pp. 4978–5002, 2007.

[AVR 08a] AVRIL S., PIERRON F., PANNIER Y., ROTINAT R., "Stress reconstruction and constitutive parameter identification in plane-stress elastoplastic problems using surface measurements of deformation fields", *Experimental Mechanics*, vol. 48, no. 5, pp. 403–420, 2008.

[AVR 08b] AVRIL S., PIERRON F., YAN J., SUTTON M., "Identification of viscoplastic parameters and characterization of Lüders behavior using digital image correlation and the virtual fields method", *Mechanics of Materials*, vol. 40, no. 9, pp. 729–742, 2008.

[AVR 10] AVRIL S., BADEL P., DUPREY A., "Anisotropic and hyperelastic identification of in vitro human arteries from full-field measurements", *Journal of Biomechanics*, vol. 43, pp. 2978–2985, 2010.

[BRO 00] BROUGHTON W., "Shear", *Mechanical Testing of Advanced Fibre Composites*, Woodhead Publishing Limited, Cambridge, pp. 100–123, 2000.

[BRU 05] BRUNO L., POGGIALINI A., "Elastic characterization of anisotropic materials by speckle interferometry", *Journal of Applied Mechanics*, vol. 45, no. 3, pp. 205–212, 2005.

[CHA 04] CHALAL H., MERAGHNI F., PIERRON F., GRÉDIAC M., "Direct identification of the damage behaviour of composite materials using the virtual fields method", *Composites Part A*, vol. 35, pp. 841–848, 2004.

[CHA 06] CHALAL H., AVRIL S., PIERRON F., MERAGHNI F., "Experimental identification of a damage model for composites using the grid technique coupled to the virtual fields method", *Composites Part A: Applied Science and Manufacturing*, vol. 37, no. 2, pp. 315–325, 2006.

[CLA 02] CLAIRE D., HILD F., ROUX S., "Identification of damage fields using kinematic measurements", *Comptes Rendus de l'Académie des Sciences*, pp. 729–734, 2002.

[DYM 73] DYM C.L., SHAMES I.H., *Solid Mechanics: A Variational Approach*, McGraw-Hill, New York, 1973.

[GEY 02] GEYMONAT G., HILD F., PAGANO S., "Identification of elastic parameters by displacement field measurement", *Comptes Rendus de l'Académie des Sciences*, pp. 403–408, 2002.

[GIR 05] GIRAUDEAU A., PIERRON F., "Identification of stiffness and damping properties of thin isotropic vibrating plates using the virtual fields method. Theory and simulations", *Journal of Sound and Vibration*, vol. 284, nos. 3–5, pp. 757–781, 2005.

[GIR 10] GIRAUDEAU A., PIERRON F., GUO B., "An alternative to modal analysis for material stiffness 411 and damping identification from vibrating plates", *Journal of Sound and Vibrations*, vol. 329, no. 10, pp. 1653–1672, 2010.

[GRÉ 89]   GRÉDIAC M., "Principe des travaux virtuels et identification", *Comptes Rendus de l'Académie des Sciences*, pp. 1–5, 1989.

[GRÉ 90]   GRÉDIAC M., VAUTRIN A., "A new method for determination of bending rigidities of thin anisotropic plates", *Journal of Applied Mechanics*, (*Transactions of the American Society of Mechanical Engineers*), vol. 57, pp. 964–968, 1990.

[GRÉ 93]   GRÉDIAC M., VAUTRIN A., "Mechanical characterization of anisotropic plates: experiments and results", *European Journal of Mechanics/A Solids*, vol. 12, no. 6, pp. 819–838, 1993.

[GRÉ 96a]   GRÉDIAC M., "On the direct determination of invariant parameters governing the bending of anisotropic plates", *International Journal of Solids and Structures*, vol. 33, no. 27, pp. 3969–3982, 1996.

[GRÉ 96b]   GRÉDIAC M., PARIS P.-A., "Direct identification of elastic constants of anisotropic plates by modal analysis: theoretical and numerical aspects", *Journal of Sound and Vibration*, vol. 195, no. 3, pp. 401–415, 1996.

[GRÉ 98a]   GRÉDIAC M., FOURNIER N., PARIS P.-A., SURREL Y., "Direct identification of elastic constants of anisotropic plates by modal analysis: experiments and results", *Journal of Sound and Vibration*, vol. 210, no. 5, pp. 645–659, 1998.

[GRÉ 98b]   GRÉDIAC M., PIERRON F., "A T-shaped specimen for the direct characterization of orthotropic materials", *International Journal for Numerical Methods in Engineering*, vol. 41, pp. 293–309, 1998.

[GRÉ 99]   GRÉDIAC M., FOURNIER N., PARIS P.-A., SURREL Y., "Direct measurement of invariant parameters of composite plates", *Journal of Composite Materials*, vol. 33, no. 20, pp. 1939–1965, 1999.

[GRÉ 01]   GRÉDIAC M., AUSLENDER F., PIERRON F., "Using the virtual fields method to identify the through-thickness moduli of thick composites with a nonlinear shear response", *Composites Part A: Applied Science and Manufacturing*, vol. 32, no. 12, pp. 1713–1725, 2001.

[GRÉ 02a]   GRÉDIAC M., TOUSSAINT E., PIERRON F., "Special virtual fields for the direct determination of material parameters with the virtual fields method. 1: Principle and definition", *International Journal of Solids and Structures*, vol. 39, pp. 2691–2705, 2002.

[GRÉ 02b]   GRÉDIAC M., TOUSSAINT E., PIERRON F., "Special virtual fields for the direct determination of material parameters with the virtual fields method. 2: Application to in-plane properties", *International Journal of Solids and Structures*, vol. 39, pp. 2707–2730, 2002.

[GRÉ 03]   GRÉDIAC M., TOUSSAINT E., PIERRON F., "Special virtual fields for the direct determination of material parameters with the virtual fields method. 3: Application to the bending rigidities of anisotropic plates", *International Journal of Solids and Structures*, vol. 40, pp. 2401–2419, 2003.

[GRÉ 04]   GRÉDIAC M., PIERRON F., "Numerical issues in the virtual fields method", *International Journal for Numerical Methods in Engineering*, vol. 59, no. 10, pp. 1287–1312, 2004.

[GRÉ 06] GRÉDIAC M., PIERRON F., "Applying the virtual fields method to the identification of plastic constitutive equations", *International Journal of Plasticity*, vol. 26, no. 4, pp. 602–627, 2006.

[GUÉ 09] GUÉLON T., TOUSSAINT E., CAM J.-B.L., PROMMA N., GRÉDIAC M., "A new characterization method for rubber", *Polymer Testing*, vol. 28, no. 7, pp. 715–723, 2009.

[HEN 91] HENDRICKS M.A.N., Identification of the mechanical properties of solid materials, Thesis, Eindhoven University of Technology, 1991.

[HOL 00a] HOLZAPFEL G.A., *Nonlinear Solid Mechanics: A Continuum Approach for Engineering*, John Wiley & Sons, New York, 2000.

[HOL 00b] HOLZAPFEL G., GASSER T., OGDEN R., "A new constitutive framework for arterial wall mechanics and a comparative study of material models", *Journal of Elasticity*, vol. 61, pp. 1–48, 2000.

[KIM 08] KIM J.-H., Identification de cartes d'endommagement de plaques composites impactées par la méthode des champs virtuels, Thesis, Ecole Nationale Supérieure d'Arts et Métiers, 2008.

[KIM 09] KIM J.-H., PIERRON F., WISNOM F., AVRIL S., "Local stiffness reduction in impacted composite plates from full-field measurements", *Composites Part A*, vol. 40, no. 12, pp. 1961–1974, 2009.

[KIM 10] KIM J.-H., NUNIO F., PIERRON F., VÉDRINE P., "Identification of the mechanical properties of 224 superconducting windings using the virtual fields method", *IEEE Transactions on Applied Superconductivity*, vol. 20, no. 3, pp. 1993–1997, 2010.

[KIM 12] KIM J.-H., AVRIL S., DUPREY A., FAVRE J.-P., "Experimental characterization of rupture in human aortic aneurysms using full-field measurement technique", *Biomechanics and Modeling in Mechanobiology*, vol. 11, no. 6, pp. 841–853, 2012.

[LAD 92] LADEVÈZE P., LEDANTEC E., "Damage modeling of the elementary ply for laminated composites", *Composites Science and Technology*, vol. 43, no. 3, pp. 257–267, 1992.

[MOO 40] MOONEY M., "A theory of large elastic deformation", *Journal of Applied Physics*, vol. 11, pp. 582–592, 1940.

[MOU 06] MOULART R., AVRIL S., PIERRON F., "Identification of the through-thickness rigidities of a thick laminated composite tube", *Composites Part A: Applied Science and Manufacturing*, vol. 59, no. 10, pp. 1287–1312, 2006.

[PAN 06] PANNIER Y., AVRIL S., ROTINAT R., PIERRON F., "Identification of elasto-plastic constitutive parameters from statically undetermined tests using the virtual fields method", *Experimental Mechanics*, vol. 46, no. 6, pp. 735–755, 2006.

[PER 63] PERZYNA P., "The constitutive equations for rate-sensitive plastic materials", *Quarterly of Applied Mathematics*, vol. 20, pp. 321–332, 1963.

[PIE 00] PIERRON F., GRÉDIAC M., "Identification of the through-thickness moduli of thick composites from whole-field measurements using the Iosipescu fixture", *Composites Part A: Applied Science and Manufacturing*, vol. 31, no. 4, pp. 309–318, 2000.

[PIE 10] PIERRON F., AVRIL S., TRAN V., "Extension of the virtual fields method to elasto-plastic material identification with cyclic loads and kinematic hardening", *International Journal of Solids and Structures*, vol. 96, no. 47, pp. 2993–3010, 2010.

[PIE 12] PIERRON F., GRÉDIAC M., *The Virtual Fields Method-Extracting Constitutive Mechanical Parameters from Full-field Deformation Measurements*, Springer, New York, 2012.

[PRO 09] PROMMA N., RAKA B., GRÉDIAC M., TOUSSAINT E., LECAM J. B., BALANDRAUD X., HILD F., "Application of the virtual fields method to mechanical characterization of elastomeric materials", *International Journal of Solids and Structures*, vol. 46, pp. 698–715, 2009.

[ROS 12a] ROSSI M., PIERRON F., "Identification of plastic constitutive parameters at large deformations from three dimensional displacement fields", *Computational Mechanics*, vol. 49, pp. 53–71, 2012.

[ROS 12b] ROSSI M., PIERRON F., "On the use of simulated experiments in designing tests for material characterization from full-field measurements", *International Journal of Solids and Structures*, vol. 49, pp. 420–435, 2012.

[SOO 93] SOONG T., GRIGORIU M., *Random Vibration of Mechanical and Structural Systems*, PTR Prentice Hall, Inc., Englewood Cliffs, NJ, 1993.

[SUR 99] SURREL Y., FOURNIER N., GRÉDIAC M., PARIS P.-A., "Phase-stepped deflectometry applied to shape measurement of bent plates", *Experimental Mechanics*, vol. 39, no. 1, pp. 66–70, 1999.

[SUT 96] SUTTON M., DENG X., LIU J., YANG L., "Determination of elastic plastic stresses and strains from measured surface strain data", *Experimental Mechanics*, vol. 36, no. 2, pp. 99–112, 1996.

[SUT 08] SUTTON M., YAN J., AVRIL S., PIERRON F., ADEEB S., "Identification of heterogeneous constitutive parameters in a welded specimen: uniform stress and virtual fields methods for material property estimation", *Experimental Mechanics*, vol. 48, no. 5, pp. 451–464, 2008.

[SYE 07] SYED-MUHAMMAD K., Sur quelques aspects numériques de la méthode des champs virtuels: optimisation de conditions d'essais et champs virtuels définis par sous-domaines, Thesis, Blaise Pascal University, 2007.

[SYE 08] SYED-MUHAMMAD K., TOUSSAINT E., GRÉDIAC M., AVRIL S., KIM J., "Characterization of composite plates using the virtual fields method using optimized loading conditions", *Composite Structures*, vol. 85, no. 1, pp. 70–82, 2008.

[TIM 59] TIMOSHENKO S., WOINOWSKY-KRIEGER S., *Theory of Plates and Shells*, McGraw-Hill, New York, 1959.

[TOU 06] TOUSSAINT E., GRÉDIAC M., PIERRON F., "The virtual fields method with piecewise virtual fields", *International Journal of Mechanical Sciences*, vol. 48, pp. 256–264, 2006.

[TRU 65] TRUESDELL C., NOLL W., *The Non-linear Field Theories of Mechanics*, Handbuch der Physik, III/3, Springer, Berlin, 1965.

[XAV 07] XAVIER J., AVRIL S., PIERRON F., MORAIS J., "Novel experimental approach for the characterisation of the LR stiffness parameters of clear wood using a single specimen", *Holzforschung*, vol. 61, pp. 573–581, 2007.

Chapter 12

Equilibrium Gap Method

Chapter 8 presented an overview of the various identification strategies. Although they may appear different, they often use similar principles. Starting from a displacement field assumed to be known, the "equilibrium gap method" (EGM) consists of determining a field of mechanical properties (and/or of loading), which best fulfills the balance equations.

This chapter first presents the basic principles within the framework of continuum mechanics to enable its numerical implementation within any numerical formulation. Sections 12.2 and 12.3 apply this method to finite differences (FDs) and finite elements (FEs), respectively. The application of this method to beam-type geometries is presented in sections 12.4 and 12.5. Returning to fundamentals, the sensitivity to noise of this method is discussed in section 12.6, and a reconditioning is proposed to limit the impact of noise at small scales. The identification of a nonlinear behavior law is then addressed in the case of damage (section 12.7), and is applied to composite materials (section 12.8). Finally, the treatment of noise covariance in kinematic measurements is discussed in section 12.9. Section 12.10 concludes the chapter with a brief summary of the main results, and outlines perspectives for future developments.

12.1. Theoretical basis

The EGM aims at allowing for the identification of mechanical properties based on measured displacement fields. The mechanical framework is *a priori* quite broad, and can be applied to either homogeneous or heterogeneous elastic media, or even nonlinear constitutive laws. The complexity of constitutive laws will progressively

---

Chapter written by Fabien AMIOT, Jean-Noël PÉRIÉ and Stéphane ROUX.

increase all along this chapter. We begin with the elementary case of homogeneous linear elasticity, considered in the framework of continuum mechanics, to highlight the basic principles. This presentation allows for a dedicated formulation within any chosen geometry (e.g. slender bodies such as beams) or a discretization method for practical implementations.

12.1.1. *Homogeneous elastic medium*

In the case of linear elasticity, the stress field $\boldsymbol{\sigma}$ follows the balance equation:

$$\operatorname{div}(\boldsymbol{\sigma}) + \boldsymbol{f} = 0 \qquad [12.1]$$

where $\boldsymbol{f}$ denotes the density of body forces. The constitutive equation is written as $\boldsymbol{\sigma} = \boldsymbol{C} : \boldsymbol{\epsilon}$. Combining these two equations, together with the introduction of the strain tensor as the symmetric part of the displacement gradient, $\boldsymbol{\epsilon} = (\boldsymbol{\nabla u})_s$, the displacement field fulfills Lamé's equation:

$$\mathfrak{Lam}(\boldsymbol{u}; \boldsymbol{C}) + \boldsymbol{f} \equiv \operatorname{div}(\boldsymbol{C} : \boldsymbol{\nabla u}) + \boldsymbol{f} = 0 \qquad [12.2]$$

where Lamé's operator, $\mathfrak{Lam}$, is a second-order differential operator acting on $\boldsymbol{u}$ and linear in $\boldsymbol{C}$. It can be seen as providing the opposite of the distribution of body forces that makes the displacement field the solution to the elastic problem. If identification is solely based on kinematic measurements, it is clear that Hooke's tensor can only be determined up to a multiplicative constant, as nothing can set the scale for stress. For an isotropic elastic solid, the Young's modulus cannot be determined from the displacement field, whereas the Poisson's ratio can be determined.

Information on the loading can be described as an imposed Neumann condition, $\boldsymbol{\sigma}.\boldsymbol{n} = \boldsymbol{t}$, where $\boldsymbol{n}$ is the unit vector normal to the boundary and $\boldsymbol{t}$ the imposed force density on a part of the domain boundary, $\partial \mathcal{D}_N$:

$$\boldsymbol{C} : \boldsymbol{\nabla u}.\boldsymbol{n} = \boldsymbol{t} \qquad [12.3]$$

This condition is well suited to free boundaries ($\boldsymbol{t} = \boldsymbol{0}$), but in most other cases, the detail of the force distribution is hardly accessible, in contrast to the kinematic information. It may be more appropriate to minimize the Neumann condition to a weak form, imposing the resulting force (or torque) on a part of the boundary, $\mathcal{C}_i$:

$$\int_{\mathcal{C}_i} \boldsymbol{C} : \boldsymbol{\nabla u}.\boldsymbol{n} \, \mathrm{d}\boldsymbol{x} = T_i \qquad [12.4]$$

The EGM is a tool to determine the Hooke tensor, $\boldsymbol{C}$ – homogeneous over the domain to start with – such that the measured displacement field fulfills "at best" equation [12.2] in the interior of the domain, $\mathcal{D}$, together with the load boundary conditions on $\partial \mathcal{D}_N$ [CLA 02]. A general expression can be proposed as the minimum of a quadratic norm:

$$
\begin{aligned}
\mathcal{R}(\boldsymbol{C}) = {} & \alpha V(\mathcal{D}) \| \mathfrak{Lam}(\boldsymbol{u}; \boldsymbol{C}) + \boldsymbol{f} \|_{\mathcal{D}}^2 \\
& + \beta S(\partial \mathcal{D}_N) \| \boldsymbol{C} : \boldsymbol{\nabla u}.\boldsymbol{n} - \boldsymbol{t} \|_{\partial \mathcal{D}_N}^2 \\
& + \gamma \sum_i \left\| \int_{\mathcal{C}_i} \boldsymbol{C} : \boldsymbol{\nabla u}.\boldsymbol{n} \, \mathrm{d}\boldsymbol{x} - T_i \right\|^2
\end{aligned}
\tag{12.5}
$$

where V is the domain volume and S the area of the boundary $\partial \mathcal{D}_N$ (in two dimensions, these two quantities are an area and a length, respectively). Symbols $\| \ldots \|$ in this equation refer to norms of volume, surface and discrete fields. Thus, $\mathcal{R}$ is homogeneous to a force squared, and prefactors α, β and γ are dimensionless. They weigh, respectively, the contributions of the bulk displacement field and the load imposed on the studied body. To set the value of these constants that weigh information coming from various sources, it is necessary to know the uncertainty that is attached to them.

The natural formulation based on the above principles uses a L_2-norm of the displacement field defined over the domain:

$$
\begin{aligned}
\mathcal{R}_0(\boldsymbol{C}) = {} & \alpha V(\mathcal{D}) \int_{\mathcal{D}} |\mathfrak{Lam}(\boldsymbol{u}; \boldsymbol{C}) + \boldsymbol{f}|^2 \, \mathrm{d}\boldsymbol{x} \\
& + \beta S(\partial \mathcal{D}_N) \int_{\partial \mathcal{D}_N} |\boldsymbol{C} : \boldsymbol{\nabla u}.\boldsymbol{n} - \boldsymbol{t}|^2 \, \mathrm{d}\boldsymbol{x} \\
& + \gamma \sum_i \left| \int_{\mathcal{C}_i} \boldsymbol{C} : \boldsymbol{\nabla u}.\boldsymbol{n} \, \mathrm{d}\boldsymbol{x} - T_i \right|^2
\end{aligned}
\tag{12.6}
$$

It is to be noted that the above functional is quadratic in $\boldsymbol{C}$, and hence its minimization with respect to $\boldsymbol{C}$ is a linear problem. To make it explicit, with index notations, when no load information is available or used, in that case, the pre-factor αV can be dropped:

$$
\mathcal{R}_0(\boldsymbol{C}) = \int (C_{ijkl} u_{k,jl} + f_i)(C_{imnp} u_{n,mp} + f_i) \, \mathrm{d}\boldsymbol{x}
\tag{12.7}
$$

$$
\frac{1}{2} \frac{\partial \mathcal{R}_0(\boldsymbol{C})}{\partial C_{ijkl}} = C_{imnp} \int u_{k,jl} u_{n,mp} \, \mathrm{d}\boldsymbol{x} + \int u_{k,jl} f_i \, \mathrm{d}\boldsymbol{x} = 0
\tag{12.8}
$$

Thus, the operator that allows for the inversion of $\boldsymbol{C}$ is the sixth-order tensor:

$$
\mathfrak{J} = \int (\boldsymbol{\nabla} \otimes \boldsymbol{\nabla} \otimes \boldsymbol{u}) \otimes (\boldsymbol{\nabla} \otimes \boldsymbol{\nabla} \otimes \boldsymbol{u}) \, \mathrm{d}\boldsymbol{x}
\tag{12.9}
$$

Let us emphasize that, locally, this operator is a projector, and hence cannot be inverted. At each point, we have a single strain and strain gradient, and hence this is insufficient to probe the entire constitutive law. However, the spatial integration may – if the test is rich enough – restore the invertibility of $\mathfrak{J}$. For an isotropic elastic solid, this notation can easily be simplified. In particular, if we introduce the shear modulus μ and Poisson ratio ν

$$\mathcal{R}_0(C) = \mu^2 \sum_i \int_{\mathcal{D}} \left| u_{i,jj} + \frac{u_{j,ji}}{(1 - 2\nu)} + \frac{f_i}{\mu} \right|^2 \mathrm{d}\boldsymbol{x} \qquad [12.10]$$

12.1.2. *Heterogeneous elastic medium*

Extension to the case where the medium is heterogeneous (but still elastic) is trivial. Attention should be paid to the fact that the Lamé operator will involve the gradient of the Hooke tensor, as shown by the following explicit expression:

$$C_{ijkl}u_{k,lj} + C_{ijkl,j}u_{k,l} + f_i = 0 \qquad [12.11]$$

If the elastic properties show discontinuities, as for a multiphase material, Lamé's equation should be rewritten as the continuity of the normal component of the stress tensor across the interface. If n is the interface normal vector, then the balance equation reads:

$$[\![(C : \nabla u).n]\!] = 0 \qquad [12.12]$$

This, however, does not involve any further changes.

12.1.3. *Incremental formulation*

The previous section opens the way to nonlinear constitutive laws. When expressed in terms of increments, a nonlinear problem, including plasticity or damage, can be expressed as an inhomogeneous elastic problem [CLA 04, CLA 07]. In these cases, a parametric representation of the constitutive law is needed, and the latter has to be injected into the functional $\mathcal{R}$ to be minimized. In most cases (although not always), this functional becomes more complex than a mere quadratic form of the constitutive parameters. However, a quadratic form is only a matter of comfort (a constant Hessian can be computed once and for all) and, based on successive linearizations, a general nonlinear identification procedure can be designed. It is worth noting that there are some specific cases (e.g. damage laws, in particular) where the linearity of the problem is preserved [ROU 08]. Such a case is presented in section 12.8.

The advantage of resorting to such a formulation is the major reduction in the number of unknowns. The tangent elastic properties do not need to be identified as a field, but rather as a function of kinematic variables (and their time record), thereby reducing the ill-posedness of the problem, and only the constitutive parameters have to be determined. The end of this chapter will consider such a situation.

12.2. Finite difference implementation

Most full-field measurement methods give access to displacement fields defined on a regular (uniform) grid. Because they are often exclusively based on image processing, these techniques lead to displacement fields that do not *a priori* respect the regularity assumptions expected in continuum mechanics. For instance, the continuity of the displacement field is usually not achieved. Digital image correlation (DIC) software (see Chapter 6), for example, is commonly subset based. In practice, the region of interest (ROI) is then split into an array of overlapping subsets (or zones of interest (ZOI)). For each ZOI, and for the assumed local kinematics, the software tries to locate the best corresponding subregion in the deformed image [SUT 83]. Once registration is achieved, the parameters of the given transformation are obtained. Most of the time, only the displacement of the ZOI center is retained. Other information, such as strains, is not used. Strains are then generally computed using an FD scheme (see Chapter 7).

Here, we focus on an FD version of the EGM inspired from the initial studies [CLA 04] but compatible with the way the information is assessed and treated in common DIC approaches. Let us consider a heterogeneous elastic solid. A part of the bulk is arbitrarily defined in square cells (or elements). Their behavior is assumed to be cell-wise homogeneous. The cell boundaries coincide with the measurement grid. For instance, the cell length corresponds here to twice the grid step h (see Figure 12.1). In the following, we consider a plane problem and neglect volume forces. The material properties being piecewise constant, they are, thus, not derivable or continuous. The equilibrium inside the bulk is then expressed by writing the continuity of the stress vector across the discontinuity surface of normal n (see Figure 12.1 and equation [12.12]). In practice, the interface between adjacent cells (denoted $\oplus$ and $\ominus$) is considered and we write that the jump in the normal stress $\boldsymbol{\sigma}.\boldsymbol{n}$ should disappear:

$$[\![\boldsymbol{\sigma}.\boldsymbol{n}]\!] = \boldsymbol{\sigma}^{\oplus}.\boldsymbol{n} - \boldsymbol{\sigma}^{\ominus}.\boldsymbol{n} = 0 \qquad\qquad [12.13]$$

At this stage, the behavior could be anisotropic (e.g. orthotropic, as in [CRO 08]). Nevertheless, for the sake of simplicity, it is in the following considered as isotropic in each cell. Furthermore, we assume that only the Young's modulus E varies from one cell to another. The Poisson's ratio ν is kept identical in all the cells. The rigidity matrix $[C]$ associated with a given cell can then be expressed simply as a function of a reference modulus E^{ref} and a contrast δ ($\delta = E/E^{\text{ref}}$). Assuming a plane stress state, for example, the stress inside a cell e may finally be expressed as a function of the contrast δ_e:

$$\{\tilde{\sigma}\}_e = \delta_e \left(\frac{E^{\text{ref}}}{1-\nu^2} \begin{bmatrix} 1 & \nu & 0 \\ \nu & 1 & 0 \\ 0 & 0 & \frac{1-\nu}{2} \end{bmatrix} \right) \{\tilde{\epsilon}\}_e = \delta_e \, [C^{\text{ref}}] \{\tilde{\epsilon}\}_e \qquad [12.14]$$

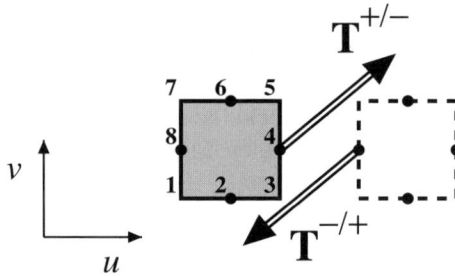

Figure 12.1. *Equilibrium for an interface between adjacent cells. Only the measurement points used for computing strains and stresses are shown*

The continuity of the normal stress vector across the interface between two adjacent cells is expressed (see equation [12.13]). It leads to two scalar equations. For the example presented in Figure 12.1 where $n = e_x$, we obtain two equations related, respectively, to the normal component and the tangential component of the stress jump:

$$\begin{cases} [\![\delta\,(\epsilon_{xx} + \nu\epsilon_{yy})]\!] = 0 \\ [\![\delta(\epsilon_{xy})]\!] = 0 \end{cases} \qquad [12.15]$$

In a conventional way, the horizontal and vertical components of the displacement of point i are denoted by u_i and v_i, respectively. With the node numbering proposed in Figure 12.1, the equilibrium of the central node finally reads:

$$\delta_\oplus \left(u_4^\oplus - u_8^\oplus + \nu \left(v_6^\oplus - v_2^\oplus \right) \right) = \delta_\ominus \left(u_4^\ominus - u_8^\ominus + \nu \left(v_6^\ominus - v_2^\ominus \right) \right)$$
$$\delta_\oplus \left(v_4^\oplus - v_8^\oplus + u_6^\oplus - u_2^\oplus \right) = \delta_\ominus \left(v_4^\ominus - v_8^\ominus + u_6^\ominus - u_2^\ominus \right) \qquad [12.16]$$

The balance of the stress across an interface thus yields two equations linking the rigidity contrasts (i.e. $\delta_\oplus$ and $\delta_\ominus$) to the *given* (measured) displacements of the grid points (i.e. u_i and v_i). When writing this jump condition for all the interfaces located inside the discretized body, we end up with a linear system of the form $[M]\{\delta\} = \{S\}$, where ${}^t\{\delta\} = \{\delta_1 \ \delta_2 \ \dots \ \delta_N\}$ corresponds to a vector collecting the rigidity contrasts of the cells, and $[M]$ and $\{S\}$ are, respectively, a matrix and a right-hand vector that depends on the measured displacements. At this stage, since the reference modulus is unknown E^{ref}, it is worth mentioning that only ratios between moduli, or rigidity contrasts δ, can be retrieved. Consequently, to solve the system, we may arbitrarily choose one reference cell for which the value of the rigidity contrast δ is defined as being unity. We could alternatively impose the value of the arithmetic mean (e.g. $\langle \delta \rangle = 1$). The value of the reference modulus E^{ref} can be estimated, but only, of course, if static data are available. The system to be solved is overdetermined. In practice, for a rectangular domain containing $N = m \times n$ cells, $2(2mn - m - n)$ linear equations are formed to determine $(mn - 1)$ unknowns. Therefore, a "best fit" (least-squares) solution to the problem is sought by using the pseudo-inverse operator. A variant to this formulation is to introduce a distinct weight to the data according to their relative reliability. In this general framework, the system to be solved becomes:

$$(^t[M][W][M])\{\delta\} = (^t[M][W])\{S\} \qquad [12.17]$$

where $^t[M][W][M]$ is a symmetric matrix – allowing the use, for example, of a conjugate gradient algorithm – and $[W]$ a weight matrix.

The proposed FD approach can be implemented easily. For each pair of images discussed, it can provide a *rigidity contrast* map. It may, for instance, reveal an initial heterogeneity in a specimen, or allow for the monitoring of a degradation mechanism. When a model is associated with the technique, or when the reference rigidity is updated, because of measured static data (loading), damage maps and damage laws may be identified. This approach allowed [CRO 08] to study a sheet-molding compound (SMC) composite material and to identify a damage law related to a carbon/carbon composite [CRO 09]. In addition, this numerical scheme is perfectly consistent with the data retrieved from many full-field measurement techniques (grid techniques, classic DIC, etc.) and leads to a simple physical interpretation of the method. However, the method is not generic, and not well suited to coping with complex geometries. The choice of identifying a contrast field without any other assumption induces a rather high sensitivity to measurement noise.

12.3. Finite element implementation

Here, we discretize the ROI of the solid into N FE. Because it offers a general framework, the finite element method (FEM) is the ideal tool for a generic numerical implementation. FEM can deal, in particular, with complex geometries without worrying about boundary condition issues. In addition, the recent development of FE-DIC [BES 06] (see Chapter 6) provides us with the opportunity to build a consistent identification approach. The issues related to reprojection in particular are avoided. The field of properties may, however, be defined on a different mesh from that used for retrieving the displacement field.

We are still considering inhomogeneous solids here. As mentioned previously, the mechanical properties are assumed to be piecewise constant, i.e. element-wise uniform. For simplicity, we limit our study to elastic solids. We adopt the classic FE notations: $\{u\}$ represents the nodal displacement vector, $[K]$ the rigidity matrix and $\{f\}$ a vector of nodal forces. As in the previous section, the rigidity matrix depends on the contrast vector $\{\delta\}$. Unlike classic FE problems, the nodal displacement field is assumed to be *known*. Here the latter displacements are those measured with the help of an FE-DIC approach, and are denoted $\{u^{\mathrm{mes}}\}$ (see Figure 12.2). The aim is now to determine the *unknown* rigidity contrast field (i.e. δ) minimizing the "equilibrium gap" vector norm $\|f^{\mathrm{res}}\|^2$, where:

$$\{f^{\mathrm{res}}\} = [K(\{\delta\})]\{u^{\mathrm{mes}}\} - \{f\} \qquad [12.18]$$

represents the residual vector associated with the measured displacement fields $\{u^{\mathrm{mes}}\}$ and the applied nodal forces $\{f\}$. We next assume that volume forces can be neglected. Therefore, the nodal force vector $\{f^{\mathrm{res}}\}$ should disappear for all interior nodes. In general, the strain energy related to element e reads:

$$W_e = \frac{1}{2}{}^t\{u_e\} \left(\int_e {}^t[B(x)][C][B(x)]d\Omega_e \right) \{u_e\} = \frac{1}{2}{}^t\{u_e\}[k^{el}]\{u_e\}$$
$$[12.19]$$

where $[k^{el}]$ is, by definition, the elementary stiffness matrix. In the following, for simplicity of presentation, we use the same assumptions as in section 12.2 (plane stress state, isotropic material and only the Young's modulus varies from one element to another). In this particular case, $[C(\delta_e)] = \delta_e[C^{\mathrm{ref}}]$ leads to:

$$W_e = \frac{1}{2}\delta_e{}^t\{u_e\}[k^{\mathrm{ref}}]\{u_e\} \qquad [12.20]$$

where $[k^{\mathrm{ref}}]$ represents a reference elementary rigidity matrix computed for the reference Young's modulus E_{ref}. Here, the elementary rigidity matrix $[k^{el}]$ is thus simply proportional to the reference elementary rigidity matrix. The contribution of element e to the nodal force f_{ie} at node i then reads (see Figure 12.2):

$$f_{ie} = \sum_j k^{el}_{ije} u_j = \delta_e \sum_j k^{\mathrm{ref}}_{ije} u_j \qquad [12.21]$$

The contribution of a unit nodal displacement u_j on nodal forces corresponds in fact to column j of $[k^{el}_e]$.

The example of a regular mesh made up of four-noded square elements (Q4) is provided next for illustration purposes. For any element e, we adopt the numbering of the eight displacement components u_j proposed in Figure 12.2. The 8×8 reference elementary rigidity matrix $[k^{\mathrm{ref}}]$ is the same for all the elements. It can then be computed once and for all, and stored.

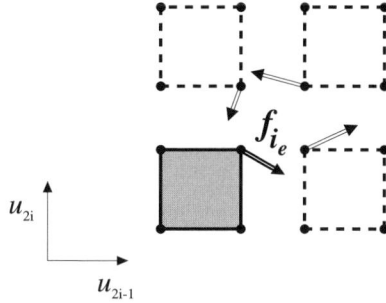

Figure 12.2. *Equilibrium of an interior node located in a regular mesh based on Q4 elements. Vector f_{ie} represents the nodal resulting force at node i generated by element e when subjected to imposed displacements*

We now seek to express the nodal residual force at an interior node. One node i inside this kind of mesh is always shared by four elements. The corresponding nodal residual f_i^{res} may be expressed as:

$$f_i^{res} = \sum_e f_{ie} = \sum_e \delta_e \sum_j k_{ij}^{\text{ref}} u_j \qquad [12.22]$$

To enforce the equilibrium in the whole solid, the quadratic norm $\mathcal{R}$ of the "equilibrium gap", $\|f^{res}\|^2$, is then minimized:

$$\mathcal{R} = \sum_i f_i^{\text{res}2} = \sum_i \left(\sum_e L_{ie} \delta_e \right)^2 \qquad [12.23]$$

where i denotes the indices of the internal nodes and $L_{ie} = \sum_j k_{ij}^{\text{ref}} u_j$ corresponds to the contribution of element e to the nodal resulting force at interior node i for the reference behavior and the imposed displacements u_j.

As mentioned previously, the minimization of $\mathcal{R}$ calls for an additional condition because the contrast is only defined up to a constant scale factor. This conventional condition may consist of imposing the mean value of the identified contrasts, i.e. $\langle \delta \rangle = 1$. By resorting to a Lagrange multiplier to enforce the latter condition α, the functional to be minimized becomes $\mathcal{R} - (\alpha/N) \sum_e \delta_e$. This finally leads to the following linear system:

$$^t[L][L]\{\delta\} = (\alpha/N)\{\mathbf{1}\} \qquad [12.24]$$

where α is determined by using the condition $\langle \delta \rangle = 1$. As in section 12.2, the problem is again overdetermined. In 2D, two equilibrium equations could be obtained for any interior node while there is only one unknown per element, or even less when the mesh used to describe the properties is coarser than the mesh used for measurements.

This implementation also allows the initial state to be studied and the damage process of a structure to be followed. Less intuitive, it is however totally generic. The technique can be extended to anisotropic behavior and can manage any kind of element. It has up to now successfully been applied to 1D problems with beam elements (see sections 12.4 and 12.5) and to 2D plane stress problems using Q4 elements (see section 12.8). It has, for example, been used to identify rigidity fields, loading fields (see section 12.5) or even damage fields and damage growth laws (see section 12.8). The extension to 3D, for example from tomographies, does not introduce particular challenges. To be consistent, it is desirable to perform the experimental determination of the displacement on the same basis as that of the FEM. As presented in this section, the approach shows a significant sensitivity to the noise intrinsic to measurements. Section 12.6 provides some elements intended to reduce this sensitivity.

12.4. Application to beam theory: local buckling

12.4.1. *Application to beam theory*

The EGM is illustrated using beam structures in the following, so that every step yields closed-form equations, thus highlighting the foundations of the method. In the current and following sections, a simple Euler–Bernoulli beam theory framework is chosen. Let us consider a plane cantilever beam, whose centroidal axis is made coincident with the x-axis of the coordinate system. Let us consider that this cantilever beam features a heterogeneous bending stiffness field $EI(x)$ and is subjected to a transverse distributed force field $p(x)$. The local equilibrium condition reads:

$$\frac{\mathrm{d}^2 M_f(x)}{\mathrm{d}x^2} = p(x) \qquad\qquad [12.25]$$

where $M_f(x)$ is the bending moment in section x. The bending moment and the transverse displacement $v(x)$ are related through the constitutive equation:

$$M_f(x) = EI(x)\frac{\mathrm{d}^2 v(x)}{\mathrm{d}x^2} \qquad\qquad [12.26]$$

so that satisfying simultaneously both the equilibrium condition and the constitutive equation yields:

$$\frac{\mathrm{d}^2}{\mathrm{d}x^2}\left\{EI(x)\frac{\mathrm{d}^2 v(x)}{\mathrm{d}x^2}\right\} = p(x) \qquad\qquad [12.27]$$

Equation [12.27] clearly shows that it is impossible, without any additional information or assumption, to recover a unique pair of loading and stiffness fields solely using the measured displacement field $v(x)$. This section will nevertheless display two examples for which considering simple (but realistic) assumptions allows the loading or stiffness fields to be recovered from the measured displacement field.

Figure 12.3 [HIL 11] shows a cantilever beam made from construction steel, with a square cross-section (ASTM-A-500) that is transversally loaded at its end section. It is 1.28 m long, 120 mm wide and 4 mm thick. The displacement field for the side of the beam is obtained throughout the experiment by using DIC (see Chapter 6), for the regions of interest described in Figure 12.3(a). The basis of the considered displacement fields is that of a beam-like FE (length l) with six degrees of freedom. These degrees of freedom are two longitudinal displacements (u_1 and u_4), two transverse displacements (u_2 and u_5) and two cross-section rotations (u_3 and u_6), described in Figure 12.4. The displacement $\mathbf{u}(x)$ of a given cross-section therefore reads:

$$\mathbf{u}(x) = [N_e(x)]\{u\} \qquad [12.28]$$

where $N_e(x)$ concatenates the shape functions $(N_1, \ldots, N_6)$ corresponding to the degrees of freedom $\{u\}^t = \{u_1, \ldots, u_6\}$

$$N_1(x) = 1 - x/l, \qquad\qquad N_2(x) = (2x^3 - 3lx^2 + l^3)/l^3$$
$$N_3(x) = (x^3 - 2lx^2 + l^2x)/l^2, \quad N_4(x) = x/l \qquad [12.29]$$
$$N_5(x) = -(2x^3 - 3lx^2)/l^3, \qquad N_6(x) = (x^3 - lx^2)/l^2$$

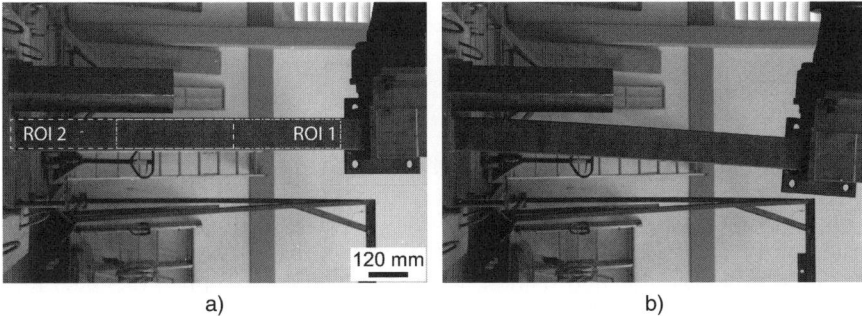

a)   b)

Figure 12.3. *The considered cantilever beam in its initial a) and deformed b) states. The two regions of interest are displayed for the initial state*

The rotation $\theta(x)$ and curvature $\kappa(x)$ fields read:

$$\theta(x) = \left[\frac{d\hat{N}}{dx}(x)\right]\{\hat{u}\}, \quad \kappa(x) = \left[\frac{d^2\hat{N}}{dx^2}(x)\right]\{\hat{u}\} \qquad [12.30]$$

with $\{\hat{u}\}^t = \{u_2, u_3, u_5, u_6\}$ and $[\hat{N}(x)] = [N_2(x), N_3(x), N_5(x), N_6(x)]$. The curvature field is a linear function of x.

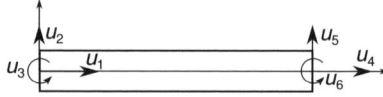

Figure 12.4. *Considered beam element with six degrees of freedom*

12.4.2. *Loading identification*

It is important to stress that this assumption regarding the kinematic description is not independent of assumptions on both the applied loading and the beam stiffness field. The chosen function basis actually correctly describes displacement fields obtained when the curvature field

$$\kappa(x) = \frac{M_f(x)}{EI(x)} \qquad [12.31]$$

is linear with x, under the assumption of elastic behavior. This is, for instance, the case when the bending stiffness is homogeneous over the element and if the external loading is localized. Under these assumptions, the curvature field reads:

$$\kappa(x) = \frac{M_f(x)}{(EI)_0} = \frac{F(x_F - x)}{(EI)_0} \qquad [12.32]$$

where F is the projection along y of the external force applied at position x_F. The curvature, therefore, disappears at $x = x_F$ and its derivative with respect to x

$$\frac{d\kappa(x)}{dx} = -\frac{F}{(EI)_0} \qquad [12.33]$$

is a measure of the applied external loading.

Figure 12.5(a) describes the overall structure behavior for the 10 loading levels considered in the following. Figure 12.5(b) shows the curvature fields measured over the ROI close to the end section, where the loading is applied (denoted ROI 1 in Figure 12.3). These fields are linear functions of x over the measurement zone (because of the chosen basis) and are extrapolated near the loading zone by assuming that the behavior remains linear elastic, so that the location where the (equivalent) point-load is applied, and its value, are estimated as a function of the loading level.

It should be noted that the identified variations for the point-load locations scale as one-tenth of the measurement window size, so that it is absolutely necessary to take these variations into account in order to accurately estimate the bending moment over the whole beam.

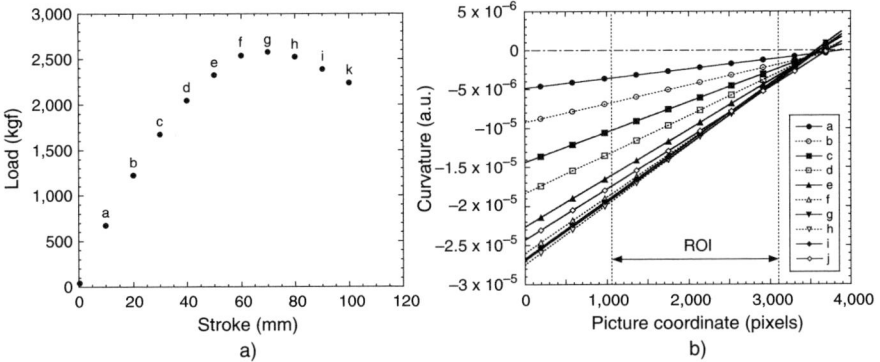

Figure 12.5. *a) Applied load versus stroke for the full experiment and b) measured curvature fields for 10 loading levels*

12.4.3. *Identification of a heterogeneous stiffness field*

As for the loading identification, the displacement field is measured for 10 loading levels and extrapolated toward the base, where the beam is welded (Figure 12.6(a)). The extrapolation again assumes that the material remains elastic. For the first five loadings, the extrapolated rotation disappears at the base, contrary to the last five loadings, where the boundary conditions are no longer satisfied, possibly because of local buckling. The beam kinematic and constitutive descriptions, therefore, have to be enriched. The measurement zone is thus shifted toward the base (denoted ROI 2 in Figure 12.3) and the displacement field measurement procedure is enriched [HIL 11] in order to correctly describe the rotation field. The curvature is imposed as being linear with x (homogeneous bending stiffness and a single point-load) only outside the local buckling zone.

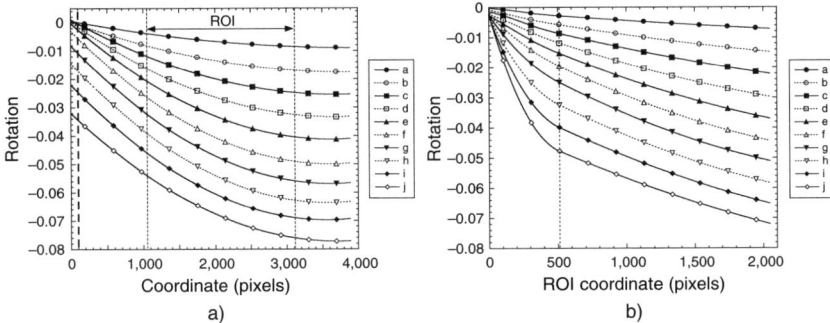

Figure 12.6. *a) Rotation field for 10 loading levels (measured over the region of interest and extrapolated). The base is indicated by the bold dashed line. b) Rotation field for 10 loading levels measured near the base. The localized deformation near the base delimits the zone affected by local buckling*

Two zones are distinguished and are described here (over the full experiment) with two bending stiffness: $\widehat{EI}$ over a distance l_b from the base and $(EI)_0$ away from this degraded element. The equilibrium condition at $x = l_b$ translates as:

$$\frac{L - l_b}{(EI)_0} = \frac{L_p - l_b}{\widehat{EI}} \qquad [12.34]$$

where L is obtained out of x_F, and the scaling $F/(EI)_0$ is deduced from equation [12.33]. L_p is introduced to ensure curvature continuity at $x = l_b$. Parameters l_b and $\widehat{EI}$ are then obtained as the minimizers of the squared gap between the measured rotation field (Figure 12.6(b)) and the rotation field computed from the heterogeneous beam model. The agreement is quite good for every loading level, and the identified l_b parameter is constant throughout the experiment, at about 500 pixels.

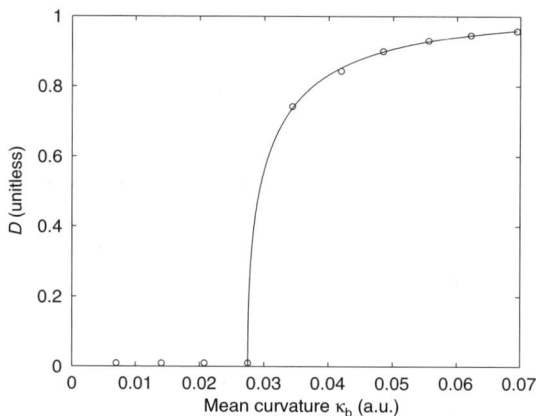

Figure 12.7. *Damage parameter as a function of the mean curvature over the degraded element*

The identified bending stiffness is shown in Figure 12.7 as a damage parameter D, defined as:

$$1 - D = \frac{\widehat{EI}}{(EI)_0} \qquad [12.35]$$

as a function of the mean curvature κ_b over the degraded zone:

$$l_b \kappa_b = \int_0^{l_b} \frac{M_f(x)}{\widehat{EI}} \, dx \qquad [12.36]$$

It is then clear that the stiffness is significantly degraded near to the base at the end of the experiment. This last example illustrates how a heterogeneous stiffness field can be recovered by using equilibrium conditions over regions with well-defined loading (i.e. the damaged and loading regions are distinct).

12.5. Simultaneous identification of stiffness and loading fields

For the case of heterogeneous mechanical properties and distributed mechanical loadings, it is convenient to make use of an FE description of the structure. This can be illustrated using micrometer-sized cantilever beams with electrostatic actuation. The cantilever beams are made from silicon dioxide and obtained by wet etching, an oxidized silicon wafer after having designed the structures in the oxide layer [AMI 05]. The cantilevers are released by under-etching so that it is rather difficult to precisely control the final cantilever thickness. This may result in heterogeneous bending stiffness. These cantilever beams are finally covered with an evaporated gold layer (150 nm).

Figure 12.8(a) shows a scanning electron microscope (SEM) view of an obtained cantilever beam. It is 70 μm long, 20 μm wide, for a 1.2 μm total thickness. It can be seen in Figure 12.8(b) that the beam thickness significantly varies near the fixed part.

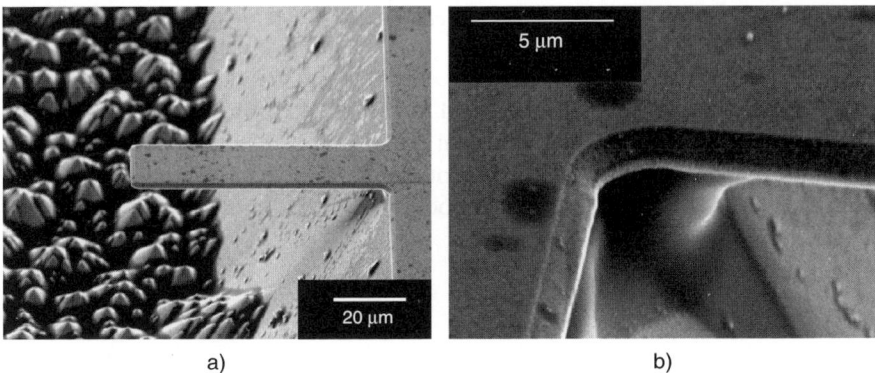

a)   b)

Figure 12.8. *SEM views of the considered cantilevers: a) a beam and b) zoom at the fixed part*

A drilled metallic plate is placed 1 mm above the cantilever. The gold layer and this metallic plate are connected to a DC generator (Figure 12.9(a)) so that an electric field orthogonal to the conductive surfaces is established. The electric field, and thus the electrostatic pressure, distributions are expected to be heterogeneous because of the various geometrical singularities (edges, corners, etc.). The results presented hereafter are obtained for a bias of about 800 V and a gap equal to 1 mm between the plate and the cantilever upper surface. A Nomarski imaging interferometer [AMI 06] is used to measure the out-of-plane displacement field of the cantilever induced by the electric field (Figure 12.9(b)).

The previously described function's basis $d\hat{N}(x)/dx$ is used in the following to describe the measured displacement field. The cantilever beam is discretized using N

elements, for which the bending stiffness is assumed to be homogeneous. The element length is denoted by l. The elastic properties field is assumed to be heterogeneous, and is described through a multiplicative contrast field δ, so that $EI\delta_n$ is the bending stiffness of element n, $n \in \{1, \ldots, N\}$.

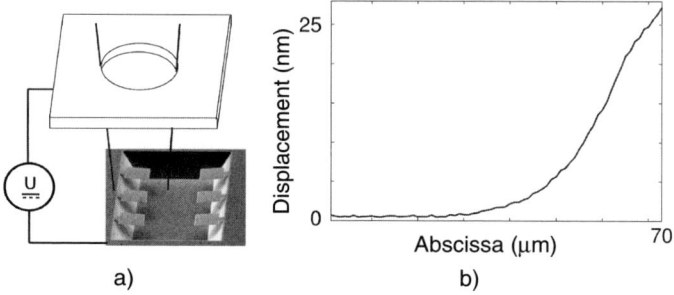

Figure 12.9. *Schematic view of the a) setup and b) measured displacement field along the cantilever beam*

The beam is assumed to be subjected to (Figure 12.10(a)): (1) a pressure field, uniform over an element. The equivalent nodal forces f_i and nodal moments M_i depend on the pressure acting on adjacent elements (energetic equivalence); (2) an additional moment C, applied to the node at the free end and described by the dimensionless parameter λ, defined by:

$$C = -\lambda p_N \frac{l}{12} \qquad\qquad [12.37]$$

where p_N is the electrostatic pressure acting over the element at the free end.

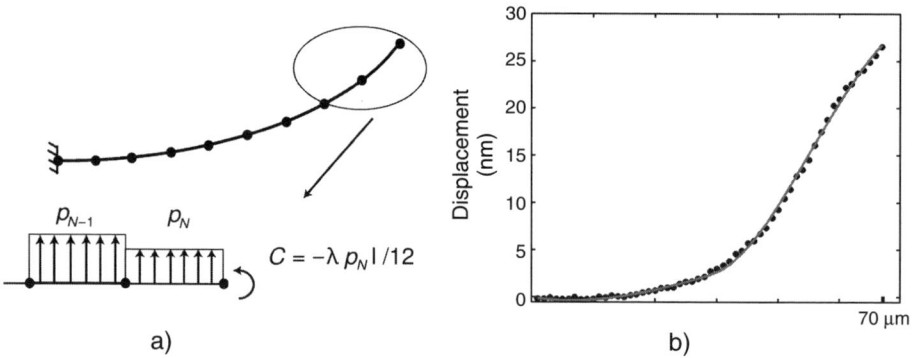

Figure 12.10. *a) Proposed modeling for the cantilever beam under electrostatic loading and b) measured (points) as well as computed statically admissible (solid line) displacement fields*

The goal is to retrieve the stiffness and loading fields when the projection $\{\hat{u}\}$ of the out-of-plane displacement field $v(x)$ (onto the kinematically admissible fields basis, $\mathrm{d}\hat{N}(x)/\mathrm{d}x$ is known (measured). The identification problem then becomes a question of recovering the mechanical action of the fixed part on the cantilever, one pressure and one stiffness per element in addition to the λ parameter; that is $2(N+1)+1$ parameters for $2(N+1)$ equilibrium conditions obtained as described in section 12.1. It has already been mentioned that without any additional static information, the pressure and stiffness fields are defined up to a multiplicative factor. Factorizing (and canceling) this unknown leads to a system with the same number of equations and unknowns. It should be noted that this corresponds to the case for which $\beta \neq 0$ in equation [12.5]. The solution is obtained here using a singular value decomposition [DEM 92].

Figure 12.10(b) shows the computed statically admissible displacement field (using the identified stiffness and loading fields) for $N = 2$. It corresponds to an attractive pressure field all over the cantilever beam, with $\lambda = -1.45$. The agreement between measured and identified fields is quite good and demonstrates the identification quality even though the number of elements used to describe the structure is rather low. This full identification procedure may be run taking into account additional noise that corrupts the measured displacement field [AMI 07]. We then obtain a probability density for the parameter λ.

12.6. Spectral sensitivity and reconditioning

It has already been mentioned that the EGM involves the displacement field through its second derivatives, increasing measurement noise at small scales. To quantify this effect, it is instructive to consider the contribution of a plane wave field of amplitude A and wave vector k, $u(x) = Ae^{i k \cdot x}$. Functional $\mathcal{R}$ introduced in equation [12.5] for an L_2 norm scales as:

$$\|\mathfrak{Lam}(u; C)\|_{\mathcal{D}}^2 \propto A^2 k^4 \qquad [12.38]$$

The quadratic dependence in A is the simple consequence of our choice of a quadratic norm. The fourth power of the wave vector results from the second-order derivative of the displacement that appears in the formulation. The exponent of k, $s = 4$, in the latter expression is called the "spectral sensitivity". To modify (or correct) this high sensitivity, it is natural to propose a different norm. For instance, if we chose the following norm in Fourier space:

$$\|\mathfrak{Lam}(u; C)\|_{\mathcal{D}}^2 = \int |k|^{-2m} |\mathcal{F}[\mathfrak{Lam}(u; C)]|^2 \, \mathrm{d}k \qquad [12.39]$$

where $\mathcal{F}[\ldots]$ denotes the Fourier transform, the spectral sensitivity would amount to $4 - 2m$. The idea is simply to filter the residual field, $\mathfrak{Lam}(u; C)$, with a power-law filter, $|k|^{-m}$.

Now that we know how to tune this sensitivity, we can specify the objective we are aiming for. The answer depends on the type of noise or uncertainty that is expected for the measurement of the displacement field. If the latter can be considered as white noise – a good approximation for DIC – its power spectrum will be uniform over all wave vectors, and hence we should use a method whose spectral sensitivity is $s = 0$, or $m = 2$ in equation [12.39]. If, on the other hand, we have access to a strain measurement, corrupted by white noise, it will be convenient to use an identification method with a spectral sensitivity $s = 2$.

Using a norm in Fourier space is suited to periodic media, but for displacement fields defined over domains of arbitrary shape, Fourier transforms are inconvenient and introduce numerous artifacts. It is, therefore, natural to look for other reconditioning operators. Multiplication of the integrand by k^{-2m} can be seen dimensionally as an integration of order m of $\mathfrak{Lam}(u; C)$. More precisely, this would correspond to a L_2-norm on a "corrected" field, w, such that:

$$\nabla^2 w = \mathfrak{Lam}(u; C) \tag{12.40}$$

However, the Laplacian operator ∇^2 is quite arbitrary here, and any other elliptical (to preserve the notion of a norm) second-order differential operator would result in the same corrected spectral sensitivity as long as dimensional analysis is concerned. Among these operators, a natural choice is Lamé's operator for a reference homogeneous elastic problem, with elastic properties C_0, so that the corrected field w would follow, in the interior of the domain:

$$\mathfrak{Lam}(w; C_0) = \mathfrak{Lam}(u; C) \tag{12.41}$$

The boundary conditions to be applied to the corrected field are naturally of Dirichlet type, with a null displacement all along the boundary, $w = 0$. This leads to a *reconditioned* EGM where the objective functional to minimize is the L_2-norm of w:

$$\tilde{\mathcal{R}}(C) = \|w\|^2 \tag{12.42}$$

The very construction of this reconditioned equilibrium gap leads to a spectral sensitivity $s = 0$. It is to be stressed that the quadratic dependence of $\tilde{\mathcal{R}}$ in C is not altered, and hence the identification problem remains linear. The reconditioning consists of turning the initial residual, interpreted as a distribution of body forces, into a field that is dimensionally a displacement field, that is of the same nature as the raw information exploited in the identification.

The reference solid to be used and its Hooke tensor C_0 have not yet been discussed. On the basis of dimensional considerations, complete freedom is still preserved. It is natural to determine C_0 so that it is close to C. This is actually the choice made in finite element method updating (FEMU), where implicitly the identity $C_0 = C$ is imposed. When this choice is inserted rigidly in the formulation of the identification, the problem becomes nonlinear. However, the spirit of the approach is very similar.

12.7. Damage

We now focus on the identification of damage growth laws describing material degradation. The material is again considered as elastic. It is now assumed to be *initially homogeneous*. Following sections 12.2 and 12.3, let us first consider isotropic behavior. The solid is still discretized using N cells or elements with piecewise-homogeneous material properties. The (piecewise constant) contrasts of the elastic properties δ now result from a damage process. We next rely on continuum damage theory [VOY 98]. The effect of the microscopic material failure is described at the representative volume element (RVE) scale by a gradual decrease in the elastic properties. The introduced damage variables represent a relative loss of modulus. These quantities can only vary from 0 (initial state) to 1 (RVE failure). Damage is in the following assumed to be isotropic. We then write that $\delta_e = (1 - d_e)$, that is $E_e = E^0(1 - d_e)$, where E^0 is the initial Young's modulus of the studied material.

The displacement fields measured between the virgin (undamaged) state and various damaged states (corresponding to a certain loading history) can be used as entries for the techniques presented in sections 12.2 and 12.3. This would provide a collection of *rigidity contrast* maps with respect to a reference modulus E^{ref}. The latter is different for each map, and *a fortiori* different from E^0. A procedure was originally proposed in [CLA 07] to go back to *damage*. All the identified contrast maps were then associated with the *same* damage growth law. An alternative method consists of using measured static data (resultants or moments) to identify the reference modulus [CRO 08]. The latter approach has the merit of requiring only basic assumptions about the material's behavior (which may be useful, for example when no continuum damage model has been chosen).

As proposed in [CLA 07], we consider hereafter that the contrasts maps come from a *unique* damage law. A continuum thermodynamics framework is used [GER 83]. For a given damage state, the state potential ψ (Helmholtz free energy density) reads:

$$\psi = \frac{1}{2}(1 - d)\epsilon : \mathbf{K}^0 : \epsilon \qquad\qquad [12.43]$$

The state laws are derived from this state potential. In particular, the driving force, or energy release rate density, Y, associated with the damage variable d is defined as:

$$Y = Y_d = -\frac{\partial \psi}{\partial d} \qquad\qquad [12.44]$$

Usually, the parameters of the damage law relating d and $\max_{0 < \tau < t}(Y)$ must be identified. However, it can be simply proved that the non-dimensional quantity $\epsilon_{eq} = \sqrt{2Y/E}$ may be interpreted as a strain that can be computed from the (indirectly) measured strains. The identification problem can then be advantageously rephrased. Finally, we try to identify a monotonically increasing function H such that $d = H(\widehat{\epsilon_{eq}})$, and where $\widehat{\epsilon_{eq}} = \max_{0<\tau<t}(\epsilon_{eq})$. To regularize the problem, instead of exploiting *a posteriori* some identified contrasts maps, this law is enforced everywhere in the ROI at the very root of the procedure. In practice, this means that two regions where the equivalent strain is the same will experience identical damage. The equilibrium gap norm $\mathcal{R}$ can now be written as:

$$\mathcal{R} = \sum_i \left(\sum_e L_{ie}\delta_e \right)^2 = \sum_i \left(\sum_e L_{ie}(1 - d_e) \right)^2$$

$$= \sum_i \left(\sum_e L_{ie}(1 - H(\widehat{\epsilon_{eq}})) \right)^2 \qquad [12.45]$$

A decomposition of the damage law H on a specific basis is then adopted:

$$d = H(\widehat{\epsilon_{eq}}, c_k) = \sum_k c_k \varphi_k(\widehat{\epsilon_{eq}}) \qquad [12.46]$$

under the constraint $c_k > 0$ for any k, and with:

$$\varphi_k(\widehat{\epsilon_{eq}}) = 1 - \exp\left(-\frac{\widehat{\epsilon_{eq}}}{\epsilon_k} \right) \quad \text{and} \quad \widehat{\epsilon_{eq}} = \max_{0<\tau<t} (\epsilon_{eq}(\tau)) \qquad [12.47]$$

The parameters ϵ_k are preset in order to select the space in which the damage function is sought. Only ϵ_k values in the range of the experimentally observed equivalent strains are considered. The choice usually made is to set a fixed ratio of 2 between two consecutive values. This corresponds to a good compromise between the conditioning of the system and the number of degrees of freedom used to describe the damage function. In practice, only three to four degrees of freedom c_k are generally required. It should be noted that it is no longer useful to set the contrast scale since $\varphi_k(0) = 0$.

However, as mentioned previously in section 12.6, the above system shows a high sensitivity to measurement noise. As a consequence, we will use a reconditioned version in order to enhance the robustness of the method. It is proposed to introduce the operator $\mathbf{S}$ (a compliance matrix) such that $[\mathbf{S}]\{\mathbf{L}\} = \{\mathbf{u}^{mes}\}$, where $\mathbf{S}$ solves an elastic problem for which the medium is assumed to be undamaged. It corresponds to a particular choice of the reference elastic body that is well suited to the problem. In theory, we should impose Dirichlet boundary conditions with null displacements at the boundaries. The specific form of the damage law, associated with the particular choice of the undamaged solid as a reference, avoids solving an elastic problem if the measured displacement fields on the boundaries are introduced as boundary conditions. This formulation is mathematically equivalent to the general approach of the "reconditioned EGM", but its numerical implementation is far simpler.

An objective function corresponding to the "reconditioned" equilibrium gap $\widetilde{\mathcal{R}}$ is finally proposed:

$$
\widetilde{\mathcal{R}}(c_k) = \sum_i \left(\sum_j S_{ij} \sum_e L_{je} \left(1 - \sum_k c_k \varphi_k(\widehat{\epsilon_{eq}^e}) \right) \right)^2
$$

$$
= \sum_i \left(u_i^{\text{mes}} - \sum_k c_k \sum_j S_{ij} \sum_e L_{ie} \varphi_k(\widehat{\epsilon_{eq}^e}) \right)^2 \qquad [12.48]
$$

It is to be noted that the minimization is to be carried out under the constraint $c_k > 0$ for any k. We can then write the objective function as:

$$
\widetilde{\mathcal{R}}(c_k) = \sum_i \left(u_i^{\text{mes}} - \sum_k c_k w_k(\widehat{\epsilon_{eq}^e})) \right)^2 \qquad [12.49]
$$

It thus remains for us to search for the c_k that minimize the latter functional. A remarkable feature of this procedure is that the identification of the entire (nonlinear) damage evolution law is reduced to the resolution of a few linear systems. This results, in particular, from the use of a specific decomposition of the damage law (equation [12.46]) and from the choice of the norm (equation [12.45]).

We have chosen here to impose a form of damage law at the very beginning of the identification procedure [ROU 08]. This means that this new equilibrium gap approach leaves much less flexibility in terms of material behavior. The method enables, however, the direct identification of the damage law parameters and of a set of damage maps. Furthermore, once combined with a reconditioning that allows for a comparison of kinematic quantities, the method becomes much more robust with respect to the noise inherent in displacement measurements.

12.8. Application to a biaxial test carried out on a composite material

In this section, it is proposed to identify a damage model developed to describe the degradation of a composite material. A 3D needled C/C composite [PÉR 02] is studied. This material is manufactured in two main steps. In the first step, a fiber preform is made up. Plies of satin layers made of carbon yarn are laid up to form a laminate (stacking sequence: $[0, 90]_n$), then needled together. In practice, fibers are transferred along the third direction (perpendicular to the layer). In the second step, the preform is densified using a chemical vapor infiltration process.

The fiber reinforcement distribution and orientation impact the elastic properties, but also the degradation modes. In most composite materials, fiber/matrix debonding, transverse cracking and micro-delamination are commonly observed. At the ply scale, we can model the effect of such degradations by inelastic phenomena and stiffness losses, often described by damage models. In the specific case of the studied needled $[0, 90]_n$, the degradation mechanisms lead to a significant nonlinear macroscopic behavior when the composite is subjected to shear loading. Usually, an identification procedure is based on a set of elementary tests (e.g. tensile tests and bend tests) performed on various stacking sequences. However, for this material, the sampling of coupons is a limiting issue. The use of identification procedures based on the use of non-conventional experiments and of full-field measurements is thus particularly appealing.

A flat cruciform specimen, considered as a $[\pm 45°]$ needled laminate, is subjected to a biaxial test. The loading involves an intense shear in the center of the specimen. Digital images of the surface are taken at various steps of the loading. These images are analyzed by using an FE-based DIC algorithm. The measured displacement fields are subsequently used to identify a shear damage law due to the procedure introduced in the previous section (digital image mechanical identification (DIMI) [ROU 08]). It should be noted that other identification techniques, such as the so-called virtual fields method [CHA 04] and the finite element model updating method [GEE 99, AVR 08], have already been used to analyze composite materials.

In the following section, the considered damage model used to describe degradation is presented. Then, the interest and the performance of the proposed identification technique are emphasized with the aid of a few results.

12.8.1. *Damage modeling*

Many models developed in the framework of continuum damage mechanics have been proposed [VOY 98]. The damage model considered herein derives from an approach developed at the mesoscopic scale (ply scale) for laminates [LAD 92]. It is assumed that damage is mainly dictated by fiber orientation and not by the applied load [LAD 95].

The material is now assumed to be initially homogeneous. Indices $(1, 2)$ refer to the ply coordinate system (i.e. material frame) here coincide with the fiber directions. The angle between this frame and that of the camera coordinate system (x, y) is denoted by θ (Figure 12.11). With these notations, E_1 and E_2 denote the initial Young's moduli (in the fiber directions), G_{12} the initial shear modulus and ν_{12} one of Poisson's ratios. At this stage, we could describe damage by using three scalar variables d_1, d_2 and d_{12} related, respectively, to the decrease in the elastic moduli E_1, E_2 and G_{12}. In the following, only one damage variable d_{12} is considered, and describes a *gradual* degradation of the shear modulus. This appears long before the fiber breakage described by the damage variables d_1 and d_2. Many [0, 90] carbon epoxy woven composites, as a first-order approximation, and at a certain scale, behave in a such a way [GAO 99]. A continuum thermodynamics framework is used [GER 83]. Gibbs' free enthalpy density Φ of woven plies reads:

$$\Phi = \frac{1}{2}\left[\frac{\sigma_{11}^2}{E_1} - 2\frac{\nu_{12}}{E_1}\sigma_{11}\sigma_{22} + \frac{\sigma_{22}^2}{E_2} + \frac{\sigma_{12}^2}{G_{12}(1-d_{12})}\right] \qquad [12.50]$$

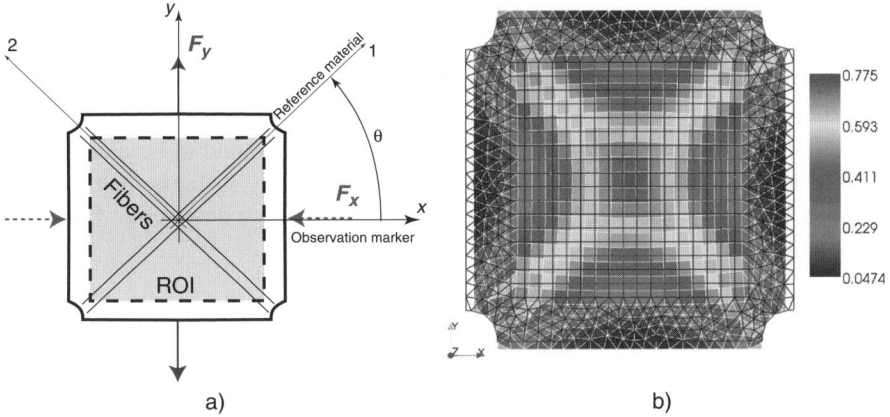

Figure 12.11. *a) Scheme of a shear test on a [$\pm 45°$] cruciform specimen and b) reference mesh used and damage map obtained at the last load level plotted over the deformed $\times$ 10 mesh*

From the state potential Φ, the state laws are derived. The driving force Y_{12}, or energy release rate density associated with the damage variable d_{12}, is written as:

$$Y_{12} = \frac{\partial\Phi}{\partial d_{12}} = \frac{1}{2}\frac{\sigma_{12}^2}{G_{12}(1-d_{12})^2} \qquad [12.51]$$

In practice, the driving force Y_{12} thus simply depends on elastic parameters and kinematic quantities. It is actually directly related to the shear strains expressed in the ply coordinate system, which are accessible due to kinematic measurements:

$$Y_{12} = 2G_{12}\epsilon_{12}^2$$

In the following, the growth law for d_{12} will be assumed to be controlled solely by its associated thermodynamic force Y_{12}. Let us note that when three damage variables are introduced (to account for fiber breakage), other expressions for the driving force of the damage variable d_{12} are proposed, for example by [HOC 07]. For the sake of simplicity, the following short-hand notations are used here $d \equiv d_{12}$ and $Y \equiv Y_{12}$. We also define an equivalent strain $\epsilon_{eq} = \sqrt{Y/2G_{12}} = |\epsilon_{12}|$. This quantity is estimated from the strains measured in the camera frame (x, y) and the angle between axes 1 and x.

$$\epsilon_{eq} = \left| \frac{1}{2}(\epsilon_{yy} - \epsilon_{xx}) \sin(2\theta) + \epsilon_{xy} \cos(2\theta) \right| \qquad [12.52]$$

Finally, we need to identify the parameters of the damage law linking the damage variable d and the maximum over the elapsed time of the equivalent strain ϵ_{eq}.

12.8.2. *Adapted expression of the reconditioned equilibrium gap*

As stated in section 12.7, the damage variable d is assumed to be element-wise uniform. Because the damage is anisotropic, the elementary stiffness matrix $[k^{el}]$ is no longer linear in d (as was the case of an isotropic damage description [ROU 08]) but rather affine:

$$k_{ij}^{el} = M_{ij}^0 + M_{ij}^1(1 - d) = k_{ij}^0 - k_{ij}^1 d \qquad [12.53]$$

where $[k^0]$ and $[k^1]$ are matrices dependent on the initial elastic parameters (i.e. of the undamaged element). These elastic constants may also be identified using full-field measurements, for example using the FEMU method [LEC 07, AVR 08] (see Chapter 9). The contribution of element e to the nodal force at the internal node j is $L_{je}^n = \sum_i k_{ije}^n u_i$ ($n = 0$ or 1). The problem is then equivalent (see section 12.8) to minimizing the quadratic norm $\mathcal{R}$ of the "equilibrium gap":

$$\mathcal{R} = \sum_i \left(\sum_e (L_{ie}^0 - L_{ie}^1 d_e) \right)^2 \qquad [12.54]$$

As explained in section 12.3, the solution to such a problem would provide a map of shear modulus contrasts for each loading step [CLA 04]. In the following, it is assumed that all the maps obtained result from the *same* damage law $d = H(\widehat{\epsilon_{eq}}, c_k)$. We adopt the decomposition of the damage law H introduced in section 12.7 and defined by equations [12.46] and [12.47]. The objective function $\mathcal{R}$ depends quadratically on the coefficients c_k defining the damage law for the given set of characteristic strains ϵ_k:

$$\mathcal{R}(c_k) = \sum_i \left(\sum_e (L_{ie}^0 - L_{ie}^1 H(\widehat{\epsilon_{eq}^e}, c_k)) \right)^2 \qquad [12.55]$$

As in the case of isotropic damage, we introduce the operator S such that $[S]\{L^0\} = \{u^{mes}\}$. The "reconditioned" equilibrium gap objective function $\widetilde{\mathcal{R}}$ finally reads:

$$\widetilde{\mathcal{R}}(c_k) = \sum_i \left(\sum_j S_{ij} \sum_e \left(L^0_{je} - L^1_{je} \sum_k c_k \varphi_k(\widehat{\epsilon^e_{eq}}) \right) \right)^2$$

$$= \sum_i \left(u^{mes}_i - \sum_k c_k \sum_j S_{ij} \sum_e L^1_{ie} \varphi_k(\widehat{\epsilon^e_{eq}}) \right)^2, \qquad [12.56]$$

which can be interpreted as the quadratic norm of a homogeneous nodal vector to a displacement field.

12.8.3. *Application to a biaxial test*

The adopted configuration, defined due to numerical simulations [PÉR 02], consists of a biaxial test on a flat cruciform specimen. The sample, considered as a [±45°] laminate, is subjected to tension with respect to y and to compression with respect to x (Figure 12.11). This loading induces high values of shear damage in the central part of the specimen.

The proposed identification procedure is first evaluated using simulated data. Various simulations are then performed with an in-house FE code using a given damage law [LEC 08]. A plane stress state is assumed. The whole cruciform specimen is meshed (Figure 12.11). The central part of the specimen is uniformly meshed using 17×17 Q4 elements to be consistent with the measured kinematic field. The arms and the material surrounding the ROI are meshed using T3 elements. The computed displacements, more or less corrupted with additional white noise, are used as input data for the identification procedure. All the analyses demonstrate the stability and robustness of the proposed approach [PÉR 09].

The "real" test was carried out on the ASTRÉE multiaxial machine. Tabs glued on the (100 mm large) arms allowed for a transmission of the load to the gauge section (Figure 12.12). Owing to the specimen thickness (i.e. 10 mm), a plane stress state was assumed. Eleven unloading/loading cycles were performed. The details of the loading path and experimental setup are given in [PÉR 02]. Digital images of the surface ($1,016 \times 1,008$ pixel resolution, 8 bit depth) were shot at various loading steps (Figures 12.12(b) and (c)). The images were analyzed using the Q4-DIC algorithm (see Chapter 6). The element size was set to 32 pixels ($\approx$3 mm). The measured displacement fields were subsequently used to identify the parameters of the proposed damage law.

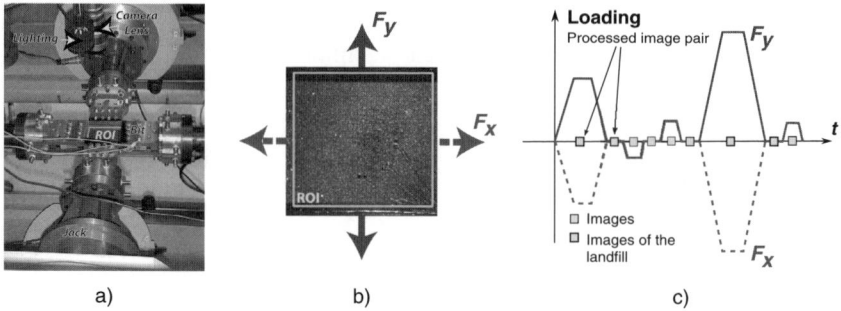

Figure 12.12. *a) Detail of the experimental setup of the shear test, b) ROI and mesh in the initial configuration and c) loading path. Note that "compressive" loads were applied along the x direction and "tensile" loads along the y direction. More details can be found in [PÉR 02]*

When using the first image as the reference, the displacement fields result not only from damage but also from inelastic-related effects. As is usually done when a classic identification procedure (based on tensile tests) is followed, we use 11 loading/unloading cycles. The hysteretic effects are here neglected and the unloading is considered with a frozen state of damage. The first image is taken at maximum shear loading and the second image at the following unloaded state in terms of resultant in each arm (Figure 12.12(c)). The entries of the EGM then correspond to the differences between the displacement fields measured between the reference image and these two images. The elastic parameters are identified using a classic procedure based on tensile tests.

Figure 12.13 represents the identified law while Figure 12.14(b) shows a damage map retrieved for the final loading step.

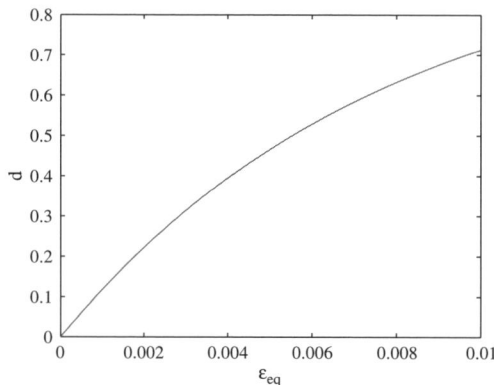

Figure 12.13. *Damage law identified using 11 pairs of images taken during loading/unloading cycles*

The damage field within the ROI can also be computed using a damage post-processor [PÉR 02]. The nonlinear parameters are identified using a classic procedure based on tensile tests. The woven ply is then modeled as a [0, 90] laminate made up of unidirectional plies. Shear damage maps computed for both types of ply are shown in Figure 12.14(a). We can note a very good agreement between these post-processed damage maps and those determined by following the present procedure (Figure 12.14(b)).

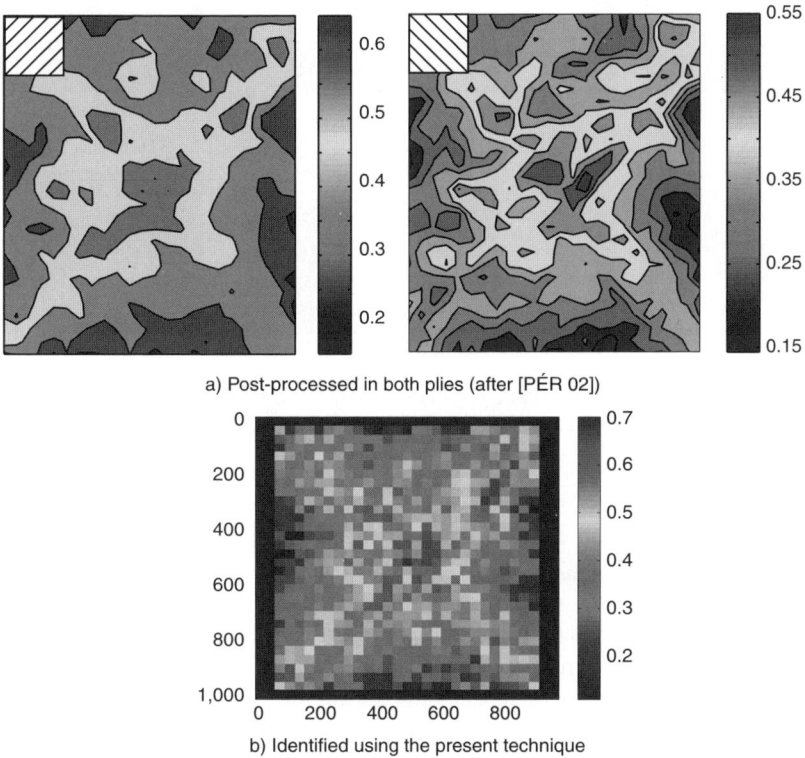

a) Post-processed in both plies (after [PÉR 02])

b) Identified using the present technique

Figure 12.14. *Comparison of shear damage maps obtained using a post-processing procedure [PÉR 02] and the technique proposed here*

It is noteworthy that the present approach enables us to identify a damage law with higher levels of damage (Figure 12.13) than those observed during classic tensile tests [PÉR 02] (typically less than 0.5). Figure 12.15 shows a comparison between the measured and reconstructed displacement fields for the final loading step. The corresponding residual, representing a normalized standard deviation of the difference between the measured and reconstructed displacement fields, is estimated here to be $R = 5 \times 10^{-2}$. This approach has been validated on a different test applied to a layered composite [BEN 11].

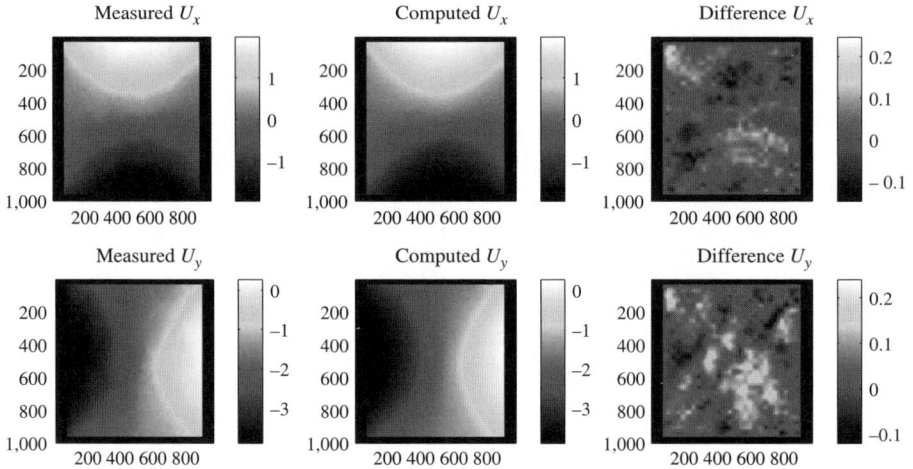

Figure 12.15. *Comparison between measured and reconstructed displacements (expressed in pixels), and corresponding differences for the final loading level*

12.9. Exploitation of measurement uncertainty

We saw in section 12.6 how to adapt the spectral sensitivity to the EGM. This can be converted into $s = 0$ through appropriate reconditioning. This objective is optimal for white noise affecting the displacement measurements.

White noise is, however, often an approximation of the actual uncertainty. Some measurement techniques, such as DIC (see Chapter 6), allow for the exact computation of the covariance of the kinematic degrees of freedom, $\{\mathbf{u}\}$, (for example the nodal displacement for a FE-DIC approach), resulting from white noise in the initial images. The covariance matrix characterizes the mathematical expectation (denoted as $\langle ... \rangle$) of quadratic forms of the fluctuating part $\breve{u}_i$ of the displacement u_i:

$$\text{Cov}(u_i, u_j) \equiv \langle \breve{u}_i \breve{u}_j \rangle \qquad [12.57]$$

Section 12.6 led to a reconditioned EGM where the objective functional to be minimized can be read as a quadratic norm on the displacement field $\{\mathbf{u}\}$ defined with the same discretization as the measured one, or $\widetilde{\mathcal{R}} = \|\mathbf{u}\|^2$. In general, in the case of an arbitrary covariance matrix, it is possible to show that the minimum sensitivity to measurement noise is obtained for a norm of the displacement field that is written as:

$$\|\mathbf{u}\|^2_{cov} = \{\mathbf{u}\}^t [\mathbf{Cov}]^{-1} \{\mathbf{u}\} \qquad\qquad [12.58]$$

The reconditioned EGM exploiting the above expression provides a robust formulation that is tailored specifically to image textures and displacement discretization [LEC 09].

12.10. Conclusions

This chapter is devoted to recent developments of the EGM, in particular, using its reconditioned formulation, and taking explicitly into account the covariance matrix of the displacement measurements. These last two ingredients jointly allow for a significant reduction in measurement noise, which is often the weak aspect in identification problems.

A number of applications to vastly different systems and materials have been presented to highlight the flexibility of the approach to different geometries (e.g. beams) and numerical implementations (FDs, FEs).

We have chosen, in this chapter, to consider identification based on previously determined kinematic fields. This presentation should not hide the fact that the reverse approach is also very fruitful: the (non-reconditioned) equilibrium gap method can equally well assist DIC or digital volume correlation, by providing a more suited kinematic basis, in the sense of being closer to mechanical requirements [RÉT 09]. Very small meshes – down to the pixel or voxel scale [LEC 11, LEC 12] – can be used, thereby providing a very good spatial resolution, and yet the minimization of the equilibrium gap will allow for convergence and mastering of small-scale fluctuations.

Beyond the use of DIC to feed the EGM, or the EGM to regularize DIC, it is clearly from the intimate coupling of kinematic and material property evaluations, performing both determinations simultaneously, that the best performances will be achieved. This emphasizes, if it were necessary, the importance of being able to take advantage of an arbitrary kinematic basis for displacement measurements (see Chapter 6) and reciprocally to exploit these kinematic measurements and uncertainties for identification. The EGM, without being exclusive, is one of the tools allowing for such a dialog [LEC 09]. In addition, the incorporation of other exogenous information (e.g. temperature field and microstructure) to supplement the available data is easily done, using a formalism similar to the formalism used for damage laws. This point will be detailed in Chapter 15. This direction, which has not often been explored, is a very promising direction for the future.

The direction to head for is thus easily readable. The success of these first attempts is a motivation to incorporate complexity, either in the description of local (elastic) properties or in the constitutive laws. This increasing complexity means more parameters to determine from the same experimental corpus of data, until the ill-posedness of the problem becomes the ultimate barrier. However, ill-posedness being dependent on the prior information available about the system, identification techniques aiming at the measurement of mechanical properties have a huge potential waiting to be exploited.

12.11. Bibliography

[AMI 05] AMIOT F., Mesure de champs à l'échelle micrométrique pour l'identification d'effets mécaniques surfaciques: vers une nouvelle instrumentation pour la biologie, PhD Thesis, ENS Cachan, 2005.

[AMI 06] AMIOT F., ROGER J.P., "Nomarski imaging interferometry to measure the displacement field of micro-electro-mechanical systems", *Applied Optics*, vol. 45, pp. 7800–7810, 2006.

[AMI 07] AMIOT F., HILD F., ROGER J.P., "Identification of elastic property and loading fields from full-field measurements", *International Journal of Solids and Structures*, vol. 44, no. 9, pp. 2863–2887, 2007.

[AVR 08] AVRIL S., BONNET M., BRETELLE A.-S., GRÉDIAC M., HILD F., IENNY P., LATOURTE F., LEMOSSE D., PAGANO S., PAGNACCO E., PIERRON F., "Overview of identification methods of mechanical parameters based on full-field measurements", *Experimental Mechanics*, vol. 48, no. 4, pp. 381–402, 2008.

[BEN 11] BEN AZZOUNA M., PÉRIÉ J.-N., GUIMARD J.-M., HILD F., ROUX S., "On the identification and validation of an anisotropic damage model by using full-field measurements", *International Journal of Damage Mechanics*, vol. 20, pp. 1130–1150, 2011.

[BES 06] BESNARD G., HILD F., ROUX S., "'Finite-element' displacement fields analysis from digital images: application to Portevin-Le Châtelier bands", *Experimental Mechanics*, vol. 46, pp. 789–803, 2006.

[CHA 04] CHALAL H., MERAGHNI F., PIERRON F., GRÉDIAC M., "Direct identification of the damage behaviour of composite materials using the virtual fields method", *Composites Part A: Applied Science and Manufacturing*, vol. 35, pp. 841–848, 2004.

[CLA 02] CLAIRE D., HILD F., ROUX S., "Identification of damage fields using kinematic measurements", *Comptes Rendus Mécanique*, vol. 330, pp. 729–734, 2002.

[CLA 04] CLAIRE, D., HILD, F., ROUX, S., "A finite element formulation to identify damage fields: the equilibrium gap method", *International Journal for Numerical Methods in Engineering*, vol. 61, no. 2, pp. 189–208, 2004.

[CLA 07] CLAIRE D., HILD F., ROUX S., "Identification of a damage law by using full-field displacement measurements", *International Journal of Damage Mechanics*, vol. 16, no. 2, pp. 179–197, 2007.

[CRO 08] CROUZEIX L., Identification de champs de propriétés mécaniques de structures composites à partir de mesures de champs de déplacement, PhD Thesis, University of Toulouse III - Paul Sabatier, 2008.

[CRO 09] CROUZEIX L., PÉRIÉ J.N., COLLOMBET F., DOUCHIN B., "An orthotropic variant of the equilibrium gap method applied to the analysis of a biaxial test on a composite material", *Composites Part A: Applied Science and Manufacturing*, vol. 40, no. 11, pp. 1732–1740, 2009.

[DEM 92] DEMMEL J., VESELIC K., "Jacobi's method is more accurate than QR", *SIAM Journal on Scientific and Statistical Computing*, vol. 11, pp. 1204–1246, 1992.

[GAO 99] GAO F., BONIFACE L., OGIN S.L., SMITH P.A., GREAVES R.P., "Damage accumulation in woven-fabric CFRP laminates under tensile loading: part 1. Observations of damage accumulation", *Composites Science and Technology*, vol. 59, no. 1, pp. 123–136, 1999.

[GEE 99] GEERS M.G.D., DE BORST R., PEIJS T., "Mixed numerical-experimental identification of nonlocal characteristics of random fiber reinforced composites", *Composites Science and Technology*, vol. 59, no. 10, pp. 1569–1578, 1999.

[GER 83] GERMAIN P., NGUYEN Q.S., SUQUET P., "Continuum thermodynamics", *ASME Journal of Applied Mechanics*, vol. 50, pp. 1010–1020, 1983.

[HIL 11] HILD F., ROUX S., GUERRERO N., MARANTE M.E., FLÓREZ-LÓPEZ J., "Calibration of constitutive models of steel beams subject to local buckling by using digital image correlation", *European Journal of Mechanics A/Solids*, vol. 30, pp. 1–10, 2011.

[HOC 07] HOCHARD C., LAHELLEC N., BORDREUIL C., "A ply scale non-local fibre rupture criterion for CFRP woven ply laminated structures", *Computers and Structures*, vol. 80, no. 3, pp. 321–326, 2007.

[LAD 92] LADEVÈZE P., LE DANTEC E., "Damage modelling of the elementary ply for laminated composites", *Composites Science and Technology*, vol. 43, pp. 257–267, 1992.

[LAD 95] LADEVÈZE P., "A damage computational approach for composites: basic aspects and micromechanical relations", *Computational Mechanics*, vol. 17, nos. 1–2, pp. 142–150, 1995.

[LEC 07] LECOMPTE D., SMITS A., SOL H., VANTOMME J., VAN HEMELRIJCK D., "Mixed numerical-experimental technique for orthotropic parameter identification using biaxial tensile tests on cruciform specimens", *International Journal of Solids and Structures*, vol. 44, no. 5, pp. 1643–1656, 2007.

[LEC 08] LECLERC H., "Toward higher performance of FEM implementations using lazy evaluation and symbolic programming", *Proceedings of WCCM-8 and ECCOMAS-8 5th European Congress on Computational Methods in Applied Sciences and Engineering*, Venice, 30 June–4 July 2008.

[LEC 09] LECLERC H., PÉRIÉ J.N., ROUX S., HILD F., "Integrated digital image correlation for the identification of mechanical properties", *Lecture Notes in Computer Science*, vol. 5496, pp. 161–171, 2009.

[LEC 11] LECLERC H., PÉRIÉ J.N., ROUX S., HILD F., "Voxel-scale digital volume correlation", *Experimental Mechanics*, vol. 51, pp. 479–490, 2011.

[LEC 12] LECLERC H., PÉRIÉ J.N., HILD F., ROUX S., *Digital Volume Correlation: What are the Limits to the Spatial Resolution?*, forthcoming.

[PÉR 02] PÉRIÉ J.-N., CALLOCH S., CLUZEL C., HILD F., "Analysis of a multiaxial test on a C/C composite by using digital image correlation and a damage model", *Experimental Mechanics*, vol. 42, no. 3, pp. 318–328, 2002.

[PÉR 09] PÉRIÉ J.-N., LECLERC H., ROUX S., HILD F., "Digital image correlation and biaxial test on composite material for anisotropic damage law identification", *International Journal of Solids and Structures*, vol. 46, pp. 2388–2396, 2009.

[RÉT 09] RÉTHORÉ J., ROUX S., HILD F., "An extended and integrated digital image correlation technique applied to the analysis fractured samples. The equilibrium gap method as a mechanical filter", *European Journal of Computational Mechanics*, vol. 18, pp. 285–306, 2009.

[ROU 08] ROUX S., HILD F., "Digital image mechanical identification (DIMI)", *Experimental Mechanics*, vol. 48, no. 4, pp. 495–508, 2008.

[SUT 83] SUTTON M.A., WOLTERS W.J., PETERS W.H., RANSON W.F., MCNEILL S.R., "Determination of displacements using an improved digital correlation method", *Image Vision and Computing*, vol. 1, no. 3, pp. 133–139, 1983.

[VOY 98] VOYIADJIS G.Z., JU J.-W., CHABOCHE J.-L. (eds), "Damage mechanics in engineering materials", *Studies in Applied Mechanics*, vol. 46, 1998.

Chapter 13

Reciprocity Gap Method

13.1. Introduction

The class of problems discussed in this chapter concerns the identification of sources, cracks, boundary conditions or physical parameters on the inner surfaces of a body for which we possess overspecified data on the outer surface of the body.

The direct problem is usually defined as the determination of a solution inside the body from knowledge of its geometry, material data and boundary conditions. The so-called "primal formulation" describes the physical phenomenon and involves the "natural boundary conditions" of the problem, which ensures the existence and uniqueness of the solution. For a given mathematical operator representing the physical phenomenon under scrutiny, for example electrical or thermal diffusion or linear elasticity, overspecified data denote pairs of fields, typically representing Neumann, Dirichlet or Robin boundary conditions when the Laplace operator is under examination. Let us remark that for a well-defined direct problem, only one of the specified fields is necessary to ensure the existence of the solution of the direct problem.

In the case of an elastic problem, knowing both tractions and displacements at the surface for a given problem setting constitutes an overspecified data pair. A similar pattern for heat diffusion would be the simultaneous knowledge of the temperature field and the heat flux on the boundary.

The classic approach to solve this type of problem is to define the unknowns (cracks, inclusions, etc.) as a finite set of parameters, denoted by a vector x. The

Chapter written by Stéphane ANDRIEUX, Huy Duong BUI and Andrei CONSTANTINESCU.

overspecified data will be split into an "input" and an "output". The direct problem setting will provide now a solution $u(\boldsymbol{x})$ using the input data and the "output" can finally be compared with the additional output data. The two output data sets are compared in almost all cases using a least squares cost functional expressed as $MC(u - u^m)$, where u and u^m are the computed and the measured output data, respectively. The identification method becomes therefore a minimization algorithm:

$$\underset{\boldsymbol{x}}{\arg\min}\, MC[u(\boldsymbol{x}) - u^m] \qquad\qquad [13.1]$$

The method is straightforward and can easily be implemented if a solver of the direct problem is available.

This method requires numerous solutions to the direct problem and does not usually enable the formulation of identifiability results, that is specifying whether or not the available data are sufficient to uniquely identify the sought unknowns.

In the particular case where the data are overspecified on the *complete* outer surface of the solid, another approach, called *reciprocity gap*, can be used. This allows us to:

– obtain   *without solving direct problems* an infinite quantity of information concerning the unknown elements;

– *identify explicitly* the elements in certain particular configurations.

The reciprocity gap method relies on a simple idea, generally denoted as *reciprocity*, which underlines the *symmetry* of the operator describing the underlying physics (steady-state diffusion, linear elasticity, etc.). Depending on the area of application, reciprocity takes different names according to the physical matter: Maxwell–Betti reciprocity in elastostatics, Rayleigh reciprocity in harmonic elastodynamics, etc.

This property states that for a solid subjected to two different "loads" C_1 and C_2 and presenting the different responses R_1 and R_2, respectively, the "work" of crossed solicitations will be equal. More specifically, the "work" defined as the loading C_1 in response R_2 equals that of C_2 in R_1. This fundamental property defines self-adjoined operators in mathematical theories and leads to symmetric "stiffness" matrices in the finite element methods.

The idea behind the *reciprocity gap method* is to analyze the gap in reciprocity between the actual field in the real solid that provided the measurements and a fictitious solution field in a fictitious solid in the absence of the unknown elements (cracks, inclusions, sources, etc.). When comparing the two fields, the property of reciprocity is not verified; however, the scalar value of this difference provides access to the "difference" between the real domain and the fictitious solid in the absence of the unknown elements. This difference will further enable the identification of the unknown elements.

As a fictitious solid, we can take the homogeneous solid that can be embedded in an infinite domain, and as such we gain access to an infinite class of closed-form solutions of the direct problem for the fictitious domain. The reciprocity gap provides information about the unknowns from a straightforward integral computation on the boundary.

In several cases, the acquired information can lead to theoretical identifiability results and provide a convenient reconstruction method based on closed-form expressions for the unknowns. One such case is the identification of flat (planar) cracks in electrostatics.

In this chapter, we present three simple examples of identification using the reciprocity gap: (1) identification of planar cracks in steady-state diffusion (thermal or electrical conduction), (2) thermal sources and (3) planar cracks in linear thermoelasticity using only displacement fields for the identification.

13.2. The reciprocity gap method

The Maxwell–Betti reciprocity theorem in elasticity is equivalent to the symmetry property of the bilinear form $a(\boldsymbol{u}, \boldsymbol{v})$, describing the weak formulation of the direct problem. Let us consider the balance of linear momentum of an elastic body Ω, with a boundary $\partial \Omega$ and unit outward normal $\boldsymbol{n}$:

$$a(\boldsymbol{u}, \boldsymbol{v}) = a(\boldsymbol{v}, \boldsymbol{u}) \qquad \forall (\boldsymbol{u}, \boldsymbol{v}) \in \mathcal{V}^2 \tag{13.2}$$

$$a(\boldsymbol{u}, \boldsymbol{v}) = \int_\Omega \varepsilon[\boldsymbol{u}] : \boldsymbol{A} : \varepsilon[\boldsymbol{v}] \, d\Omega \tag{13.3}$$

where $\boldsymbol{A}$ denotes the fourth-order elastic stiffness tensor and $\mathcal{V}$ denotes the set of displacement fields with finite deformation energy.

Using a similar pattern, we can access a Maxwell–Betti-type reciprocity property for all operators describing a physical setting, provided the underlying operator has the previous symmetry. The following list provides a series of examples of symmetric bilinear operators and the corresponding operators of natural boundary conditions σ:

– harmonic elastodynamics at fixed frequency:

$$a(\boldsymbol{u}, \boldsymbol{v}) = \int_\Omega \varepsilon[\boldsymbol{u}] : \boldsymbol{A} : \varepsilon[\boldsymbol{v}] \, d\Omega - \omega^2 \int_\Omega \rho \boldsymbol{u} \cdot \boldsymbol{v} \, d\Omega \tag{13.4}$$

$$\sigma[\boldsymbol{u}] \cdot \boldsymbol{n} = (\boldsymbol{A} : \varepsilon[\boldsymbol{u}]) \cdot \boldsymbol{n} \tag{13.5}$$

where $\boldsymbol{u}, \boldsymbol{v}$ are the real and virtual displacement fields, ε is the small strain operator and σ is the stress operator.

– steady-state diffusion (electric conduction heat problem):

$$a(u, v) = \int_{\Omega} \nabla u \cdot \boldsymbol{k} \cdot \nabla v \, d\Omega \qquad\qquad\qquad [13.6]$$

$$\sigma[u] \cdot \boldsymbol{n} = (\boldsymbol{k} \cdot \nabla u) \cdot \boldsymbol{n} \qquad\qquad\qquad [13.7]$$

where u, v are the real and virtual potential temperature or electrical potential fields, respectively, and σ is the operator describing the boundary flux.

– harmonic acoustics (Helmholtz equation):

$$a(u, v) = \int_{\Omega} \nabla u \cdot \boldsymbol{k} \cdot \nabla v \, d\Omega - \omega^2 \int_{\Omega} u \cdot v \, d\Omega \qquad\qquad [13.8]$$

$$\sigma[u] \cdot \boldsymbol{n} = (\boldsymbol{k} \cdot \nabla u) \cdot \boldsymbol{n} \qquad\qquad\qquad [13.9]$$

where u, v are the real and virtual acoustic potential fields, and σ is the acoustic flux operator.

The general setting is therefore a problem that admits a weak formulation of the type $a(u, v) = l(v)$, where $l(v)$ is the linear application of the work of surface forces. The work $l(v)$ is associated with the natural boundary conditions of the problem under consideration and a is a symmetric bilinear application.

The reciprocity property states that two linear applications l^1 and l^2, expressing the work of external forces producing the solutions u^1 and u^2, respectively, of the underlying problems:

$$a(u^i, v) = l^i(v) \qquad \forall v \in \mathcal{V} \quad i = 1, 2 \qquad\qquad [13.10]$$

will equate the crossed work of the corresponding solutions:

$$l^1(u^2) = l^2(u^1) \qquad\qquad\qquad [13.11]$$

The boundary integrals involve the operator σ of the natural boundary conditions associated with the bilinear application a, defined in the next equation:

$$\forall u \in \mathcal{V}: \qquad a(u, v) = \int_{\partial\Omega} (\sigma[u] \cdot \boldsymbol{n}) \cdot v \, ds \quad \forall v \in \mathcal{V} \qquad [13.12]$$

The precise definition of σ for a series of physical models was presented in the previous list.

The identification problems discussed in this chapter can be expressed using this general framework under the following form:

The measured experiment or the applied load leads to a field u verifying:

$$a_0(u, w) + a_1(u, w) = l(w) + \int_{\partial\Omega_{\text{ext}}} f \cdot w \, ds \qquad \forall w \in V \qquad\qquad [13.13]$$

In this expression, the linear application l (denoting sources or boundary conditions on interior boundaries) and the bilinear application a_1 (denoting material characteristics, inclusions, etc.) are unknown. We possess complete field measurements of $u = U$ on the complete boundary of the domain.

The inverse problem seeks to determine l and a_1 from the overspecified boundary data (U, f).

13.2.1. *Definition of the reciprocity gap*

The reciprocity gap is a linear application defined over the space of potentials with finite energy by the following expression:

$$\mathcal{R} = \int_{\partial\Omega_{\text{ext}}} (f \cdot v - U\sigma_0[v] \cdot \boldsymbol{n}) \, ds \qquad\qquad [13.14]$$

This linear application is computed by integration on the external boundary of the domain and involves only known quantities. Hence, for its computation, we do not need to solve any direct problem.

The interest of this definition lies within the fundamental property of the reciprocity gap, which involves as fictitious fields v of V "in balance" under the known bilinear form A_0 with surface fluxes and tractions on the outside boundary. Let us further denote this space by $V_{A_0}^{eq}$. A complete definition is given by:

$$V_{a_0}^{eq} = \left\{ v \in V \, \Big| \, a_0(v, w) = \int_{\partial\Omega_{\text{ext}}} (\sigma_0[v] \cdot \boldsymbol{n}) \cdot w \, ds \quad \forall w \in V \right\} \qquad [13.15]$$

13.2.2. *Fundamental property of the reciprocity gap*

For all potential fields, the reciprocity gap can also be expressed using the following expression:

$$\mathcal{R}(v) = \ell(v) - a_1(u, v) \qquad \forall v \in V_{a_0}^{eq} \qquad\qquad [13.16]$$

The proof of this property is straightforward. Indeed, taking as a test field $w = u$ in the characterization of v, and $w = v$ in the equilibrium condition of u, we obtain only:

$$a_0(v, u) = \int_{\partial\Omega_{\text{ext}}} (\sigma_0[v] \cdot \boldsymbol{n}) \cdot u \, ds \qquad [13.17]$$

and

$$a_0(u, v) + a_1(u, v) = \ell(v) + \int_{\partial\Omega_{\text{ext}}} f \cdot v \, ds \qquad [13.18]$$

The sought equality follows from the difference of the previous equation, simple algebra and the symmetry of the bilinear application a_0, which is the essential property here.

For each equilibrated potential field v, we can obtain, by computing the reciprocity $\mathcal{R}(v)$ directly from its definition, that is a boundary integration, valuable information about the main unknowns of the problems ℓ and a_1. The technique of the reciprocity gap method is reduced essentially to the wise choice of the fictitious potentials v. This technique leads to either theoretical identification results or constructive partial or complete reconstruction results. The theoretical identification results answer the following question: does a unique pair (ℓ, a_1) exist? In other words, does there exist a single set of sources, cracks, inclusions for which the response of the solid coincides with the measured boundary data? The reconstruction techniques provide methodologies for completely or partially characterizing inclusions, cracks, sources, etc., contained in the solid.

The chapter next presents a series of examples of such results. Note that no direct problem solving is required to apply the reciprocity gap method if analytical expressions of the auxiliary fields v are available. The calculations are limited to integrations on the surface of the body under scrutiny. This specificity of the method gives rise to an extremely fast and efficient identification technique. However, we should take into consideration its two major limitations: (1) the need for data over the entire boundary of the solid and (2) the lack of a systematic method for choosing the fictitious potential fields v.

Some proposals for reducing both limitations have been made, but they are beyond the scope of this chapter.

13.3. Identification of cracks in electrostatics

In the case of flat planar cracks, it is possible to formulate closed-form expressions of the normal of the plane containing one or more cracks, the complete position of the plane and even the complete extension of the crack in this plane. Let us now analyze the case of the steady-state heat conduction problem.

In thermal balance, if the heat flux over the boundary is known to be equal to f, the partial differential equations of the underlying problem are described for a thermal conductivity k and a temperature field T as:

$$\mathrm{div}(k\nabla T) = 0 \quad \text{in } \Omega, \qquad k\nabla T \cdot \boldsymbol{n} = f \quad \text{on } \partial\Omega_{\text{ext}} \tag{13.19}$$

If the body under discussion, Ω, contains a crack Γ, its boundary $\partial\Omega$ is defined by the external surface $\partial\Omega_{\text{ext}}$ and the crack lips Γ_j, $j = 1, N$ (see [13.1]).

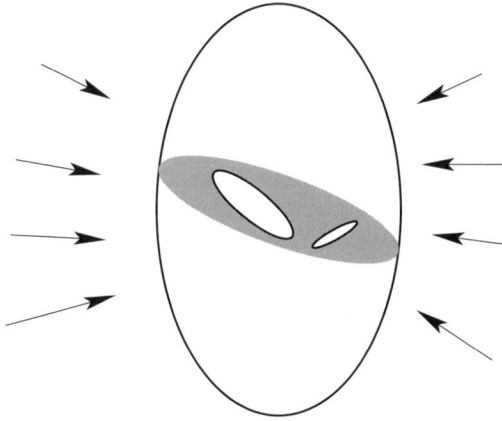

Figure 13.1. *Flat planar cracks in the interior of a body*

Let us assume that on the crack lips Γ_j, $j = 1, N$, the boundary conditions are defined as a nonlinear contact resistivity, that is a relation representing the normal flux with the temperature jump over the lips: $[\![T]\!]$.

$$k\nabla T \cdot \boldsymbol{N}_j = r_j([\![T]\!])\text{on } \Gamma_j, \quad \forall j = 1, \text{etc.}, N \tag{13.20}$$

Let us finally assume that the temperature is measured on the outer surface of the body and the field T^m is known.

Under these conditions, the reciprocity gap is defined as:

$$\mathcal{R}(v) = \int_{\Omega_{\text{ext}}} [f \cdot v - (k\nabla v \cdot \boldsymbol{n}) \cdot T^m]\, ds \tag{13.21}$$

A series of straightforward computations conclude that the fundamental property of the reciprocity gap is given by:

$$\mathcal{R}(v) = \sum_{j=1,N} \nabla v \cdot \boldsymbol{N}_j [\![T]\!] \, ds \quad \forall v \text{ such that } \mathrm{div} k \nabla v = 0 \quad \text{in} \quad \Omega \qquad [13.22]$$

13.3.1. Identification formulas for the plane of the crack(s)

Let us suppose that the cracks are situated within the same plane Π of unit normal $\boldsymbol{N}$. The choice of linear fictitious auxiliary functions v, more precisely coordinate projections of the current point $\boldsymbol{x}$:

$$x_i(\boldsymbol{x}) = x_i \qquad i = 1, 2, 3 \qquad [13.23]$$

leads to the following formulas defining the unit normal of the plane and the mean value of the temperature jump over the crack:

$$\boldsymbol{N} = \frac{\boldsymbol{L}}{\|\boldsymbol{L}\|} \qquad \boldsymbol{L} = \mathcal{R}(x_i)\boldsymbol{e}_i \quad i = 1, 3 \qquad [13.24]$$

and

$$\left\| \sum_{j=1}^{N} \boldsymbol{e}_j \int_{\Gamma_i} [\![T]\!] \, ds \right\| = \frac{1}{k} \|\boldsymbol{L}\|, \qquad [13.25]$$

respectively.

Similarly, we can completely determine the plane Π using quadratic fields. In this case, it is important to be aware of the fact that not all quadratic fields are harmonic and, therefore, do not satisfy the balance conditions over a homogeneous body, that is an area without cracks. However, by effecting a change of reference frame x_k $k = 1, 2, 3$ with coordinate x_3 directed along the normal $\boldsymbol{N}$ that has just been determined, the equation for plane Π is given by $x_3 + C = 0$. Constant C is obtained by using the auxiliary field:

$$v(\boldsymbol{x}) = x_2^2 - x_3^2 \qquad [13.26]$$

As the normal to the crack plane is oriented toward x_3, we obtain:

$$\nabla v \cdot \boldsymbol{N} = -2x_3$$

The integration of reciprocity for the heat equation [13.22] over the plane of the crack defined by $x_3 + C = 0$, for the particular field v defined previously, leads to the following value of the constant:

$$C = \frac{1}{2\|\boldsymbol{L}\|} \mathcal{R}(x_2^2 - x_3^2) \qquad [13.27]$$

These formulas show that a single measurement identifies the plane including the cracks and that its determination requires only four simple calculations:

(i–iii) for i=1,2,3:

$$\mathcal{R}(x_i) = \int_{\Omega_{\text{ext}}} \left[f x_i - k n_i T^m \right] ds$$

(iv)

$$\mathcal{R}(x_2^2 - x_3^2) = \int_{\Omega_{\text{ext}}} \left[f(x_2^2 - x_3^2) - k(x_2 n_2 - x_3 n_3) T^m \right] ds$$

An illustration of an indentified crack plane is shown in Figure 13.3. It is interesting to note that better accuracy for the plane position is obtained in a finite element computation when it is quadratic in the calculation of the temperature T and the auxiliary fields. The error observed for linear elements derives directly from the approximation error of quadratic fields with such elements.

The identifiability condition derives from the previous arguments:

$$\int_\Gamma [\![T]\!] \, ds \neq 0 \qquad\qquad [13.28]$$

This condition is interpreted simply by the requirement that the stress applied to a solid generally "excites" the cracks. The cracks are detectable only if they are illuminated by a temperature discontinuity (because the source absence imposes the continuity of heat flow). If the loading does not lead to these discontinuities, the test is not relevant. Figure 13.2 shows such a situation.

13.3.2. *Complete identification of cracks*

Once the crack plane is completely determined, it is possible to choose further auxiliary fields to completely identify the extension of the crack, and more precisely to determine their position and shape in the plane. The shape in the plane will be reconstructed from the Fourier decomposition of the temperature jump in plane Π. The reconstruction is based on a mathematical result that ensures that the support of the cracks coincides with the support of the temperature discontinuities. In other words, it is physically impossible for the temperature discontinuity to vanish on a measureable part of the finite field occupied by the cracks (see Figure 13.4).

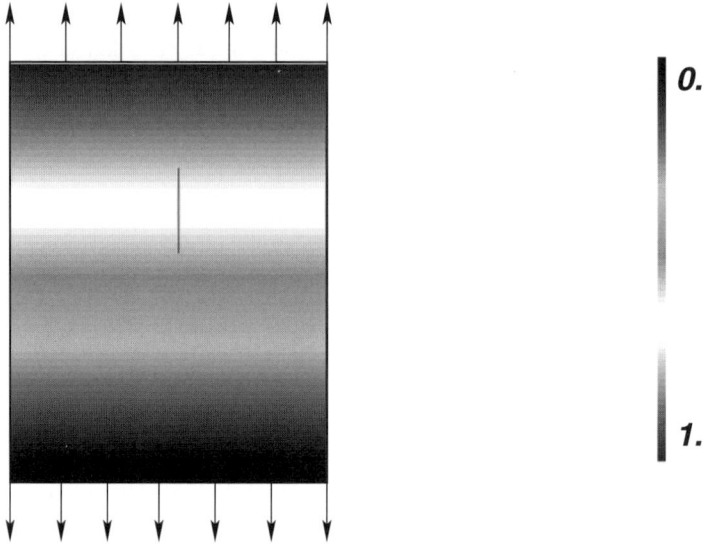

Figure 13.2. *Example of a loading that does not enable the identification of planar cracks. Gray levels are isovalues of the potential* u

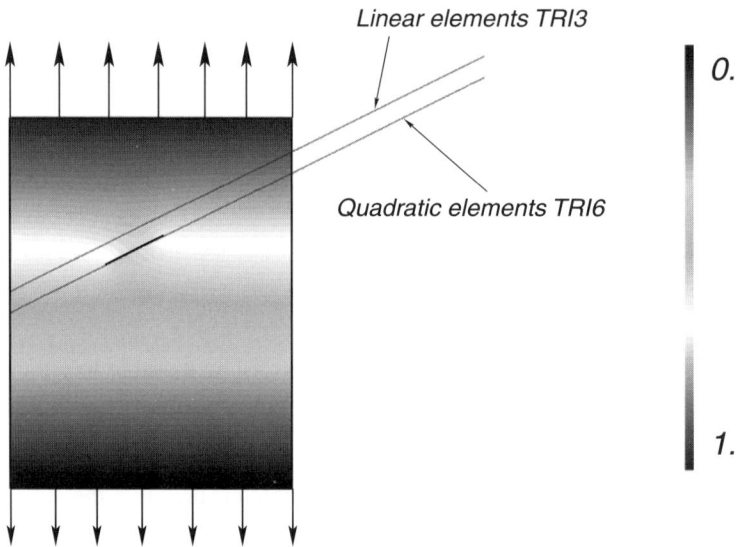

Figure 13.3. *Identification example of the position of the plane of the crack for reconstructions using linear or quadratic finite elements for the representation of the temperature* T *and the auxiliary function* v

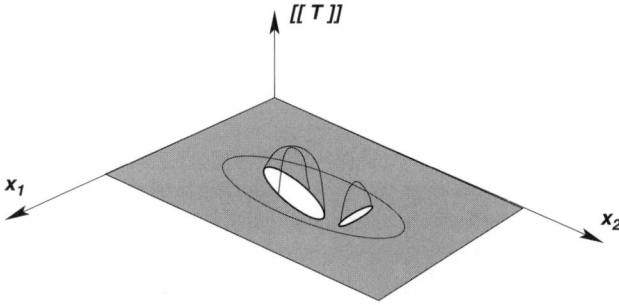

Figure 13.4. *Extension of the temperature discontinuity $[\![T]\!]$ with a zero field in a rectangle R containing the intersection of plane Π with body Ω*

By extending the temperature jumps $[\![T]\!]$ with zero outside the cracks in the rectangle $R = [0, L] \times [0, H]$, which contains the intersection of plane Π with body Ω, and by using auxiliary fictitious functions w_{pq}^{cc}, w_{pq}^{cs}, w_{pq}^{sc}, which are a combination of sine and cosine functions, defined as:

$$w_{pq}^{ss}(X_1, X_2, X_3) = \frac{1}{\sqrt{\lambda_{pq}}} \sin\left(p\pi \frac{X_1}{L}\right) \sin\left(p\pi \frac{X_2}{H}\right) \sinh\left(\sqrt{\lambda_{pq}} X_3\right)$$

$$\lambda_{pq} = \pi^2 \left(\frac{p^2}{L^2} + \frac{q^2}{H^2}\right)$$

we obtain the following expression of the reciprocity gap:

$$\mathcal{R}(w_{pq}^{ss}) = \int_R [\![T]\!] \sin\left(p\pi \frac{X_1}{L}\right) \sin\left(p\pi \frac{X_2}{H}\right) dX_1 dX_2 \qquad [13.29]$$

This expression enables us to recover the Fourier expansion of the temperature discontinuity $[\![T]\!]$.

Similar results have equally been obtained in linear isotropic elasticity (1) for the Lamé operator in [AND 92] and (2) for the Helmholtz operator in [BEN 05]. The result of the reconstruction using the Fourier expansion is shown in Figure 13.5.

13.4. Crack identification in thermoelasticity using displacement measurements

Let us consider a heterogeneous, isotropic and linear elastic body subjected only to thermal loading. The stress–strain relation is, in this case, expressed as:

$$\boldsymbol{\sigma} = \boldsymbol{A} : (\varepsilon - \alpha T \boldsymbol{I}_2) \qquad \boldsymbol{A} \frac{\nu E}{(1+\nu)(1-2\nu)} \boldsymbol{I}_2 \otimes \boldsymbol{I}_2 + \frac{E}{1+\nu} \boldsymbol{I}_4 \qquad [13.30]$$

where α is the thermal expansion coefficient, E is the Young's modulus, ν is the Poisson coefficient and T is the temperature increase with respect to the reference temperature of the material. $\boldsymbol{I}_2$ and $\boldsymbol{I}_4$ will denote the fourth- and second-order identity tensors, respectively.

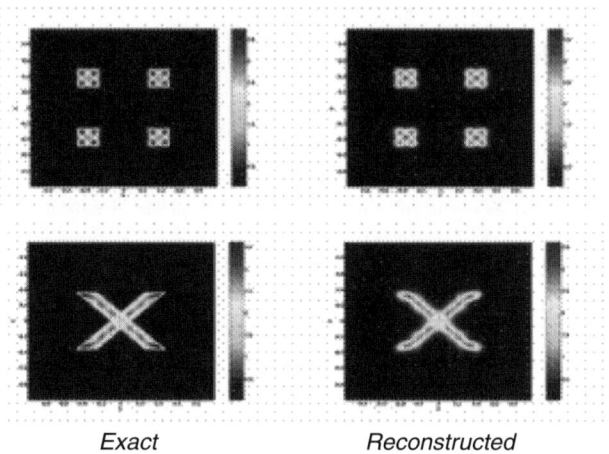

Exact   Reconstructed

Figure 13.5. *Multiple cracks and non-convex cracks: temperature discontinuities, original fields and their reconstructions [BEN 05]*

An interesting test case is that of imposed, but not measured, thermal loading on a traction-free solid. The measurement will imply *only* the displacement field $\boldsymbol{u}^r$ on the exterior surface.

In the case of small strain and neglecting inertial forces, the thermomechanical balance equations and the transient heat equation are expressed by the following expressions:

	Thermal evolution	Thermoelastic balance
in $\Omega\backslash\Gamma$	$\rho c \dot{T} - \mathrm{div}k\nabla T = s$	$\mathrm{div}\boldsymbol{\sigma} = 0$
on Γ	$k\nabla T \cdot \boldsymbol{N} = r_T([\![T]\!])$	$\boldsymbol{\sigma} \cdot \boldsymbol{n} = r([\![u]\!])$
on $\partial\Omega$	$k\nabla T \cdot \boldsymbol{n} = \Phi$	$\boldsymbol{\sigma} \cdot \boldsymbol{n} = \boldsymbol{0}$
on $\partial\Omega$		$\boldsymbol{u} = \boldsymbol{u}^D$
in $\Omega\backslash\Gamma$	$T = T^0$	

where s is the heat source, Φ is the surface flux, T_0 is the initial temperature, r_T is the thermal resistance and r is the contact stiffness of the cracks. All these parameters are considered to be unknown.

The reciprocity gap in terms of mechanical fields is defined for each time instant $t \in (0, D)$ in the now usual sense for the displacements and traction fields by the formula:

$$\mathcal{R}_t(v) = - \int_{\partial \Omega} \boldsymbol{u}^\tau \cdot (\boldsymbol{A} : \varepsilon((v))) \cdot \boldsymbol{n} \; ds \qquad [13.31]$$

where the traction-free condition at the exterior surface of the body has already been taken into account.

For auxiliary displacement fields v, with the additional properties of *assuring the balance of forces and zero divergence*, the interpretation of the reciprocity gap is given by the following equation:

$$\mathcal{R}_t(v) = \int_\Gamma [\![u]\!] \cdot (\boldsymbol{A} : \varepsilon(v) \cdot \boldsymbol{N}) \; ds \qquad [13.32]$$

for all auxiliary displacement fields of the set:

$$\mathcal{V}_{\text{div}} = \left\{ v \mid \int_\Omega \varepsilon(v) : \boldsymbol{A} : \varepsilon(w) = \int_{\partial \Omega} (\boldsymbol{A} : \varepsilon(v)) \cdot \boldsymbol{n} \; \forall w \text{ with div} w = 0 \right\}$$

$$[13.33]$$

The zero divergence property of the auxiliary fields partly erases the coupling with the thermal problem. As a result, we can exploit only the displacement measurements at the surface of the body and neglect the information about its thermal state.

Let us define the following auxiliary displacement fields with zero divergence:

$$v_i(\boldsymbol{x}) = 2(3x_i e_i - \boldsymbol{x}) \quad i = 1, 3 \qquad [13.34]$$

$$w(\boldsymbol{x}) = 2(x_3 x_2 e_1 + x_1 x_3 e_2 + x_2 x_1 e_3) \qquad [13.35]$$

If the cracks are included in the plane Π of normal $\boldsymbol{N}$, we obtain the following expressions for the reciprocity gap tensor:

$$\hat{\boldsymbol{R}} = \text{dev } \boldsymbol{R} = \boldsymbol{R} - \frac{1}{3}(\text{tr } \boldsymbol{R}) \boldsymbol{I}_2 \qquad [13.36]$$

$$\boldsymbol{R} = \frac{1}{2}(\boldsymbol{N} \otimes \boldsymbol{U} + \boldsymbol{U} \otimes \boldsymbol{N}) \qquad [13.37]$$

$$\boldsymbol{U} = \int_\Gamma [\![u]\!] \; ds \qquad [13.38]$$

The components of the tensor are related to the reciprocity gap as expressed in the following relations:

– If $i \neq j$, then for k different from indices i and j, we have:

$$\hat{R}_{ij} = \frac{1}{8\mu} \mathcal{R}_t(\boldsymbol{w}_{,k}) \quad k \neq j \, k \neq j$$

– If $i = j$, then:

$$\hat{R}_{ii} = \frac{1}{12\mu} \mathcal{R}_t(\boldsymbol{v}_i) \qquad \text{(without summing)}$$

The main question is how to retrieve the normal unit vector $\boldsymbol{N}$ and the vector field of the displacement dicontinuity $\boldsymbol{U}$ from the knowledge of $\hat{\boldsymbol{R}}$.

Let us consider the eigenvalues $(\lambda_1, \lambda_2, \lambda_3)$ of $\hat{\boldsymbol{R}}$, in decreasing order of corresponding unit eigenvectors $(\boldsymbol{g}_1, \boldsymbol{g}_2, \boldsymbol{g}_3)$. The vectors $\boldsymbol{N}$ and $\boldsymbol{U}$ are in one of the two cases: they are either (1) colinear or (2) not colinear.

If $\boldsymbol{N}$ and $\boldsymbol{U}$ are colinear vectors, then we have:

$$\boldsymbol{U} = U\boldsymbol{N}$$

and as a necessary condition, we further have $\boldsymbol{N} = \boldsymbol{g}_1$. Because $\boldsymbol{R} = U\boldsymbol{N} \otimes \boldsymbol{N}$, the eigenvalues of $\hat{\boldsymbol{R}}$ are related to U by the following relations:

$$\lambda_1 = \frac{2}{3}U \qquad \lambda_2 = \lambda_3 = -\frac{1}{3}U$$

if $U > 0$.

If $\boldsymbol{N}$ and $\boldsymbol{U}$ are not colinear vectors, let us note:

$$\boldsymbol{U} = U\boldsymbol{d}_u$$

Then, we can easily prove that $\boldsymbol{u} \times \boldsymbol{N}$ is colinear with $\boldsymbol{g}_3$. As a result, we obtain:

$$\boldsymbol{N} = n_1\boldsymbol{g}_1 + n_3\boldsymbol{g}_3 \qquad \boldsymbol{d}_u = n_1\boldsymbol{g}_1 - n_3\boldsymbol{g}_3$$

A complete discussion of the order of eigenvalues [AND 06] and the assumption that $U > 0$ lead to the following relationship for directions 1 and 3:

$$n_1 = \frac{\sqrt{\lambda_1 - \lambda_2}}{\sqrt{\lambda_1 - \lambda_3}} \qquad n_3 = \frac{\sqrt{\lambda_2 - \lambda_3}}{\sqrt{\lambda_1 - \lambda_3}}$$

It is important to note that it is impossible to distinguish between the directions N and U.

In a practical application, it must be considered that N will stay constant over time while u varies, in general, with the load and therefore with time.

Finally, the amplitude of the displacement jump U is given by:

$$U = \left\| \int_\Gamma [\![u]\!]\, ds \right\| = \sqrt{\lambda_1 - \lambda_3}$$

Once the normal is known (using the above formulas at least two different time instants), the following change of coordinates can be performed: the vector E_3 is taken parallel to the normal to plane Π, which is now defined by $x \cdot e_3 + C = 0$.

We only need to identify constant C locating the affine plane containing the cracks. This follows from the reciprocity gap computation for the auxiliary displacement field h:

$$C = -\frac{1}{6\mu \displaystyle\int_\Gamma [\![u_x]\!]\, ds} \mathcal{R}_t(h) \quad \text{with} \quad h = 3(x_3^2 - x_2^2)e_1 \qquad [13.39]$$

13.5. Conclusions and perspectives

The reciprocity gap technique has a great advantage: that is its simplicity. A number of identification problems have been solved from a theoretical or numerical point of view. The identification of planar cracks for different families of auxiliary fields has been discussed in the following cases:

– scalar diffusion equation, that is *steady-state heat equation* or *electric conductivity* in [AND 92, AND 99, AND 96];

– *transient heat equation* in [BEN 01];

– *transient acoustics* (wave propagation with a scalar acoustic potential) in [BUI 99]:

– *elastostatics* (diffusion equation with vector potential) in [BAN 97, BAN 99]

– *elastodynamics* (wave propagation with vector potential) in [BUI 04, BUI 05];

– *small strain coupled thermoelasticity* in [AND 06];

– *damped elastodynamics* with Zehner-type viscoelasticity (Helmholtz equation) in [BUI 10];

– *acoustic dispersion* (Helmholtz equation with a scalar potential) [COL 05, BEN 05];

The disadvantage of the method, as indeed for a number of inverse problems, is the necessity of knowing the measured fields over the complete outer boundary of the body. The search for suitable ways to overcome this difficulty remains an open problem for the years to come.

13.6. Bibliography

[AND 92] ANDRIEUX S., BEN ABDA A., "Identification des fissures planes par une donnée de bord unique: un procédé direct de localisation et d'identification", *Comptes Rendus de l'Académie des Sciences (Série I)*, vol. 315, pp. 1323–1328, 1992.

[AND 96] ANDRIEUX S., BEN ABDA A., "Identification of planar cracks by complete overdetermined data: inversion formulae", *Inverse Problems*, vol. 12, no. 5, pp. 553–563, 1996.

[AND 99] ANDRIEUX S., BEN ABDA A., BUI H.D., "Reciprocity principle and crack identification", *Inverse Problems*, vol. 15, no. 1, pp. 59–65, 1999.

[AND 06] ANDRIEUX S., BUI H.D., "Écart à la réciprocité et identification de fissures en thermo-élasticité isotrope transitoire", *Comptes Rendus de l'Académie des Sciences (Série I)*, vol. 334, no. 4, pp. 225–229, 2006.

[BAN 97] BANNOUR T., BEN ABDA A., JAOUA M., "A semi-explicit algorithm for the reconstruction of 3D planar cracks", *Inverse Problems*, vol. 13, no. 4, pp. 899–917, 1997.

[BAN 99] BANNOUR T., BEN ABDA A., JAOUA M., "Identification of 2D cracks by elastic boundary measurements", *Inverse Problems*, vol. 15, no. 1, pp. 67–77, 1999.

[BEN 01] BEN ABDA A., BUI H.D., "Reciprocity principles and crack identification in transient thermal problems", *Inverse and Ill-Posed Problems*, vol. 9, no. 1, pp. 1–6, 2001.

[BEN 05] BEN ABDA A., DELBARY F., HADDART H., "On the use of the reciprocity-gap functional in inverse scattering from planar cracks", *Mathematical Models and Methods in Applied Sciences*, vol. 15, no. 10, pp. 1553–1574, 2005.

[BUI 99] BUI H.D., CONSTANTINESCU A., MAIGRE H., "Diffraction acoustique inverse de fissure plane: solution explicite pour un solide borné", *Comptes Rendus de l'Académie des Sciences (série II)*, vol. 327, pp. 971–976, 1999.

[BUI 04] BUI H., CONSTANTINESCU A., MAIGRE H., "Numerical identification of planar cracks in elastodynamics using the instantaneous reciprocity gap", *Inverse Problems*, vol. 20, pp. 993–1001, 2004.

[BUI 05] BUI H.D., CONSTANTINESCU A., MAIGRE H., "An exact inversion formula for determining a planar fault from boundary measurements", *Inverse and Ill-Posed Problems*, vol. 13, no. 6, pp. 553–565, 2005.

[BUI 10] BUI H.D., CHAILLAT S., CONSTANTINESCU A., GRASSO, E., "Identification of a planar crack in Zener type viscoelasticity", *Annals of Solid and Structural Mechanics*, vol. 1, pp. 3–8, 2010,

[COL 05] COLTON D., HADDAR H., "An application of the reciprocity gap functional to inverse scattering theory", *Inverse Problems*, vol. 21, no. 1, pp. 383–398, 2005.

Chapter 14

Characterization of Localized Phenomena

14.1. Introduction

Localization (the progressive or sudden process of the concentration of physical phenomena in narrow structures in which their intensity is amplified) is a vast subject whose study in this chapter is limited to the *localization of strains* in deformable bodies. We show that, in this case, localization can take many, and sometimes complex, configurations. In particular, it can take on the form of strain or displacement discontinuities, for example shear bands and cracks. The theoretical study and the modeling of these essential phenomena, which are related to fracture, require the maximum possible knowledge of their causes. This chapter illustrates how, over the last three decades (and particularly in the last few years), tools based on quantitative image analysis of the deforming object have led to a better understanding of the physics of localization. These studies have pinpointed the existence of sequences of mechanisms, which would have gone unnoticed in a simple examination of the final configurations. Today, some developments of measurement techniques through digital image correlation (DIC) are following the same trend as modeling, and the same discretization techniques are being used to measure the kinematics in one case and to model it in the other. Another major breakthrough in the experimenter's toolbox was the switch to three-dimensional (3D) due to a combination of tomography that enables us to "see" inside 3D solids and digital correlation extended to these 3D images. In the following sections, we present these methodological advances and discuss them by giving some representative examples.

Chapter written by Jacques DESRUES and Julien RÉTHORÉ.

14.2. Definitions and properties of the localized phenomena being considered

The localized phenomena, which are the subject of this chapter, can be of very diverse natures. Let us start with a crude definition of a *localization structure* as the concentration of the intensity of a quantity within one or several zones that are well defined geometrically and are observable (given the necessary technical tools). Thus, even from the viewpoint of solid mechanics alone, due to the many coupling mechanisms that exist between mechanics and other phenomena (physical, chemical, and even biological), numerous quantities relevant to a given mechanical problem, such as temperature, electrical conductivity, ultrasound wave velocity, density and permeability, can take on localization structures. These coupling mechanisms can play a role at the very source of the localization or merely occur beside it. Here, we will focus on the localization of kinematic quantities, that is on the *localization of deformations*, which is often simply called *localization* in solid mechanics. This can be defined as the transition from a spatially continuous deformation process to a discontinuous process. A first examination distinguishes between *cracks* and *shear bands*. A more detailed inspection, especially by changing the scale of observation, may render this distinction less obvious; this point will be addressed later. However, if we retain this first level of analysis for the sake of conceptualization, the two classes will differ in the nature of the discontinuity, which concerns either the displacement field or the strain field. The first case corresponds to the loss of continuity by the solid itself, leading to what is commonly referred to as the *crack*. This is what Thomas [THO 61] calls a *strong discontinuity* of the displacement field. Conversely, the second case is that of a *weak discontinuity* in which the displacement field remains continuous, but one or several of its derivatives are discontinuous. The kinematic object that results from this situation is a *shear band*.

The simplest situation in which localization of the deformation can be observed is that of the *elementary test*. This refers to a test carried out on an initially homogeneous specimen that is loaded mechanically in such a way that it could, ideally, remain homogeneous. Although such ideal experimental conditions are practically unattainable, they can reasonably be approximated through proper care in the preparation of the specimens and in the boundary conditions that produce the loading. Then, at the beginning of the loading, the strain is not only diffuse *a priori*, but also homogeneous (i.e. constant over the domain being studied). In addition, the characteristic being recorded, such as a force versus a displacement, can be interpreted directly in terms of a constitutive law of the stress–strain type (which can be nonlinear and even incrementally nonlinear). This confers on these tests, which are called *laboratory tests*, a special importance in the characterization of materials. Nevertheless, very often, this type of test results in a loss of homogeneity of the deformation, which becomes localized in the form of cracks or shear bands (or for some media, especially in the case of geological materials, compaction). This phenomenon is associated with fracture because in many cases, it is almost immediately followed by an often catastrophic loss of strength in the structure that has replaced the (no longer homogeneous) specimen.

Among the elementary tests, depending on the domains of investigation and materials, we encounter different specimen geometries, primarily plane (especially for metals) or cylindrical (various types of materials). Such types of tests are either "uniaxial" (simple tension or compression) or "triaxial" (in cylindrical coordinates) with two identical principal stresses over the lateral sides of the specimen. We also find the so-called "biaxial" tests, which are no less triaxial than the previous tests, especially when they are performed in plane strain, and lead to three different principal stresses. There are also, much more rarely, truly triaxial tests in which the three loading directions are controlled independently. Finally, for the sake of completeness, let us mention that elementary, or quasi-elementary, tests have been developed in order to introduce loading cases with rotating principal axes (shearing): these tests involve specific geometries and loading conditions, such as hollow tubes loaded in tension–torsion or compression–torsion, along with various types of homogeneous "shear boxes" capable of transferring shear stresses at the boundaries of the specimens.

All of these elementary tests, without exception, can have localized responses. The case of a strain that remains diffuse throughout the test cannot entirely be ruled out[1], but experimentally established cases of non-localized fracture are rare, and the lack of mention of strain (or the mention of a strain that is considered to remain homogeneous until the end of the test) is often an indication of too cursory an observation, or even an absence of observation, of the strain field. Using bifurcation theory [RIC 76], it can be theoretically proved that the occurrence of a shear band in a homogeneous medium subjected to a homogeneous loading history is rooted in the behavior of the material itself, given that it has reached its strength limitation and its behavior is nonlinear and dependent on the loading direction (e.g. the difference in incremental response between loading and unloading). Thus, localization is not associated with imperfect test conditions or uncontrolled perturbations, but it is part of the material behavior. Therefore, it deserves to be studied because it is of highest practical importance when, due to the overloading of a structure or building, or to forced fracture in forming processes (manufactured materials) or civil engineering works (geological materials), the limit state of the material is reached. Let us discuss some examples of localized strain.

Figure 14.1 shows the distinction between *diffuse heterogenous deformation modes* and *localized deformation*. The figure shows the final deformations of two

---

1 The diffuse failure cases mentioned in the literature correspond to specific loading paths that can lead to failure within the resistance domain defined by the localization criterion; such a case can be found, for example, in the liquefaction of granular media as a result of forced isochoric behavior due to the trapping of interstitial water in the pores (see [DES 06] among many references).

identical clay specimens tested under triaxial axisymmetric loading with confinement, one in extension (a) and the other in compression (b). In both tests, all three principal stresses are compressive, but the axial stress in the extension test, contrary to the compression test, is less than the two radial stresses and, therefore, becomes the so-called minor principal stress. We can observe some necking (in the extension case) and some barreling (in the compression case). These two geometric effects are comparable and they result from radial compressive strain in the first case and from the radial extension strain in the second case, accompanied by a limitation of the radial strain in the vicinity of the heads due to the friction of the material against them. Generally, necking is self-catalytic because a reduction in the cross section increases the axial stress in the middle section of the specimen and, therefore, increases the stress deviator locally; the opposite occurs in barrel deformation. Thus, extension could possibly lead to structural instability. However, localization traces on the surface can be observed with both specimens: several parallel shearing planes are visible in the extension case, and there is a single, very visible and much more pronounced one in the compressive case. These minor differences cannot be generalized; for example, only one plane can be observed in extension tests on sand. However, the common fact in both cases is that although the diffuse heterogeneous strains occurred before localization, they did not hamper its development. It is also interesting to note that the orientations of the shearing planes, although very different in the two cases when measured with respect to the axes of the specimens, are quite comparable if measured with respect to the major principal stress direction, which is along the axis in the compression case, but perpendicular to the axis in the extension case.

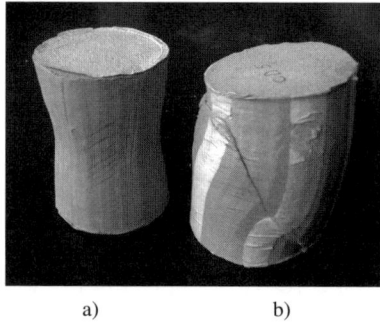

a)   b)

Figure 14.1. *Two clay specimens tested under triaxial axisymmetric loading with confinement, one in extension a) and the other in compression b). Both specimens present localized strains as indicated by several parallel planes in extension and a single shearing plane in compression*

Figure 14.1 also shows an important aspect of localization phenomena, that is they depend on the material microstructure: in this clay, the fineness of the particles compared to the size of the specimen is such that the shear bands can be viewed, under direct observation, as zero thickness lines. On this observation scale, the distinction between Mode-II fracture and a shear band is not obvious. However, on a microstructural scale, the qualitative observation under microscopic imaging enables us to distinguish displacement discontinuity from strain discontinuity (see, e.g. [RAY 08]). On the contrary, in coarser granular media such as sands, we can easily pinpoint the effect of the particle size on the thickness of the shear bands: Figure 14.2, from studies by Colliat-Dangus [COL 86, COL 88], gives a clear illustration in the case of two dense specimens of Hostun sand tested under triaxial compression loading with 100 kPa confinement pressure. The two specimens were of the same size (10 cm initial diameter) and made up of two kinds of sand with similar mineral composition, but different particle sizes: 0.32 mm average diameter on the left and 3.2 mm on the right. These examples clearly show that the thickness of the shear bands increases with particle size.

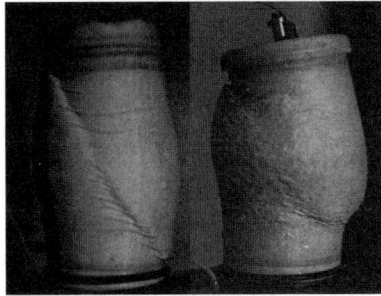

Figure 14.2. *Localized deformations in two sand specimens tested under triaxial axisymmetric loading with confinement: effect of the sand particle size on the thickness of the shear band (adapted from [COL 86, COL 88])*

Thus, the distinction between the two kinematic discontinuity modes is more complex than anticipated. In particular, we can observe *the effect of the average stress* on the occurrence of shear bands rather than cracks, and vice versa. This is shown in Figure 14.3, from the works of Bésuelle [BÉS 00]: on a series of red Vosges sandstone specimens subjected to triaxial axisymmetric compression tests with confinement pressures from 0 to 60 MPa, we can observe a variety of localized strain patterns, from quasi-brittle fracture with axial cracking in the absence of confinement – uniaxial compression – through single shearing plane patterns under intermediary pressures, to multiple intersecting shearing planes under high confinement. It can be observed that the slope of the shearing planes with respect to the specimen axis increases with the confinement pressure level. The shearing planes are directly visible to the naked eye due to the whitening of the material in the zones with high shear stresses. This phenomenon does not appear at the level of the cracks in specimens tested under low confinement pressure, which confirms that the traces observed at higher pressures correspond to shear bands of a certain thickness within which the material has undergone large enough strains to induce changes in its visual aspect.

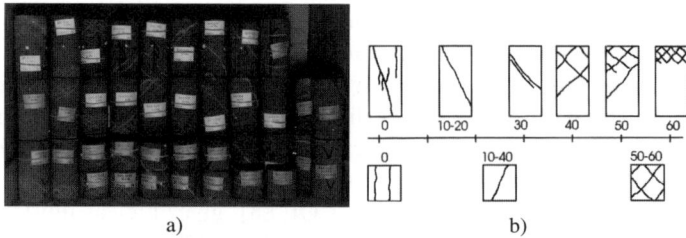

Figure 14.3. *A series of triaxial axisymmetric compression tests on Vosges sandstone with confinement pressures from 0 to 60 MPa: a) photograph of the specimens after the tests and b) diagram of the localization patterns observed (adapted from [BÉS 00])*

Another interesting aspect of localized deformations is their *multiscale character*, which is made particularly obvious by geophysicists who carry out observations of localized deformation modes on widely varying scales, from approximately 100 km to 1 m or less. The interested reader is referred, for example, to the works of Marsan and Weiss *et al.* [WEI 04, MAR 04] on the deformation of the Arctic pack ice using satellite images analyzed on different scales; these authors characterized the heterogeneity of the crack networks and strain fields in the pack ice – an ideal two-dimensional (2D) body – and derived multifractal scale laws which, according to them, invalidate the use of homogenization techniques in the modeling of scale changes. Figure 14.4 shows measurements of the amplitude of the strain analyzed on two different scales (left, 360 km; right, 18 km) in a window of about 2,000 km × 2,500 km, that is a sizeable portion of the Arctic Ocean basin.

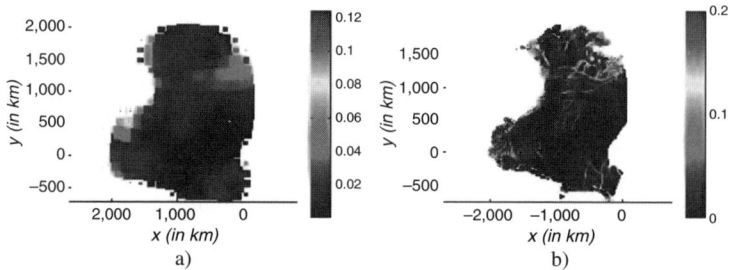

Figure 14.4. *Deformation of the Arctic pack ice observed on two scales (left, 360 km; right, 18 km) in the same 2,000 km × 2,500 km zone (courtesy of J. Weiss). For a color version of this figure, see www.iste.co.uk/gh/solidmech.zip*

Granular materials represent an interesting limit case from the viewpoint of multiscale aspects: under mean stress levels in which the particles remain unaffected during the deformation process[2], the scale on which the particles can be observed as individual grains represents the ultimate characterization scale of the deformation of the granular medium; that scale is within the experimental reach with current observation tools, including in 3D, as shown in Figure 14.5, obtained at the 3SR Laboratory in Grenoble with a fine Hostun sand specimen (average particle diameter 0.32 mm) using X-ray microtomography.

Figure 14.5. *A 3D image of a specimen (diameter 10 mm) of fine Hostun sand (average particle diameter: 0.32 mm) obtained using the X-ray microtomography of 3SR Laboratory (image courtesy of [LEN 06])*

One last aspect of localization phenomena should be mentioned here: these phenomena, in essence, are *locally* 2D phenomena because, as a first approximation, they involve the displacement of rigid blocks over discontinuity surfaces that can be viewed locally as planes. From the material viewpoint, these are unstretchable surfaces (they are called "zero extension lines" in 2D). This is an obvious result in the case of strong discontinuities that can also be easily proved in the case of weak discontinuities. Therefore, at the onset of localization, *a plane kinematics scheme develops* within a process that originally, except for the special case of 2D tests, was a 3D kinematics. This is true at least locally. However, on the structure level, when the 3D boundary conditions being applied are incompatible with a global 2D kinematics[3], they are satisfied through the occurrence of several concurrent deformation mechanisms. This type of phenomenon has been clearly illustrated in the case of a triaxial axisymmetric sand test using X-ray tomography [DES 96]. Figure 14.6 shows the various plane deformation mechanisms that could be observed in successive cutting planes perpendicular to the axis. Their organization in pairs of oblique planes is revealed by a cut parallel to the axis and perpendicular to one

---

2 Which is not always the case, but that is a different question that we are not addressing here.

3 Or even, in the case of situations that are indifferent due to the absence of kinematic constraints, the 3D boundary conditions considered do not present a sufficient bias to enable the structure to "choose" (or, more appropriately, to *align spontaneously with*) a given plane deformation mode.

of the mechanisms, which was obtained by reconstruction from the tomographic sections. The explanation of this complex mechanism is that, in the axisymmetric context of the test, the good quality of the preparation of the sample and of the application of the boundary conditions left no preferential direction for breaking the symmetry of the structure (i.e. the specimen) once the mechanical conditions for local bifurcation were met; therefore, several mechanisms occurred concurrently. These mechanisms were probably initiated by minute imperfections of similar magnitudes; their geometric distribution does not appear to follow any regularity rule. This type of structure was subsequently observed by other authors in various types of materials, including metallic alloys [XUE 09]. Thus, localization does lead to 2D strain patterns, but there can be many of these, in which case their global organization is markedly 3D. In the case of the specimen illustrated here, a naked-eye observation of the external surface could very well have failed to pinpoint the localization of the internal deformation. This example shows that, in order to characterize localization phenomena in specimens and structures objectively, advanced measurement techniques are mandatory, especially in 3D.

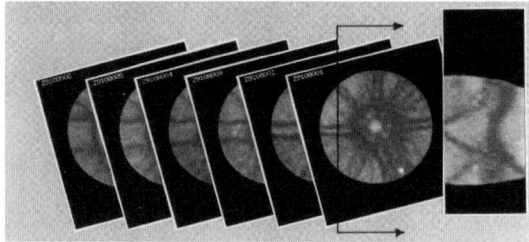

Figure 14.6. *The emergence of concurrent plane strain mechanisms revealed by tomography during an axisymmetric triaxial sand test (adapted from [DES 96])*

14.3. Available methods for the experimental characterization of localized phenomena

Our rapid survey of localized phenomena and their connection with the behavior of the materials involved show that it is extremely important to be able to identify the initiation of localization phenomena, which begins by observing them. Various experimental approaches have been used for this purpose, and extremely promising developments have appeared recently.

14.3.1. *Direct observation*

Although naked-eye observation of laboratory specimens had enabled the characterization of shear bands, cracks and fracture phenomena for many years, the information could not be collected at the very beginning of the process (inception of the phenomena) and was only qualitative. It was in the 1960s that the first attempts to measure displacement fields using photographic and radiographic techniques appeared.

14.3.2. *Recording the coordinates of predefined markers*

The first displacement fields recorded were characterized rather coarsely, using *recordings of the coordinates of predefined markers*, from photographs or radiographs taken during the tests. The Cambridge (England) school of soil mechanics was a noted pioneer in this field [BUT 70]. The sets of displacement vectors thus obtained led to the first calculations of incremental strains, which showed (or, more precisely, confirmed) the prevalence of localized fracture strains in civil engineering structures, such as embankments and retaining walls. However, the resolution was poor and the characterization was coarse.

14.3.3. *False relief photogrammetry*

This method made it possible to visualize displacement fields by the direct comparison of successive photographs, thus increasing the relevance of photography-based information. Initially, a simple observation tool [BUT 70], false relief photogrammetry, was developed from the viewpoints of both the practical measurement technique and numerical processing in order to take best advantage of the principle of tracking, image after image, material points in a deforming body [DES 84]. Improvements concerned first the use of high-resolution instruments, professional photography equipment and stereocomparators, then the point acquisition protocols – not necessarily following a regular grid, but concentrated on the high-gradient zones observed by the operator – and, finally, the general processing of the data from the kinematic viewpoint: finite transformations, expression of the strains in terms of their invariants and associated mapping. This approach was used intensively between the 1970s and 2000 to study localization in soils and rocks, especially in connection with biaxial plane strain tests [DES 04].

14.3.4. *Digital image correlation*

After 2000, due to considerable progress in 2D image acquisition instruments and the regular increase in the computational power of individual workstations, DIC, which was previously limited to a few specialists, saw a sharp surge in popularity and rapidly superseded classical photogrammetry techniques in the domain we are interested in. Today, image correlation enables the treatment of 2D stereophotographic image pairs automatically, either to calculate the 3D geometry of an object photographed at various angles or to calculate the displacement field in a 2D object photographed at several stages of a deformation process. The two aspects, that is the 3D recovery of the relief of a surface from pairs of 2D images and the deformation of 2D images with time, can even be coupled in order to track the 3D deformation of the surface (see Chapter 6).

14.3.5. *Digital volume correlation*

It is only quite recently, however, that the development of high-resolution tomography, especially using X-ray microtomographs, led to the first operational applications of DIC to 3D volumes, a relatively straightforward extension of 2D in principle, but a very demanding extension in terms of computer resources. Today, the method uses principally 3D digital images from tomographs in the domain of mechanics, but it was developed concurrently for other types of images, such as 3D images resulting from seismic data collection campaigns in geophysics, or MRI images in biomechanics. The use of laboratory tomographs can lead to some difficulties because, contrary to synchrotron-type sources that produce a parallel beam, the former use a divergent beam. The source, whose temperature as well as that of its environment varies in operation, may move due to thermal expansion and may lead to artificial magnification or contraction in the radiographs [LIM 09]. If these phenomena cause little disturbance in the qualitative examination of these images, they can affect quantitative image correlation to a much greater extent.

14.3.6. *X-ray tomography*

For a very long time, prior to focusing on digital volume correlation, X-ray tomography was one of the field measurement techniques used to study the localization of strains [COL 88, DES 96]. This method enabled the geometry of shear bands to be characterized due to the dilatancy phenomenon, which leads to density variations in granular media subjected to shear and, therefore, is naturally intensified in the bands. This description of the localized kinematics was purely qualitative. However, starting with the works cited above, the density variations observed in the bands began to be used quantitatively to collect new information about the behavior of granular media, particularly regarding the critical void ratio toward which they are considered to evolve when subjected to large strains. Digital volume correlation has enriched this approach considerably by enabling the measurement of deviatoric strains directly through the analysis of the complete 3D kinematics, as opposed to indirectly using dilatancy. The improvement in characterization capability (which will be illustrated in section 14.4.3) is considerable because not all materials are dilatant, and even when they are dilatancy is a function of the local stress state and current density of the material, which drastically limits the possibilities of an analysis based on density images alone.

To conclude this rapid overview of kinematic methods, we should mention some field measurement methods that take advantage of multiphysics coupling phenomena associated with the deformation. The localization of strains leads to an intensification and an acceleration of the deformation in some zones of the structures and specimens being studied. This is accompanied by a series of other phenomena, which may be related, in the form of various multiphysics coupling effects that are specific to the materials considered. We can take advantage of these associated effects to detect, and even visualize, the localization of the strain field. The techniques involved depend on the coupling mechanisms, and we will only give a few examples:

Acoustic emission: In some materials, the microcracks associated with the deformation may send acoustic waves that can be detected by microphones placed at the boundaries of the domain being studied, typically along the lateral faces of the specimens. This method has long been in use in the mechanics of rocks and concrete to record with a single sensor the accumulated number of acoustic events. A sudden acceleration of the number of events indicates a change in the behavior and signals imminent fracture. A more advanced use of acoustic emission consists of seeking, thanks to several sensors, the location of the source by comparing the arrival times of the same identified signal at the sensors. Under homogeneity and isotropy assumptions on the acoustic wave propagation velocity in the structure being considered, this localization is possible. This method has been used by numerous authors, including Lockner and Byerlee [LOC 77a, LOC 77b], Falls *et al.* [FAL 92] and Zang *et al.* [ZAN 98]. Now, however, progress in instrumentation in terms of acquisition rate, sensitivity and the ability to accommodate multiple sensors has enabled the development of measurement systems with increasing numbers of sensors, leading to a drastic improvement in the spatial resolution of the localization of acoustic events, for example by Stanchits [STA 03].

Acoustic tomography: This technique relies on a recording of the propagation times of an acoustic wave across a specimen between two specific points, usually located on its boundary. By carrying out many such measurements along carefully distributed paths, it is possible to define a map of the acoustic wave velocities within the specimen. The practical interest of such methods has benefited greatly from the recent appearance, in 2008, of miniature acoustic sensor-transmitter modules of typically 64 sensors per board (similar to those found in medical ultrasound equipment) and also, as in many other cases, from the increase in computational power that has made the processing of the massive amounts of data possible. The combination of many transceivers enables the specimen to be associated with a network of paths, each characterized by a mean velocity, leading to velocity maps with rapidly improving resolutions (see [VIG 08] for recent results and references).

Infrared tomography: At first glance, this method bears a strong resemblance to photographic measurements because it uses infrared cameras as imaging devices in order to record 2D pictures of the surface temperatures of a specimen. Thus, it is through thermomechanical coupling, that is the association of heat sources with plastic strains or with phase changes related to the deformation (especially in the case of shape memory alloys), that thermographic images are related to the strain fields [BOU 85]. However, heat diffuses through materials in relation to their thermal conductivity; therefore, heat flow pours out of the sources to be localized, and the temperature map observed on the surface must be viewed as a result of a complex process of heat production by the sources followed by diffusion (see Chapter 16). Then, the processing of these data calls for a specific treatment in which it is often necessary to associate a pure kinematic measurement (2D or 3D surface) with the surface temperature measurement. By taking into account the two fields together, it is possible to locate the heat sources (see, e.g., [FAV 07]).

14.4. Localization kinematics: a case study

To illustrate the methods presented above, we will examine some examples in which localization can be observed. The first set of examples concerns geomaterials, and the second set focuses on metallic materials.

14.4.1. *Emergence and development of shear bands in a sand specimen under plane strain revealed by stereophotogrammetry*

Figure 14.7 shows the maps of incremental displacements V, deviatoric strain ε_s and volume strain ε_v for eight strain increments of a biaxial test on a sand specimen whose stress–strain curve is shown in Figure 14.8 along with the numbering of the states (photos) defining the increments. These maps show that:

– Localization began at increment three to four, before the peak of the stress–strain curve.

– It coincided with the occurrence of two parallel shear bands (see the middle line ε_s), which indicates that the possible singularity represented by the corner of the specimen was not the triggering factor, but that *band orientation* was a significant parameter of the phenomenon.

– The localization pattern changed with the increments and eventually led, as a first order approximation, to a mechanism of kinematically admissible blocks, which was not the case for the initial mechanism.

– The deviatoric strain was accompanied by a volume change (dilatancy), especially at the beginning of the localization; subsequently, volume change measurements become dominated by significant measurement noise, especially in the bands.

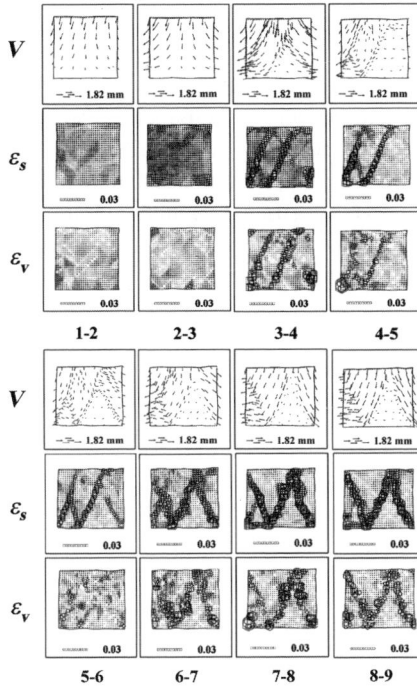

Figure 14.7. *Stereophotogrammetric characterization of strain localization in a biaxial test of sand (adapted from [DES 04])*

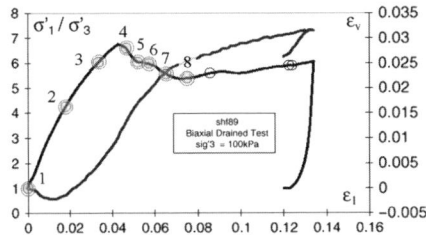

Figure 14.8. *The stress–strain curve of test shf89*

A summary of the stereophotogrammetric localization studies performed at IMG Laboratory, then at 3SR in Grenoble, was published in [DES 04] and [MOK 99].

14.4.2. *Comparison of stereophotogrammetry and digital image correlation for a biaxial test of a soft clay-rock specimen*

Both stereophotogrammetry and DIC enable displacement and strain fields to be measured. During the 1990s, the evolution of acquisition and processing equipment allowed the latter to catch up with, then surpass the former. Today, not only the resolution, but also the ease of use is far superior with image correlation. An illustration is shown in Figure 14.9, from Viggiani and Desrues [VIG 04], which compares the results of an analog stereophotogrammetric treatment of high-resolution silver negatives on 4" × 5" sheet film to the results of a digital treatment of images scanned from the same negatives. The test consisted of three increments of localized deformation of a soft clay-rock specimen (Beaucaire marl) under biaxial loading. The same overall localization structures can be identified in both cases, but the description is much more refined in the correlation case. Furthermore, the maps obtained by correlation show secondary structures that certainly did not go unnoticed by the stereophotogrammetry operator, but whose (necessarily manual) detailed recording would have been too time consuming.

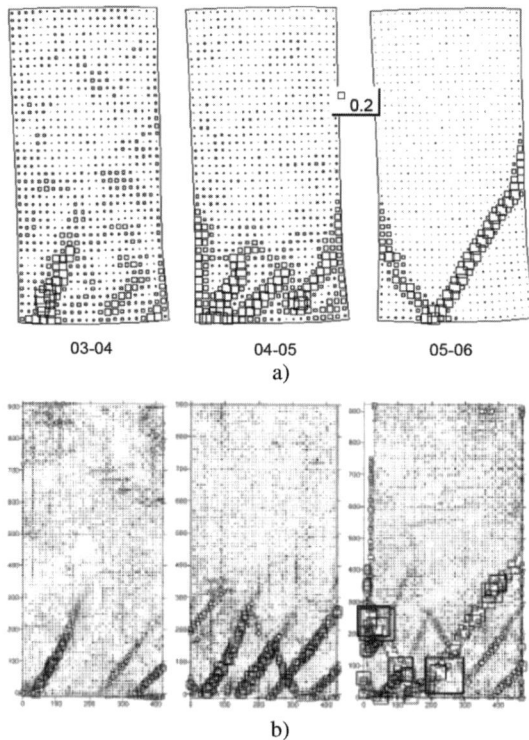

Figure 14.9. *Comparison of the results obtained by: a) stereophotogrammetry; and b) digital image correlation (adapted from [VIG 04])*

14.4.3. *The contribution of digital volume correlation to the detection of localization in isochoric shearing*

X-ray tomography enables volume images of specimens and structures to be recorded. For a chemically homogeneous medium, the gray levels of the voxels that form the images are proportional to the material local mass density. In the case of sands, whose straining is accompanied by significant dilating or contracting volume variations, the mapping of densities obtained by X-ray tomography is very enlightening qualitatively regarding the localized deformation structures being formed, and even quantitatively concerning the volume variation if a prior calibration of the mass density versus the absorption coefficient measured by the tomograph has been performed. Similarly, in the case of cracked media, the opening of cracks can be observed provided the resolution of the instrument is adapted to the crack scale. However, there are cases in which strain is isochoric and, therefore, does not result in local density variations. Similarly, cracking may occur in Mode II, that is with no relative displacement normal to the crack, in which case nothing can be observed through X-ray tomography. Conversely, in these two cases, digital volume correlation enables, as in 2D situations, the measurement of the displacement field, and therefore the calculation of the strain field or the detection of displacement discontinuities (which are tangent displacement jumps in the crack plane in Mode II). This is shown in Figure 14.10 that shows, on the left, a 3D representation of the incremental deviatoric strain in a block of a hard clay subjected to an axisymmetric triaxial test with 10 MPa confinement. This test was performed using tomography on beam line 15 of the ESRF synchrotron in Grenoble by a team from the 3SR Laboratory [LEN 06], and processed through image correlation with M. Bornert at the LMS Laboratory in Palaiseau, France. Figure 14.10(a) also shows a sagittal section of the tomographic image in the same location. We can observe that the localization of the deviatoric strain is perfectly characterized by image correlation, whereas on the image itself only a thorough examination with an experienced eye would reveal signs of the phenomenon: a slight shift at mid-height of the right edge of the cut and a hardly visible thin dark mark in the lower left quarter. The reason is that, in this case, the discontinuity produced no dilatancy or pore opening, except in the lower left where a few minor undulations of the sliding surface led to a very slight normal opening in the relative displacement of the two blocks [LEN 07].

14.4.4. *Characterization of severe discontinuities: stereophotogrammetry and correlation*

The previous example also takes us back to the problem of the distinction between strong discontinuities (cracks) and weak discontinuities (shear bands). It is not always easy to choose between these two analysis frameworks because, as mentioned previously, experimental reality can assume either aspect depending on the observation scale. When the "strong discontinuity" framework seems to be essential, specific strategies must be defined in order to characterize the location of the discontinuity and the displacement jump it involves. The DIC framework offers solutions, which will be illustrated in the examples presented hereafter concerning metallic materials. Some development work was done in the context of stereophotogrammetry: due to the necessary manual data gathering in photogrammetry, a collection strategy was defined for the geometric (path) and kinematic (displacement jump) characterization of cracks through the measurement of the displacements of points on both sides of the crack mouth [DES 95]. An adaptation of this principle to image correlation was recently proposed by Nguyen *et al.* [NGU 10].

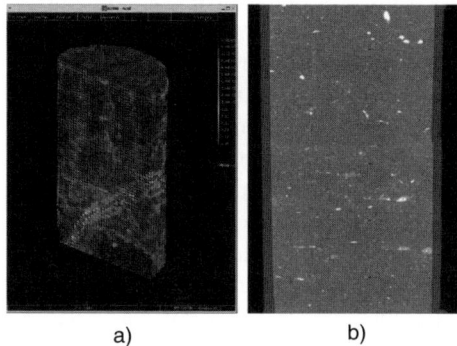

a)   b)

Figure 14.10. *Characterization of Mode-II discontinuity in a hard clay-rock specimen using digital volume correlation (adapted from [LEN 07]). For a color version of this figure, see www.iste.co.uk/gh/solidmech.zip*

14.4.5. *Localization on the grain scale: the contribution of discrete DVC*

Image correlation can also be used for the discrete mechanical analysis of granular materials, even in 3D. Figure 14.11 shows the field of the vertical displacement component in a cross section of a sand specimen during a triaxial test studied by tomography. The analysis was based on a prior segmentation of the 3D image that enabled 45,000 or so grains of the specimen to be labeled, followed by a grain-by-grain correlation to define the grain kinematics (rigid-body displacement rotation). The grain displacement was calculated from these data. The representation of the displacement components reveals, on the one hand, the zones with high global gradients (shear band localization) and enables, on the other hand, the identification of the contacts in which sliding occurred during the strain increment: for these contacts, the displacement field at the contact point is discontinuous. These recent results [HAL 10] give access to information that was previously unavailable, especially in the general case of 3D volume specimens.

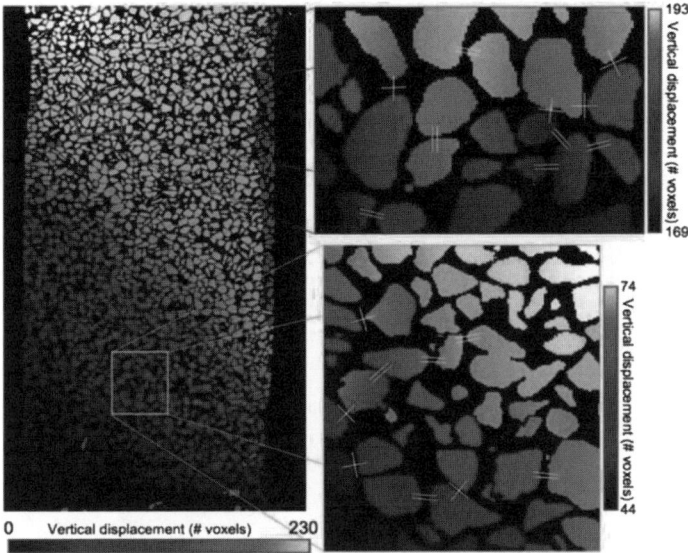

Figure 14.11. *Discrete volume correlation of a sand specimen: incremental displacement map (vertical component). The shear band localization and the mechanism of contact between grains (sliding for some, non-sliding for others) can be observed. Cooperation 3SR-LMS, ANR MicroMODEX (adapted from [HAL 10])*

14.4.6. *A fatigue crack in steel*

The experimental data for this example comes from [HAM 06]. A CCT specimen made of XC48 (or C45) steel was subjected to fatigue loading with a load ratio $R = 0.4$. Here, we are focusing on the state corresponding to the maximum load after about 300,000 cycles. At this stage of the experiment, the crack length was about $2a = 14.5$ mm. The use of an image correlation algorithm with a finite element description of the displacement field [BES 06] (see Chapter 6) led to the results shown in Figure 14.12(a). We clearly observe a discontinuity of the vertical displacement component blurred by the use of continuous functions. The amplitude of that discontinuity seems to decrease between the right edge of the region of interest and a point located approximately one-third of the length of the region from its left edge.

14.4.7. *Piobert–Lüders band in steel*

This is a tension test whose complete data can be found in [FAY 07]. The material tested was a steel with elastic/perfectly plastic behavior. Therefore, localized deformation was expected to take place. Figure 14.13 shows the reference image used for correlation as well as the region of interest and the mesh used for the analysis. Results obtained with 32-pixel elements are shown in Figure 14.13. Examination of the displacement fields shows the existence of a quasi-discontinuity. By plotting the equivalent strain obtained from the measured displacement field, we can see that on the image scale (86 μm physical pixel size) the strain was localized along a thin band that can be viewed as a line. This example was documented in [RÉT 07].

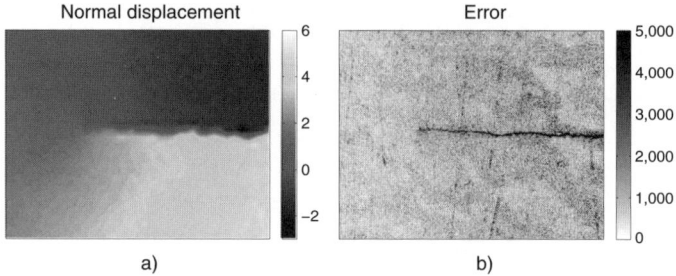

Normal displacement   Error

a)   b)

Figure 14.12. *The vertical displacement field, in pixels (1 pixel = 2.08 μm), obtained with classical finite element functions (32-pixel elements), and the correlation residual in gray levels*

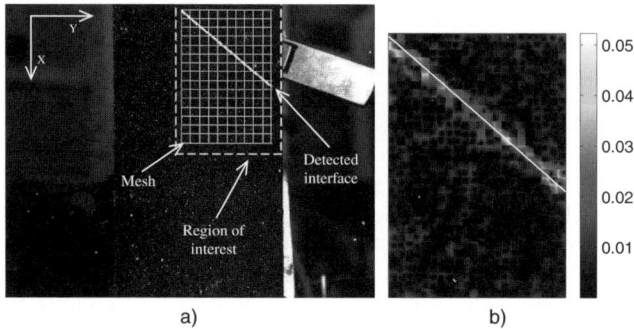

a)   b)

Figure 14.13. *The reference image for the Piobert–Lüders band case along with the equivalent strain map. The region of interest and the mesh used for image correlation can also be seen. Pixel size: 86 μm*

14.4.8. *Portevin–Le Châtelier band*

In this case, the zone of localized deformation, on the image scale, has a finite thickness. This is a tension test on 5005 aluminum. The authors [BES 06] used a spray-painted speckled pattern, as shown in Figure 14.14. The pixel size was 25 μm. The loading rate was 10 μm/s and images were taken every 60 μm. Figure 14.15 shows the displacement gradients obtained through finite element kinematics supported by 16-pixel elements. The strain maps clearly show a band in which the strains are localized. These strains are discontinuous because they are concentrated within the band. The displacement field is slightly discontinuous.

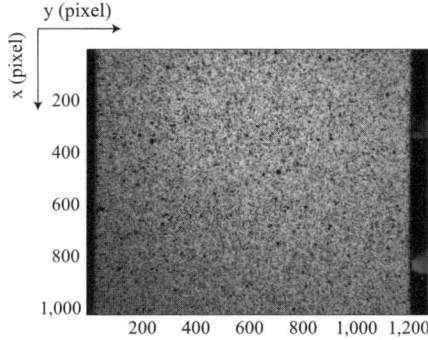

Figure 14.14. *Reference image of the 5005 aluminum alloy specimen in which Portevin–Le Châtelier bands developed. The loading axis was vertical.*
Pixel size: 25 μm

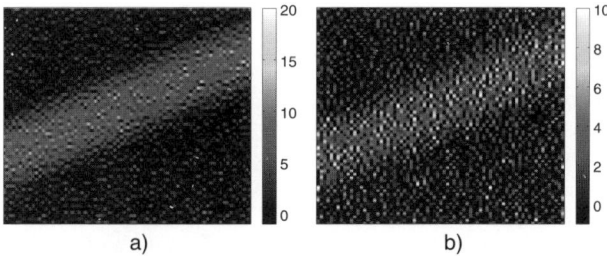

Figure 14.15. *Gradients of the vertical displacement component ($\times 10^{-3}$) in the vertical and horizontal directions obtained with 16-pixel finite elements*

14.5. The use of enriched kinematics

As mentioned earlier, the localization of strains (whether in a hyperplane or not) leads to distinctive kinematics: strain discontinuities or displacement discontinuities. Borrowing from recently developed numerical techniques [MOË 99], we can enrich a finite element description basis (see Chapter 6) with specific enrichment functions in order to capture particular shapes of the field of interest.

Let $\mathcal{N}$ be the set of the nodes of the domain being considered, and let $(N_n)_{n \in \mathcal{N}}$ be a set of finite element shape functions forming a partition of unity of the domain:

$$\sum_{n \in \mathcal{N}} N_n(\mathbf{x}) = 1 \qquad\qquad [14.1]$$

Then, this approximation can be enriched by an arbitrary function F (see [BAB 97]):

$$\mathbf{u}(\mathbf{x}) = \sum_{j=1,2} \sum_{n\in\mathcal{N}} a_{nj} N_n(\mathbf{x})\mathbf{X}_j + \sum_{j=1,2} \sum_{n\in\mathcal{N}_e} d_{nj} N_n(\mathbf{x})F(\mathbf{x})\mathbf{X}_j \qquad [14.2]$$

Here, the approximation is expressed for a vector field: $\mathbf{X}_j$ are the basis vectors of the image plane, a_{nj} the standard degrees of freedom and d_{nj} the degrees of freedom supported by the subset of the nodes associated with the enrichment function F. It can be noted that, due to the property of the partition of unity [14.1], the enrichment function can be captured exactly by that enriched description. In practice, this function is chosen such that it describes a predetermined shape of the solution. For example, we can use a Heaviside step function in order to be able to measure a discontinuity, or a weak discontinuous distance function in order to evaluate a gradient discontinuity. In the following case studies, we illustrate the use of such enrichment functions by applying them to the examples presented above.

Let us note that these functions are defined based on a weak or strong geometric discontinuity that is also an unknown of the problem. In the following section, we will show what strategies can be implemented in order to determine the geometry of the support of these discontinuities.

14.5.1. *Displacement discontinuity*

For the fatigue crack example [RÉT 08d], as discussed above, we enriched the approximation of the two components of the displacement field by a discontinuous function. The results are presented in Figure 14.16, which shows the component of the displacement field normal to the median line of the crack. At large distance from the crack, the results are exactly the same as in Figure 14.12. However, the discontinuity of the displacement field is now described explicitly. Due to the enriched degrees of freedom associated with the discontinuous enrichment function, the value of the displacement jump across the crack mouth is accessible. Figure 14.16 also shows the evolution of this displacement jump for a straight or optimized crack geometry (see section 14.5.2). The displacement jumps are of the order of 10 pixels or so at the edge of the zone of interest. The signal/noise ratio is high, but the $\sqrt{r}$ shape predicted by linear elastic fracture mechanics is recognizable nevertheless.

In the case of the Piobert–Lüders band, Figure 14.17 shows the displacement field obtained by enriching both of its components [RÉT 07]. For such measurements, although we would expect the tangent component alone to be discontinuous, dilatancy effects in the band or 3D effects can create a discontinuity of the normal component. In this case, the normal discontinuity levels were about 0.1 pixel with an uncertainty level of 0.1 pixel. Therefore, it is difficult to assess the relevance of an enrichment of the normal component. For the tangent component, the levels obtained were on average 0.9 pixel.

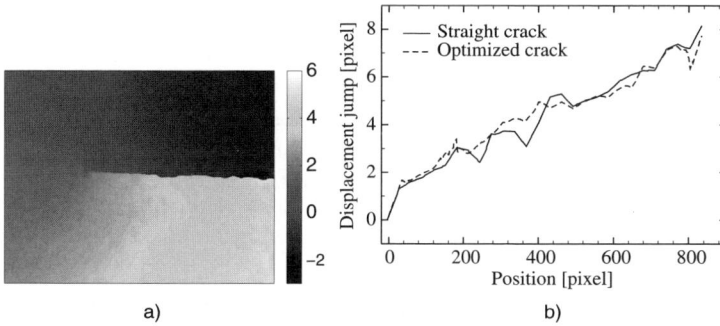

Figure 14.16. *The vertical component of the displacement field (in pixels) obtained with kinematics enriched by a discontinuous function: a) On the right, evolution of the opening along the crack mouth for a straight crack or a crack with optimized geometry*

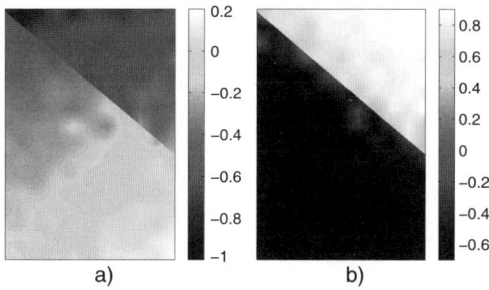

Figure 14.17. *The normal and tangent components with respect to the displacement band*

14.5.2. *Strain discontinuity*

For the Portevin–Le Châtelier band example, we chose an enrichment function of the distance type. This required a regularization, as explained in [RÉT 08a]. Thus, the elements used were much larger (128 pixels) than for a non-enriched approach. Figure 14.18 shows the normal strain in the band local coordinate system for three successive image pairs. We can observe the propagation of the localization band, whose 65° orientation remained constant, but whose thickness decreased progressively, indicating a slight deceleration. The strain discontinuity levels were comparable: 0.35% normal strain and 0.27% shear strain.

14.6. Localization of the discontinuity zone

The use of enriched kinematics, or even the characterization of localized phenomena, in general, inevitably requires the detection of their geometric support.

In some cases, these are simple geometric shapes (e.g. lines, bands, planes), but it may be desirable to capture some irregularities of these elements that are often characteristics of the material microstructure. To detect these geometric entities, we must define a criterion for the localization of the discontinuity zone and develop a technique for the extraction of geometric information. Regarding the latter point, whenever we are interested in complex geometries, the level set function method is a suitable tool [SET 99]. As for the localization criterion, we will go through the various possible solutions for the examples described previously.

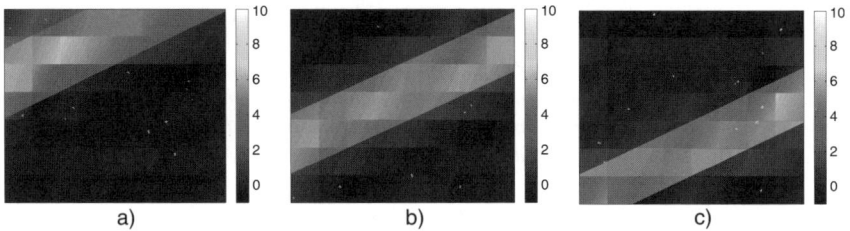

Figure 14.18. *Normal strain* $(\times 10^{-3})$ *in the band local coordinate system for three successive image pairs. These results were obtained using 128-pixel elements enriched by a slightly discontinuous function*

14.6.1. *The use of strain fields*

For the Piobert–Lüders band example, the equivalent strain map in Figure 14.13 provides a good indication of the geometric support of the localization. For this particular work, a Radon transform can be used to extract from that map the straight line along which we can find the maximum average strain [RÉT 07].

14.6.2. *The use of correlation residuals*

We saw previously that cracking or strain localization leads to specific types of kinematics. In a first measurement of a displacement field by image correlation with no enriched kinematics, the map of correlation residuals (i.e. the difference between the reference image and the deformed image corrected by the measured displacement field) provides worthwhile information. The continuous types of kinematics (see Chapter 6) used in this first determination of the displacement field do not satisfy the actual kinematics. This inconsistency leads inevitably to high error levels along the geometric support of the localization.

This is especially true in the case of cracks: Figure 14.12 shows the map of correlation residuals obtained with a standard finite element kinematics. We can observe a concentration of the error along the crack mouth. As a first approximation, this concentration can be viewed as a line. In this case again, the Radon transform is a suitable tool. Then, we may wish to describe the crack irregularities more precisely. To do this, an algorithm for moving the line describing the crack to the set of the points that maximize the correlation residual (using the capabilities of the level set function method) was presented in [RÉT 08d]. This was the crack geometry used to obtain the displacement field in Figure 14.16. Let us note that for cracks, we must also determine the location of the tip. To do that, it is possible to consider several strategies that will not be developed here (see [HAM 06, RÉT 09]).

For Portevin–Le Châtelier bands, the correlation residuals are used to define the position, thickness and orientation of the localization band precisely. Figure 14.19 shows the evolution of the average residual over the whole region of interest as a function of the parameters defining the geometry of the localization band. There is a relatively well-defined minimum that characterizes the optimum position, thickness and orientation from an image correlation viewpoint. As shown in Figure 14.18, this strategy allows the propagation of the band to be tracked.

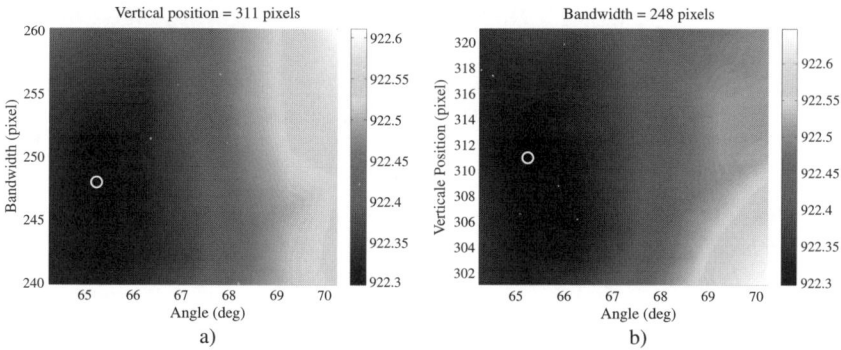

Figure 14.19. *Evolution of the average correlation error over the whole region of interest a) as a function of band thickness and orientation for a fixed position; and b) as a function of the position and orientation for a fixed band thickness*

14.7. Identification of fracture parameters

The presence of a crack in an elastic solid creates singularities in the stress and strain fields [BUI 78]. The parameters that characterize the mechanical state at a crack tip are the stress intensity factors that quantify the amplitude of the singularity at the tip. For Mode I (opening), the factor is defined as:

$$K_I = \lim_{r \to 0} \sqrt{2\pi r}\sigma_{22}(\theta = 0)$$

where σ_{22}, r and θ denote, respectively, the stress normal to the direction defined by the crack, the distance to the crack front and the angle with respect to the crack direction. A similar definition holds for Mode II (inplane shear). Such definitions in limit form make the identification of the stress intensity factors very difficult. Therefore, the aim of most developments is to extend the support of the data that can be used for this identification. For an illustration of this point, let us use the example of Figure 14.20, which was presented in [FOR 04]. This is a three-point bend test of a pre-notched specimen made of silicon carbide. No crack is visible to the naked eye in the deformed image, although we can see from the correlation results in Figure 14.20 that a discontinuity is indeed present. Nevertheless, the amplitudes of the normal displacement jump reported in Figure 14.21 are much smaller than one pixel, that is a physical size of 1.85 μm, on the order of 1 μm. It can also be noted that for this example the signal/noise ratio is relatively high, which makes it a discriminating case for assessing the various stress intensity factor extraction techniques.

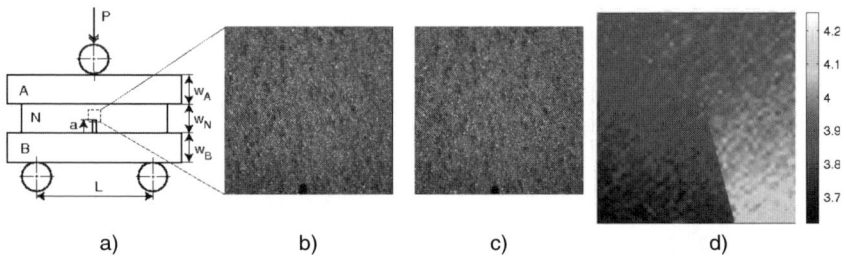

Figure 14.20. *Experimental setup and images for a silicon carbide specimen in bending. The machined pre-notch is visible at the bottom of the images. Pixel size: 1.85 μm. d) the horizontal displacement field in pixels*

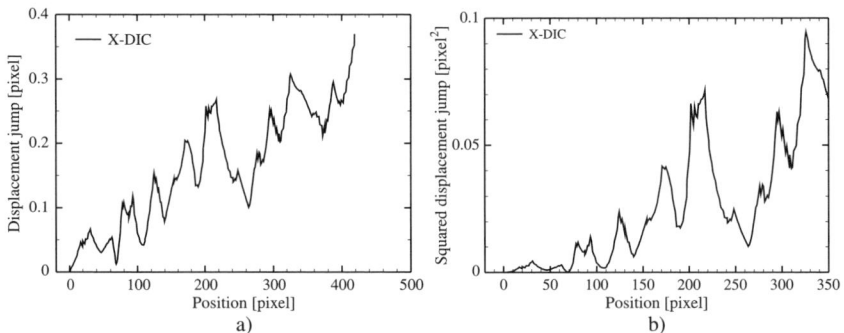

Figure 14.21. *a) Evolution of the normal displacement jump along the crack mouth; and b) the profile of the displacement jump squared is plotted*

Extending the support of the data to a line leads to the crack opening displacement (COD) family of methods. These methods consist of identifying the stress intensity factors from the crack opening profile. The opening values are considered to be proportional to the stress intensity factors and to vary as the square root of the distance to the crack tip. Thus, we could consider using the curves in Figure 14.21 to identify the square of the slope of the opening, leading to a value of K_I. From the data, we can immediately conclude that such a result would be very unreliable.

The J integral [RIC 68] is a contour integral that characterizes the energy flow through the contour during the virtual propagation of the crack. Thus, this integral uses the data over a support of the same dimension as in COD methods. The J integral can also be related to stress intensity factors through Irwin's formula [IRW 57], and its independence from the integration contour makes its use interesting. However, it involves the strain and stress fields and, therefore, the displacement derivation operation may make its use delicate because of its sensitivity to measurement noise.

In the context of the measurement of displacement fields, it is unfortunate that all the available information cannot be used, even if we have access to a measurement of the discontinuity due to an enrichment. Therefore, it is desirable to be able to use the measurement over the whole image, or at least part of the image. Then the objective becomes the extraction of the stress intensity factors from surface data. Extraction methods can be categorized according to the underlying "norm": kinematic "norm" or mechanical "norm". Methods derived for COD-type approaches belong to the kinematic category: given a basis of theoretical functions known to be solutions to the problem (Williams' series [WIL 57]), the measured field is projected onto that basis in the least squares sense (see [MCN 87, ANB 02, HAM 06]). Methods derived from the J integral belong to the "mechanical" category: tools such as the virtual extension field and auxiliary fields are used to transform the contour integral into a surface integral and uncouple the fracture modes (see [BUI 83, DES 83]). An example would be the interaction integral that was used along with image correlation measurements in [RÉT 05].

Taking the silicon carbide example, we show results obtained with the interaction integral, along with a variant that is optimized with respect to measurement noise [RÉT 08c], and with a least squares method. Figure 14.22 shows the evolution of K_I as a function of the size of the zone within which the data of Figure 14.20 are used. Although there is no bias in the evolution of K_I, significant fluctuations can be observed. The property of independence of the interaction integral with respect to the domain, which has made this integral popular in the numerical simulation world, is lost because the assumptions of equilibrium of the displacement fields and the free surface along the crack mouth are not satisfied. To some extent, the optimization of the sensitivity to measurement noise helps reduce the amplitude of the oscillations and measurement uncertainties on the curve in Figure 14.22. Nevertheless, the least squares method appears to be more robust and, indeed, it can be shown that it is optimal when the noise affecting the displacement field is white (i.e. spatially uncorrelated), which, unfortunately, is not the case. Another point is that the interaction integral involves derivatives of the displacement fields, which explain that the results thus obtained are less robust. In terms of sensitivity to noise, the optimum is obtained by using an image correlation method known as "integrated", which was proposed in [ROU 06], in which there is no post-processing, but the stress intensity factors are obtained directly during the measurement of the displacement field. To obtain this result, the basis of the functions used to describe the displacement in the correlation algorithm is the same as that used for the least squares method.

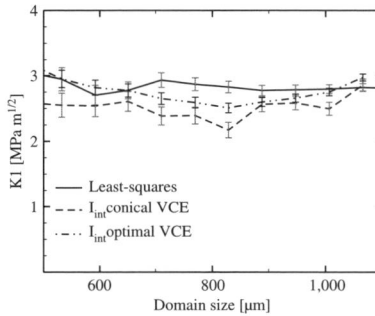

Figure 14.22. *Evolution of the stress intensity factor K_I as a function of the size of the domain over which the data are integrated. The error bars correspond to uncertainties obtained by considering sensor noise with two gray level standard deviation*

As already mentioned, localization phenomena often induce 3D mechanical states. As illustrated above, with the development of 3D imaging techniques such as tomography, the use of these new tools for the exploration of localized phenomena seems inevitable.

Here, in order to use image correlation under favorable conditions (texture, autocorrelation function), we chose to use a spheroidal graphite cast iron whose nodules served as "markers". As in 2D, we used the correlation residuals of an analysis performed with a continuous kinematics to define the geometric support of the crack, which was described by level set functions. The enrichment strategy for the description of the displacement was also identical to the 2D case [RÉT 08b]. We obtained the results in Figure 14.23 for an *in situ* tensile test of a precracked specimen. Concerning the extraction of the stress intensity factors, it is possible to use the same techniques as in 2D [LIM 09, RAN 10]. Figure 14.24 shows the evolution of the three stress intensity factors obtained along the crack front using a least squares technique.

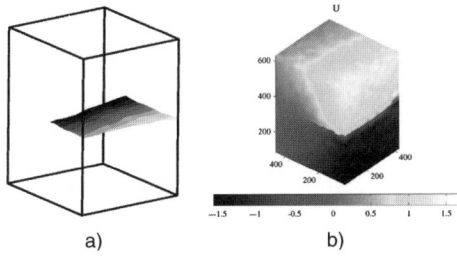

Figure 14.23. *a) The crack geometry detected and used to enrich the approximation of the displacement field; and b) the displacement field along the loading direction obtained with 32-voxel elements and a discontinuous enrichment*

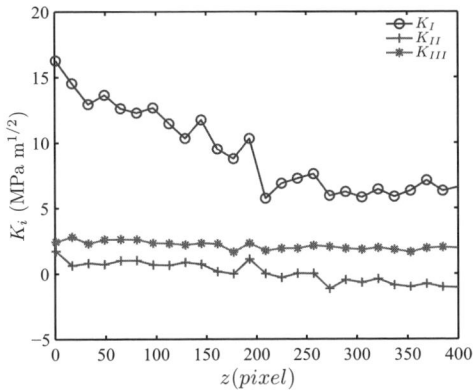

Figure 14.24. *Evolution of stress intensity factors along the crack front for the three fracture modes*

14.8. Conclusion

This chapter, dedicated to the study of what are called localized phenomena, showed the considerable improvements that have been made possible by the progress of full-field measurement techniques. The first section offered a rapid overview of the onset of strain localization. We discussed a number of case studies in the domain of geomaterials, whose microstructure and behavior are prone to localization phenomena. Then, the various optical and non-optical field methods that have been used in this context over the past 20 years were briefly described. These advances were illustrated by case studies of both 2D and 3D field methods. The use of global correlation schemes involving the displacement discontinuity explicitly was also introduced. Finally, for the particular case of cracks, we discussed recent progress concerning the robust extraction of stress intensity factors from 2D and 3D experimental displacement fields. Today, with the field measurement methods we have described, experimental mechanics has at its disposal effective and stimulating new tools for the characterization of localized phenomena.

14.9. Bibliography

[ANB 02] ANBANTO-BUENO J., LAMBROS J., "Investigation of crack growth in functionally graded materials using digital image correlation", *Engineering Fracture Mechanics*, vol. 69, pp. 1695–1711, 2002.

[BAB 97] BABUSKA I., MELENK J., "The partition of unity method", *International Journal for Numerical Methods in Engineering*, vol. 40, pp. 727–758, 1997.

[BÉS 00] BÉSUELLE P., DESRUES J., RAYNAUD S., "Experimental characterisation of the localisation phenomenon inside a Vosges sandstone in a triaxial cell", *International Journal Rock Mechanics and Mining Science*, vol. 37, pp. 1223–1237, 2000.

[BES 06] BESNARD G., HILD F., ROUX S., "'Finite-element' displacement fields analysis from digital images: application to Portevin-Le Châtelier bands", *Experimental Mechanics*, vol. 46, no. 6, pp. 789–803, 2006.

[BOU 85] BOUC R., NAYROLES B., "Methods and results in infrared thermography of solids", *Journal de mécanique théorique et appliquée*, vol. 4, no. 1, pp. 27–58, 1985.

[BUI 78] BUI H., *Mécanique de la rupture fragile*, Masson, Paris, 1978.

[BUI 83] BUI H., "Associated path independent *J*-integral for separating mixed modes", *Journal of Mechanics and Physics of Solids*, vol. 31, pp. 439–448, 1983.

[BUT 70] BUTTERFIELD R., HARKNESS R., ANDRAWES K., "A stereo-photogrammetric method for measuring displacement fields", *Géotechnique*, vol. 20, no. 3, pp. 308–314, 1970.

[COL 86] COLLIAT-DANGUS J., Comportement des matériaux granulaires sous fortes contraintes, PhD thesis, USMG - INPG, 1986, PhD Thesis, University of Grenoble, available at http://l3sphnum.hmg.inpg.fr/HomepageS1/etagere/theses/These_JLCD.pdf.

[COL 88] COLLIAT-DANGUS J., DESRUES J., FORAY P., "Triaxial testing of granular soil under elevated cell pressure", *Advanced Triaxial Testing for Soil and Rocks – ASTM STP 977*, ASTM, pp. 290–310, 1988.

[DES 83] DESTUYNDER P., DJAOUA M., LESCURE S., "Quelques remarques sur la mécanique de la rupture élastique", *Journal de Mécanique Théorique et Appliquée*, vol. 2, no. 1, 1983.

[DES 84] DESRUES J., DUTHILLEUL B., "Mesure du champ de déformation d'un objet plan par la méthode stéréophotogrammétrique de faux relief", *Journal de Mécanique Théorique et Appliquée*, vol. 3, no. 1, pp. 79–103, 1984.

[DES 95] DESRUES J., "Analyse stéréophotogrammétrique de la fissuration progressive", in BERTHAUD Y., PARASKEVAS D.T.M.E. (eds), *Etudes du comportement des matériaux et des structures, Photomécanique 95*, Eyrolles, Cachan, pp. 149–162, 14–16 March 1995.

[DES 96] DESRUES J., CHAMBON R., MOKNI M., MAZEROLLE F., "Void ratio evolution inside shear bands in triaxial sand specimens studied by computed tomography", *Géotechnique*, vol. 46, no. 3, pp. 529–546, 1996.

[DES 04] DESRUES J., VIGGIANI G., "Strain localization in sand: an overview of the experimental results obtained in Grenoble using stereophotogrammetry", *International Journal for Numerical and Analytical Methods in Geomechanics*, vol. 28, no. 4, pp. 279–321, 2004.

[DES 06] DESRUES J., GEORGOPOULOS I., "An investigation of diffuse failure modes in undrained triaxial tests on loose sand", *Soils and Foundations*, vol. 46, no. 5, pp. 587–596, 2006.

[FAL 92] FALLS S., YOUNG R., CARLSON S., CHOW T., "Ultrasonic tomography and acoustic emission in hydraulically fractured Lac du Bonnet grey granite", *Journal of Geophysical Research-Solid Earth*, vol. 97, no. B5, pp. 6867–6884, 1992.

[FAV 07] FAVIER D., LOUCHE H., SCHLOSSER P., ORGEAS L., VACHER P., DEBOVE L., "Homogeneous and heterogeneous deformation mechanisms in an austenitic polycrystalline Ti-50.8 at.% Ni thin tube under tension. Investigation via temperature and strain fields measurements", *Acta Materialia*, vol. 55, no. 16, pp. 5310–5322, 2007.

[FAY 07] FAYOLLE X., CALLOCH S., HILD F., "Controlling testing machines with digital image correlation", *Experimental Techniques*, vol. 31, no. 3, pp. 57–63, 2007.

[FOR 04] FORQUIN P., ROTA L., CHARLES Y., HILD F., "A method to determine the toughness scatter of brittle materials", *International Journal of Fracture*, vol. 125, no. 1, pp. 171–187, 2004.

[HAL 10] HALL S., BORNERT M., DESRUES J., PANNIER Y., LENOIR N., VIGGIANI G., BÉSUELLE P., "Discrete and continuum analysis of localised deformation in sand using X-ray micro CT and volumetric digital image correlation", *Géotechnique*, vol. 60, no. 5, pp. 315–322, 2010.

[HAM 06] HAMMAM R., HILD F., ROUX S., "Stress intensity factor gauging by digital image correlation: application in cyclic fatigue", *Strain*, vol. 43, no. 3, pp. 181–192, 2006.

[IRW 57] IRWIN G., "Analysis of stress and strains near the end of a crack traversing a plate", *Journal of Applied Mechanics*, vol. 24, no. 3, pp. 361–364, 1957.

[LEN 06] LENOIR N., Comportement mécanique et rupture dans les roches argileuses étudiés par tomographie à rayons X, PhD Thesis, UJF Grenoble-1, 2006.

[LEN 07] LENOIR N., BORNERT M., DESRUES J., BÉSUELLE P., VIGGIANI G., "Volumetric digital image correlation applied to X-ray microtomography images from triaxial compression tests on argillaceous rock", *Strain*, vol. 43, no. 3, pp. 193–205, 2007.

[LIM 09] LIMODIN N., RÉTHORÉ J., BUFFIÈRE J., GRAVOUIL A., HILD F., ROUX S., "Crack closure and stress intensity factor measurements in nodular graphite cast iron using 3D correlation of laboratory X-ray microtomography images", *Acta Materialia*, vol. 57, no. 14, pp. 4090–4101, 2009.

[LOC 77a] LOCKNER D., BYERLEE J., "Acoustic-emission and creep in rock at high confining pressure and differential stress", *Bulletin of the Seismological Society of America*, vol. 67, no. 2, pp. 247–258, 1977.

[LOC 77b] LOCKNER D., BYERLEE J., "Hydrofracture in Weber sandstone at high confining pressure and differential stress", *Journal of Geophysical Research*, vol. 82, no. 14, pp. 2018–2026, 1977.

[MAR 04] MARSAN D., STERN H., LINDSAY R., WEISS J., "Scale dependence and localization of the deformation of Arctic sea ice", *Physical Review Letters*, vol. 93, no. 17, 2004.

[MCN 87] MCNEILL S., PETERS W., SUTTON M., "Estimation of stress intensity factor by digital image correlation", *Engineering Fracture Mechanics*, vol. 28, no. 1, pp. 101–112, 1987.

[MOË 99] MOËS N., DOLBOW J., BELYTSCHKO T., "A finite element method for crack growth without remeshing", *International Journal for Numerical Methods in Engineering*, vol. 46, no. 1, pp. 133–150, 1999.

[MOK 99] MOKNI M., DESRUES J., "Strain localisation measurements in undrained plane-strain biaxial tests on Hostun RF sand", *Mechanics of cohesive-frictional materials*, vol. 4, no. 4, pp. 419–441, 1999.

[NGU 10] NGUYEN T., HALL S., VACHER P., VIGGIANI G., "Fracture mechanisms in soft rock: identification and quantification of evolving displacement discontinuities by digital image correlation", *Tectonophysics*, vol. 503, nos. 1–2, pp. 117–128, 2010.

[RAN 10] RANNOU J., LIMODIN N., RÉTHORÉ J., GRAVOUIL A., LUDWIG W., BAIETTO-DUBOURG M., BUFFIÈRE J., COMBESCURE A., HILD F., ROUX S., "Three dimensional experimental and numerical analysis of a fatigue crack", *Computer Methods in Applied Mechanics and Engineering*, vol. 199, nos. 21–22, pp. 1307–1325, 2010.

[RAY 08] RAYNAUD S., NGAN-TILLARD D., DESRUES J., MAZEROLLES F., "Brittle-to-ductile transition in Beaucaire marl from triaxial tests under the CT-scanner", *International Journal of Rock Mechanics and Mining Sciences*, vol. 45, no. 5, pp. 653–671, 2008.

[RÉT 05] RÉTHORÉ J., GRAVOUIL A., MORESTIN F., COMBESCURE A., "Estimation of mixed-mode stress intensity factors using digital image correlation and an interaction integral", *International Journal of Fracture*, vol. 132, no. 1, pp. 65–79, 2005.

[RÉT 07] RÉTHORÉ J., HILD F., ROUX S., "Shear-band capturing using a multiscale extended digital image correlation technique", *Computer Methods in Applied Mechanics and Engineering*, vol. 196, nos. 49–52, pp. 5016–5030, 2007.

[RÉT 08a] RÉTHORÉ J., BESNARD G., VIVIER G., HILD F., ROUX S., "Experimental investigation of localized phenomena using digital image correlation", *Philosophical Magazine*, vol. 88, nos. 28–29, pp. 3339–3355, 2008.

[RÉT 08b] RÉTHORÉ J., TINNES J., ROUX S., BUFFIÈRE J., HILD F., "Extended three-dimensional digital image correlation (X3D-DIC)", *Comptes Rendus Mécanique*, vol. 336, pp. 643–649, 2008.

[RÉT 08c] RÉTHORÉ J., ROUX S., HILD F., "Noise-robust stress intensity factor determination from kinematic field measurements", *Engineering Fracture Mechanics*, vol. 75, no. 13, pp. 3763–3781, 2006.

[RÉT 08d] RÉTHORÉ J., HILD F., ROUX S., "Extended digital image correlation with crack shape optimization", *International Journal for Numerical Methods in Engineering*, vol. 73, no. 2, pp. 248–272, 2007.

[RÉT 09] RÉTHORÉ J., HILD F., ROUX S., "An extended and integrated digital image correlation technique applied to the analysis of fractured samples", *European Journal of Computational Mechanics*, vol. 18, pp. 285–306, 2009.

[RIC 68] RICE J., "A path independant integral and the approximate analysis of strain concentration by notches and cracks", *Journal of Applied Mechanics*, vol. 35, pp. 379–386, 1968.

[RIC 76] RICE J., "The localization of plastic deformation", *Theoretical and Applied Mechanics*, North-Holland Publishing Company, pp. 207–220, 1976.

[ROU 06] ROUX S., HILD F., "Stress intensity factor measurement from digital image correlation: post-processing and integrated approaches", *International Journal of Fracture*, vol. 140, nos. 1–4, pp. 141–157, 2006.

[SET 99] SETHIAN J., *Level Set Methods and Fast Marching Methods: Evolving Interfaces in Computational Geometry, Fluid Mechanics, Computer Vision, and Materials Science*, Cambridge University Press, New York, NY, 1999.

[STA 03] STANCHITS S., LOCKNER D., PONOMAREV A., "Anisotropic changes in P-wave velocity and attenuation during deformation and fluid infiltration of granite", *Bulletin of the Seismological Society of America*, vol. 93, no. 4, pp. 1803–1822, 2003.

[THO 61] THOMAS T., *Plastic Flow and Fracture in Solids*, Academic Press, New York, NY, 1961.

[VIG 04] VIGGIANI G., DESRUES J., "Experimental observation of shear banding in stiff clay", *Geotechnical Innovations*, vol. 1, Verlag Glückauf Essen, pp. 649–658, 2004.

[VIG 08] VIGGIANI G., HALL S., "Full-field measurements, a new tool for laboratory experimental geomechanics", *Fourth Symposium on Deformation Characteristics of Geomaterials*, IOS Press, Amsterdam, vol. 1, pp. 3–26, 2008.

[WEI 04] WEISS J., MARSAN D., "Scale properties of sea ice deformation and fracturing", *Comptes Rendus Physique*, vol. 5, no. 7, pp. 735–751, 2004.

[WIL 57] WILLIAMS M., "On the stress distribution at the base of a stationary crack", *ASME Journal Applied Mechanics*, vol. 24, pp. 109–114, 1957.

[XUE 09] XUE L., WIEBZBICKI T., "Ductile fracture characterization of aluminum alloy 2024-T351 using damage plasticity theory", *International Journal of Applied Mechanics*, vol. 1, no. 2, pp. 267–304, 2009.

[ZAN 98] ZANG A., WAGNER F., STANCHITS S., DRESEN G., ANDRESEN R., HAIDEKKER M., "Source analysis of acoustic emissions in Aue granite cores under symmetric and asymmetric compressive loads", *Geophysical Journal International*, vol. 135, no. 3, pp. 1113–1130, 1998.

Chapter 15

From Microstructure to Constitutive Laws

15.1. Introduction

The previous chapters have presented different methods for the identification of mechanical behavior (see also [AVR 08]). For most of them, this behavior can display a spatial heterogeneity. Thus it is appealing to exploit such a possibility to analyze a mechanical test. However, if mechanical material properties are modeled as a heterogeneous field down to very small scales, identification becomes an ill-posed – or at least ill-conditioned – problem.

The objective of this chapter is to propose an alternative to allow for both "rich" local mechanical behavior and a decent conditioning of the problem. The main tool to achieve this goal is to exploit exogenous information, in the form of one or more auxiliary fields used to parameterize local constitutive laws. Although examples of such a methodology are as yet still scarce, it appears to be a very promising strategy, the development of which is expected to be fast as well as useful.

15.2. General problem

15.2.1. *How can we appreciate spatial heterogeneity?*

Characterizing a field of elastic properties requires a large amount of information. Depending on the generality (anisotropic behavior) or assumed local symmetries, the number of unknown independent elastic parameters, m, can vary widely, but for

---

Chapter written by Jérôme CRÉPIN and Stéphane ROUX.

isotropic local behavior at least two moduli are needed. To evaluate such fields of elastic properties, it is natural to resort to spatial discretization, assuming implicitly that the variation in these parameters at a scale smaller than that of the discretization, ξ, is negligible. Thus, in two dimensions, for a domain of size L, the number of unknowns to be identified amounts to $m(L/\xi)^2$. As has been made explicit in the previous chapters, this identification can use a displacement field measurement, giving $2N^2$ elements of kinematic information with $N = (L/\ell)$ and ℓ being the discretization scale of the displacement field. In the case where L and N are fixed parameters of the problem, it is observed that the resolution ξ is severely constrained: indeed, decreasing ξ leads to a fast increase in the number of unknowns ($\propto \xi^{-2}$), and thus to a reduction in the amount of kinematic information available per unknown ($\alpha \propto \xi^2$). Identification techniques aim at reducing the required ratio α, yet a reasonable uncertainty on local elastic properties calls for a large redundancy of information, and thus $\alpha \gg 1$. In conclusion, the resolution that we can expect from such a direct approach is poor.

It is possible to argue that detailed information is often unnecessary, and indeed homogenization theory has shown that many different microstructures could result in similar macroscopic behavior [BRE 01]. Thus, it is possible to avoid a very detailed spatial description of the material, compensating for the lack of information by a homogenization step. Then, in a second step, the relationship between the local microstructure and the local elastic behavior can be addressed, based on the identified macroscopic (homogeneous) elastic stiffness. This route may be envisioned, but it requires detailed microstructural information and the fulfillment of necessary assumptions for the homogenization step, and the latter are most often satisfied only approximately. Thus, some information loss is to be anticipated. Therefore the objective of the present chapter is to use additional microstructural information but without the need to resort to homogenization, and hence with the minimum loss of information.

In many interesting cases, the above presentation is only a caricature: indeed, each volume element of size ξ is not independent from its surroundings. For instance, in a multiphase material, each phase will be characterized by the same behavior. For a porous material at a small scale, it may be legitimate to assume that the local elastic properties are given by functions of the local pore volume fraction. For a material that has been corroded from its surface, the elastic properties may depend on the distance to this surface, but two points located at the same distance from the surface may be expected to display identical mechanical properties. Formulating such hypotheses allows for a drastic reduction in the number of unknowns, and hence for a much more faithful picture of the geometry of the specimen, together with a much more accurate determination of the constitutive parameters. This is the path that we propose to follow in this chapter.

15.2.2. *Phase segmentation*

The segmentation of an image into different components, corresponding to different phases, is a classic image analysis problem, for which many powerful techniques exist [NAJ 08]. Most often, these techniques use gray (or color) levels, which can be seen as a signature of a specific phase. They may account for topological constraints, correct imaging bias, and so on.

Once this segmentation has been performed, the hypothesis that each point within a phase shares the same mechanical behavior allows for a very significant number of unknowns to be identified. The latter is equal to the number of phases to be identified, N_ϕ, times the number of moduli per phase, m. The scale of resolution, ξ, now becomes irrelevant. The quantity of available information per unknown is of the order of $\alpha = (2/m)(N^2/N_\phi)$. A limited number of phases allows for a rich description of their behavior (large m) to be considered.

A particular case of multiphase materials concerns those whose constituents are of a similar nature, characterized by anisotropic behavior. In such a situation, each specific orientation is to be considered as a phase of its own. The problem reduces to determining the orientation of the grains after they have been segmented, or to using an orientation field possibly obtained using another instrument/procedure, and determining the elastic properties of a single reference grain representative of the phase. If the orientation is known, the number of degrees of freedom is thus limited to m. The full field of elastic properties is then deduced from the segmentation and an appropriate rotation of the reference elastic Hooke's tensor (or more complex mechanical properties).

15.2.3. *Inverse problem*

Identification is generally considered in this book with reference to mechanical behavior. It is, however, possible to enlarge this scope: for a known set of mechanical properties per phase, and a predetermined spatial distribution obtained from segmentation, we can wonder if this identification would not enable the segmentation to be refined, reaching a better determination of the grain shapes, or correcting the orientation field itself.

As an illustration, let us mention a simple example in the case of a cracked elastic solid (Chapter 14). Segmentation here is limited to the determination of the crack path in two dimensions or crack surface in three dimensions. This segmentation is difficult because the crack tip or front is rather imprecise (since the opening vanishes at the tip). In this case, identification allows the crack tip to be determined very precisely, because the knowledge of the displacement field over the entire domain (provided the latter is elastic) is much richer than the sole contrast of a static picture [RÉT 08, RAN 10].

Interfacial crack decohesion in a composite medium can be tackled this way, although a direct image analysis technique will not be able to detect it.

15.2.4. *Statistical description/morphological model*

An independent determination of the phase geometry and/or orientation over the same domain of analysis as the one used for the measurement of the displacement field is certainly one of the most favorable cases. Is it possible to relax this assumption? Is it possible to use *statistical* rather than deterministic information? Such a situation is quite common, since the statistical information is more easily accessible than an actual map. This is a field that has been very poorly explored up to now, and the following section proposes, in a rather speculative way, some directions that could be explored.

Let us consider some simple examples where a morphological model is also available. For instance, we may know that a two-phase material consists of spherical inclusions of a known radius randomly dispersed in a homogeneous matrix. Such information is easily exploitable in the identification context. An inclusion in an elastic homogeneous matrix will give rise to a displacement field that can be decomposed onto a multipolar basis. The dominant term will be a perturbative displacement field (with respect to the homogeneous case) decaying as $1/r^{d-1}$, where r is the distance to the inclusion center, and d the space dimension (Eshelby problem [ESH 57]). The prefactor of this perturbation term will be a combination of the inclusion size and the contrast of elastic properties, and it will depend on the particular inclusion shape. Knowing that the latter is a sphere (or ellipsoid) with a known radius enables us to determine the elastic properties of the inclusion. Moreover, it is not necessary to have a precise estimate of the inclusion center, as contour integrals in two dimensions may provide the answer [TAL 08]. In the same spirit, looking for microcracks of a known orientation but with unknown length is realistic, even in the unfavorable case of a small (subpixel) opening, and where an image analysis treatment is doomed to failure [RUP 11]. This approach, however, involves a dilution assumption (small volume fraction of inclusions) to resort to the Eshelby or crack influence function. Although not very demanding in practice, this criterion may not always be fulfilled.

Getting away from this dilute limit, the same type of approach may be applied if we have at least an (approximate) representation of the microstructure. Indeed, still starting from the assumption of a known inclusion type, and a known spatial distribution, the computation of the gain in the objective function – which would be a measurement of the adequacy between computed and measured displacements – resulting from the repositioning of an inclusion center gives access to a "configuration force", analogous to what would be obtained in traditional solid mechanics, but where energy is converted into the objective function [MAU 93]. Successive relaxations allow inclusions to be located as precisely as possible. In a similar manner, we can compute the benefit brought about by the introduction of an additional inclusion. This would be the direct analog of a volume energy release rate in damage mechanics [LEM 05]. Such a strategy enables an initial configuration to be corrected and a more satisfactory solution to be obtained [ESC 94]. However, it is impossible to avoid the ill-posed nature of the problem at small scales. Incorporating structural information may push back the limits of the smallest resolvable scales, but these limits do not disappear (unless the microstructural information precludes the presence of arbitrarily small features).

Purely statistical information, without morphological elements, is more difficult to incorporate. To mention examples, the simplest case is when a phase has a known volume fraction. A more complicated example would be the incorporation of a known pair correlation function. The major issue here is a matter of representation.

An attractive approach consists of using a "phase field" approach [STE 09]. For instance, for a two-phase composite, it is possible to account for the microstructure through the indicator function, $\psi(\boldsymbol{x})$, of one phase, say A. This function is valued 1 for the support of phase A, and 0 outside. The phase field approach consists of dealing with a smooth approximation of ψ constrained to the $[0, 1]$ interval. This function allows the spatial phase distribution to be approximated without being constrained to a fixed topology, hence allowing for a progressive mapping of the microstructure. ψ will be sought as the minimum of an objective functional. In order to favor pure phases, $\psi = 0$ or 1, a "free energy" density is introduced as

$$e_1 = A_1\psi^2(1 - \psi^2) \qquad\qquad [15.1]$$

The prefactor A_1 sets the energy barrier separating the two phases, and induces steep variations between these two values. (At this stage all binary functions are in the ground state of this energy.) To introduce a smooth transition, we have introduced a second term into the free energy

$$e_2 = A_2(\boldsymbol{\nabla}\psi(\boldsymbol{x}))^2 \qquad\qquad [15.2]$$

This expression tends to smooth out the ψ field, introducing a "line tension" term. Finally, the phase field is coupled with other data, and a final contribution is introduced as

$$e_3 = A_3 H(\boldsymbol{x}).\psi(\boldsymbol{x}) \qquad\qquad [15.3]$$

For instance, starting from a gray-level image f with two different reference gray levels g_A and g_B for the two phases A and B, we may wish to retrieve the phase distribution in a case where the image is severely polluted by noise. In this case, we can design a field H such that

$$H(\boldsymbol{x}) = -\frac{2f(\boldsymbol{x}) - (g_A + g_B)}{(g_A - g_B)} \qquad [15.4]$$

where H would ideally assume the value -1 (respectively, $+1$) on A (respectively, B).

The phase field is simply computed from the minimization of the total energy

$$E = \int (e_1 + e_2 + e_3) \, d\boldsymbol{x} \qquad [15.5]$$

In order to compute a known volume fraction C of phase A, it is sufficient to introduce a Lagrange multiplier (analog of a chemical potential) associated with the constraint $\int \psi(\boldsymbol{x}) \, d\boldsymbol{x} = C$. The pair correlation function can be expressed as

$$C[\psi](\boldsymbol{r}) \equiv \int \psi(\boldsymbol{x}) \psi(\boldsymbol{x} + \boldsymbol{r}) \, d\boldsymbol{x} \qquad [15.6]$$

and we can see that the sought field also minimizes $C[\psi](\boldsymbol{r}) - C(\boldsymbol{r}) = 0$. However, it is to be emphasized that this constraint is no longer linear in ψ and requires a specific treatment.

For the segmentation of the original image into two well-defined phases, the algorithm consists of making A_1 progressively larger. The transition between the two phases will occur over a specific length scale, $\ell \propto (A_2/A_1)^{1/2}$. Note that the length scale over which the transition takes place between the two phases is determined by $\xi \sim (A_2/A_1)^{1/2}$. Along a curved interface between the two phases, a Laplace pressure originates from the line tension. The latter will have a tendency to erase small isolated clusters of a phase within a matrix of the other constituent when their size is smaller than a critical value (or order ξ) in a region where only e_1 and e_2 are active.

Up to this point, the phase field approach is not coupled with mechanical identification, but may appear in the brief description above as a technical treatment of image processing, among many others. The approaches discussed in the previous chapters are, however, very easily used. In order to give a simple mechanical meaning to the phase segmentation, it is natural to attribute to each phase different elastic properties, through their Hooke's tensors, $\boldsymbol{C}_A$ and $\boldsymbol{C}_B$. It is also necessary to extend these elastic properties continuously between the two pure phases, and hence an interpolating function $\boldsymbol{C}(\psi)$ is introduced, following $\boldsymbol{C}(1) = \boldsymbol{C}_A$ and $\boldsymbol{C}(0) = \boldsymbol{C}_B$.

For a given phase field, $\psi(\boldsymbol{x})$, seeking the best elastic properties $\boldsymbol{C}_A$ and $\boldsymbol{C}_B$ is a problem that fits exactly within the scope of the previous chapters, up to the fact that the two Hooke's tensors are involved at each point where ψ is different from 0 or 1.

This is, however, not a real difficulty. In addition, the role of the interpolating function will soon become negligible when the transition between the phases is steep (large A_1 term).

The most interesting question for a fixed pair of C_A and C_B is the adjustment of the phase field. As suggested earlier, this has to be done from the third energy term, e_3, and the field H now to be defined. If the chosen identification method results from the minimization of a functional, denoted generically, $\mathcal{T}[C(x)]$ (constitutive law error, equilibrium gap, etc.; see Chapter 8), the latter can be substituted directly for the spatial integral of e_3. A linearization would then lead to the identification

$$H = \frac{\delta \mathcal{T}}{\delta \psi(x)} \qquad [15.7]$$

but this linearization will only be legitimate for an infinitesimal variation of ψ, which may not always be safe for a quasi-binary ψ field. Another option is to introduce the finite variation

$$H(x) = \mathcal{T}(\psi(x) = 1) - \mathcal{T}(\psi(x) = 0) \qquad [15.8]$$

The interpretation is quite straightforward. The H field is a configuration force being applied along each (smoothed) interface between the two phases. If a pointwise interpretation makes little sense, the smoothing of the interface enables us to integrate this configuration force across the interface and restore a meaningful configuration force, delocalized over the regularization length ξ.

15.2.5. *Coupling of identification with an exogenous field*

Many identification techniques aim at estimating a field of elastic properties from a functional minimization, $\mathfrak{J}[C(x)]$. The latter can be a simple quadratic form, resulting in a linear problem, or a more complex form that can still be tackled from successive linearizations. In both cases, the elementary step is the solution of a linear system of equations leading to an estimate of the discretized elastic properties, C_i, where the index i refers both to the different components of the Hooke's tensor and to the spatial discretization of the element endowed with these properties.

$$M_{ij}C_j = B_i \qquad [15.9]$$

Classically, the field $C(x)$ has to be strongly regularized (using for instance a very coarse mesh for the elastic properties) so that the number of final unknowns C_i is limited, otherwise the system will not be invertible or will at best be poorly conditioned.

What is proposed in the present chapter is another way of regularizing the problem. At variance with an arbitrary reduction in the number of unknowns through a limited spatial variability, the proposed strategy is to relate the mechanical properties to an exogenous field, say $D(\boldsymbol{x})$. An example of such a strategy was used for the identification of a damage law in Chapter 12. The (tangent) elastic properties are thus parameterized by a (potentially nonlinear) relation between $\boldsymbol{C}$ and $D(\boldsymbol{x})$

$$\boldsymbol{C} = \sum_{i=1}^{N} \alpha_i \varphi_i(D) \tag{15.10}$$

The choice of the appropriate functions φ_i and the knowledge of the field D allow for the reduction in the number of unknown to the N amplitudes α_i, without losing the spatial resolution accounting for fine scale details. Let us introduce the rectangular matrix $\boldsymbol{P}$ such that

$$C_i = P_{ij}\alpha_j \tag{15.11}$$

$\boldsymbol{P}$ can easily be computed from D and φ, and it allows the identification problem to be rephrased as

$$P_{ij}M_{jk}P_{km}\alpha_m = P_{ij}B_j \tag{15.12}$$

The exogenous field D can originate from different sources:

– It can contain simple geometrical information. An interface from which diffusion/corrosion has altered the mechanical properties can justify the choice of D as a simple distance to the interface.

– It can be based on morphological properties such as the examples of a local orientation field developed in the following sections.

– It can result from a colorimetric analysis allowing for the differentiation of the constituents, or other more sophisticated imaging facilities enabling a spatial resolution compatible with the images/displacement fields used for the identification step. For instance, infrared or Raman microscopy could be considered. Hyperspectral imaging (resolved in wavelength) could also be very promising.

– It can finally result from digital image correlation (DIC) itself. Indeed, the knowledge of the total or incremental strain field allows the formulation of an identification problem for nonlinear behavior laws. The tangent elastic operator and the plastic strain rate will result from the strain history with an assumed simple algebraic dependence, even though the strain history may itself be quite complex. This opens the path to nonlinear behavior identification.

15.3. Examples of local field characterization

During the past two decades, developments in computation resources and advances in electronics have given rise to new observation and characterization technologies in the field of material sciences. Indeed, the spatial resolution of measurements is now less than the size of the microstructure heterogeneities (grain size, for instance) that are typical for polycrystals used as structural materials. Currently, the spatial resolution given for a commercial transmission electron microscope (TEM) is less than 1 Å, and the energy dispersion of the electron beam is less than 0.1 eV. The same improvement is observed for field emission gun (FEG) scanning electron microscopes, which present a spatial resolution equal to a few nanometers, meaning a size less than the grain size of polycrystals. This kind of microscope, coupled with energy-dispersive X-ray spectroscopy (EDX analysis), wavelength dispersive X-ray spectroscopy (WDS analysis) or crystallographic orientation imaging microscopy (OIM) obtained from Electron Back Scattering Diffraction (EBSD) analysis, gives access to the characterization of microstructures at a resolution scale compatible with those of their chemical or physical heterogeneities, whereas beforehand only average measurements of the distribution of these sought properties were possible. Indeed, because of the increase in data analysis speed and signal processing, it is now possible to access hundreds of thousands of measurement points in a reasonable data acquisition time. As a consequence, a multiscale analysis is now possible and allows us to estimate simultaneously the size of the representative volume element (RVE) that defines the elementary volume of matter that is necessary to characterize the macroscopic properties of the investigated material and the spatial distribution of its heterogeneities inside the RVE. EBSD systems [DIN 86, RAN 00] are a good illustration of the use of such field measurements.

15.3.1. *EBSD analysis and orientation imaging microscopy*

The EBSD system is based on the exploitation of Bragg's law, and consists of measuring the rotation matrix from the crystallographic to the microscope frame. This matrix is defined by the Euler angles triplet (Φ_1, Φ, Φ_2), with a spatial resolution less than the cubic micrometer and is identified by an automated indexation of the Kikuchi bands, based on DIC between experimental Kikuchi patterns (see Figure 15.1) and simulated patterns for different theoretical crystallographic orientations. The Kikuchi pattern is the result of the in-plane projections of conic diffractions associated with crystallographic plane families that are in diffraction conditions (Bragg's law conditions) regarding the monochromatic electron beam of the microscope. Even if this property of interaction between the crystal and the monochromatic beam has been known since the 1920s, and the indexation technique was developed by Venables [VEN 73] in 1973, only a few points could be manually indexed in a reasonable time. The recent speed increase in the simulation processes gives access to hundreds of thousands of points, allowing us to define crystallographic orientation maps. The result of such an analysis is presented in Figure 15.2.

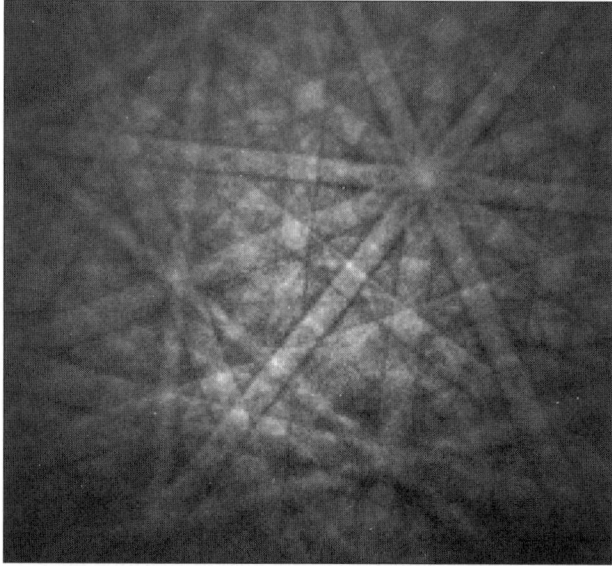

Figure 15.1. *Kikuchi pattern obtained on nickel alloy for an accelerator tension equal to 20 kV*

a)   b)   c)

Figure 15.2. *Microstructure maps obtained by EBSD for an austeno-ferritic stainless steel. The analyzed surface is equal to 600 × 470 μm² with a spatial resolution of 1 μm. (a) Quality index map. (b) Grain orientation map. (c) Crystallographic phase map: light gray: austenite (FCC structure), dark gray: ferrite (BCC structure) [RUP 07]*

Figure 15.2 shows the distribution of the grains for a two-phase stainless steel. The first phase is austenite and presents a face-centered cubic (FCC) crystallographic structure; the second phase is ferrite and presents a body-centered cubic (BCC) crystallographic structure. Figure 15.2(a) corresponds to the material microstructure observed using an SEM microscope. From this figure, it is easy to identify the grains and to perform a statistical analysis of morphological parameters such as grain size and grain aspect ratio. However, it is quite difficult to associate these grains with one or the other crystallographic phases. With this kind of microstructure representation, the information about the distribution of the chemical phases is missing. But, as already mentioned, Kikuchi patterns are a signature of the chemical phase and crystallographic orientation. Figures 15.2(b) and (c) represent the same microstructure as Figure 15(a), but in these cases grain orientations and chemical phases can be determined. From these data, it is now possible to define a criterion for grain boundaries, and usually a misorientation criterion between two adjacent points equal to $5°$ is used. Thus it is now possible to characterize the morphological parameters of the microstructure from a grain point of view (Figure 15.2(b) and Table 15.1) or from a phase point of view (Figure 15.2(c) and Table 15.2). It appears that the morphological parameters of the microstructure depend on the properties, namely grain or phase. These properties could modify the size of the RVE, depending on the distribution of the phases inside the microstructure (percolation or homogeneous distribution, for instance). This example is an illustration of multiscale analysis from the scale of the grains to the scale of the phases or the scale of the RVE. Depending on the kind of heterogeneities that occur, due to contrasts in the mechanical behavior of the phases or due to their anisotropy, one scale or the other could be more relevant to the study. Finally, the averaging of the local crystallographic orientation for each phase gives access to the crystallographic texture of the material, as presented in Figure 15.3. Note that only this last characterization could be performed with classic X-ray diffraction, due to the spot size of the beam that is commonly used (a few millimeters), which is greater than the grain size.

Phase	Diameter (μm)		Area (μm^2)		Aspect ratio	
	$\langle d \rangle$	$\sigma(d)$	$\langle A \rangle$	$\sigma(A)$	$\langle \lambda \rangle$	$\sigma(\lambda)$
FCC	16.2	17.3	479	1003	1.75	0.70
BCC	11.5	9.5	184	308	1.80	0.68

This table corresponds to Figure 15.2(b).

Table 15.1. *Morphological characterization of the grains, average value $\langle \ldots \rangle$, standard deviation $\sigma(\ldots)$, for the two crystallographic phases*

Phase	Diameter (μm)		Aspect ratio	
	$\langle d \rangle$	$\sigma(d)$	$\langle \lambda \rangle$	$\sigma(\lambda)$
FCC	62.6	169.9	2.38	1.00
BCC	32.7	29.7	2.15	0.96

This table corresponds to Figure 15.2(c).

Table 15.2. *Morphological characterization of the two crystallographic phases, average value $\langle \ldots \rangle$, standard deviation $\sigma(\ldots)$*

a)

b)

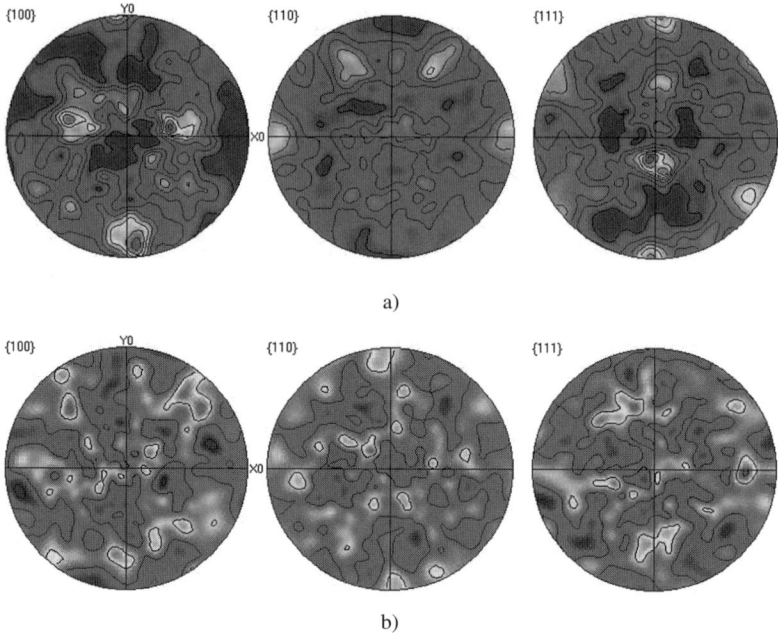

Figure 15.3. *Crystallographic textures of the ferritic phase (a) and austenite phase (b), corresponding to the microstructure of Figure 15.2 [RUP 07]*

Moreover, the analysis of the misorientation angle between adjacent grains can reveal some special relations between them, due to thermal treatment and allotropic transformation mechanisms such as mother (high temperature) and sister (room temperature) phases, as observed in steels, zirconium or titanium alloys or between twins and the surrounding matrix. For instance, in the FCC structure, the twins–matrix misorientation angle is equal to 60° around a $\langle 111 \rangle$ crystallographic direction. Figure 15.4 shows, for the same microstructure, two different distributions for misorientation angles. One corresponds to uncorrelated points, obtained from a random choice of pairs of grains, and the other corresponds to correlated distributions between adjacent grains. The comparison between these two distributions reveals that, in this case, a special arrangement of the grains occurs inside the microstructure. This arrangement corresponds to the occurrence of twins.

The different examples presented here represent two-dimensional characterization. However, such examples could not claim to characterize a three-dimensional material microstructure, especially if some phases are connected in the bulk. As a consequence in this case, three-dimensional analysis seems to be essential, and can now be performed for chemical multiphase materials and polycrystals using X-ray sources such as those supplied by synchrotron facilities. This technique, called 3DXRD tomography, is presented in [LUD 07], [LUD 08] and [LUD 09]. It enables the characterization of a polycrystal microstructure in three dimensions as presented in Figure 15.5. Using this data, we can use the same methodology as presented before to determine the characteristics of the microstructure.

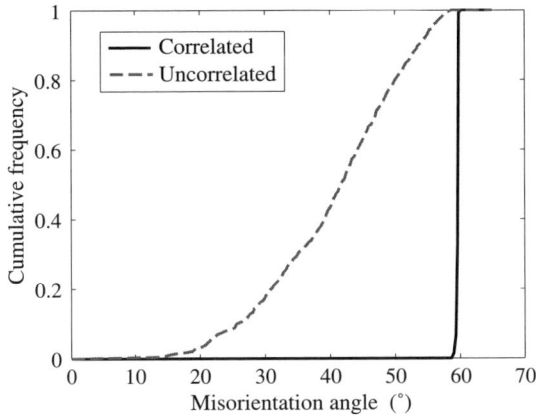

Figure 15.4. *Cumulative distribution of misorientations in the field shown in Figure 15.2. The jump at 60° of the distribution labeled "correlated" is a signature of the favored relative orientation between twins and matrix in a FCC structure [RUP 07]. The distribution labeled "uncorrelated" corresponds to the misorientation between two grains chosen at random within the field*

To conclude this section, we note that the same improvements concerning the increase in data acquisition speed are possible to be applied to TEM analysis with the introduction of numerical images. Now chemical composition cartography or crystallographic orientation cartography [FUN 03, RAU 05] are also available with this technique. However, it is important to keep in mind that the volume of matter contained in a thin foil is not representative of the dimension of the RVE, which should be characterized at a larger scale.

15.4. First example: elastic medium with microstructure

15.4.1. *Glass wool*

Mineral wools are fibrous materials used for their excellent thermal insulation performances. Typically, mechanical properties are not expected from such low-density products. However there are a few applications that demand some mechanical strength (such as resisting the weight of a roofer and providing shear resistance for sandwich panels). For these applications, higher-density products are used. For a given density, better mechanical performances can be obtained from a process called "crimping". Its goal is to reorient the fibers (or fiber bundles) in either a more isotropic direction or along the thickness, as shown in Figure 15.6.

Figure 15.5. *Three-dimensional microstructure characterization of β-titanium 21S alloy obtained by X-ray diffraction contrast tomography (DCT) performed on ESRF beam line [LUD 09]*

An algorithm designed to estimate the local preferential alignment direction from images is now introduced. The domain is first segmented into zones of interest (ZOI) [BER 07b]. An anisotropic texture is expected to be characterized by an autocorrelation function that decreases more slowly along the preferential oriental than perpendicular to it. Thus, a natural choice to identify these directions is to consider the curvature tensor of the autocorrelation function at the origin, and to extract its principal directions. If $f(x)$ is the local ZOI image, its autocorrelation function is simply expressed from its Fourier transform, $\tilde{f}(k)$ of f,

$$C(x) = \mathcal{F}^{-1}(|\tilde{f}(k)|^2) \qquad [15.13]$$

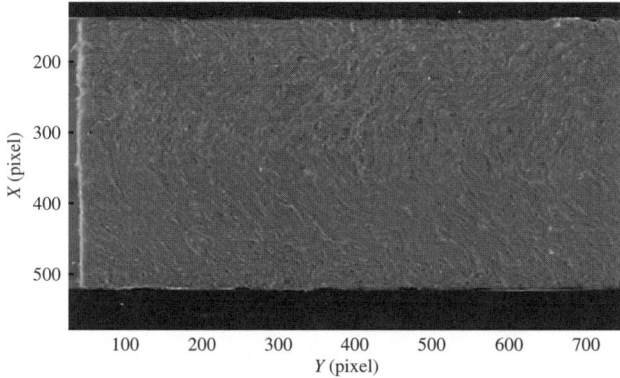

Figure 15.6. *Microstructure of a sample of crimped glass wool: cross-section view. A local orientation analysis provides a field of local directions shown as segments superimposed on the image. The thickness (vertical in the image) corresponds to 8 cm. For a color version of this figure, see www.iste.co.uk/gh/solidmech.zip*

Its curvature tensor at the origin is thus simply

$$\left.\frac{\partial^2 C}{\partial x \partial x}\right|_{x=0} = \int |\tilde{f}(k)|^2 (k \otimes k) \, dk \qquad [15.14]$$

The difficulty we have to face with mineral wool (experienced in many case studies) is that the texture is not smooth enough to provide a well-defined curvature. The consequence is that this curvature tensor strongly depends on the observation scale, and is very sensitive to discretization. A way to remedy this lack of regularity, and turn the curvature into an objective quantity independent of the discretization scale, is to filter the ZOI image so that $C(x)$ becomes twice differentiable. A power law of the wave-number modulus (or fractional integration) allows this, since the original image typically shows a power-law power spectrum (revealing an underlying self-similarity). Let α be the exponent of the power-law filter. The regularized "curvature" assumes the following expression

$$\mathcal{C} = \int |\tilde{f}(k)|^2 |k|^{2\alpha} (k \otimes k) \, dk \qquad [15.15]$$

The appropriate value of α has to be chosen from the power spectrum of the original image. From the above tensor $\mathcal{C}$, the eigenvector associated with the smallest eigenvalue gives the direction of the preferential orientation. The norm of the deviatoric part of this tensor, which is sensitive to the contrast of the two eigen curvatures, provides a measurement of the anisotropy *amplitude*.

This procedure, when applied to the segmentation of the entire image into ZOIs, the size of which being of the order of a few tens of pixels, gives access to a preferential orientation map, as well as a measurement of its amplitude. Figure 15.6 shows the result of such an analysis superimposed on the original image where the local preferential orientation is indicated by a segment whose length is proportional to the anisotropy amplitude. It is this microstructural information that will now be used for the elastic identification.

15.4.2. *Identification*

Glass wool is a very heterogeneous material. However, for medium-density products, the initial compression of the material, prior to the polymerization of the binder, considerably reduces this density heterogeneity. At the same time, crimping induces the local reorientation of the initial stratification of the wool. Describing the mechanical behavior in the range of a few percents is compatible with simple linear elastic behavior; however, it is imperative to account for the local orientation, because of the high contrast of elastic stiffness along and perpendicular to the preferential orientation. The proposed model thus assumes that the same anisotropic elastic Hooke's tensor rules the behavior, but with spatially varying local orientation of the principal axes. At the product scale, the orientation map is used to set the orientation of the elastic principal axes with those of the microstructure. Thus heterogeneity is reduced in the orientation field, which is assumed to be known. Identification will then consist of determining the four elastic constants of the assumed homogeneous orthotropic behavior. We can see clearly that the incorporation of *a priori* information allows us to reduce drastically the number of unknowns as compared with the complexity of the microstructure. Finally, for this specific material, the fibrous constituents imply that the transverse Poisson ratio is approximately 0, hence reducing the effective number of unknowns to three.

The identification procedure is now simply the result of this particular description. For an arbitrary choice of the three elastic constants, it is easy to compute from, say, a finite-element code the displacement field expected in the bulk from the specific microscopic description. Determining the constitutive parameters now proceeds naturally with the chosen strategy: finite-element updating, constitutive law error or equilibrium gap. These different techniques have been applied to the present case and provide similar results [WIT 08].

Using the identified elastic properties, and the Dirichlet boundary conditions as measured from DIC, we can compute the displacement field for the entire domain.

The difference between the measured and computed displacement fields, as shown in Figure 15.7, enables the quality of the model to be appraised.

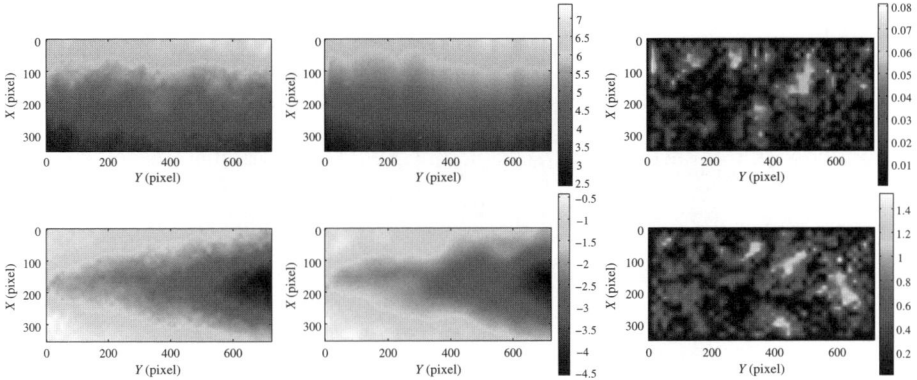

Figure 15.7. *Measured (left column) and computed (central column) displacement fields and their difference (right column). Vertical (respectively, horizontal) components are shown in the top (respectively, bottom) line*

The dimensionless residual is quantified by the norm of the difference in displacement fields, normalized by the standard deviation of the displacement to cancel the influence of rigid body motion. The order of magnitude of the final residuals obtained with a series of images is consistently of order 3%–4%.

15.5. Second example: crystal plasticity

We have already seen in section 15.3 that there is not necessarily only a single possibility to represent the microstructure of materials, and that are in fact it depends on the properties that are in question, such as chemical and crystallographic phase(s). These phases correspond to different physical or chemical fields, named images of properties or cartographies of properties, associated with the corresponding microstructure. However, one of the key points of the crystal plasticity research field does not concern only microstructure characterization but also observations of plastic deformation mechanisms (such as slip systems, twinning systems and grain boundary sliding) and the identification of the critical resolved shear stresses associated with these mechanisms. All these data allow us to propose multiscale mechanical models that give a better prediction of the mechanical behavior of the materials, or that provide some answers corresponding to process improvement to optimize the microstructure (microstructure design) regarding specific properties such as an increase in operating temperature, fatigue design, irradiation and corrosion resistance.

It is interesting to note that often one of the reasons for the fracture of industrial components is linked to plastic strain heterogeneities inside the microstructure, as shown in Figure 15.8, obtained after a tensile test performed at room temperature on α-grade 702 polycrystal zirconium alloy. As shown in this figure, plastic strain localization causes localized voids that lead to cracks and the premature failure of the material.

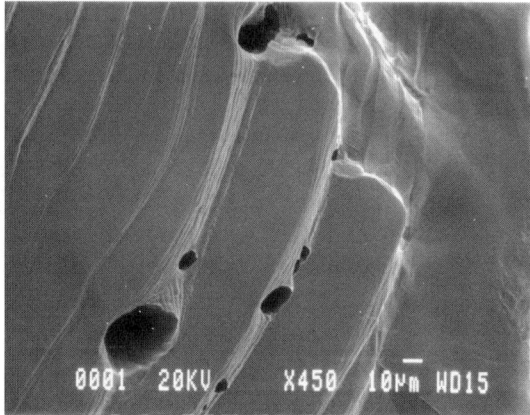

Figure 15.8. *After tensile test, scanning electron microscopy observation of cavity alignments parallel to the plastic strain localizations along grain boundaries for β-treated grade 702 zirconium alloy*

To improve the mechanical behavior of this type of industrial component, it is necessary to study the distribution of such strain heterogeneities, often linked with microstructure heterogeneities.

15.5.1. *Multiscale approach for identification of material mechanical behavior*

What does "multiscale approach" mean? Figure 15.9 represents, for a tensile test performed on α-grade 702 zirconium alloy, the principal component of the strain field, depending on the scale of the considered gauge length. The total area under investigation is $450 \times 450 \ \mu m^2$ and the chosen gauge lengths are 150, 70, 20 and 4 μm. This means that for the 4-μm case, the gauge length is less than the grain size (equal to 15 μm). From this figure, it is clear that the greater the gauge length, the more homogeneous the strain field and the closer to the average value of the strain (macroscopic strain, equal to 5% for this example). This strain homogeneity variation allows us to estimate a lower bound for the RVE of the microstructure. In this case, $300 \times 300 \ \mu m^2$ corresponds to an appropriate estimate, meaning that the size of the RVE is 20 times the grain size. From this figure we can also notice that when the gauge length decreases, the strain field is no longer homogeneous and a localization scheme occurs. First, some small-amplitude fluctuations around the average value of the macroscopic strain are observed, then a more accurate localization pattern occurs with maximum and minimum values that depend on the chosen gauge length. This gauge length defines the spatial resolution of the strain field heterogeneities. It is important to emphasize that strain values cannot be dissociated from gauge length (see Chapter 7).

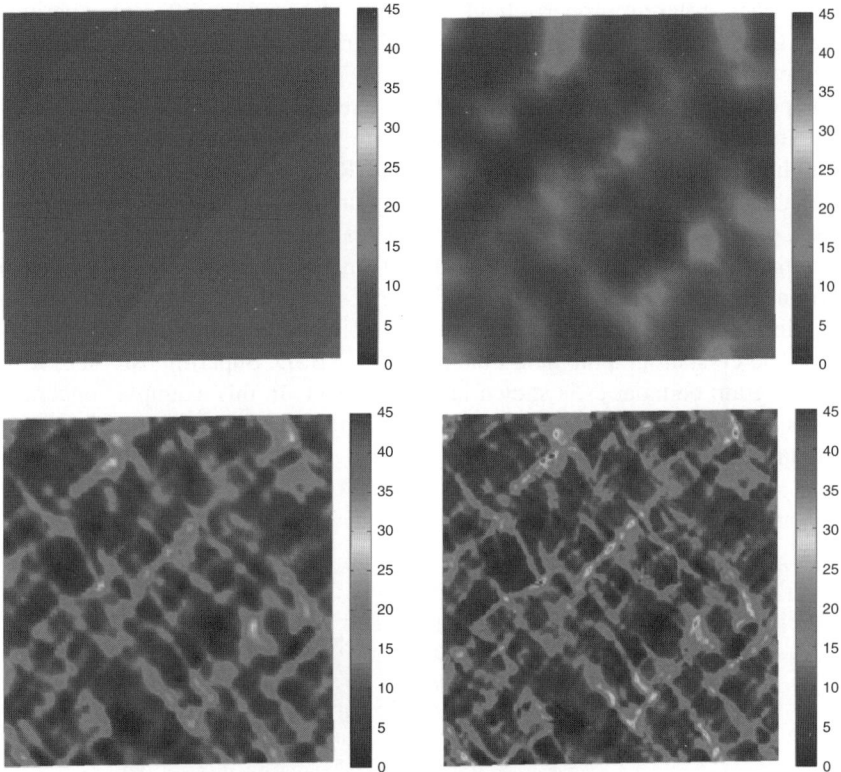

Figure 15.9. *Tensile test performed on grade 702 zirconium alloy for a macroscopic strain equal to 5%. Von Mises strain field map established on a 450×450 μm² with different gauge lengths (150, 70, 20 and 4 μm). The same scale range is used from 0% to 45% equivalent strain. For a color version of this figure, see www.iste.co.uk/gh/solidmech.zip*

This simple example highlights the fact that care must be taken with respect to the gauge length, especially when we want to compare different results coming from different experiments, or to compare experiments with numerical simulations. As is already the case for computer simulations, the strain (stress) fields are mesh dependent, and for the identification of constitutive law parameters caution is necessary concerning the correlation between experimental data (such as microstructure fields and strain fields) and kinematic fields obtained by computer simulation, for instance.

15.5.2. *Methodology*

The methodology proposed here is based on the recognition of the same geometric marks in each different property field, generally obtained by different experimental devices. In each map (chemical phases, crystallographic phases, strain heterogeneities, etc.), the spatial coordinates of the marks are obtained, enabling the data to be superimposed with special attention paid to the relative geometrical distortion of the fields. This point is not so trivial when we are interested in field sizes of around the millimeter scale to cover the size of the RVE and with a spatial resolution smaller than the grain size (microscale) to capture the intragranular behavior of the grains. Consequently, the ratio of the scale sizes that we have to take into account in this multiscale approach is close to 1,000, and more often 10,000, so that field distortions cannot be neglected. Figure 15.10 gives the different steps of the procedure to superimpose, for instance, the principal component of a strain field, obtained using DIC, on the crystallographic field obtained from EBSD. Superimposition of a strain field with grain boundaries is shown in Figure 15.11. In this example concerning α polycrystal grade 702 zirconium alloy, it is clear that strain localization is correlated with grain boundaries.

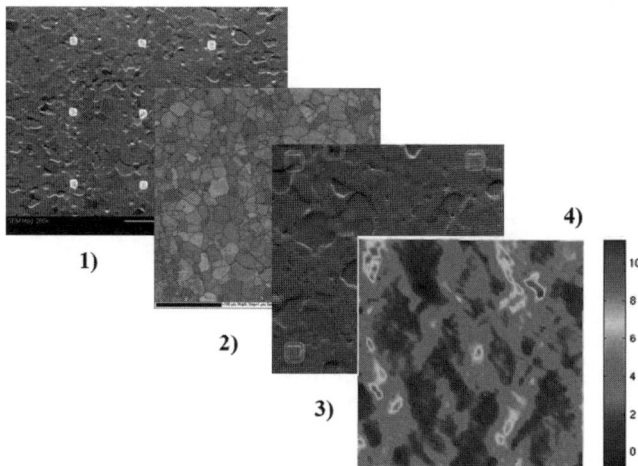

Figure 15.10. *Procedure for the superimposition of microstructural and kinematic fields [HÉR 06]: (1) after electrolytic etching, gold marks are deposited on the material; (2) EBSD mapping with spatial resolution less than 1 µm; (3) gold grids deposited using an electrolithographic technique with a spatial resolution of around 1 µm [ALL 94] (4) kinematic field associated with the grid (gauge length: 4 µm). For a color version of this figure, see www.iste.co.uk/gh/solidmech.zip*

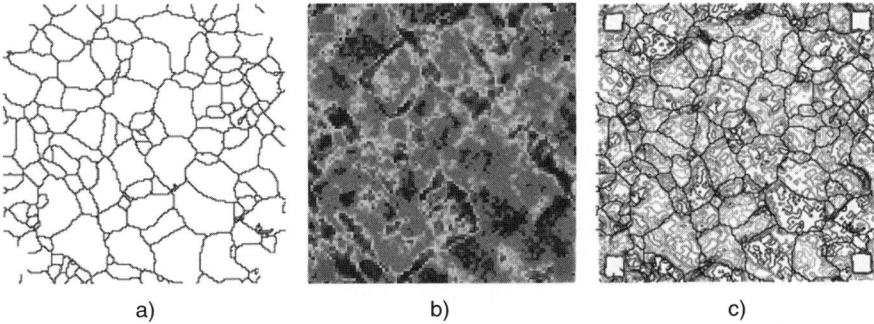

a)   b)   c)

Figure 15.11. *Polycrystal of α-grade 702 zirconium alloy: (a) grain boundary cartography obtained by EBSD mapping; (b) von Mises equivalent strain cartography obtained by DIC; (c) superimposition of the two previous fields [HÉR 06]*

15.5.3. *Numerical simulation of mechanical behavior*

Even if the previous sections remind us that some correlations occur between microstructure and mechanical fields depending on the gauge length representation, it is only a qualitative comparison, which cannot allow us to identify the constitutive equations or their parameters. The identification procedure for the constitutive equation parameters consists of minimizing a cost function F_k [HOC 03, HÉR 06]. This function is built from experimental mechanical fields, such as stress or strain fields, and their counterparts obtained from numerical simulations. The algorithm for the minimization procedure could be based on the Levenberg–Marquardt algorithm [LEV 44, MAR 63] or a genetic algorithm [GOL 89] (see Chapter 8 for more details).

First, we need to postulate the mathematical formalism of the constitutive equations for each identified phase, and in this case the constitutive equation used is an elastoviscoplastic crystalline constitutive law, as already proposed by [CAI 88, TEO 93, CAI 03] with

$$\dot{\sigma} = C(\dot{\varepsilon} - \dot{\varepsilon}^{vp}) \tag{15.16}$$

where C, $\dot{\sigma}$, $\dot{\varepsilon}$ and $\dot{\varepsilon}^{vp}$ are the tensors of elasticity, stress rate, total strain rate and viscoplastic strain rates, respectively. The viscoplastic strain rate tensor is given by

$$\dot{\varepsilon}^{vp} = \sum_{s} \dot{\gamma}_s R_s \tag{15.17}$$

where R_s corresponds to the orientation matrix of system (s) with

$$R_s = \frac{1}{2}(m_s \otimes n_s + n_s \otimes m_s) \tag{15.18}$$

and $\dot{\gamma}_s$ is the shear strain rate on slip system (s). $\boldsymbol{m}_s$ is the slip direction of system (s) and $\boldsymbol{n}_s$ is the direction normal to its plane. The shear rate follows a law that presents a hyperbolic sinus according to [MAS 98, BRE 02]

$$\dot{\gamma}_s = \dot{\gamma}_{s0} \left| \sinh \left(\frac{\tau_s}{\tau_{s0}} \right) \right|^{n_s} \mathrm{sgn}(\tau_s) \qquad [15.19]$$

τ_s is the resolved shear stress on slip system (s). Each slip system is associated with a reference shear rate $\dot{\gamma}_{s0}$, a critical resolved shear stress τ_{s0} and a coefficient of stress sensitivity n_s. The hardening law is limited here to linear hardening

$$\dot{\tau}_{s0} = \sum_k h_{sk} \left(\frac{\tau_{sat}^k - \tau_0^k}{\tau_{sat}^k - \tau_{ini}^k} \right) |\dot{\gamma}_s| \qquad [15.20]$$

where h_{sk} is the matrix of interaction between the slip systems. This matrix has been simplified by using

$$h_{sk} = H_0 Q_{sk} \qquad [15.21]$$

where H_0 is the hardening parameter and Q_{sk} is given by

$$Q_{sk} = Q_{sk0} + (1 - Q_{sk0})\delta_{sk} \qquad [15.22]$$

This implies that the hardening matrix is reduced to two values, one for diagonal terms expressing self-hardening and one for off-diagonal terms expressing latent hardening. This expression means that hardening (self- or latent hardening) does not depend on nature of the considered slip systems.

After performing an optimization procedure and minimizing the cost function (see Chapter 8), Table 15.3 gives the set of parameters that have been identified. Figure 15.12 shows a comparison between (1) the DIC experimental axial strain field and (2) the simulated field obtained with the optimized set of constitutive equation parameters. The boundary conditions applied in this example correspond to experimentally measured kinematic conditions.

If this methodology allows us to identify the parameters of the postulated constitutive equations, it also allows us to predict the activity of each plastic slip system. This plastic strain activity must be confronted with experimental observations (identification of dislocation density and Bürgers vectors) such as those made, because of TEM analysis. However, in the case of TEM observations, the volume of explored matter (volume of a thin foil) is quite small in comparison with the size of the RVE. This means that from a statistical point of view, we should bear in mind this possible artifact in the analysis.

Coefficient	$\dot{\gamma}_0$ (s^{-1})	H_0 (MPa)	Q_0	
Value	5×10^{-5}	375	1.9	
Coefficient	n_p	$n_{\langle a\rangle}$	$n_{\text{pyr}\langle c+a\rangle}$	n_b
Value	3	7.5	7.7	7.1
Coefficient	τ_{ini}^{p} (MPa)	$\tau_{\text{ini}}^{\text{pyr}\,\langle a\rangle}$ (MPa)	$\tau_{\text{ini}}^{\text{pyr}\,\langle c+a\rangle}$ (MPa)	τ_{ini}^{b} (MPa)
Value	44	44	175	175
Coefficient	τ_{sat}^{p} (MPa)	$\tau_{\text{sat}}^{\text{pyr}\,\langle a\rangle}$ (MPa)	$\tau_{\text{sat}}^{\text{pyr}\,\langle c+a\rangle}$ (MPa)	τ_{sat}^{b} (MPa)
Value	74	74	205	205

Table 15.3. *Optimized values of the constitutive equation parameters*

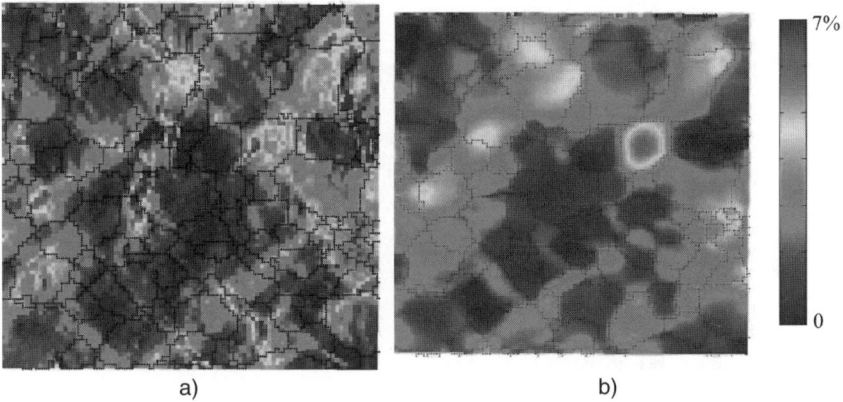

a)   b)

Figure 15.12. *Tensile test performed on grade 702 zirconium alloy.
Cartography of (a) experimental (b) simulated axial strain field component.
For a color version of this figure, see www.iste.co.uk/gh/solidmech.zip*

In our example, it is clear that even if the identification procedure presents the same parameter values for pyramidal $\langle a\rangle$ slip systems and prismatic slip systems, the activity maps are quite different (Figure 15.13). This is a consequence of the crystallographic orientation of the grains inside the microstructure (texture effect). Moreover, the activity map for pyramidal $\langle c + a\rangle$ slip systems shows that this activity is limited to grain boundaries, areas where strong mechanical incompatibilities are usually observed. This example illustrates the interest of such a coupling approach between microstructure and kinematic fields with the aim of identifying the constitutive equations and their parameters. This approach reveals the fact that this kind of study is a multiscale study, with special attention drawn to the definition of the gauge length to avoid inappropriate comparisons due to mesh dependency.

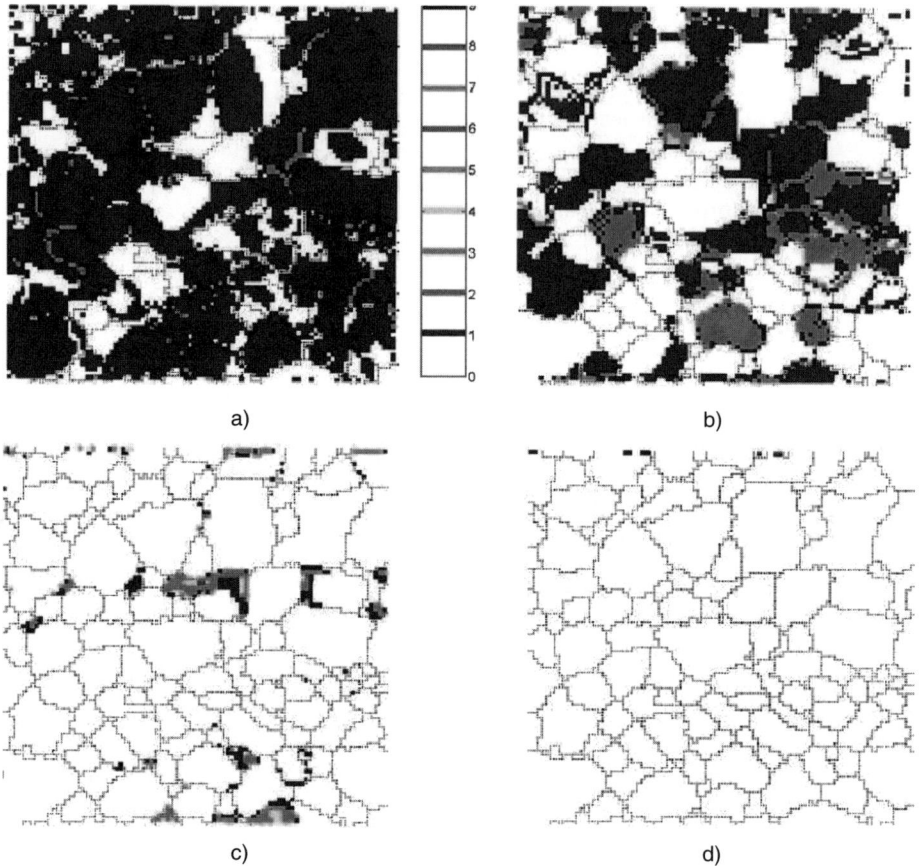

Figure 15.13. *Maps of activated slip systems for (a) prismatic, (b) pyramidal*
⟨a⟩, *(c) pyramidal* ⟨c + a⟩ *and (d) basal slip systems. For a color version of*
this figure, see www.iste.co.uk/gh/solidmech.zip

15.6. Conclusions

The aim of this chapter was to emphasize the very high potential of adding
local characterization fields to kinematic field measurements to identify mechanical
behavior. Coupling experiments to numerical simulation is the most promising and
reliable route for the evaluation of constitutive parameters. Even if such associations
of fields coming from different methods are still not numerous today, it is reasonable
to invest in such techniques because they open new horizons for problems such as
identification.

Mixing information of different natures requires special care with respect to the
question of scale, especially concerning strain. The systematic size dependence of

micromechanical fields calls for a consistent selection of the fields to be knitted together. Similarly, geometrical distortion and corrections necessitate strict calibration protocols – only briefly mentioned throughout this chapter – but whose importance should not be overlooked.

Finally, the incorporation of *statistical* rather than deterministic information into the identification process is essentially unexplored. Some directions have been pointed out in this chapter, for example the use of phase fields, but most of these techniques have to be invented and promoted within a proper probabilistic framework.

15.7. Bibliography

[ALL 94] ALLAIS L., BORNERT M., BRETHEAU T., CALDEMAISON D., "Experimental characterization of the local strain field in a heterogeneous elastoplastic material", *Acta Metallurgica and Materalia*, vol. 42, no. 11, pp. 3865–3880, 1994.

[AVR 08] AVRIL S., BONNET M., BRETELLE A.S., GREDIAC M., HILD F., IENNY P., LATOURTE F., LEMOSSE D., PAGANO S., PAGNACCO E., PIERRON F., "Overview of identification methods of mechanical parameters based on full-field measurements", *Experimental Mechanics*, vol. 48, pp. 381–402, 2008.

[BER 05] BERGONNIER S., HILD F., ROUX S., "Digital image analysis used for mechanical tests on crimped glass wool samples", *Journal of Strain Analysis*, vol. 40, pp. 185–197, 2005.

[BER 07a] BERGONNIER S., Relation entre microstructure et propriétés mécaniques de matériaux enchevêtrés, Doctoral Thesis, University of Paris 6, 2007.

[BER 07b] BERGONNIER S., HILD F., ROUX S., "Local anisotropy analysis for non-smooth images", *Pattern Recognition*, vol. 40, pp. 544–556, 2007.

[BRE 01] BRETHEAU T., BORNERT M., GILORMINI P., *Homogénéisation en mécanique des matériaux*, vol. 1, Hermès, Paris, 2001.

[BRE 02] BRENNER R., BÉCHADE J.L., CASTELNAU O., BACROIX B., "Thermal creep of Zr-Nb1%-O alloys: experimental analysis and micromechanical modelling", *Journal of Nuclear Materials*, vol. 305, pp. 175–186, 2002.

[CAI 88] CAILLETAUD G., "Une approche micromécanique du comportement des polycristaux", *Rev. Phys. App.*, vol. 23, pp. 353–363, 1988.

[CAI 03] CAILLETAUD G., FOREST S., JEULIN D., FEYEL F., GALLIET I., MOUNOURY V., QUILICI S., "Some elements of microstructural mechanics", *Computational Materials Science*, vol. 27, pp. 351–374, 2003.

[DEX 06] DEXET M., Méthode de couplage entre expérimentation et simulations numériques en vue de l'identification de loi de comportement intracristallin; application aux alliages de zirconium, Doctoral Thesis, Ecole Polytechnique, France, 2006.

[DIN 86] DINGLEY D.J., BABA-KISHI K., "Use of backscattered electron diffraction patterns for determination of crystal symmetry elements", in JOHARI O. (ed.), *Scanning Electron Microscopy*, SEM Inc., Chicago, vol. II, pp. 383–391, 1986.

[ESC 94] ESCHENAUER H.A., KOBELEV V.V., SCHUMACHER A., "Bubble method for topology and shape optimization of structures", *Structural Optimization*, vol. 8, pp. 42–51, 1994.

[ESH 57] ESHELBY J.D., "The determination of the elastic field of an ellipsoidal inclusion and related problems", *Proceedings of the Royal Society London, Series A, Mathematical and Physical Sciences*, vol. 241, pp. 376–396, 1957.

[FU 03] FU X., POULSEN H.F., SCHMIDT S., NIELSEN S.F., LAURIDSEN E.M., JUUL JENSEN D., "Non-destructive mapping of grains in three dimensions", *Scripta Materialia*, vol. 49, pp. 1093–1096, 2003.

[FUN 03] FUNDENBERGER J.J., MORAWIEC A., BOUZY E., LECOMTE J.S., "Polycrystal orientation maps from TEM", *Ultramicroscopy*, vol. 96, pp. 127–137, 2003.

[GÉR 08] GÉRARD C., Mesure de champs et identification de modèles de plasticité cristalline, Doctoral Thesis, University of Paris 13, 2008.

[GOL 89] GOLDBERG D.E., *Genetic Algorithms in Search, Optimization, and Machine Learning*, Addison-Wesley, Reading, MA, 1989.

[HÉR 06] HÉRIPRÉ E., Méthode de couplage multi-échelles entre simulations numériques polycristallines et mesures de champs pour l'identification de paramètres de lois de comportement et de fissuration des matériaux métalliques. Application à l'étude des alliages TiAl, Doctoral Thesis, Ecole Polytechnique, Paris, France, 2006.

[HOC 03] HOC T., CRÉPIN J., GÉLÉBART L., ZAOUI A., "A procedure for identifying the plastic behavior of single crystals from the local response of polycrystals", *Acta Materialia*, vol. 51, no. 18, pp. 5479–5490, 2003.

[JUU 06] JUUL JENSEN D., LAURIDSEN E.M., MARGULIES L., POULSEN H.F., SCHMIDT S., SØRENSEN H.O., VAUGHAN G.B.M., "X-ray microscopy in four dimensions", *Materials Today*, vol. 9, pp. 18–25, 2006.

[LEM 05] LEMAITRE J., DESMORAT R., *Engineering Damage Mechanics: Ductile, Fatigue and Brittle Failures*, Springer, Berlin, 2005.

[LEV 44] LEVENBERG K., "A method for the solution of certain problems in least squares", *Quarterly of Applied Mathematics*, vol. 2, pp. 164–168, 1944.

[LUD 07] LUDWIG W., SCHMIDT S., LAURIDSEN E.M., POULSEN H.F., BARUCHEL J., "High-resolution three-dimensional mapping of individual grains in polycrystals by topotomography", *Journal of Applied Crystallography*, vol. 40, pp. 905–911, 2007.

[LUD 08] LUDWIG W., LAURIDSEN E.M., SCHMIDT S., POULSEN H.F., "X-ray diffraction contrast tomography: a novel technique for three-dimensional grain mapping of polycrystals. I. Direct beam case", *Journal of Applied Crystallography*, vol. 41, pp. 302–309, 2008.

[LUD 09] LUDWIG W., KING A., REISCHIG P., HERBIG M., LAURIDSEN E.M., SCHMIDT S., PROUDHON H., FOREST S., CLOETENS P., ROLLAND DU ROSCOAT S., BUFFIÉRE J.Y., MARROW T.J., POULSEN H.F., "New opportunities for 3D materials science of polycrystalline materials at the micrometre lengthscale by combined use of X-ray diffraction and X-ray imaging", *Materials Science and Engineering A*, vol. 524, pp. 69–76, 2009.

[MAR 63] MARQUARDT D., "An algorithm for least-squares estimation of nonlinear parameters", *SIAM Journal on Applied Mathematics*, vol. 11, pp. 431–441, 1963.

[MAS 98] MASSON R., Estimations non linéaires du comportement global de matériaux hétérogènes en formulation affine. Application aux alliages de zirconium, Doctoral Thesis, Ecole Polytechnique, Paris, France, 1998.

[MAU 93] MAUGIN G.A., *Material Inhomogeneities in Elasticity*, Series Applied Mathematics and Computation 3, Chapman & Hall, London, 1993.

[NAJ 08] NAJMAN L., TALBOT H., *Morphologie mathématique 1: approches déterministes*, Traité IC2, série signal et image, Hermès, Paris, 2008.

[RAN 00] RANDLE V., ENGLER O., *Introduction to Texture Analysis: Macrotexture, Microtexture and Orientation Mapping*, CRC Press, Taylor & Francis, 2000.

[RAN 10] RANNOU J., LIMODIN N., RÉTHORÉ J., GRAVOUIL A., LUDWIG W., BAÏETTO-DUBOURG M.-C., BUFFIÈRE J.-Y., COMBESCURE A., HILD F., ROUX S., "Three dimensional experimental and numerical multiscale analysis of a fatigue crack", *Computer Methods in Applied Mechanics and Engineering*, vol. 199, pp. 1307–1325, 2010.

[RAU 05] RAUCH E.K., DUPUY L., "Rapid spot diffraction patterns identification through template matching", *Archives of Metallurgy and Materials*, vol. 50, pp. 87–99, 2005.

[RÉT 08] RÉTHORÉ J., HILD F., ROUX S., "Extended digital image correlation with crack shape optimization", *International Journal for Numerical Methods in Engineering*, vol. 73, pp. 248–272, 2008.

[RUP 07] RUPIN N., Modélisation micromécanique du comportement à chaud de polycristaux biphasés, Doctoral Thesis, Ecole Polytechnique, Paris, France, 2007.

[RUP 11] RUPIL J., ROUX S., HILD F., VINCENT L., "Fatigue microcrack detection from digital image correlation", *Journal of Strain Analysis*, vol. 46, pp. 492–509, 2011.

[STE 09] STEINBACH I., "Phase-field models in materials science", *Modelling and Simulation in Materials Science and Engineering*, vol. 17, pp. 073001, 2009.

[TAL 08] TALAMALI M., PETÄJÄ V., VANDEMBROUCQ D., ROUX S., "Path independent integrals to identify localized plastic events in two dimensions", *Physical Review E*, vol. 78, pp. 016109, 2008.

[TEO 93] TEODOSIU C., RAPHANEL J.L., TABOUROT L., in TEODOSIU C., RAPHANEL J.L., SIDOROFF F. (eds), *Large Plastic Deformations*, Balkema, Rotterdam, p. 153, 1993.

[VEN 73] VENABLES J.A., HARLAND C.J., "Electron backscattering patterns – a new technique for obtaining crystallographic information in the scanning electron microscope", *Philosophical Magazine*, vol. 27, pp. 1193–1200, 1973.

[WIT 08] WITZ J.F., ROUX S., HILD F., RIEUNIER J.B., "Mechanical properties of crimped mineral wools: identification from digital image correlation", *ASME Journal of Engineering Materials and Technology*, vol. 130, pp. 021016/1-6, 2008.

[WIT 09] WITZ J.-F., Relations entre texture et propriétés thermomécaniques de laines minérales, Doctoral Thesis, University of Paris 6, 2009.

Chapter 16

Thermographic Analysis of Material Behavior

16.1. Introduction

Infrared (IR) techniques have been used increasingly over the last 15 years in the field of mechanics of materials and structures in university laboratories and industrial research and development centers. This has been facilitated by the progress achieved in signal digitization, data transfer rates and image-processing techniques, thus making new thermographic workstations more user-friendly. A key asset of thermographic systems is that they generate contactless full-field surface temperature measurements. In addition, current IR cameras allow the storage of thermal images of about 80,000 pixels at frame rates of up to several hundred hertz, and even higher under certain conditions. New sensors enable the estimation of temperature variations ranging from several hundredths to several hundreds of degrees celsius.

Thermal imaging is of interest in mechanics because material deformation is nearly always accompanied by heat release and thus temperature variations. Strain-induced heat sources have several origins – we have to distinguish between sources that are associated with energy dissipation and those that are introduced by thermomechanical coupling mechanisms. The first sources reflect the irreversibility of deformation processes. When the material behavior is plastic or viscoplastic, when aging or damage processes occur, part of the deformation energy is irreversibly transformed into heat. The second sources, which are induced by thermomechanical couplings, also reveal the material behavior. They occur because

Chapter written by Jean-Christophe Batsale, André Chrysochoos, Hervé Pron and Bertrand Wattrisse.

the microstructural state is sensitive to temperature variations, inducing a strong interaction between the mechanical and thermal states. Thermoelasticity is surely the most famous example of coupling mechanisms [THO 53]. A thermoelastic material warms up under compression or expands when warmed. Of course, other coupling mechanisms exist. Another example concerns rubber elasticity [CHA 84] or mechanisms associated with austenite–martensite phase changes [PEY 98]. In the two latter cases, the materials cool under compression or contract when heated.

The quantitative assessment of evolved heat, caused by deformation, remains problematic despite progress achieved in the field of IR data measurement, acquisition and processing. This analysis is based on measuring effects (temperature changes) to trace causes (release or stored heat). The relationship between temperature and source is complex because temperature changes depend not only on the intensity and distribution of sources but also on the nature of the material (diffusivity) and the thermal boundary conditions (external exchanges). Temperature variations are thus not definitely intrinsic to material behavior. Finally, the heat diffusion equation involves several thermophysical parameters (e.g. emissivity, specific heat, conduction tensor) that must be known in order to properly assess the temperature field and estimate the heat source distribution.

IR thermography applications in the mechanics of materials have primarily been focused on non-destructive testing. Thermal stress analysis is the oldest and most widespread application domain. For a material having linear thermoelastic behavior (isotropic), surface temperature variations – with a magnitude of several tenths of a degree Celsius – can be linked, with some additional assumptions, to variations in the stress tensor trace [SAT 07]. Another currently booming application area is the use of thermography to analyze the fatigue behavior of components. For monochromatic cyclic stress tests, the principle used to identify the most loaded zones of components is to analyze the time spectrum content of temperature fields, particularly the component associated with twice the loading frequency ($2f$ method [BRE 82]). This latter frequency generally does not correspond to thermoelastic effects and can be associated with the dissipative effects of fatigue. Self-heating ranges from a few degrees to several tens of degrees. Finally, we should also mention applications where temperature variations reach several tens to several hundreds of degrees. A few examples of this may be found in studies of material dynamic behavior [RAN 08], machining processes [SUT 07], forming and welding operations [PIN 04], friction mechanisms.

This chapter describes the main steps of a source field estimation protocol. It begins with a brief overview of the theoretical framework used to interpret thermal and calorimetric measurements and provides a link between heat sources and constitutive equations. Difficulties, associated with IR measurements and their conversion in terms of temperature, are then discussed. Finally, various classic diffusion models linking temperature fields to heat source distributions, and allowing the estimation of thermophysical parameters from thermal data, are introduced.

16.2. Thermomechanical framework

Thermal measurements cannot be interpreted without a reference to a specific thermodynamic framework, which is necessary for the definition of thermophysical parameters and the classification of heat sources. This section presents the framework of classic thermodynamics of irreversible processes (TIP), whose applications to material behavior modeling began increasing in the 1960s. For further information, readers should refer to the work of [COL 67, HAL 75, GER 83] and more recently to [FRÉ 02]. TIP offers a consistent and flexible framework to describe material behavior. It is flexible because a set of state variables may be chosen to describe the macroscopic effects of numerous, complex and often combined microstructural phenomena.

16.2.1. Constitutive equations

The deformation process is considered to be quasi-static in a thermodynamic sense, that is consisting of a continuous series of equilibrium states. The equilibrium state of each elementary volume of matter is then described using a set of N state variables. The temperature T, a strain tensor ε and N-2 additional state variables are generally used to complete the description of the material state. α denotes the vector that symbolically groups these state variables, often called "internal" state variables.

Let $\psi = e - Ts$ define the Helmholtz specific free energy, where e and s are the internal energy and entropy, respectively. It is assumed that internal energy is a monotonically increasing function of entropy and absolute temperature is by definition $T = e_{,s}$.

The material constitutive equations are divided into two sets:

State equations: these are the first derivatives of the free energy with respect to state variables. These state laws read:

$$\begin{cases} s = -\psi_{,T} & (a) \\ \sigma^r = \rho\psi_{,\varepsilon} & (b) \\ A = \rho\psi_{,\alpha} & (c) \end{cases} \qquad [16.1]$$

where σ^r is the reversible part of the stress tensor and A is a vector that groups the conjugated variables associated with internal state variables.

The local expression of the second principle of thermodynamics leads to the Clausius–Duhem inequality and allows the dissipation to be defined as:

$$d = \sigma : D - \rho\psi_{,\varepsilon} : \dot{\varepsilon} - \rho\psi_{,\alpha}\cdot\dot{\alpha} - \frac{q}{T}\cdot\mathrm{grad}T \geq 0 \qquad [16.2]$$

where σ is the Cauchy stress tensor, D the Eulerian strain rate tensor, ρ the mass density and q the heat influx vector.

Evolution equations: in the particular context of generalized standard materials [HAL 75], evolution equations are derived from the dissipation potential $\varphi(\dot{\varepsilon},\dot{\alpha},q;T)$, depending on the flow of state variables, temperature T (or any other state variable) acting as a parameter. The dissipation potential must be convex, positive, minimum and equal to zero in $(\dot{\varepsilon}=0,\dot{\alpha}=0,q=0)$ so that the Clausius–Duhem inequality can be satisified during any thermodynamic process. When the dissipation potential is differentiable, the complementary behavioral equations can be formulated as:

$$\left\{\begin{array}{ll} -\dfrac{\mathrm{grad}\,T}{T} = \varphi_{,q} & (a)\\[2mm] \sigma^{\mathrm{ir}} = \varphi_{,\dot{\varepsilon}} & (b)\\[2mm] X = \varphi_{,\dot{\alpha}} & (c) \end{array}\right. \qquad [16.3]$$

These equations define the thermodynamic forces and allow the dissipation to be reformulated as a sum of force-flux products. Under the small strain assumption, the strain rate tensor *is identified with* the time derivative of the strain tensor, and the dissipation becomes:

$$d = \underbrace{(\sigma-\sigma^{\mathrm{r}})}_{\sigma^{\mathrm{ir}}} : \dot{\varepsilon}\; \underbrace{-A}_{X}.\dot{\alpha} - \frac{\mathrm{grad}T}{T}.q \geq 0 \qquad [16.4]$$

The identity $d \equiv 0$ defines a reversible thermodynamic process. Dissipation is usually written as the sum of two terms that are supposed to be positive separately: and that are the intrinsic dissipation, $d_1 = \sigma^{\mathrm{ir}}:\dot{\varepsilon}+X.\dot{\alpha}$, characterizing material degradation, and thermal dissipation, $d_2 = -\mathrm{grad}T/T.q$, reflecting the irreversible character of heat diffusion.

16.2.2. *Heat equation*

Under the small strain hypothesis, the two principles of thermodynamics can be written locally as:

$$\begin{cases} \rho \dot{e} = \sigma : \dot{\varepsilon} - \mathrm{div}q + r_e \\ \rho T \dot{s} = d_1 - \mathrm{div}q + r_e \end{cases} \qquad [16.5]$$

where r_e is the external heat supply (e.g. radiation energy). Once it is noted that:

$$\rho T \dot{s} = -\rho T \frac{d\psi_{,T}}{dt} = \underbrace{-\rho T \psi_{,TT}}_{\rho C_{\varepsilon,\alpha}} \dot{T} - \underbrace{\rho T \psi_{,T\varepsilon} : \dot{\varepsilon}}_{-w^{\circ}_{the}} - \underbrace{\rho T \psi_{,T\alpha} . \dot{\alpha}}_{-w^{\circ}_{thc}} \qquad [16.6]$$

the local form of the heat diffusion equation can be directly deduced. Based on Fourier's conduction law, the heat equation can be written as follows:

$$\rho C_{\varepsilon,\alpha} \dot{T} \underbrace{-\mathrm{div}(k\,\mathrm{grad}T)}_{q = -k\,\mathrm{grad}\,T} = \underbrace{d_1 + T\,\sigma^r_{,T} : \dot{\varepsilon} + T\,A_{,T} . \dot{\alpha} + r_e}_{w^{\circ}_h} \qquad [16.7]$$

where $C_{\varepsilon,\alpha}$ is the specific heat[1] at constant ε and α, and k is the conduction tensor.

The notation $(w)^{\circ}$ indicates that (w) is dependent on the thermodynamic path followed. The term w°_h stands for the overall heat rate accompanying the deformation process. The notation w°_h means that w_h is not, *a priori*, a state function.

On the left-hand side of equation [16.7], the term $\rho C_{\varepsilon,\alpha} \dot{T}$ represents the heat rate stored or released per unit volume, while the term $-\mathrm{div}(k\,\mathrm{grad}T)$ corresponds to the local heat losses by conduction. On the right-hand side, we successively find the intrinsic dissipation d_1, the heat sources associated with the thermomechanical coupling mechanisms (in brief, the coupling sources), taking into account the possible temperature sensitivity of reversible stress σ^r and conjugated variables A, and finally the external heat supply r_e.

---

1 The specific heat involved here is thus defined for isothermal processes during which all state variables except the temperature remain constant. The reader may question the relevance of the existence of such processes. Note also that the classic C_p found in *Handbooks of Chemistry and Physics* implies that temperature and pressure are the only state variables used to describe the thermomechanical behavior of a material.

The coupling sources can be divided into two parts: the thermoelastic source w_{the}°, induced by material (thermo)dilatability, and other coupling sources w_{thc}°, such as the latent heat rate of a first-order solid–solid phase change. It should be noted that, whatever the thermomechanical model, w_{the}° and w_{thc}° do not correspond to $T\sigma_{,T}^{r}:\dot{\varepsilon}$ and $TA_{,T}.\dot{\alpha}$, respectively. The only general relationship that may be written is:

$$w_{the}^{\circ} + w_{thc}^{\circ} = T\sigma_{,T}^{r}:\dot{\varepsilon} + TA_{,T}.\dot{\alpha} \qquad [16.8]$$

If ρ and $C_{\varepsilon,\alpha}$ are material constants and if the conduction tensor also remains constant and isotropic, and where k denotes the isotropic conduction coefficient, the heat equation can be simplified as:

$$\rho C_{\varepsilon,\alpha}\dot{T} - k\Delta T = w_{h}^{\circ} + r_{e} \qquad [16.9]$$

If we finally assume that the equilibrium temperature field T_0 and r_e are time-independent, with $\theta = T - T_0$ denoting the temperature variation with respect to the local thermal equilibrium, then $-k\Delta T_0 = r_e$ and:

$$\rho C_{\varepsilon,\alpha}\dot{\theta} - k\Delta\theta = w_{h}^{\circ} \qquad [16.10]$$

16.2.3. Energy balance over a load–unload cycle

Let us now consider a load–unload cycle, where $A = (T_A, \varepsilon_A, \alpha_A)$ and $B = (T_B, \varepsilon_B, \alpha_B)$ denote the thermomechanical states of the material at the extremities of the process. Let us then consider the energy balance corresponding to the three following situations:

Case i: $A \neq B$, this general situation can be illustrated by the schematic stress–strain diagram proposed in Figure 16.1(a).

Case ii: $A \neq B$ and $\varepsilon_A = \varepsilon_B$, a mechanical cycle is then associated with the load-unload test. The stress–strain diagram shows a hysteresis loop.

Case iii: $A = B$, the mechanical cycle is now a thermodynamic cycle.

In the small strain framework, the deformation energy rate is defined by:

$$w_{def}^{\circ} = \sigma:\dot{\varepsilon} \qquad [16.11]$$

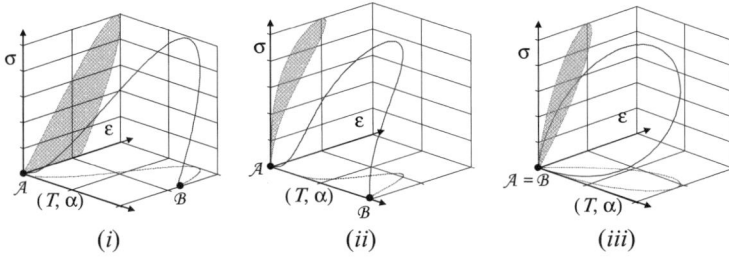

Figure 16.1. *Stress–strain curve constructed as the projection of a thermodynamic path onto the mechanical plane (source: [CHR 10])*

Equation [16.4] shows that the intrinsic dissipation d_1 is then the difference between the deformation energy rate and the sum of the elastic and stored energy rates.

$$\begin{cases} d_1 = w_{\text{def}}^{\circ} - (w_{\text{e}}^{\circ} + w_{\text{s}}^{\circ}) & \text{(a)} \\ w_{\text{e}}^{\circ} + w_{\text{s}}^{\circ} = \sigma^{\text{r}}:\dot{\varepsilon} + A \cdot \dot{\alpha} & \text{(b)} \end{cases} \qquad\qquad [16.12]$$

Note again that the systematic correspondence between the two terms of equation [16.12b] is possible only for particular thermomechanical models.

When $t_B - t_A$ denotes the duration of the test and is based on equations [16.5] and [16.12], the deformation energy associated with a load–unload cycle may be formulated as follows:

$$w_{\text{def}} = \int_{t_A}^{t_B} d_1 \, dt + \int_{t_A}^{t_B} (\rho\dot{e} - \rho C_{\varepsilon,\alpha}\dot{T} + w_{\text{the}}^{\circ} + w_{\text{thc}}^{\circ}) \, dt \qquad\qquad [16.13]$$

where e is the specific internal energy. Equation [16.13] shows that:

Case *i*: in the general case, the deformation energy balance during a load–unload test involves energy dissipation, internal energy variations (variations in heat stored in the material) and coupling heat sources.

Case *ii*: the deformation energy then corresponds to the generalized area A_h of the different hysteresis loops in planes $\sigma_{\text{ij}}-\varepsilon_{\text{ij}}$, $i, j = 1,2,3$.

Case *iii*: for a thermodynamic cycle, these hysteresis loops are only due to dissipation and couplings, because the heat capacity ρC is assumed to be constant. The energy balance then takes the following simplified form:

$$w_{\text{def}} = \int_{t_A}^{t_B} \left(\sum_{\text{i,j}=1}^{3} \sigma_{\text{ij}} \, \dot{\varepsilon}_{\text{ij}} \right) dt = \int_{t_A}^{t_B} d_1 \, dt + \int_{t_A}^{t_B} (w_{\text{the}}^{\circ} + w_{\text{thc}}^{\circ}) \, dt \qquad\qquad [16.14]$$

The energy balance then gives the restricted conditions for which the dissipated energy can be estimated by computing the hysteresis area of a uniaxial load–unload cycle. Equation [16.13] shows that coupling sources or internal energy variations can also influence the size of the hysteresis area. This underlines the necessity of analyzing the thermal effects to verify if a mechanical cycle is also a thermodynamic cycle, and to check the relative energy importance of coupling effects [CHR 10].

16.3. Metrological considerations

IR radiation corresponds to the electromagnetic spectrum ranging from 0.8 to 1,000 μm wavelengths. IR thermography generally uses shortwave (SW) and longwave (LW) bands associated with the spectral domains 2–5 μm and 7–15 μm, respectively. Radiation from the observed thermal scene crosses the ambient atmosphere, then the camera lenses before reaching the IR sensor (in most cases, the sensor uses InSb or HgCdTe technology). The IR radiation modifies the electrical state of the sensor during the integration time of the image acquisition process. When using IR focal plane array (IRFPA) cameras, the electrical charges of each sensor (pixel) are then transferred to the acquisition system via a multiplexor before being stored in the computer memory. The quantity to be measured using these systems is rarely IR radiation but rather temperature. It is thus necessary to convert the digitized radiation into temperature fields of the observed scene as accurately as possible. Each of these steps may generate its own biases, and the combination of these different perturbations will decrease the accuracy and quality of the thermal measurement.

In numerous applications – heat sources or thermophysical property estimation – it is necessary to estimate the spatial and temporal derivatives of the thermal fields obtained by IR thermography. Consequently, the metrological considerations related to temporal reliability, spatial resolution and calibration remain fundamental whatever the chosen analysis e.g. nodal, modal, impulse, modulation. For instance, the accuracy of the gradient computation is directly linked to the accuracy of the thermal calibration and to the spatial resolution of the thermal measurement, while the accuracy of the time derivative stems from the calibration procedure and the stability of the time base. In most cases, these three aspects are combined when dealing with thermophysical property identification, except for a few particular cases such as the thermal diffusivity determination, which "only" requires precise knowledge of the spatial and temporal parameters.

In this section, the basic laws of thermal radiation are briefly reviewed. They enable the description of the radiation path from the observed object to the camera [GAU 99, PAJ 89, PAP 97]. The main calibration methods used for IRFPA are then briefly presented, followed by several metrological considerations regarding IR measurements. The purely geometrical effects linked to image formation, e.g. distortions, aberrations, are not within the scope of this section, and they will not be discussed hereafter.

16.3.1. *Physics of radiation preliminaries*

The luminance L (unit: $W.m^{-2}.sr^{-1}$) is the power emitted ($d\Phi$) per unit of solid angle ($d\Omega$) and per unit of apparent surface ($dS \cos\theta$) of an extended source in a given direction defined by the angles (θ,φ):

$$L = \frac{d^2\Phi}{d\Omega\, dS\, \cos\theta} \qquad [16.15]$$

The emittance M (unit: $W.m^{-2}$) is the power emitted $d\Phi$ by a unit surface of an extended source emitting in a whole hemisphere. Consequently, it corresponds to the luminance integrated over all directions of a hemisphere. If the luminance is constant in all directions (θ, φ), it follows Lambert's law and may be formulated as: $M = \pi\, L$. Monochromatic luminance L_λ and emittance M_λ are defined as the luminance and emittance derivatives with respect to the wavelength λ. When incident radiation encounters an object, a fraction of this radiation is reflected Φ_r, another fraction is absorbed by the object Φ_a and the rest is transmitted Φ_t. The following factors are then introduced: reflection $\rho_r(\lambda, T, \theta, \varphi)$, absorption $\alpha_r(\lambda, T, \theta, \varphi)$ and transmission $\tau_r(\lambda, T, \theta, \varphi)$. These factors depend on the wavelength λ, the object temperature T and the radiation direction (θ, φ). Energy conservation implies that the sum of these three factors must always be equal to unity. When the object is at thermodynamic equilibrium, it re-emits the absorbed radiation such that the emissivity factor ε_r is equal to α_r.

IR sensors are sensitive to IR radiation from an object. Black bodies have the ability to re-emit all received radiation whatever the wavelength (i.e. $\alpha_r = \varepsilon_r = 1$). In this case, according to Planck's law, the spectral luminance of an observed black body L_λ^{CN} can be associated with its temperature T:

$$L_\lambda^{CN} = \frac{c_1}{\pi\lambda^5 \left(e^{\frac{c_2}{\lambda T}} - 1\right)} \qquad [16.16]$$

where $c_1 = 3.741832.10^{-16}$ $W.m^2$ and $c_2 = 1.438786.10^{-2}$ $K.m$. Stefan's law: $L^{CN} = \pi^{-1}\sigma_b\, T^4$ is obtained by integrating Planck's law over the entire wavelength spectrum. Assuming Lambertian behavior of the black body, Stefan's law can also be written: $M_{CN} = \sigma_b\, T^4$, with $\sigma_b = 5.67032.10^8$ $W.m^{-2}.K^{-4}$ being Stefan–Boltzmann's constant. The latter two expressions allow the radiated flux, measured by an IR sensor, to be associated with the temperature of the observed black body. In practice, only a few objects behave like black bodies, and the relationship between radiation and temperature is established by integrating the monochromatic luminance L_λ of the observed body on the wavelength spectrum with $L_\lambda = \varepsilon_r \left(\lambda, T, \theta, \varphi\right) L_\lambda^{CN}$. Several sets of bodies can be defined according to the specific values taken by parameters α_r, ρ_r and τ_r: opaque bodies ($\tau_r \approx 0$), brilliant bodies ($\rho_r \gg 0$ and $\varepsilon_r \approx 0$), and gray bodies ($\varepsilon_r \approx \varepsilon_{r0}$ = constant). Measurements are often performed on plane samples, coated with matt black paint in order to obtain high emissivity gray bodies (i.e. ε_{r0} close to 1), with quite a uniform emissivity distribution, independent of the observation direction. The temperature variation should also remain small enough to neglect the temperature-dependence of the emissivity.

16.3.2. *Calibration*

IR cameras often have their own setting and acquisition applications, along with basic data-processing applications, e.g. conversion from digitized flux to temperature, average and standard deviation computations, display. Calibration laws proposed by manufacturers are based on the assumption that the sensor's response is linear in terms of emitted flux versus digitized flux over the whole range between the lower and upper saturation levels of the sensor (i.e. the pixel). The differences between the pixels' responses are interpreted as distributions of gains and offsets to apply to the mean response of the sensor matrix. This operation is often referred to as non-uniformity correction (NUC). The gains and offsets of each pixel are computed to obtain uniform distributions of digitized fluxes for two specific images of two uniform thermal scenes taken at different temperatures. To that end, it is advised to use a precision extended black body. Recently, some manufacturers have proposed to relate gain and offset values to the "internal temperature" of the camera in order to account for thermal drifts associated with heat losses that occur between the camera and its environment (extended non-uniformity correction (ENUC)). The temporal drift of IR cameras will be addressed thoroughly in section 16.3.3.

Once the response of each pixel is brought back to the mean response of the sensor matrix, a global calibration law is used to relate the average digitized flux to the average temperature. In most cases, the calibration law is taken in the form of a two- or three-degree polynomial.

Sensor matrices always include a certain number of defective pixels (in most cases, less than 0.5%). For instance, they correspond to saturated pixels or to "dead" pixels. They are localized using criteria dealing mainly with the discrepancy with respect to the mean response (in terms of digitized flux, gain, offset). Manufacturers propose to replace the digitized flux value of these pixels by that of their nearest non-defective neighbor (bad pixel replacement, or *BPR*, operation).

The validity of the "factory" calibration can easily be verified by observing a given thermal scene with a single camera, but using different calibration laws associated with different acquisition settings (integration time, or measurement range). Figure 16.2 illustrates different observations on a black body using different IRFPA cameras. They show that it is appropriate to use the center of the matrix and the middle of the calibration range when using the manufacturer's calibration laws. Furthermore, if the specifications concerning measurement accuracy are more stringent than 1 K, it is advisable to perform a complete customized calibration of the sensor matrix with testing conditions and camera configuration (integration time, windowing) similar to those used for the application.

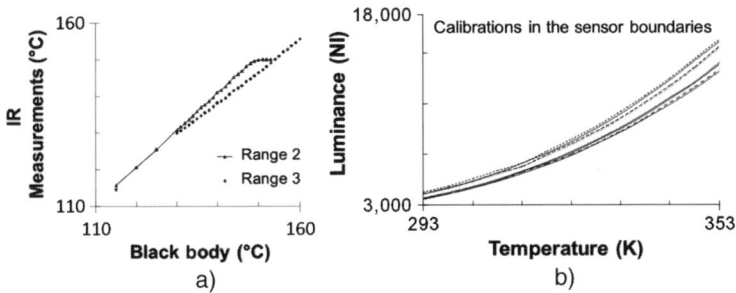

Figure 16.2. *Verification of factory calibration using an extended black body:*
(a) comparison between two ranges of a single camera (FLIR SC1000);
(b) comparison between several pixel responses (CEDIP IRC 320-4LW)

This calibration overcomes the limitations inherent to the *NUC–BPR* procedures (linearity assumption valid far enough from saturation for the *NUC*, introduction of a strong spatial correlation between neighboring pixels for the *BPR* operation). However, this calibration requires an extended black body in order to have a uniform radiation source at different temperature levels. The calibration law of each pixel can be chosen as a polynomial or as a Planck-like function. The calibration coefficients are obtained by approximating, generally in the least squares sense, the couples (digitized radiation–temperature) by the chosen calibration function.

Defective pixels are then localized using a criterion for measuring the mismatch between the calibrated and specified temperature. The BPR operation is not performed: temperatures of the defective pixels are not taken into account in the subsequent data processing. A specific pixel-to-pixel calibration is detailed in [HON 05].

16.3.2.1. *Temporal analysis*

The integration time sets the duration during which radiation resulting from the thermal scene will be captured by the camera. For this reason, it determines the temporal resolution of the device. As the image transfer time to the storage memory is often much higher than the integration time (a few milliseconds vs. several microseconds), the camera thus does not see the scene most of the time, which is particularly penalizing for observing fast phenomena.

Once the integration time is chosen, the temporal analysis can be disturbed by the absence of some images in the stored sequence. Depending on the equipment, a temporal shift of one or two images can occur at the beginning of the sequence. This is due to the fact that the first stored image corresponds to the one that was captured when the starting order occurred, not the actual image at the beginning of the sequence. A simple sequence shift is sufficient to correct this edge effect.

The second, more penalizing, problem is the absence of some images in a film. This problem is relatively unimportant in terms of visualization, but can become critical in image processing when time is involved. Algorithms that are compatible with variable acquisition frequencies are then required. To count and isolate times from the missing images, it is possible to directly read time information in the files from the camera, provided that they have been accurately stored, that is, sufficient with respect to the acquisition frequencies used. Depending on the camera model, the number of missing images can thus range from one to several dozen.

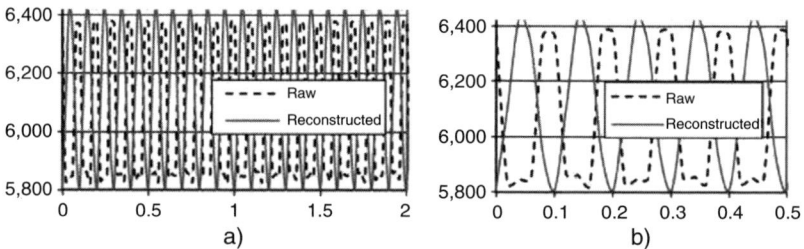

Figure 16.3. *Errors induced by the loss of images: in lock-in thermography, the amplitude is not really affected (a) but the phase is strongly distorted (b)*

Figure 16.3 presents the artifacts observed in the case of a numerical lock-in procedure applied to a series of 500 images in which only two images are missing. If the amplitude is not significantly affected, the phase has a completely erratic behavior, and takes a value that depends directly on the number and the phase of the missing images.

16.3.2.2. *Spatial resolution*

The adoption of focal plane array technology has led to clear improvements in image quality (Figure 16.4).

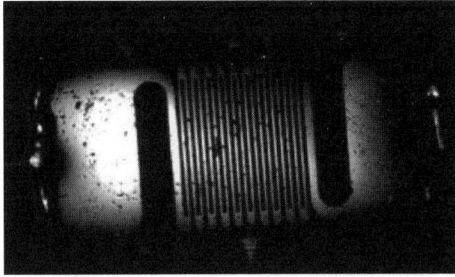

Figure 16.4. *Extensometric gauge (tracks of approximately 20 μm) observed with an M1 lens*

However, the independence of each sensor relative to its neighbors must be checked. One of the most current tests for characterizing such equipment is the slit response function (SRF) test: the camera focuses on a thermal slit (side-cooled slit of variable width placed in front of a hot plate), and the following function is studied (contrast):

$$\text{FRF} = \frac{V(x) - V_{\min}}{V_{\max} - V_{\min}} \qquad [16.17]$$

where $V(x)$ is the value recorded for a slit width equal to x, $V_{\max}$ is the value recorded when the slit is wide open ($x \to \infty$) and $V_{\min}$ is the recorded value on the cooled part (Figure 16.5(a)). In general, it is assumed that, for 320×240 pixel cameras, to obtain a good measurement, the object must be projected on at least two detectors. Thus, with a lens magnification of 1 (M1) and a matrix periodicity of 30 μm, truly independent information can only be obtained at each step of 60 μm.

A study of this SRF for different positions clearly shows (Figure 16.5(b)) that the pixels are definitely more correlated on the edges of the image than in the center. Note that there is indeed a problem of correlation between close measurement points, that is, on the one hand, only the contrast (and by no means the average value) is affected and, on the other hand, there is convolution of the thermal scene by this response function. Consequently, a simple geometrical correction (e.g. of repositioning of the points in the image, or amplification and/or offsets applied to each pixel) is necessary to recover the real quantitative image of the scene, in addition to a deconvolution procedure.

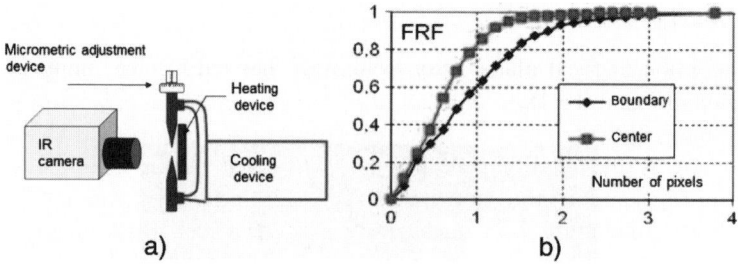

Figure 16.5. *(a) Slit response function; (b) SRF near the edge of the array compared to SRF at the center (CEDIP IRC 320-4 LW camera)*

16.3.3. *Thermal noise and thermal drift*

16.3.3.1. *Thermal stability of devices*

In order to reduce radiation in the vicinity of the IR sensors, the array is cooled at approximately 80 K. In new-generation IR cameras, cooling is done using a Stirling cycle engine, which has replaced the liquid nitrogen cooling systems of older cameras. However, cameras have gained in portability, but with a loss of performance: the cooling that was quasi-instantaneous with nitrogen now requires at least 10 min before any measurement is possible (Figure 16.6(a)).

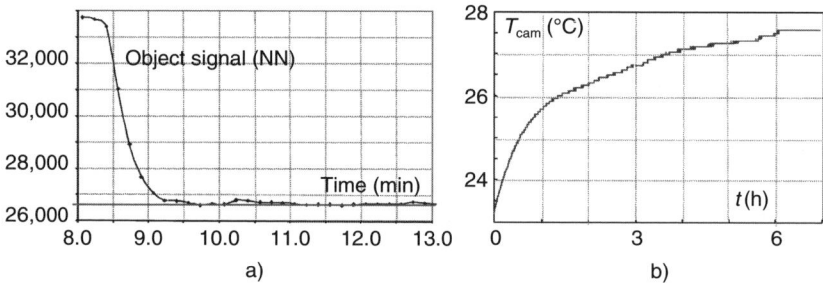

Figure 16.6. *(a) Cooling CEDIP IRC320-4LW; (b) thermal drift CEDIP JADE III*

In addition, once cooling is achieved, a slow drift of about 1–5 mK/s can occur with certain equipment, sometimes over durations reaching a few hours (Figure 16.6(b)). Because these internal temperature drifts in the camera modify the sensor responses, it is appropriate in many situations to wait until the camera temperature is stabilized, or to take this internal drift into account in the conversion of the digitized signal into temperature (compensated NUC). In addition, certain lower-quality devices have instabilities of 0.5 or even 1 K, which is incompatible with quantitative measurements.

16.3.3.2. *Environmental stability*

The signal measured by a camera (which is roughly an intensity) comes primarily from the object (which is assumed to be opaque and gray in the camera's spectral range), and also, to a lesser extent, from the environment and atmosphere (Figure 16.7). If the environment can be considered as an integral radiator of temperature T_{env} and if the atmosphere between the target and the camera is isothermal at temperature T_{atm}, considering a coefficient of transmission τ_{atm}, the measured intensity L_{mes} can be formulated as a function of the intensity of a black body at the object temperature L^0:

$$L_{mes} = \tau_a.\varepsilon.L^0(T_{obj}) + \tau_a(1-\varepsilon)L^0(T_{env}) + (1-\tau_a)L^0(T_{atm})$$

[16.18]

For short distance measurements (about a few dozen centimeters), the atmosphere can reasonably be considered as being transparent, and thus:

$$L_{mes} = \varepsilon L^0(T_{obj}) + (1-\varepsilon)L^0(T_{env})$$

[16.19]

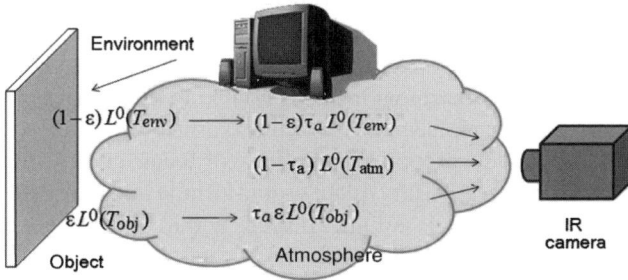

Figure 16.7. *Simplified radiometric balance*

This equation shows that the environment must be reasonably well controlled in order to limit the influence of spurious radiation (reflection from a radiator or any other radiative IR source, or even from the operator). This precaution is all the more important when the measured temperature increases are minor. In addition, using a high emissivity coating (thus of low reflectivity) is obviously advantageous to minimize the parasitic flow/object flow ratio.

Along the same lines, note also the presence of the Narcissus effect (reflection of the cold detector on the scene), which is often observed when using a macro lens (e.g. lens magnification of 1 [PRO 04]). Usually, this is only an offset map that is superimposed on the scene, and that can thus be offset by subtraction of a reference image.

Last but not least, possible environmental instabilities could modify the exchange conditions between the sample and its environment and thus must be taken into account, especially when there are marked temperature variations over time. Consequently, it is useful to check that room temperature variations at the different hours of the day (Figure 16.8) do not significantly influence the measurements.

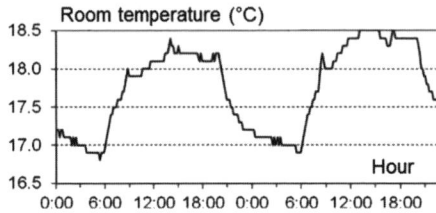

Figure 16.8. *Environmental temperature variations over 24 h*

16.4. Heat diffusion models and identification methods

An ultimate goal of thermal image processing could be the determination of source fields at any point in a 3D structure while taking variations in thermophysical properties of the material into account throughout the process. However, this is (currently) out of reach for metrology reasons related to the physics of radiation and IR techniques, and also for mathematical reasons (surface measurements, inverse ill-posed problem). The models proposed hereafter will focus on diffusion problems in simplified systems, either 2D (thin plate, semi-infinite medium) or 1D (straight beam) and even 0D (material volume placed within a homogeneous medium uniformly cooled or heated). We will, however, introduce, through integral transforms, direct relationships between full-field temperature measurements and certain types of sources in the case of thick 3D media. The following section puts forward equations that enable us to estimate thermophysical parameters or heat sources using thermographic images. The many possibilities offered by integral transformations of temperature fields, in relation to the heat transfer model, are the cause of numerous IR image issues. In what follows, we will refer to operational application examples to highlight this.

16.4.1. *Diffusion equation for thin plates*

The medium in question is a thin plate of length L, width l and thickness e. The corresponding coordinates are (x, y, z), respectively. Within the framework of our experimental tests, the following hypotheses are often accepted:

– the specific heat and conduction coefficient are constant material characteristics, independent of the internal state;

– as the tests are quasi-static, the convective terms are neglected in the particular time derivative. Note that the combination of IR and digital image correlation (DIC) techniques enables us to monitor the temperature of material surface elements (SEM) and to consider the convected heat [BER 08, CHR 09, CHR 10]. In some cases (e.g. necking lips of polymers [MUR 08]), the convected heat may reach up to 40% of the heat source.

To reduce it to a 2D problem, the idea here is to consider that the averaged value of the source $s(x,y,z,t)$ over the thickness is sufficiently representative of this distribution throughout this thickness. We classically define:

$$\overline{f}(x,y,t) = \frac{1}{e}\int_{-\frac{e}{2}}^{+\frac{e}{2}} f(x,y,z,t)\,dz \tag{16.20}$$

as the averaging operation. Assuming constant thermophysical parameters and applying the averaging operation to the heat diffusion equation [16.10], we obtain:

$$\rho C\left(\frac{\partial\overline{\theta}}{\partial t} + \frac{\overline{\theta}}{\tau_{th}^{2D}}\right) - k\left(\frac{\partial^2\overline{\theta}}{\partial x^2} + \frac{\partial^2\overline{\theta}}{\partial y^2}\right) = \overline{s} \text{ with } \rho C\frac{\overline{\theta}}{\tau_{th}^{2D}} \approx -\frac{k}{e}\left[\frac{\partial\theta}{\partial z}\right]_{-\frac{e}{2}}^{+\frac{e}{2}} \tag{16.21}$$

where out-of-plane heat losses are estimated under the assumption of linear heat fluxes (Fourier boundary conditions) and where the time constant τ_{th}^{2D} characterizes the heat fluxes through the surfaces defined by $z = \pm e/2$.

16.4.2. *Diffusion equation for straight beams*

For some sample shapes, it may be interesting to work only with temperature and source profiles along a straight beam. This represents a gain with respect to data storage, processing speed and the quality of the signal-to-noise ratio of thermal data. This approximation is all the more acceptable because the heat source distribution slightly fluctuates within a cross-section S, inducing a representative average value. Assuming that thermophysical parameters remain constant over each cross-section and then integrating equation [16.10] over S, we obtain a 1D version of the heat diffusion equation. Noting:

$$\overline{\overline{f}}(x,t) = \frac{1}{S}\int_{-\frac{l}{2}}^{+\frac{l}{2}}\int_{-\frac{e}{2}}^{+\frac{e}{2}} f(x,y,z,t)\,dz\,dy \tag{16.22}$$

the diffusion equation becomes:

$$\rho C \left(\frac{\partial \bar{\bar{\theta}}}{\partial t} + \frac{\bar{\bar{\theta}}}{\tau_{th}^{1D}} \right) - k \left(\frac{\partial^2 \bar{\bar{\theta}}}{\partial x^2} \right) = \bar{\bar{s}} \quad \text{with} \quad \frac{\bar{\bar{\theta}}}{\tau_{th}^{1D}} \approx -\frac{k}{\rho C} \overline{\left(\frac{\partial^2 \theta}{\partial y^2} + \frac{\partial^2 \theta}{\partial z^2} \right)} \qquad [16.23]$$

The approximation of lateral heat losses is still based on the assumption of linear heat exchange temperature at the boundary ∂S of each cross-section S. The heat exchange coefficient h between the specimen and the surrounding air is assumed to be constant:

$$-k \, \text{grad}\theta . n = h\theta \qquad [16.24]$$

where n is the outward-pointing normal vector to ∂S.

Application examples on the necking of steel and polymers have been presented in [CHR 00] and [MUR 08].

16.4.3. *Diffusion equation for a monotherm material volume element*

In the classic framework of tensile loading on homogeneous plates, which is of practical importance, we can reduce the heat diffusion problem in the center of the sample gauge part to an ordinary differential problem with respect to time. For this, the material must have constant thermophysical characteristics with a uniform load, thus resulting in a uniform heat source distribution. Moreover, the initial temperature field is uniform, and finally the thermal boundary conditions are linear (Fourier condition) and symmetrical on each pair of facing surfaces. Under these conditions, it can be shown [CHR 00] that the solution to the 3D diffusion problem is almost completely borne by the first eigenfunction of the spectral solution. The 3D diffusion problem reduces to the following differential equation:

$$\rho C \left(\frac{d\theta}{dt} + \frac{\theta}{\tau_{eq}} \right) \approx s \qquad [16.25]$$

where θ is the temperature variation recorded in the center of the specimen gauge part, and with the time constant τ_{eq} characterizing the local heat losses [CHR 95]. This simple heat equation of ODE form has been used to study the behavior of materials subjected to high cycle fatigue tests [BOU 04, DOU 09].

16.4.4. *Integral transforms and quadrupole method related to thick media*

A direct analytical relation between the observable temperature field (from an IR camera) and the heating source field is convenient for implementing estimation methods related to thick media. The characterization of thermophysical properties requires calibrated heat excitation, often imposed at the sample surface. Instead of thermophysical properties, when the characterization of an unknown internal heating source field is investigated, the transient point-source response (Green's function) can be conveniently considered.

The main idea is to consider the general heat equation reduced to a case depending only on the space variable x orthogonal to the observed sample surface. A 1D heat transfer versus the x-direction is then considered, when the heat losses are assumed to be negligible ($\tau_{th}^{1D} \rightarrow +\infty$). An expression very similar to [16.23] is then obtained. Then, based on the Laplace transformation, suitable direct expressions of the temperature response to localized heating may be formulated. The general 1D heat equation related to a homogeneous wall of thickness e is then:

$$\frac{\partial^2 \theta}{\partial x^2} = \frac{1}{a}\frac{\partial \theta}{\partial t} \qquad [16.26]$$

where $a = k/\rho C$ is the thermal diffusivity of a homogeneous medium. This thermophysical parameter (in $m^2\ s^{-1}$) is crucial for estimating the spatial distribution of source terms. The previous expression is obtained with a Laplace transform such that:

$$\tilde{\theta}(x,p) = \int_0^\infty \exp(-pt)\theta(x,t)\,dt \qquad [16.27]$$

the Laplace variable p can be related to a frequency in s^{-1}. If $\theta(x, t=0)=0$ whatever x, the Laplace transform of expression [16.26] then yields:

$$\frac{d^2\tilde{\theta}}{dx^2} = \frac{p}{a}\tilde{\theta} \qquad [16.28]$$

The definition of the Laplace transform of the heat flux at a given depth x is then:

$$\tilde{q}(x,p) = -k\frac{d\tilde{\theta}}{dx}(x,p) \qquad [16.29]$$

The following ordinary differential system expressed with the [temperature, flux] vectors in the Laplace space yields:

$$\frac{d}{dx}\begin{bmatrix} \tilde{\theta} \\ \tilde{q} \end{bmatrix} = \begin{bmatrix} 0 & -1/k \\ -\rho Cp & 0 \end{bmatrix}\begin{bmatrix} \tilde{\theta} \\ \tilde{q} \end{bmatrix}$$

[16.30]

Integration on domain $[0, e]$ gives the solution of system [16.30], such that:

$$\begin{bmatrix} \tilde{\theta}(x = 0, p) \\ \tilde{q}(x = 0, p) \end{bmatrix} = \begin{bmatrix} A & B \\ C & D \end{bmatrix}\begin{bmatrix} \tilde{\theta}(x = e, p) \\ \tilde{q}(x = e, p) \end{bmatrix}$$

[16.31]

where $r_a(p) = \sqrt{\frac{p}{a}}$:

$$A = D = cosh(r_a e), \ B = \frac{sinh(r_a e)}{kr_a} \ \text{and} \ C = kr_a \ sinh(r_a e)$$

[16.32]

This kind of approach was previously used for the analogical description of electrical circuits called quadrupoles. This name is related to such approaches, even involving heat transfer [CAR 59, MAI 00].

In the case of pulse heating at $x=0$ ($\tilde{q}(x = 0, p) = Q$, with Q in $J.m^{-2}$) for a given sample assumed to be adiabatic, it gives:

$$\begin{bmatrix} \tilde{\theta}(x = 0, p) \\ Q \end{bmatrix} = \begin{bmatrix} A & B \\ C & D \end{bmatrix}\begin{bmatrix} \tilde{\theta}(x = e, p) \\ 0 \end{bmatrix}$$

[16.33]

The Laplace transform of the temperature response, measured by the IR camera, on the front face ($x = 0$) gives:

$$\tilde{\theta}(x = 0, p) = Q\frac{A}{C} = Q\frac{cosh(r_a e)}{kr_a \ sinh(r_a e)}$$

[16.34]

and on the rear face ($x = e$):

$$\tilde{\theta}(x = e, p) = \frac{Q}{C} = \frac{Q}{kr_a \ sinh(r_a e)}$$

[16.35]

A simple expression between the front face and the rear face temperatures, at short time or high frequency (large p values), can be deduced such that:

$$\tilde{\theta}(x = e, p) = \tilde{\theta}(x = 0, p).exp(-r_a \ e)$$

[16.36]

In addition to having the advantage of reflecting explicit relationships in a Laplace space, expressions [16.34] to [16.36] can be tailored to problems with lateral heat losses and/or in 2D or 3D situations. This analytical method allows us to consider very thick layers (semi-infinite media, see section 16.4.4.2) or very thin layers (see section 16.4.4.3) with asymptotic expansions. Transient periodic phenomena are also conveniently considered by similar expressions when the Laplace transform is replaced by a Fourier transform. The main advantage is that heat transfer can be represented through multilayered samples by a simple matrix product (see [BAT 94, MAI 00]).

Such expressions and the resulting asymptotic expansions are also the starting point for designing instrumental methods for measuring thermophysical properties (diffusivity, or thermal contact resistances). 1D analytical considerations have provided the basis for flash methods (see [HAY 04] for a review of these methods).

16.4.4.1. *Multidimensional cases*

1D cases are often inadequate when the front face heat flux (at $x = 0$) is spatially non-uniform, for instance when heating is obtained with a laser spot. A 2D or 3D transient transfer assumption must then be considered, even with a non-isotropic sample. The general equation in the principal axes of anisotropy is then:

$$\frac{\partial^2\theta}{\partial x^2} + \frac{k_y}{k_x}\frac{\partial^2\theta}{\partial y^2} + \frac{k_z}{k_x}\frac{\partial^2\theta}{\partial z^2} = \frac{1}{a_x}\frac{\partial\theta}{\partial t} \qquad [16.37]$$

where k_x, k_y, k_z are the thermal conductivities following the principal directions x, y, z with $a_x = k_x/\rho C$: the thermal diffusivity related to direction x.

A simple situation can be reduced to a sample that is adiabatic on the lateral faces (at $y = 0$ and $y = L$, and at $z = 0$ and $z = l$):

$$\frac{\partial\theta}{\partial y}\bigg|_{y=0} = \frac{\partial\theta}{\partial y}\bigg|_{y=L} = \frac{\partial\theta}{\partial z}\bigg|_{z=0} = \frac{\partial\theta}{\partial z}\bigg|_{z=l} = 0 \qquad [16.38]$$

A Fourier integral transform related to the space variables y and z and a Laplace transform related to the time variable t allow us to obtain a formal expression similar to the previous 1D case. Here, the adapted Fourier–Laplace transform is as follows:

$$\theta(x, \alpha_n, \beta_m, p) = \int_0^L \int_0^l \int_0^\infty \theta(x, y, z, t) \cos(\alpha_n y) \cos(\beta_m z) \exp(-pt)\, dx\, dy\, dt \qquad [16.39]$$

where pulsations α_n and β_m are defined by $\alpha_n = \frac{n\pi}{L}$ and $\beta_m = \frac{m\pi}{l}$.

The general equation in the transformed space, where $\theta(x, y, z, t = 0) = 0$, is then:

$$\frac{d^2\tilde{\theta}}{dx^2} - \frac{k_y}{k_x}\alpha_n^2\tilde{\theta} - \frac{k_z}{k_x}\beta_m^2\tilde{\theta} = \frac{p}{a_x}\tilde{\theta} \qquad [16.40]$$

The method described previously in the 1D case can be extended with the same formal expressions, by then considering: $r_a^2 = \frac{p}{a_x} + \frac{k_y}{k_x}\alpha_n^2 + \frac{k_z}{k_x}\beta_m^2$.

The simultaneous estimation of parameters a_x, k_y/k_x and k_z/k_x is then obtained by considering the marginal averaging of temperature images related to zero pulsations ($\alpha_n = 0$, $\beta_n = 0$). These expressions enable the modeling of many experimental situations related to the use of IR thermography devices. Simultaneous estimations of thermal diffusivity tensor terms are available, for instance in [PHI 95, KRA 04].

16.4.4.2. Semi-infinite medium

3D transient cases in semi-infinite media are easily deduced from the previous expressions. At the front face (at $x = 0$), point source pulse heating at $y = z = 0$ is expressed by a Laplace–Fourier transformation by: Q_0 (in J), which is a real constant in the Laplace–Fourier space. This kind of heating can nowadays be conveniently implemented with a focused laser spot (at $y = z = 0$) in a pulsed mode. The temperature response, when e tends to infinity in expression [16.34] extended to the 3D case (Laplace–Fourier transform), yields:

$$\tilde{q}\left(x, a_n, b_m, p\right) = \frac{Q_0}{k_x r_a}\exp(-r_a x) \text{ and } \tilde{q}\left(x, a_n, b_m, p\right) = \frac{Q_0}{k_x r_a} \qquad [16.41]$$

This expression is very important in order to assess temperature behavior at the surface (at $x = 0$) of any thick medium at short times.

If the inverse expression of the heat flux at $x = 0$ is considered versus the observed temperature at $x = e$, it then yields:

$$Q_0 = \tilde{\theta}\left(x = e, \alpha_n, \beta_m, p\right)k_x r_a \exp\left(+r_a e\right) \qquad [16.42]$$

This expression is not stable and is very sensitive to measurement noise when directly applied to experimental temperature observations. It amplifies the high frequency measurement noise exponentially even if this noise amplitude is lower than the signal, especially when the observation depth e is large. This example illustrates the ill-posed nature of the attempt to estimate the characteristics of an in-depth heating source (e.g. depth, size, form, intensity) from surface temperature observations and analysis (see, for example [BEC 77]).

Expression [16.41] can also be considered in time–space by a Laplace inverse transform of the previous expression and remaining the Fourier transform:

$$\tau(x,\alpha_n,\beta_m,t) = (Q_0/\rho c)\left(e^{\left(-x^2/(4a_x t)\right)}/\sqrt{\pi a_x t}\right)e^{\left(-a_y\alpha_n^2 t\right)}e^{\left(-a_z\beta_m^2 t\right)} \quad [16.43]$$

A suitable approximation of the inverse Fourier transform can then be considered when l and L are large:

$$\theta(x,y,z,t) = (Q_0/\rho C)\left(e^{\left(-x^2/(4a_x t)\right)}/\sqrt{\pi a_x t}\right)\left(e^{\left(-y^2/(4a_y t)\right)}/\sqrt{\pi a_y t}\right)\left(e^{\left(-z^2/(4a_z t)\right)}/\sqrt{\pi a_z t}\right)$$

$$[16.44]$$

This expression is then considered to be separable or presented as a product of three terms depending on the x, y and z directions. Expression [16.44] then contains the main characteristics of the 3D diffusive heat transfer.

If the observed temperature field is summed versus the y and z directions (or if both $\alpha_n = 0$ and $\beta_m = 0$ are considered in expression [16.43]), a 1D expression related to a purely diffusive transfer versus the x direction and the parameter to be estimated is the thermal effusivity related to the x direction: $\sqrt{k_x \rho C}$:

$$\int_0^L \int_0^l \theta(x=0,y,z,t)\,dy\,dz = \frac{Q_0}{\sqrt{k_x \rho C}\sqrt{\pi t}} \quad [16.45]$$

This result provides the basis of the hot plane thermal effusivity estimation method. Uniform heating is applied (generally with the contact of a uniformly deposited heating resistor network) on the front face of the sample and the transient front face temperature response is then analyzed (see [GUS 91, FUD 05]). Generally, the inertia of the heating device and temperature probe must be taken into account with this method. Instead of a temperature probe, an IR camera may be used to conveniently consider not only one sensor (or one image pixel) in the case of uniform heating at $x = 0$ but also the whole sensor field, which may be averaged (simultaneously averaging all pixels of the image) in the case of point source heating. If heating is generated by a laser spot, the inertia of this heating and of the temperature sensor is avoided, and such optical methods may be used for investigations related to very short times or high frequencies.

If the medium is heated with a line source in the y direction (instead of a point source), this is equivalent to the consideration of averaged pixel signals versus y (or $\alpha_n = 0$) in the case of a point source. At $z = 0$, the expression is then:

$$\int_0^L \theta(x=0,y,z=0,t)\,dy = \frac{Q_0}{\sqrt{k_x \rho C}\,\sqrt{\pi t}}\,\frac{1}{\sqrt{a_z}\,\sqrt{\pi t}} = \frac{Q_0}{\sqrt{k_x k_z}\,\pi t} \qquad [16.46]$$

The analysis of the temperature response to the line source is also called the hot wire method. This method has led to many experimental applications and variants with solid sensors (platinum wire for both heating and temperature measurement, thermocouple-wire systems, see [CAR 59, FUD 05]). Contrary to the hot plane method, this method is sensitive to thermal conductivity rather than thermal effusivity. In the case of orthotropic media, the estimated thermal conductivity is related to the harmonic averaging of the two-directional transfer, then the estimated thermal conductivity is $\sqrt{k_x k_z}$. The development of IR cameras has enabled the simultaneous implementation of hot plane, hot wire and point source methods by a simple directional averaging of the temperature response mapped with a single point heating source. The prime advantage with a laser spot and an IR camera is to avoid the inertia of solid sensors. The main problem is then to manage the radiative heating of the sample. Even if this condition is not fulfilled, IR camera measurement is useful to compare the ratio between the parameters considered here.

16.4.4.3. *Thin plate*

If the studied medium is very thin (*e* small), expression [16.34] can be reduced to a 2D case, which justifies the Laplace–Fourier transform of approximation [16.16] presented in section 1.3.1:

$$\tilde{\theta}(x=0,\alpha_n,\beta_m,p) = \frac{Q_0}{e(\rho Cp + k_y \alpha_n^2 + k_z \beta_m^2)} \qquad [16.47]$$

Such analytical expressions can be studied and implemented under many different forms and represent the first steps (forward models) for thermal diffusivity estimation (even thermal diffusivity mapping) or for identifying functions related to source terms inside the medium. Such solutions can be considered in real space with inverse transforms, or directly in the transformed space [DEG 96, BAT 01].

16.4.4.4. *Inverse Laplace transform–Stehfest algorithm*

Numerical inversion algorithms for the Laplace transform are very useful if the previous expressions are considered. No universal algorithm exists, but the very convenient Stehfest algorithm is presented here and is suitable for the calculation of temperature responses in the case of pulse heating (see [MAI 00] for further details related to such algorithms).

The method consists of seeking the time–space function $f(t)$ as a linear expression such that:

$$f(t) = \frac{\text{Ln}(2)}{t} \sum_{j=0}^{j=N} V_j F\left(\frac{j\,\text{Ln}(2)}{t}\right) \tag{16.48}$$

where $F(p)$ is the Laplace transform of $f(t)$. N depends on the floating-point precision of the computer. With single precision, $N = 10$ is often the most suitable choice, with the following V_j coefficients: $V_1 = 1/12$, $V_2 = -385/12$, $V_3 = 1279$, $V_4 = -46871/3$, $V_5 = 505465/6$, $V_6 = -473915/2$, $V_7 = 1127735/3$, $V_8 = -1020215/3$, $V_9 = 328125/2$, $V_{10} = -65625/2$.

16.4.4.5. *Inverse Fourier transform*

A substantial body of scientific literature is related to Fourier transform applications (see for example [PRE 07]). Generally, for heat transfer experiments, classic solutions related to cosine Fourier transforms are adapted to adiabatic boundary conditions at $y = 0$ and $y = L$ and at $z = 0$ and $z = L$:

$$\tilde{\theta}(\alpha_n) = \int_0^L \theta(y)\cos(\alpha_n y)\,\mathrm{d}y \tag{16.49}$$

The inverse transform is then given by the following expression:

$$\theta(y) = \frac{1}{L}\hat{\theta}(0) + \frac{2}{L}\sum_{n=1}^{\infty} \cos(\alpha_n y)\tilde{\theta}(\alpha_n) \text{ with } \alpha_n = \frac{n\pi}{L} \tag{16.50}$$

Integral [16.44], applied to discrete IR camera observations, is approximated by trapezoidal rules [PRE 07]. Usually, only a finite number N of pixels provides information about the temperature distribution and then allows access to a finite number N of transformed temperatures $\tilde{\theta}(\alpha_n)$. An extension to 2D transient problems is then possible.

16.5. Concluding comments and prospects

In conclusion, research into the mechanics of materials domain concerned by IR thermography has a broad scope. The thermomechanical behavior of material under mechanical and thermal loadings is very often assessed by identifying the material temperature and comparing it to that of the close environment (e.g. room temperature). When its role is not reduced to this parameter, temperature is often referred to as a control variable. These approximations surely mask important phenomena that cannot be identified without adopting a true thermomechanical approach to material behavior. Advances in thermographic imaging now enable access to quantitatively reliable temperature measurement fields. These fields can then be used to assess certain physical coefficients, such as the local diffusivity or conduction of the studied material. In the near future, these space distributions of thermophysical properties could be used to detect material irregularities that are natural or induced by mechanical and thermal loadings. Once these thermophysical characteristics are known, temperature fields can then be used to estimate heat sources accompanying the deformation, while reflecting the irreversibility of the deformation mechanisms and/or thermomechanical coupling effects.

The examples mentioned here are limited to quasi-static transformations at constant temperature. Although much remains to be done, it should not be considered that this is the only application area for IR techniques. Increasingly, thermographic analysis can address the problem of modeling the behavior of materials at high temperatures and dynamics. During physical treatments (e.g. hardening, annealing), or welding processes (melting, solidification), thermomechanical coupling phenomena (phase transitions, precipitation kinetics) are of critical importance. The constant development of faster IR cameras already allows mechanical engineers to address the dynamic behavior of materials in which, again, couplings play an important role in the instability propagation mode (e.g. adiabatic shear band, autocatalytic cycle).

16.6. Bibliography

[BAT 94] BATSALE J.C., MAILLET D., DEGIOVANNI A., "Extension de la méthode des quadripôles thermiques à l'aide de transformations intégrales – application au défaut plan bidimensionnel", *International Journal of Heat and Mass Transfer*, vol. 37, no. 1, pp. 111–127, 1994.

[BAT 01] BATSALE J.C., LE NILIOT C., "Estimation de paramètres par thermographie IR", *Métrologie Thermique et Techniques Inverses*, vol. 1, Chapter 4, Presses Universitaires de Perpignan, pp. 179–236, 2001.

[BAT 04] BATSALE J.C., BATTAGLIA J.L., FUDYM O., "Autoregressive algorithms and spatially random flash excitation for 3D non destructive evaluation with infrared cameras", *International Journal on Quantitative InfraRed Thermography (QIRT)*, vol. 1, pp. 5–20, 2004.

[BEC 77] BECK J.V., ARNOLD K.-J., *Parameters Estimation in Engineering and Science*, Wiley, NY, 1977.

[BER 08] BERTHEL B., CHRYSOCHOOS A., WATTRISSE B., GALTIER A., "Infrared image processing for the calorimetric analysis of fatigue phenomena", *Experimental Mechanics*, vol. 48, no. 1, pp. 79–90, 2008.

[BOU 04] BOULANGER T., CHRYSOCHOOS A., MABRU C., "Calorimetric analysis of dissipative and thermoelastic effects associated with the fatigue behavior of steels", *International Journal of Fatigue*, vol. 26, no. 3, pp. 221–229, 2004.

[BRE 82] BREMOND P., Développement d'une instrumentation infrarouge pour l'étude des structures mécaniques, Doctoral Thesis, University of Marseille, France, 1982.

[CAR 59] CARSLAW H.S., JAEGER J.C., Conduction of Heat in Solids, Clarendon Press, Oxford, 1959.

[CHA 84] CHADWICK P., CREASY C.F.M., "Modified entropic elasticity of rubberlike materials", Journal of the Mechanics and Physics of Solids, vol. 32, pp. 337–357, 1984.

[CHR 95] CHRYSOCHOOS A., "Analyse du comportement thermomécanique des matériaux par thermographie infrarouge", Actes du colloque Photomécanique 95, Cachan, pp. 203–211, 1995.

[CHR 00] CHRYSOCHOOS A., LOUCHE H., "An infrared image processing to analyze the calorific effects accompanying strain localization", International Journal of Engineering Science, vol. 38, pp. 1759–1788, 2000.

[CHR 09] CHRYSOCHOOS A., WATTRISSE B., MURACCIOLE J.-M., EL KAÏM Y., "Fields of stored energy associated with localized necking of steel", Journal of the Mechanics of Materials and Structures, vol. 4, no. 2, pp. 245–262, 2009.

[CHR 10] CHRYSOCHOOS A., HUON V., JOURDAN F., MURACCIOLE J.-M., PEYROUX R., WATTRISSE B., "Use of full-field DIC & IRT measurements for the thermomechanical analysis of material behaviour", Strain, vol. 46, pp. 117–130, 2010.

[COL 67] COLEMAN B., GURTIN M., "Thermodynamics with internal state variables", The Journal of Chemical Physics, vol. 47, no. 2, pp. 597–613, 1967.

[DEG 96] DEGIOVANNI A., BATSALE J.C., MAILLET D., "Mesure de la diffusivité longitudinale de matériaux anisotropes – panorama des techniques développées au LEMTA", Revue Générale de Thermique, vol. 35, pp. 141–147, 1996.

[DOU 09] DOUDARD C., CALLOCH S., "Influence of hardening type on self-heating of metallic materials under cyclic loadings at low amplitude", European Journal of Mechanics, vol. 28, no. 2, pp. 233–240, 2009.

[FRÉ 02] FRÉMOND M., Non-Smooth Thermomechanics, Springer, Berlin, 2002.

[FUD 02] FUDYM O., BATSALE J.C., LECONTE D., "A seminumerical approach for heat diffusion in heterogeneous media: one extension of the analytical quadrupole method", Numerical Heat Transfer: Part B, vol. 42, pp. 325–348, 2002.

[FUD 04] FUDYM O., LADEVIE B., BATSALE J.C., "Heat diffusion at the boundary of stratified media: homogenized temperature field and thermal constriction", International Journal of Heat and Mass Transfer, vol. 47, pp. 2437–2447, 2004.

[FUD 05] FUDYM O., BATTAGLIA J.L., BATSALE J.C., "Measurement of thermophysical properties in semi-infinite media by random heating and fractional model identification", Review of Scientific Instruments, vol. 76, no. 2, pp. 1–10, 2005.

[FUD 07] FUDYM O., BATSALE J.C., BATTAGLIA J.L., "Thermophysical properties mapping in semi-infinite longitudinally cracked plates by temperature image processing", Inverse Problem in Engineering, vol. 15, no. 2, pp. 163–176, 2007.

[GAU 99] GAUSSORGUES G., *La Thermographie Infrarouge – Principes, Technologie, Applications*, 4th ed., Tec & Doc Lavoisier, 1999.

[GER 83] GERMAIN P., NGUYEN Q.S., SUQUET P., "Continuum thermo-mechanics", *Journal of Applied Mechanics*, vol. 50, no. 4B, pp. 1010–1020, 1983.

[GUS 91] GUSTAFSON S.E., "Transient plane source techniques for thermal conductivity and thermal diffusivity measurements of solid materials", *Review of Scientific Instruments*, vol. 62, p. 797-804, 1991.

[HAL 75] HALPHEN B., NGUYEN Q.S., "Sur les matériaux standards généralisés", *Journal de Mécanique*, vol. 14, no. 1, pp. 39–63, 1975.

[HAY 04] HAY B., FILZ J.R., BATSALE J.C., "Mesure de diffusivité thermique par méthode flash", *Techniques de l'ingénieur*, R 2955, pp. 1–12, 2004.

[HON 05] HONORAT V., MOREAU S., MURACCIOLE J.-M., B. WATTRISSE B., CHRYSOCHOOS A., "Calorimetric analysis of polymer behaviour using a pixel calibration of an IRFPA camera", *International Journal on Quantitative IR Thermography*, vol. 2, no. 2, pp. 153–172, 2005.

[KRA 04] KRAPEZ J.C., SPAGNOLO L., FRIE B.M., MAIER H.P. NETTER F., "Measurement of in-plane diffusivity in non-homogeneous slabs by applying flash thermography", *International Journal of Thermal Sciences*, vol. 43, pp. 967–977, 2004.

[LEP 96] LEPOUTRE F., LEFEBVRE J., LHERMITTE T., AINOUCH L., DELPECH P., FORGE P., HIRCHI S., JOULAUD J.L., "Mesures thermiques microscopiques", *Revue Générale de Thermique*, vol. 413, no. 35, pp. 344–354, 1996.

[MAI 00] MAILLET D., ANDRE S., BATSALE J.C., DEGIOVANNI A., MOYNE C., *Thermal Quadrupoles: Solving the Heat Equation through Integral Transforms*, John Wiley, Chichester, 2000.

[MUR 08] MURACCIOLE J.M., WATTRISSE B., CHRYSOCHOOS A., "Energy balance of a semicrystalline polymer during local plastic deformation", *Strain*, vol. 44, no. 6, pp. 468–474, 2008.

[PAJ 89] PAJANI D., Mesure Par Thermographie Infrarouge, ADD Editeur, 1989.

[PAP 97] PAPINI F., GALLET P., *Thermographie Infrarouge – Image et Mesure*, Masson, 1997.

[PEY 98] PEYROUX R., CHRYSOCHOOS A., LICHT C., LÖBEL M., "Thermo-mechanical couplings and pseudoelasticity of shape memory alloys", *International Journal of Engineering Science*, vol. 36, no. 4, pp. 489–509, 1998.

[PHI 95] PHILIPPI I., BATSALE J.C., MAILLET D., DEGIOVANNI A., "Measurement of thermal diffusivity trough processing of infrared images", *Review of Scientific Instruments*, vol. 66, pp. 182–192, 1995.

[PIN 04] PINA V., RANC N., WAGNER D., HERVE P., SUTTER G., PHILIPPON S., "Étude des comportements statique et dynamique de matériaux métalliques par pyrométrie ultraviolet, visible ou infrarouge", *Photonique*, vol. 16, pp. 1–4, 2004.

[PRE 07] PRESS W.H., TEUKOLSKY S.A., WETTERLING W.T., FLANNERY B.P., *Numerical Recipes: The Art of Scientific Computing*, 3rd ed., Cambridge University Press, 2007.

[PRO 04] PRON H., BISSIEUX C., "3D thermal modelling applied to stress-induced anisotropy of thermal conductivity", *International Journal of Thermal Sciences*, vol. 43, no. 12, pp. 1161–1169, 2004.

[RAN 08] RANC N., TARAVELLA L., PINA V., HERVÉ P., "Temperature field measurement in titanium alloy during high strain rate loading – adiabatic shear bands phenomenon", *Mechanics of Materials*, vol. 40, no. 4–5, pp. 255–270, 2008.

[SAT 07] SATHON N., DULIEU-BARTON J.M., "Evaluation of sub-surface stresses using thermoelastic stress analysis", *Applied Mechanics and Materials*, vol. 7–8, pp. 153–158, 2007.

[SUT 07] SUTTER G., RANC N., "Temperature fields in a chip during high-speed orthogonal cutting – an experimental investigation", *International Journal of Machine Tools & Manufacture*, vol. 47, pp. 1507–1517, 2007.

[THO 53] THOMSON W. (LORD KELVIN), "On the thermoelastic and thermo-magnetic properties of matter", *Transactions of the Royal Society of Edinburgh*, vol. 20, no. 161, pp. 57–77, 1853.

List of Authors

Stéphane ANDRIEUX
EDF R & D
Clamart
France

Fabien AMIOT
FEMTO-ST
Besançon
France

Stéphane AVRIL
Center for Health Engineering
École Nationale Supérieure des Mines de Saint-Étienne (ENSM-SE)
France

Jean-Christophe BATSALE
I2M
Arts et Métiers ParisTech
Bordeaux
France

Marc BONNET
UMA
ENSTA ParisTech
France

Michel BORNERT
Navier Laboratory
Ecole des Ponts ParisTech
Paris Est University
France

Dan Borza
LOFIMS
INSA de Rouen
France

Fabrice Brémand
Pprime Institute
University of Poitiers
France

Huy Duong Bui
Laboratoire de Mécanique des Solides
École Polytechnique
Palaiseau
France

Anne-Sophie Caro-Bretelle
Centre CMGD
Ecole des Mines d'Alès
France

André Chrysochoos
Mechanics and Civil Engineering Laboratory
University of Montpellier
France

Andrei Constantinescu
LMS École Polytechnique
Palaiseau
France

Jérôme Crépin
Centre des Matériaux
Mines ParisTech
Evry
France

Jacques Desrues
Laboratoire 3SR
Grenoble INP
France

Jean-Christophe DUPRÉ
Pprime Institute
University of Poitiers
France

Pierre FEISSEL
Laboratoire Roberval
Compiègne University of Technology
France

Michel GRÉDIAC
Institut Pascal
Blaise Pascal University
Clermont-Ferrand
France

François HILD
LMT
Cachan
France

Patrick IENNY
Centre CMGD
Ecole des Mines d'Alès
France

Pierre JACQUOT
École Polytechnique Fédérale de Lausanne
Switzerland

Jérôme MOLIMARD
Center for Health Engineering
École Nationale Supérieure des Mines de Saint-Étienne (ENSM-SE)
France

Jean-José ORTEU
ICA
École des Mines d'Albi
France

Stéphane PAGANO
Mechanics and Civil Engineering Laboratory
University of Montpellier
France

Emmanuel PAGNACCO
INSA
Rouen
France

Jean-Noël PÉRIÉ
ICA
University of Toulouse
France

Pascal PICART
LAUM
École Nationale Supérieure d'Ingénieurs du Mans
France

Fabrice PIERRON
University of Southampton
UK

Hervé PRON
GRESPI
University of Reims Champagne-Ardenne
France

Julien RÉTHORÉ
LaMCoS
INSA
Lyon
France

Marco ROSSI
Universita Politecnica Delle Marche
Ancona
Italy

Stéphane ROUX
LMT
Cachan
France

Pierre SLANGEN
École des Mines d'Alès
France

Paul SMIGIELSKI
University of Strasbourg
France

Yves SURREL
VISUOL Technologies
Saint-Étienne
France

Evelyne TOUSSAINT
Institut Pascal
Blaise Pascal University
Clermont-Ferrand
France

Bertrand WATTRISSE
Mechanics and Civil Engineering Laboratory
University of Montpellier
France

Index

3D-DIC, 176